Second Edition

Introduction to College Mathematics

CUSTOMIZED EDITION OF BASIC COLLEGE MATHEMATICS, THIRD EDITION

1999-2000

SINCLAIR COMMUNITY COLLEGE

Pearson Custom Publishing

Cover: "Sinclair Community College," by Sinclair / Gatland Photography

Taken from:

Basic College Mathematics, Third Edition
by Jeffrey Slater and John Tobey

A Pearson Education Company
Upper Saddle River, New Jersey 07458

This special edition published in cooperation with Pearson Custom Publishing.

Printed in the United States of America

10 9 8 7 6 5 4 3

Please visit our web site at www.pearsoncustom.com

ISBN 0–536–02788–9

BA 990420

PEARSON CUSTOM PUBLISHING
75 Arlington Street, Boston, MA 02116
A Pearson Education Company

Contents

CHAPTER 3

Decimals 175

CHAPTER 4

Ratio and Proportion 239

CHAPTER 5

Percent 281

CHAPTER 6

Measurement 335

CHAPTER 8

Statistics 463

CHAPTER 1
Whole Numbers

Just ten years ago hardly anyone in the United States leased a car for private use. New car leasing ads appear everywhere and growing numbers of Americans are leasing their own private car. But is this really a good decision? Could you figure out the mathematics to determine if leasing or buying would be best for you? Turn to the Putting Your Skills to Work Problem on page 57 and find out.

Pretest Chapter 1

1. ______
2. ______
3. ______
4. ______
5. ______
6. ______
7. ______
8. ______
9. ______
10. ______
11. ______
12. ______
13. ______
14. ______
15. ______

If you are familiar with the topics in this chapter, take this test now. Check your answers with those in the back of the book. If an answer was wrong or you couldn't do a problem, study the appropriate section of the chapter.

If you are not familiar with the topics in this chapter, don't take this test now. Instead, study the examples, work the practice problems, and then take the test.

This test will help you identify which concepts you have mastered and which you need to study further.

Section 1.1

1. Write in words. 78,310,436

2. Write in expanded notation. 38,247

3. Write in standard notation: five million, sixty-four thousand, one hundred twenty-two

Look at the table on the right.

4. How many students were enrolled in elementary and secondary schools in 1990–1991?

5. How many high school graduates were there in 1991?

EDUCATION	1990–1991 (thousands)
Elementary and secondary	41,217
High school graduates	2,237

Section 1.2

Add.

6. $\begin{array}{r} 13 \\ 31 \\ 88 \\ 43 \\ +\ 69 \\ \hline \end{array}$

7. $\begin{array}{r} 28{,}318 \\ 5{,}039 \\ +\ 17{,}213 \\ \hline \end{array}$

8. $\begin{array}{r} 7148 \\ 500 \\ 19 \\ +\ 7062 \\ \hline \end{array}$

Section 1.3

Subtract.

9. $\begin{array}{r} 6439 \\ -\ 4328 \\ \hline \end{array}$

10. $\begin{array}{r} 100{,}450 \\ -\ 24{,}139 \\ \hline \end{array}$

11. $\begin{array}{r} 45{,}861{,}413 \\ -\ 43{,}879{,}761 \\ \hline \end{array}$

Section 1.4

Multiply.

12. $9 \times 6 \times 1 \times 2$

13. $\begin{array}{r} 2658 \\ \times\ \ \ 7 \\ \hline \end{array}$

14. $\begin{array}{r} 91 \\ \times\ 74 \\ \hline \end{array}$

15. $\begin{array}{r} 365 \\ \times\ 908 \\ \hline \end{array}$

Section 1.5

Divide. If there is a remainder, be sure to state it as part of the answer.

16. $8\overline{)84{,}840}$ **17.** $7\overline{)51{,}633}$ **18.** $42\overline{)5838}$

Section 1.6

19. Write in exponent form. $6 \times 6 \times 6 \times 6$

20. Evaluate. 3^4

Perform the operations in their proper order.

21. $2 \times 3^3 - (4 + 1)^2$

22. $2^4 + (6^2 + 36) \div 2$

23. $6 \times 7 - 2 \times 6 - 2^3 + 17 \div 17$

Section 1.7

24. Round to the nearest thousand. 270,612

25. Round to the nearest hundred. 26,539

26. Round to the nearest million. 59,540,000

Estimate by first rounding each number so that there is only one nonzero digit and then performing the calculation.

27. $187 + 458 + 375 + 803 + 666$

28. $56{,}784 \times 459{,}202$

Section 1.8

Solve each of the following applied problems.

29. Emily planned a trip of 1483 miles from her home to Chicago. She traveled 317 miles on her first day of the trip. How many more miles must she travel?

30. Betsey purchased 84 shares of stock at a cost of $1848.00. How much was the cost per share?

31. Dr. Alfonso bought three chairs at $46 each, three lamps at $29 each, and two tables at $37 each for the waiting room of his office. How much did his purchases cost?

POPULATION FOR STATE OF ALASKA, U.S. CENSUS REPORT

Year	1970	1980	1990	2000
Population	302,583	401,851	550,043	?

32. Using the table above, answer the following questions.

(a) By how much did the population in Alaska increase from 1980 to 1990?

(b) Some officials predict that between 1990 and 2000 the population of Alaska will increase by 134,890. If that figure is correct, what is the projected population of Alaska for the year 2000?

16. ____

17. ____

18. ____

19. ____

20. ____

21. ____

22. ____

23. ____

24. ____

25. ____

26. ____

27. ____

28. ____

29. ____

30. ____

31. ____

32. (a) ____

(b)

1.1 Understanding Whole Numbers

MathPro Video 1 SSM

After studying this section, you will be able to:

1. *Write numbers in expanded form.*
2. *Write a word name for a number and write a number for a word name.*
3. *Read numbers in tables.*

1 *Writing Numbers in Expanded Form*

To count a number of objects or to answer the question "How many?" we use a set of numbers called **whole numbers.** These whole numbers are as follows.

0, 1, 2, 3, 4, 5, 6, 7, 8, 9, 10, 11, 12, 13, 14, 15, . . .

There is no largest whole number. The three dots . . . indicate that the set of whole numbers goes on indefinitely. Our number system is based on tens and ones and is called the *decimal system* (or the *base 10 system*). The numbers 0, 1, 2, 3, 4, 5, 6, 7, 8, 9 are called **digits.** The position, or placement, of the digits in the number tells the value of the digits. For example, in the number 521, the "5" means 5 hundreds (500). In the number 54, the "5" means 5 tens (50).

521 — 5 means 5 hundred or 500

54 — 5 means 5 tens or 50

For this reason, our number system is called a *place-value system.*

Consider the number 5643. We will use a place-value chart to illustrate the value of each digit in the number 5643.

millions			thousands			ones		
Hundred millions	Ten millions	Millions	Hundred thousands	Ten thousands	Thousands	Hundreds	Tens	Ones
					5	6	4	3

The value of the number is 5 thousands, 6 hundreds, 4 tens, 3 ones.

The place-value chart above shows the value of each place, from ones on the right to hundred millions on the left. When we write very large numbers, we place a comma after every group of three digits, moving from right to left. This makes the number easier to read. It is usually agreed that a four-digit number does not have a comma, but that numbers with five or more digits do. So 32,000 would be written with a comma but 7000 would not.

To show the value of each digit in a number, we sometimes write the number in expanded notation. For example, 56,327 in expanded notation is

$$50{,}000 + 6000 + 300 + 20 + 7$$

EXAMPLE 1 Write each number in expanded notation.

(a) 2378 **(b)** 508,271 **(c)** 980,340,654

(a) Sometimes it helps to say the number to yourself.

	two thousand		three hundred		seventy		eight
2378 =	2000	+	300	+	70	+	8

(b) When 0 is used as a placeholder, you do not include it in the expanded form.

$$508{,}271 = \overbrace{500{,}000 + 8000 + 200 + 70 + 1}^{\text{Expanded notation}}$$

(c) $980{,}340{,}654 = \overbrace{900{,}000{,}000 + 80{,}000{,}000 + 300{,}000 + 40{,}000 + 600 + 50 + 4}^{\text{Expanded notation}}$

Practice Problem 1 Write each number in expanded notation.

(a) 3182 **(b)** 520,890 **(c)** 709,680,059 ■

The number as you usually see it is called the **standard notation.** 980,340,654 is the standard notation for the number nine hundred eighty million, three hundred forty thousand, six hundred fifty-four.

EXAMPLE 2 Write each number in standard notation.

(a) 500 + 30 + 8 **(b)** 300,000 + 7000 + 40 + 7

(a) 538

(b) Be careful to keep track of the place value of each digit. You may need to use 0 as a placeholder.

$$\overbrace{300{,}000 + 7000 + 40 + 7 = 3}^{\text{3 hundred thousand}}07{,}047$$

(7000 → 7 thousand)

We needed to use 0 in the ten thousands place and in the hundreds place.

Practice Problem 2 Write each number in standard notation.

(a) 400 + 90 + 2 **(b)** 80,000 + 400 + 20 + 7 ■

EXAMPLE 3 Last year the population of Central City was 1,509,637. In the number 1,509,637

(a) How many ten thousands are there?

(b) How many tens are there?

(c) What is the value of the digit 5?

(d) In what place is the digit 6?

The place-value chart will help you to identify the value of each place.

(a) Look at the digit in the ten thousands place. There are 0 ten thousands.

(b) Look at the digit in the tens place. There are 3 tens.

(c) The digit 5 is in the hundred thousands place. The value of the digit is

5 hundred thousand or 500,000

(d) The digit 6 is in the hundreds place.

Practice Problem 3 The campus library has 904,759 books.

(a) What digit tells the number of hundreds?

(b) What digit tells the number of hundred thousands?

(c) What is the value of the digit 4?

(d) What is the value of the digit 9? Why does this question have two answers? ■

2 *Writing Word Names for Numbers and Numbers for Word Names*

A number has the same *value* no matter how we write it. For example, "a million dollars" means the same as "$1,000,000." In fact, any number in our number system can be written in several ways or forms:

• Standard notation	521
• Expanded notation	500 + 20 + 1
• In words	five hundred twenty-one

You may want to write a number in any of these ways. To write a check, you need to use both standard notation and words.

To write a word name for a number, think about how you might say the number.

EXAMPLE 4 Write the word name for each number.

(a) 1695 **(b)** 200,470 **(c)** 7,003,038

Look at the place-value chart if you need help in identifying the place for each digit.

(a) Note that this number does not represent a date. It is a number. The number begins with 1 in the thousands place. The word name is

One thousand six hundred ninety-five
↑
We use a hyphen here.

(b) The number begins with 2 in the hundred thousands place.

Two hundred thousand, four hundred seventy

(c) The number begins with 7 in the millions place.

Seven million, three thousand, thirty-eight

Practice Problem 4 Write the word name for each number.

(a) 2736 **(b)** 980,306 **(c)** 12,000,021 ■

Very large numbers are used to measure quantities in some disciplines, such as distance in astronomy and the national debt in macroeconomics. We can extend the place-value chart to include these large numbers.

The national debt for the United States as of December 1996 was $5,311,192,809,000. This number is indicated in the place-value chart below.

Trillions			Billions			Millions			Thousands			Ones		
		5	3	1	1	1	9	2	8	0	9	0	0	0

EXAMPLE 5 Write the national debt for the United States as of December 1996 in the amount of $5,311,192,809,000 using a word name.

The national debt in December 1996 was five trillion, three hundred eleven billion, one hundred ninety-two million, eight hundred nine thousand dollars.

Practice Problem 5 The U.S. Treasury reported that the national debt in 1989 was $2,857,406,300,000. Write the amount of this debt using a word name. ■

Occasionally you may want to write a word name as a number.

EXAMPLE 6 Write each number in standard notation.

(a) four hundred ninety-two

(b) twenty-six thousand, eight hundred fifty-four

(c) two billion, three hundred eighty-six million, five hundred forty-seven thousand, one hundred ninety

(a) 492

(b) 26,854

(c) 2,386,547,190

Practice Problem 6 Write in standard notation.

(a) eight hundred three **(b)** thirty thousand, two hundred twenty-nine ■

3 *Reading Numbers in Tables*

Sometimes numbers found in charts and tables are abbreviated. Look at the chart below from the *World Almanac*. Notice the third line tells us the numbers represent thousands. To understand what these numbers mean, think thousands. If the number 23 appears under 1740 for New Hampshire, the 23 represents 23 thousand. 23 thousand is 23,000. Note that census figures for some colonies are not available for certain years.

Estimated population of the American colonies from 1650 to 1780 (in thousands)

Source: U.S. Bureau of the Census

Colony	1650	1670	1690	1700	1720	1740	1750	1770	1780
Maine	1	★	★	★	★	★	★	31	49
New Hampshire	1	2	4	5	9	23	28	62	88
Vermont	★	★	★	★	★	★	★	10	48
Plymouth and Massachusetts	16	35	57	56	91	152	188	235	269
Rhode Island	1	2	4	6	12	25	33	58	53
Connecticut	4	13	22	26	59	90	111	184	207

EXAMPLE 7 Refer to the chart above to answer the following questions. Write each number in standard notation.

(a) What was the estimated population of Maine in 1780?

(b) What was the estimated population of Plymouth and Massachusetts in 1720?

(c) What was the estimated population of Rhode Island in 1700?

(a) To read the chart, first look for Maine on the left. Read across to the column under 1780. The number is 49. In this chart 49 means 49 thousands.

$$49 \text{ thousands} \Rightarrow 49{,}000$$

(b) Read the line of the chart for Plymouth and Massachusetts. The population of Plymouth and Massachusetts in 1720 was 91. This means 91 thousands. We will write this as 91,000.

(c) Read the line of the chart for Rhode Island. The population of Rhode Island in 1700 was 6. This means 6 thousands. We will write this as 6000.

To Think About Why do you think Plymouth and Massachusetts had the largest population for the years shown in the table?

Practice Problem 7 Refer to the chart above to answer the following questions. Write each number in standard notation.

(a) What was the estimated population of Connecticut in 1670?

(b) What was the estimated population of New Hampshire in 1780?

(c) What was the estimated population of Vermont in 1770? ■

1.1 Exercises

Write each number in expanded notation.

1. 6731

2. 9519

3. 108,276

4. 350,765

5. 23,761,345

6. 46,198,253

7. 103,260,768

8. 820,310,574

Write each number in standard notation.

9. $600 + 70 + 1$

10. $500 + 90 + 6$

11. $9000 + 800 + 60 + 3$

12. $7000 + 600 + 50 + 2$

13. $70{,}000 + 6000 + 30 + 6$

14. $80{,}000 + 2000 + 300 + 5$

15. $700{,}000 + 6000 + 200$

16. $900{,}000 + 50{,}000 + 40 + 7$

Verbal and Writing Skills

17. In the number 56,782
- **(a)** What digit tells the number of hundreds?
- **(b)** What is the value of the digit 5?

18. In the number 123,769
- **(a)** What digit tells the number of tens?
- **(b)** What is the value of the digit 2?

19. In the number 984,328
- **(a)** What digit tells the number of ten thousands?
- **(b)** What is the value of the digit?

20. In the number 6,789,345
- **(a)** What digit tells the number of thousands?
- **(b)** What is the value of the digit?

Write a word name for each number.

21. 53

22. 46

23. 8936

24. 4629

25. 36,118

26. 55,742

27. 105,261

28. 370,258

29. 23,561,248

30. 19,376,584

31. 4,302,156,200

32. 7,436,210,400

Write each number in standard notation.

33. Three hundred seventy-five

34. Seven hundred thirty-six

35. Fifty-six thousand, two hundred eighty-one

36. Seventy-nine thousand, three hundred forty-six

37. One hundred million, seventy-nine thousand, eight hundred twenty-six

38. Four hundred fifty million, three hundred thousand, two hundred forty-nine

Applications

When writing a check, a person must write the word name for the dollar amount of the check.

Alex J. Writer
10 Main Street
Westwood, Texas
6848
April 1 19 91
53-235/113
PAY TO THE ORDER OF
$ 1,965 00/100
DOLLARS
Last Chance Bank
Westwood, Texas
MEMO
⑆011302357⑆ 011 08115⑈ 6848

39. Alex bought new equipment for his laboratory for \$1965. What word name should he write on the check?

40. Alex later bought a new personal computer for \$6383. What word name should he write on the check?

In problems 41–44 use the following chart prepared with data from the World Almanac. *Notice the second line tells us that the numbers represent millions. These values are only approximate values representing numbers written to the nearest million. They are not exact census figures.*

Estimated population of four states from 1900 to 1990 (in millions)

Source: U.S. Bureau of the Census

State	1900	1910	1920	1930	1940	1950	1960	1970	1980	1990
California	1	2	3	6	7	11	16	20	24	30
Florida	1	1	1	1	2	3	5	7	10	13
Illinois	5	5	6	8	8	9	10	11	11	11
New York	7	9	10	13	13	15	17	18	18	18

41. What was the estimated population of New York in 1910?

42. What was the estimated population of Florida in 1970?

43. What was the estimated population of California in 1960?

44. What was the estimated population of Illinois in 1950?

In problems 45–48 use the following chart prepared with data from the World Almanac. *Notice the third line that tells us the numbers represent billions of dollars. These values are only approximate values representing numbers written to the nearest billion. They are not exact financial figures.*

PERSONAL CONSUMPTION EXPENDITURES IN THE UNITED STATES FROM 1985 TO 1991

Source: Bureau of Economic Analysis, U.S. Department of Commerce (billions of dollars)

Category	1985	1986	1987	1988	1989	1990	1991
Transportation	360	366	380	413	437	454	438
Medical care	328	358	399	488	536	596	656
Recreation	186	201	223	247	266	281	290
Religious and welfare activities	57	63	68	86	93	102	108

45. How much did people in the United States spend in 1986 on recreation?

46. How much did people in the United States spend in 1989 on medical care?

47. How much did people in the United States spend in 1987 on transportation?

48. How much did people in the United States spend in 1991 on religious and welfare activities?

To Think About

Specify the digit.

49. The speed of light is approximately 29,979,250,000 centimeters per second.
(a) What digit tells the number of ten thousands?
(b) What digit tells the number of ten billions?

50. The radius of the earth is approximately 637,814,000 centimeters.
(a) What digit tells the number of ten thousands?
(b) What digit tells the number of ten millions?

51. In 1994, more than 4,300,000 immigrants came to the United States from Mexico.
(a) Which digit tells the number of hundred thousands?
(b) Which digit tells the number of millions?

52. The world's population is expected to reach 7,900,000,000 by the year 2020, according to the U.S. Bureau of the Census.
(a) Which digit tells the number of hundred millions?
(b) Which digit tells the number of billions?

53. Write in standard notation: six hundred thirteen trillion, one billion, thirty-three million, two hundred eight thousand, three.

54. Write a word name for 3,682,968,009,931,960,747. (*Hint:* The digit 1 followed by 18 zeros represent the number *1 quintillion.* 1 followed by 15 zeros represents the number *1 quadrillion.*)

1.2 Addition of Whole Numbers

MathPro Video 1 SSM

After studying this section, you will be able to:

1 *Master basic addition facts.*
2 *Add several single-digit numbers.*
3 *Add numbers of several digits when carrying is not needed.*
4 *Add numbers of several digits when carrying is needed.*
5 *Apply addition to real-life situations.*

1 *Mastering Basic Addition Facts*

We see the addition process time and time again. Carpenters add to find the amount of lumber they need for a job. Auto mechanics add to make sure they have enough parts in the inventory. Bank tellers add to get cash totals.

What is addition? We do addition when we put sets of objects together.

□□□□□ □□□□□□□ □□□□□□□□□□□□
5 objects + 7 objects = 12 objects

$$5 + 7 = 12$$

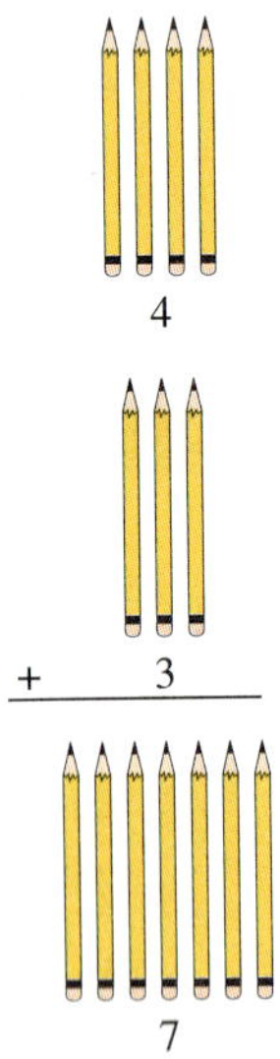

Usually when we add numbers, we put one number under the other in a column. The numbers being added are called **addends.** The result is called the **sum.**

Suppose that we have four pencils in the car and we bring three more pencils from home. How many pencils do we have with us now? We add 4 and 3 to obtain a value of 7. In this case, the numbers 4 and 3 are the addends and the answer 7 is the sum.

$$\begin{array}{rl} 4 & \text{addend} \\ +\ 3 & \text{addend} \\ \hline 7 & \text{sum} \end{array}$$

Think about what we do when we add 0 to another number. We are not making a change, so whenever we add zero to another number, that number will be the sum. Since this is always true, this is called a *property.* Since the sum is identical to the number added to zero, this is called the **identity property of zero.**

EXAMPLE 1 Add.

(a) 8 + 5 **(b)** 3 + 7 **(c)** 9 + 0

(a) $\begin{array}{r} 8 \\ +\ 5 \\ \hline 13 \end{array}$ **(b)** $\begin{array}{r} 3 \\ +\ 7 \\ \hline 10 \end{array}$ **(c)** $\begin{array}{r} 9 \\ +\ 0 \\ \hline 9 \end{array}$ ← *Note:* When we add zero to any other number, that number is the sum.

Practice Problem 1 Add.

(a) $\begin{array}{r} 7 \\ +\ 5 \\ \hline \end{array}$ **(b)** $\begin{array}{r} 9 \\ +\ 4 \\ \hline \end{array}$ **(c)** $\begin{array}{r} 3 \\ +\ 0 \\ \hline \end{array}$ ■

The table below shows the basic addition facts. You should know these facts. If any of the answers don't come to you quickly, now is the time to learn them. To check your knowledge try Exercises 1.2, problems 3 and 4.

BASIC ADDITION FACTS

+	0	1	2	3	4	5	6	7	8	9
0	0	1	2	3	4	5	6	7	8	9
1	1	2	3	4	5	6	7	8	9	10
2	2	3	4	5	6	7	8	9	10	11
3	3	4	5	6	7	8	9	10	11	12
4	4	5	6	7	8	9	10	11	12	13
5	5	6	7	8	9	10	11	12	13	14
6	6	7	8	9	10	11	12	13	14	15
7	7	8	9	10	11	12	13	14	15	16
8	8	9	10	11	12	13	14	15	16	17
9	9	10	11	12	13	14	15	16	17	18

To use the table to find the sum 4 + 7, read across the top of the table to the 4 column, and then read down the left to the 7 row. The box where the 4 and 7 meet is 11, which means that 4 + 7 = 11. Now read across the top to the 7 column and down the left to the 4 row. The box where these numbers meet is also 11. We can see that the order in which we add the numbers does not change the sum. 4 + 7 = 11, and 7 + 4 = 11. We call this the **commutative property of addition.**

This property does not hold true for everything in our lives. When you put on your socks and then your shoes, the result is not the same as if you put on your shoes first and then your socks! Can you think of any other examples where changing the order in which you add things would change the result?

2 *Adding Several Single-Digit Numbers*

If more than two numbers are to be added, we usually add from the first number to the next number and mentally note the sum. Then we add that sum to the next number, and so on.

EXAMPLE 2 Add. 3 + 4 + 8 + 2 + 5

We rewrite the addition problem in a column format.

$$\begin{array}{r} 3 \\ 4 \\ 8 \\ 2 \\ +\ 5 \\ \hline 22 \end{array}$$

Mentally, we do these steps.

3 + 4 = 7

7 + 8 = 15

15 + 2 = 17

17 + 5 = 22

Practice Problem 2 Add. 7 + 6 + 5 + 8 + 2 ■

Because the order in which we add numbers doesn't matter, we can choose to add from the top down, from the bottom up, or in any other way. One shortcut is to add first any numbers that will give a sum of 10, or 20, or 30, and so on.

EXAMPLE 3 Add.

$$\begin{array}{r} 3 \\ 4 \\ 8 \\ 2 \\ +\,6 \\ \hline \end{array}$$

We mentally group the numbers into tens.

$$\begin{array}{l} 3 \\ 4 \\ 8 \\ 2 \\ 6 \end{array}$$

8 and 2 → 8 + 2 = 10

4 and 6 → 4 + 6 = 10

The sum is 10 + 10 + 3 or 23

Practice Problem 3 Add. 1 + 7 + 2 + 9 + 3 ■

3 *Adding Numbers of Several Digits When Carrying Is Not Needed*

Of course, many numbers that we need to add have more than one digit. In such cases, we must be careful to first add the digits in the ones column, then the digits in the tens column, then those in the hundreds column, and so on. Notice that we move from *right to left.*

EXAMPLE 4 Add. 4304 + 5163

$$\begin{array}{r} 4\;3\;0\;4 \\ +\;5\;1\;6\;3 \\ \hline 9\;4\;6\;7 \end{array}$$

7: sum of 4 ones + 3 ones = 7 ones

6: sum of 0 tens + 6 tens = 6 tens

4: sum of 3 hundreds + 1 hundred = 4 hundreds

9: sum of 4 thousands + 5 thousands = 9 thousands

Practice Problem 4 Add.

$$\begin{array}{r} 8246 \\ +\,1702 \\ \hline \end{array}$$ ■

4 Adding Numbers of Several Digits When Carrying Is Needed

When you add several whole numbers, often the sum in a column is greater than 9, but we can only use *one* digit in any one place. What do we do with a two-digit sum? Look at the following example.

EXAMPLE 5 Add. 45 + 37

tens ones

$$\begin{array}{rr} & 1 \\ 4 & 5 \\ +\,3 & 7 \\ \hline & 2 \end{array}$$

5 ones and 7 ones = 12.
We rename 12 in expanded notation: 1 ten + 2 ones.
We place the 2 ones in the ones column.
We carry the 1 ten over to the tens column.

Note: Placing the 1 in the next column is often called "carrying the one."

$$\begin{array}{rr} 1 & \\ 4 & 5 \\ +\,3 & 7 \\ \hline 8 & 2 \end{array}$$

Now we can add the digits in the tens column.

Thus, 45 + 37 = 82.

Practice Problem 5 Add.

$$\begin{array}{r} 56 \\ +\,36 \\ \hline \end{array}$$ ■

Often you must use carrying several times by bringing the left digit into the next column to the left.

EXAMPLE 6 Add. 257 + 688 + 94

Thousands Column, Hundreds Column, Tens Column, Ones Column

$$\begin{array}{rrrr} & 2 & 1 & \\ & 2 & 5 & 7 \\ & 6 & 8 & 8 \\ + & & 9 & 4 \\ \hline 1 & 0 & 3 & 9 \end{array}$$

In the ones column we add 7 + 8 + 4 = 19. Because 19 is 1 ten and 9 ones, we place 9 in the ones column and carry 1 to the top of the tens column.

In the tens column we add 1 + 5 + 8 + 9 = 23. Because 23 tens is 2 hundreds and 3 tens, we place the 3 in the tens column and carry 2 to the top of the hundreds column.

In the hundreds column we add 2 + 2 + 6 = 10 hundreds. Because 10 hundreds is 1 thousand and 0 hundreds, we place the 0 in the hundreds column and place the 1 in the thousands column.

Practice Problem 6 Add. 789 + 63 + 297 ■

We can add numbers in more than one way. To add $5 + 3 + 7$ we can first add the 5 and 3. We do this by using parentheses to show the first operation to be done. This shows us that $5 + 3$ is to be grouped together.

$$\begin{aligned} 5 + 3 + 7 &= (5 + 3) + 7 = 15 \\ &= \quad 8 \quad + 7 = 15 \end{aligned}$$

We could add the 3 and 7 first. We use parentheses to show that we group $3 + 7$ together and that we will add these two numbers first.

$$\begin{aligned} 5 + 3 + 7 &= 5 + (3 + 7) = 15 \\ &= 5 + \quad 10 \quad = 15 \end{aligned}$$

The way we group numbers to be added does not change the sum. This property is called the **associative property of addition.**

Look again at the three properties of addition we have discussed in this section.

1. Associative Property of Addition When we add three numbers, we can group them in any way.	$(8 + 2) + 6 = 8 + (2 + 6)$ $10 + 6 = 8 + 8$ $16 = 16$
2. Commutative Property of Addition Two numbers can be added in either order with the same result.	$5 + 12 = 12 + 5$ $17 = 17$
3. Identity Property of Zero When zero is added to a number, the sum is that number.	$8 + 0 = 8$ $3 + 0 = 3$

Because of the commutative and associative properties of addition, we can check our addition by adding the numbers in the opposite order.

EXAMPLE 7 **(a)** Add the numbers. $39 + 7284 + 3132$
(b) Check by reversing the order of addition.

(a)

$$\begin{array}{r} {}^{1\,1}\\ 39 \\ 7284 \\ +\ 3132 \\ \hline 10455 \end{array}$$

Addition

(b)

$$\begin{array}{r} {}^{1\,1}\\ 3132 \\ 7284 \\ +\quad 39 \\ \hline 10455 \end{array}$$

Check by reversing the order.

The sum is the same in each case.

Practice Problem 7

(a) Add.

$$\begin{array}{r} 127 \\ 9876 \\ +\ 342 \\ \hline \end{array}$$

(b) Check by reversing the order.

$$\begin{array}{r} 342 \\ 9876 \\ +\ 127 \\ \hline \end{array}$$

■

5 *Applying Addition to Real-Life Situations*

We use addition in all kinds of situations. There are several key words in word problems that imply addition. For example, it may be stated that there are 12 math books, 9 chemistry books, and 8 biology books on a book shelf. To find the *total* number of books implies that we add the numbers 12 + 9 + 8. Other key words are *how much, how many,* and *all.*

Sometimes a problem will have more information than you will need to answer the question. If you have too much information, to solve the problem you will need to separate out the facts that are not important. The following three steps are involved in the problem-solving process.

Step 1 Understand the problem

Step 2 Calculate and state the answer

Step 3 Check

We may not write all of these steps down, but they are the steps we use to solve all problems.

EXAMPLE 8 The bookkeeper for Smithville Trucking was examining the following data for the company checking account.

Monday: \$23,416 was deposited and \$17,389 was debited.

Tuesday: \$44,823 was deposited and \$34,089 was debited.

Wednesday: \$16,213 was deposited and \$20,057 was debited.

What was the total of all deposits during this period?

Step 1 *Understand the problem.*
Total implies that we will use addition. Since we don't need to know about the debits to answer this question, we use only the *deposit* amounts.

Step 2 *Calculate and state the answer.*

Monday:	\$23,416 was deposited.	$\overset{1}{2}\overset{1}{3},4\overset{1}{1}6$
Tuesday:	\$44,823 was deposited.	44,823
Wednesday:	\$16,213 was deposited.	+ 16,213
		84,452

A total of \$84,452 was deposited on those three days.

Step 3 *Check.*
You may add the numbers in reverse order to check. We leave the check up to you.

Practice Problem 8 North University has 23,413 men and 18,316 women. South University has 19,316 men and 24,789 women. East University has 20,078 men and 22,965 women. What is the total enrollment of *women* at the three schools? ■

EXAMPLE 9 Mr. Ortiz has a rectangular field whose length is 400 feet and width 200 feet. What is the total number of feet of fence that would be required to fence in the field?

1. *Understand the problem.*
   To help us to get a picture of what the field looks like, we will draw a diagram.

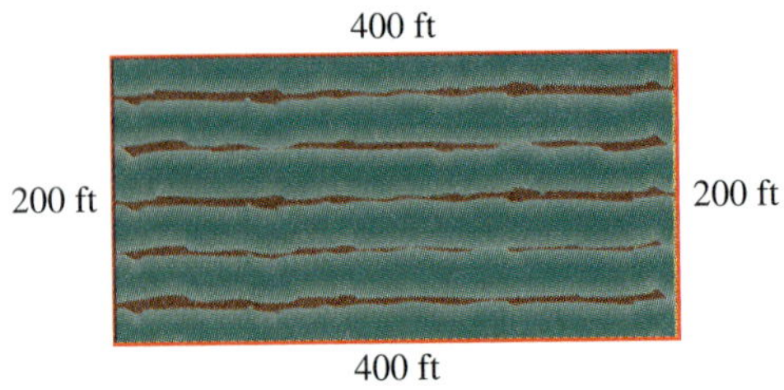

Note that ft is the abbreviation for feet. ft means feet.

2. *Calculate and state the answer.*
   Since the fence will be along each side of the field, we add the lengths all around the field.

$$\begin{array}{r} 200 \\ 200 \\ 400 \\ +\ 400 \\ \hline 1200 \end{array}$$

The amount of fence that would be required is 1200 feet.

3. *Check.*
   Regroup the addends and add.

$$\begin{array}{r} 200 \\ 400 \\ 200 \\ +\ 400 \\ \hline 1200 \end{array} \checkmark$$

Practice Problem 9 In Vermont, Gretchen fenced the rectangular field on which her sheep graze. The length of the field is 2000 feet and the width of the field is 1000 feet. What is the perimeter of the field? (*Hint:* The ''distance around'' an object [such as a field] is called the *perimeter.*) ■

Developing Your Study Skills

CLASS ATTENDANCE

You will want to get started in the right direction by choosing to attend class every day, beginning with the first day of class. Statistics show that class attendance and good grades go together. Classroom activities are designed to enhance learning, and therefore you must be in class to benefit from them. Each day vital information and explanations are given that can help you understand concepts. Do not be deceived into thinking that you can just find out from a friend what went on in class. There is no good substitute for firsthand experience. Give yourself a push in the right direction by developing the habit of going to class every day.

1.2 Exercises

Verbal and Writing Skills

1. Explain in your own words.

(a) the commutative property of addition

(b) the associative property of addition

2. When zero is added to any number, it does not change that number. Why do you think this is called the identity property of zero?

Complete the addition facts for each table. Strive for total accuracy, but work quickly. Allow a maximum of 5 minutes for each table.

3.

+	3	5	4	8	0	6	7	2	9	1
2										
7										
5										
3										
0										
4										
1										
8										
6										
9										

4.

+	1	6	5	3	0	9	4	7	2	8
3										
9										
4										
0										
2										
7										
8										
1										
6										
5										

Add.

5. $\begin{array}{r} 4 \\ 2 \\ 8 \\ +\,9 \\ \hline \end{array}$

6. $\begin{array}{r} 4 \\ 6 \\ 2 \\ +\,7 \\ \hline \end{array}$

7. $\begin{array}{r} 3 \\ 8 \\ 9 \\ +\,6 \\ \hline \end{array}$

8. $\begin{array}{r} 1 \\ 4 \\ 3 \\ +\,8 \\ \hline \end{array}$

9. $\begin{array}{r} 12 \\ 45 \\ +\;\;3 \\ \hline \end{array}$

10. $\begin{array}{r} 46 \\ 20 \\ +\;\;2 \\ \hline \end{array}$

11. $\begin{array}{r} 63 \\ 24 \\ +\,12 \\ \hline \end{array}$

12. $\begin{array}{r} 54 \\ 21 \\ +\,23 \\ \hline \end{array}$

13. $\begin{array}{r} 2847 \\ 1634 \\ +\;\;\;98 \\ \hline \end{array}$

14. $\begin{array}{r} 4816 \\ 3015 \\ +\;\;798 \\ \hline \end{array}$

15. $\begin{array}{r} 6908 \\ 2173 \\ +\,4255 \\ \hline \end{array}$

16. $\begin{array}{r} 5017 \\ 2984 \\ +\,1328 \\ \hline \end{array}$

17. $\begin{array}{r} 8235 \\ +\,5626 \\ \hline \end{array}$

18. $\begin{array}{r} 5673 \\ +\,3572 \\ \hline \end{array}$

19. $\begin{array}{r} 18{,}718 \\ +\,24{,}021 \\ \hline \end{array}$

20. $\begin{array}{r} 99{,}023 \\ +\,47{,}325 \\ \hline \end{array}$

Add from the top. Then check by adding in the reverse order.

21. $\begin{array}{r} 36 \\ 41 \\ 25 \\ 6 \\ +\ 13 \\ \hline \end{array}$

22. $\begin{array}{r} 24 \\ 39 \\ 16 \\ 14 \\ +\ 9 \\ \hline \end{array}$

23. $\begin{array}{r} 106 \\ 13 \\ 4 \\ 28 \\ +\ 981 \\ \hline \end{array}$

24. $\begin{array}{r} 463 \\ 27 \\ 8 \\ 41 \\ +\ 507 \\ \hline \end{array}$

Add.

25. $\begin{array}{r} 126 \\ 8142 \\ 37 \\ +\ 9604 \\ \hline \end{array}$

26. $\begin{array}{r} 223 \\ 7021 \\ 89 \\ +\ 7634 \\ \hline \end{array}$

27. $\begin{array}{r} 1{,}362{,}214 \\ 7{,}002{,}316 \\ +\ 3{,}214{,}896 \\ \hline \end{array}$

28. $\begin{array}{r} 4{,}002{,}983 \\ 2{,}134{,}702 \\ +\ 3{,}592{,}001 \\ \hline \end{array}$

29. $\begin{array}{r} 837{,}241{,}000 \\ +\ 298{,}039{,}240 \\ \hline \end{array}$

30. $\begin{array}{r} 982{,}306{,}000 \\ +\ 583{,}215{,}320 \\ \hline \end{array}$

31. $\begin{array}{r} 516{,}208 \\ 24{,}317 \\ +\ 1{,}763{,}295 \\ \hline \end{array}$

32. $\begin{array}{r} 32{,}500 \\ 763{,}420 \\ +\ 2{,}837{,}667 \\ \hline \end{array}$

33. $12 + 8 + 156 + 72$

34. $85 + 3 + 407 + 26$

35. $15{,}216 + 485 + 5208$

36. $26{,}002 + 599 + 3500$

Applications

Answer each question.

37. When Diana went shopping for holiday presents for her family, she spent $224 on Thursday, $387 on Friday, and $183 on Saturday. What is the total amount of money she spent on gifts?

38. Peter is building an extra room onto the house to accommodate the new baby. At the building supply store he made purchases of $421 on Friday, $218 on Saturday, and $79 on Sunday. How much did the new room cost to build?

39. Andy's annual vacation two years ago cost him $2311. Last year he spent $2502. This year his vacation cost him $3173. What is the total amount he spent on vacations for the three years?

40. The taxes on Jack's house two years ago were $4658. Last year they were $5222. This year they are $6027. What is the total amount for three years?

41. Anthony has a field with the length of each side as labeled on the sketch. What is the total number of feet of fence that would be required to fence in the field? (Find the perimeter of the field.)

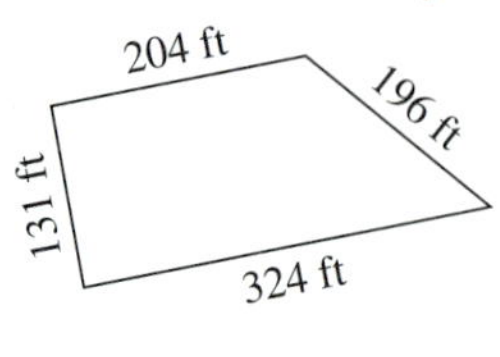

42. Jessica has a field with the length of each side as labeled on the sketch. What is the total number of feet of fence that would be required to fence in the field? (Find the perimeter of the field.)

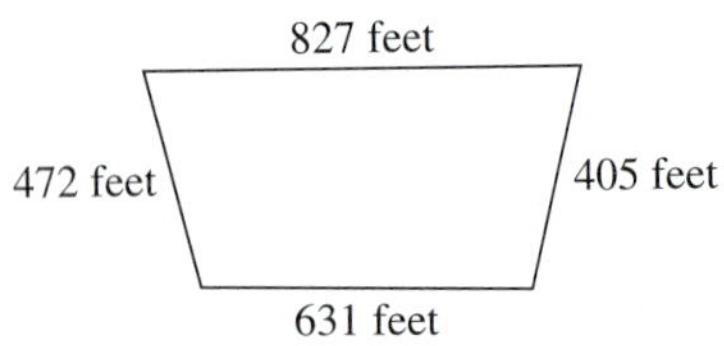

In problems 43, 44, and 46, you will find information about the relative sizes of different bodies of water found on Earth. For each problem write the total number of square miles in each grouping.

43. The Pacific Ocean, the world's largest, has an area of 64,000,000 square miles. The Atlantic Ocean has an area of 31,800,000 square miles. The Indian Ocean has an area of 25,300,000 square miles. What is the total area for these oceans?

44. The Arctic Ocean has an area of 5,400,000 square miles. The Mediterranean Sea has an area of 1,100,000 square miles. The Caribbean Sea has an area of 1,000,000 square miles. What is the total area for these oceans?

45. 2,144,856 people in California, 307,244 people in Oregon, and 470,239 people in Washington voted for Ross Perot for president in the year 1992. What was the total vote for Perot in those three states?

46. Lake Huron, which is bordered by the United States and Canada measures 23,010 square miles. Lake Tanganyika, which borders Tanzania and Zaire, measures 12,700 square miles. Lake Baikal in Russia measures 12,162 square miles. What is the total area for these lakes?

In problems 47–49, you will have to choose the numbers you need in order to answer each question. Then calculate the correct answer.

47. The admissions department of a competitive university is reviewing applications to see whether students are *eligible* or *ineligible* for student aid. On Monday, 415 were found eligible and 27 ineligible. On Tuesday, 364 were found eligible and 68 ineligible. On Wednesday, 159 were found eligible and 102 ineligible. On Thursday, 196 were found eligible and 61 ineligible.

(a) How many eligible students qualified for student aid during the four days?

(b) How many students were considered in all?

48. The quality control division of a motorcycle company classifies the final assembled bike as *passing* or *failing* final inspection. In January 14,311 vehicles passed whereas 56 failed. In February, 11,077 passed and 158 failed. In March, 12,580 passed and 97 failed.

(a) How many motorcycles passed the inspection during the three months?

(b) How many motorcycles were assembled during the three months in all?

49. Answer using the information in the following Western University expense chart for the academic year 1996–1997.

Western University Yearly Expenses	In-State Student, U.S. Citizen	Out-of-State Student, U.S. Citizen	Foreign Student
Tuition	\$3640	\$5276	\$8352
Room	1926	2437	2855
Board	1753	1840	1840

How much is the total cost for tuition, room, and board for
(a) *an out-of-state U.S. citizen?*
(b) *an in-state U.S. citizen?*
(c) *a foreign student?*

To Think About

Add.

50. 2,368,521,788 + 5,721,368,701 + 4,027,399,206

51. 89 + 166 + 23 + 45 + 72 + 190 + 203 + 77 + 18 + 93 + 46 + 73 + 66

52. What would happen if addition were not commutative?

53. What would happen if addition were not associative?

Cumulative Review Problems

Write the word name for each number.

54. 76,208,941

55. 121,000,374

Write each number in standard notation.

56. Eight million, seven hundred twenty-four thousand, three hundred ninety-six

57. Nine million, fifty-one thousand, seven hundred nineteen

Developing Your Study Skills

CLASS PARTICIPATION

People learn mathematics through active participation, not through observation from the sidelines. If you want to do well in this course, get involved in classroom activities. Sit near the front where you can see and hear well, where your focus is on the instruction process. Ask questions, be ready to contribute toward solutions, and take part in all classroom activities. Your contributions are valuable to the class and to yourself. Class participation requires an investment of yourself in the learning process, which you will find pays huge dividends.

1.3 Subtraction of Whole Numbers

After studying this section, you will be able to:

1 *Master basic subtraction facts.*
2 *Subtract whole numbers when borrowing is not necessary.*
3 *Subtract whole numbers when borrowing is necessary.*
4 *Check the answer to a subtraction problem.*
5 *Apply subtraction to real-world situations.*

MathPro Video 1 SSM

1 *Mastering Basic Subtraction Facts*

Subtraction is used day after day in the business world. The owner of a bakery placed an ad for his cakes in a local newspaper to see if this might increase his profits. To learn how many cakes had been sold, at closing time he subtracted the number of cakes remaining from the number of cakes the bakery had when it opened. To figure his profits, he subtracted his costs (including the cost of the ad) from his sales. Finally, to see if the ad paid off, he subtracted the profits he usually made in that period from the profits after advertising. He needed subtraction to see whether it paid to advertise.

What is subtraction? We do subtraction when we take objects away from a group. If you have 12 objects and take away 3 of them, 9 objects remain.

12 objects − 3 objects = 9 objects

$12 - 3 = 9$

If you earn \$400 per month, but have \$100 taken out for taxes, how much do you have left?

\$400	−	\$100	=	\$300
salary	subtraction symbol	amount withheld		amount left

We can use addition to help with a subtraction problem.

To subtract: $200 - 196 =$ what number

We can think: $196 +$ what number $= 200$

Usually when we subtract numbers, we put one number under the other in a column. When we subtract one number from another, the answer is called the **difference.**

$$\begin{array}{r} 9 \\ -\ 2 \\ \hline 7 \end{array} \qquad \begin{array}{r} 8 \\ -\ 3 \\ \hline 5 \end{array} \qquad \begin{array}{r} 12 \\ -\ \ 6 \\ \hline 6 \end{array} \qquad \begin{array}{r} 17 \\ -\ \ 9 \\ \hline 8 \end{array}$$

Each of these is called the difference of the two numbers.

The other two parts of a subtraction problem have labels, although you will not often come across them. The number being subtracted is called the **subtrahend.** The number being subtracted from is called the **minuend.**

$$\begin{array}{rl} 17 & \text{minuend} \\ -\ 9 & \text{subtrahend} \\ \hline 8 & \text{difference} \end{array}$$

In this case, the number 17 is called the *minuend.* The number 9 is called the *subtrahend.* The number 8 is called the *difference.*

Quick Recall of Subtraction Facts

It is helpful if you can subtract quickly. See if you can do Example 1 correctly in 15 seconds or less. Repeat again with Practice Problem 1. Strive to obtain all answers correctly in 15 seconds or less.

EXAMPLE 1 Subtract. **(a)** $8 - 2$ **(b)** $13 - 5$ **(c)** $12 - 4$ **(d)** $15 - 8$ **(e)** $16 - 0$

(a) $\begin{array}{r} 8 \\ -\ 2 \\ \hline 6 \end{array}$ **(b)** $\begin{array}{r} 13 \\ -\ 5 \\ \hline 8 \end{array}$ **(c)** $\begin{array}{r} 12 \\ -\ 4 \\ \hline 8 \end{array}$ **(d)** $\begin{array}{r} 15 \\ -\ 8 \\ \hline 7 \end{array}$ **(e)** $\begin{array}{r} 16 \\ -\ 0 \\ \hline 16 \end{array}$

Practice Problem 1 Subtract.

(a) $\begin{array}{r} 9 \\ -\ 6 \\ \hline \end{array}$ **(b)** $\begin{array}{r} 12 \\ -\ 5 \\ \hline \end{array}$ **(c)** $\begin{array}{r} 17 \\ -\ 8 \\ \hline \end{array}$ **(d)** $\begin{array}{r} 14 \\ -\ 0 \\ \hline \end{array}$ **(e)** $\begin{array}{r} 18 \\ -\ 9 \\ \hline \end{array}$ ■

2 *Subtracting Whole Numbers When Borrowing Is Not Necessary*

When we subtract numbers with more than two digits, in order to keep track of our work, we line up the ones column, the tens column, the hundreds column, and so on. Note that we begin with the ones column, and move from right to left.

EXAMPLE 2 Subtract. $9867 - 3725$

$$\begin{array}{r} 9\ 8\ 6\ 7 \\ -\ 3\ 7\ 2\ 5 \\ \hline 6\ 1\ 4\ 2 \end{array}$$

- 7 ones − 5 ones = 2 ones
- 6 tens − 2 tens = 4 tens
- 8 hundreds − 7 hundreds = 1 hundred
- 9 thousands − 3 thousands = 6 thousands

Practice Problem 2 Subtract. $7695 - 3481$ ■

3 *Subtracting Whole Numbers When Borrowing Is Necessary*

In the subtraction that we have looked at so far, each digit in the upper number (the minuend) has been greater than the digit in the lower number (the subtrahend) for each place value. Many times, however, a digit in the lower number is greater than the digit in the upper number for that place value.

$$\begin{array}{r} 42 \\ -28 \\ \hline \end{array}$$

The digit in the ones place in the lower number—the 8 of 28—is greater than the number in the ones place in the upper number—the 2 of 42. To subtract, we must *rename* 42, using place values. This is called **borrowing.**

EXAMPLE 3 Subtract. 42 − 28

$$\begin{array}{r} \overset{3}{\cancel{4}}\,\overset{12}{\cancel{2}} \\ -\ 2\ 8 \\ \hline 1\ 4 \end{array}$$

To subtract 8 ones from 2 ones, we need to borrow. Since 1 ten is 10 ones, we can rename 42 as 3 tens and 12 ones, writing the 3 in the tens column and the 12 in the ones column.

Now we subtract 8 ones from 12 ones to obtain 4 ones.

We then subtract 2 tens from 3 tens to obtain 1 ten.

Practice Problem 3 Subtract. 34 − 16 ■

EXAMPLE 4 Subtract. 864 − 548

To subtract 8 ones from 4 ones, we borrow 1 ten from the 6 tens and write 5 in the tens column to show what is left. Since we are borrowing 1 ten, which is 10 ones, we now have 14 ones, and write 14 in the ones column.

$$\begin{array}{r} 8\,\overset{5}{\cancel{6}}\,\overset{14}{\cancel{4}} \\ -\ 5\ 4\ 8 \\ \hline 3\ 1\ 6 \end{array}$$

Now we subtract 8 ones from 14 ones to obtain 6 ones.

We continue to subtract from right to left.

Practice Problem 4 Subtract.

$$\begin{array}{r} 693 \\ -\ 426 \\ \hline \end{array}$$ ■

EXAMPLE 5 Subtract. 8040 − 6375

$$\begin{array}{r} \overset{7}{\cancel{8}}\,\overset{\overset{9}{\cancel{10}}}{\cancel{0}}\,\overset{\overset{13}{\cancel{3}}}{\cancel{4}}\,\overset{10}{\cancel{0}} \\ -\ 6\ 3\ 7\ 5 \\ \hline 1\ 6\ 6\ 5 \end{array}$$

To subtract 5 from 0, we borrow 1 ten from the 4 tens to make 3 tens and 10 ones. 10 − 5 = 5

To subtract 7 tens from the 3 tens, we need to borrow 1 hundred to make 10 tens. Since we find a 0 in the hundreds column, first we borrow 1 thousand to make 10 hundreds. We show the number of thousands that are left, and write the 10 in the hundreds column. Now we borrow 1 hundred, show the number of hundreds that are left, and add the 10 tens to the 3 tens. We now do the subtraction.
13 tens − 7 tens = 6 tens

9 hundreds − 3 hundreds = 6 hundreds

7 thousands − 6 thousands = 1 thousand

Practice Problem 5 Subtract.

$$\begin{array}{r} 9070 \\ -\ 5886 \\ \hline \end{array}$$ ■

EXAMPLE 6 Subtract.

(a) 9521 − 943 **(b)** 40,000 − 29,056

(a)

$$\begin{array}{r} \overset{8}{\cancel{9}}\,\overset{\overset{14}{\cancel{4}}}{\cancel{5}}\,\overset{\overset{11}{\cancel{1}}}{\cancel{2}}\,\overset{11}{\cancel{1}} \\ -\quad 9\,4\,3 \\ \hline 8\,5\,7\,8 \end{array}$$

(b)

$$\begin{array}{r} \overset{3}{\cancel{4}}\,\overset{9}{\cancel{0}},\overset{9}{\cancel{0}}\,\overset{9}{\cancel{0}}\,\overset{10}{\cancel{0}} \\ -2\,9,0\,5\,6 \\ \hline 1\,0,9\,4\,4 \end{array}$$

Practice Problem 6 Subtract.

(a)

$$\begin{array}{r} 8964 \\ -\ 985 \\ \hline \end{array}$$

(b)

$$\begin{array}{r} 50{,}000 \\ -\ 32{,}508 \\ \hline \end{array}$$ ■

4 *Checking the Answer to a Subtraction Problem*

We observe that when 9 − 7 = 2 it follows that 7 + 2 = 9. Each subtraction problem is equivalent to a corresponding addition problem. This gives us a convenient way to check our answers to subtraction.

EXAMPLE 7 Check this subtraction problem.

5829 − 3647 = 2182

$$\begin{array}{r} 5\,8\,2\,9 \\ -3\,6\,4\,7 \\ \hline 2\,1\,8\,2 \end{array} \quad \text{then} \quad \begin{array}{r} 3\,6\,4\,7 \\ +2\,1\,8\,2 \\ \hline 5\,8\,2\,9 \end{array}$$

The sum should equal 5829, *which it does. We have checked our work, and it is correct.*

Practice Problem 7 Check this subtraction problem by adding.

9763 − 5732 = 4031 ■

EXAMPLE 8 Subtract and check your answers.

(a) 156,000 − 29,326 **(b)** 1,264,308 − 1,057,612

	Subtraction	Checking by addition	
(a)	156,000 − 29,326 126,674	29,326 + 126,674 156,000	It checks.
(b)	1,264,308 − 1,057,612 206,696	1,057,612 + 206,696 1,264,308	It checks.

Practice Problem 8 Subtract and check your answers.

(a)

$$\begin{array}{r} 284{,}000 \\ -\ 96{,}327 \\ \hline \end{array}$$

(b)

$$\begin{array}{r} 8{,}526{,}024 \\ -\ 6{,}397{,}518 \\ \hline \end{array}$$ ■

Subtraction can be used to solve word problems. Some problems can be expressed (and solved) with an *equation*. An equation is a number sentence with an equal sign, such as

$$10 = 4 + x$$

Here we use the letter x to represent a number we do not know. When we write $10 = 4 + x$, we are stating that 10 is equal to 4 added to some other number. Since $10 - 4 = 6$, we would assume that the number is 6. If we substitute 6 for x in the equation, we have two values that are the same.

$10 = 4 + x$

$10 = 4 + 6$ *Substitute 6 for x.*

$10 = 10$ *Both sides of the equation are the same.*

We can write an equation when one of the addends is not known, then use subtraction to solve for the unknown.

EXAMPLE 9 The librarian knows that he has eight world atlases and that five of them are in full color. How many are not in full color?

We represent the number that we don't know as x and write an equation, or mathematical sentence.

$$8 = 5 + x$$

To solve an equation means to find those values that will make the equation true. We solve this equation by reasoning and by a knowledge of the relationship between addition and subtraction.

$$8 = 5 + x \text{ is equivalent to } 8 - 5 = x$$

We know that $8 - 5 = 3$. Then $x = 3$. We can check the answer by substituting 3 for x in the original equation.

$8 = 5 + x$

$8 = 5 + 3$ True ✓

We see that $x = 3$ checks, so our answer is correct. There are 3 atlases not in full color.

Practice Problem 9 Solve.

(a) $x + 12 = 17$ **(b)** $10 + x = 22$ ■

5 *Applying Subtraction to Real-Life Situations*

We use subtraction in all kinds of situations. There are several key words in word problems that imply subtraction. Words that involve comparison, such as *how much more, how much greater,* or how much a quantity *increased* or *decreased,* all imply subtraction.

EXAMPLE 10 Look at the following population table.

ESTIMATED POPULATION OF FOUR STATES FROM 1960 TO 1990

Source: U.S. Bureau of the Census

State	1960	1970	1980	1990
California	15,717,204	19,971,069	23,667,947	29,760,021
Texas	9,579,677	11,198,655	14,227,799	16,986,510
Arizona	1,302,161	1,775,399	2,716,598	3,665,228
New Mexico	951,023	1,017,055	1,303,302	1,515,069

(a) In 1980 how much greater was the population of Texas than that of Arizona?

(b) How much did the population of California increase from 1960 to 1990?

(c) How much greater was the population of California in 1990 than the other three states combined?

(a)

$$\begin{array}{rl} 14{,}227{,}799 & \text{1980 population of Texas} \\ -\ 2{,}716{,}598 & \text{1980 population of Arizona} \\ \hline 11{,}511{,}201 & \text{difference} \end{array}$$

The population of Texas was greater by 11,511,201.

(b)

$$\begin{array}{rl} 29{,}760{,}021 & \text{1990 population of California} \\ -\ 15{,}717{,}204 & \text{1960 population of California} \\ \hline 14{,}042{,}817 & \text{difference} \end{array}$$

The population of California increased by 14,042,817 in those 30 years.

(c) First we need to find the total population in 1990 of Texas, Arizona, and New Mexico.

$$\begin{array}{rl} 16{,}986{,}510 & \text{1990 population of Texas} \\ 3{,}665{,}228 & \text{1990 population of Arizona} \\ +\ 1{,}515{,}069 & \text{1990 population of New Mexico} \\ \hline 22{,}166{,}807 & \end{array}$$

We use subtraction to compare this total with the population of California.

$$\begin{array}{rl} 29{,}760{,}021 & \text{1990 population of California} \\ -\ 22{,}166{,}807 & \\ \hline 7{,}593{,}214 & \end{array}$$

The population in California in 1990 is 7,593,214 more than the population in the other three states combined.

Practice Problem 10

(a) In 1980 how much greater was the population of California than the population of Texas?

(b) How much did the population of Texas increase from 1960 to 1970? ■

EXAMPLE 11 The number of real estate transfers in several towns during the years 1994 to 1996 is given in the following bar graph.

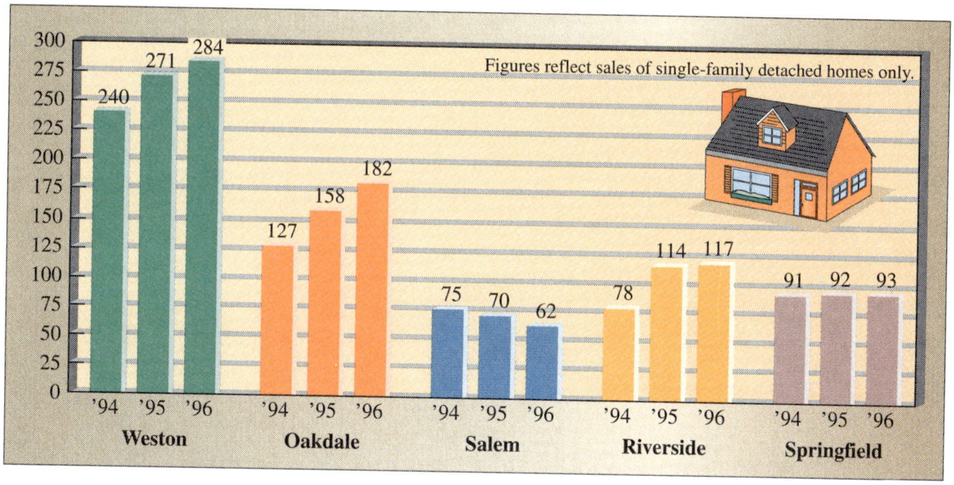

(a) What was the increase in homes sold in Weston from 1995 to 1996?

(b) What was the decrease in homes sold in Salem from 1994 to 1996?

(c) Between what two years did Oakdale have the greatest increase in sales?

(a) From the labels on the bar graph we see that 284 homes were sold in 1996 in Weston and 271 homes were sold in 1995. Thus the increase can be found by subtracting $284 - 271 = 13$. There was an increase of 13 homes sold in Weston from 1995 to 1996.

(b) In 1994, 75 homes were sold in Salem. In 1996, 62 homes were sold in Salem. The decrease in the number of homes sold is $75 - 62 = 13$. There was a decrease of 13 homes sold in Salem from 1994 to 1996.

(c) Here we will need to make two calculations in order to decide where the greatest increase occurs.

158	1995 sales	182	1996 sales
− 127	1994 sales	− 158	1995 sales
31	Sales increase from 1994 to 1995	24	Sales increase from 1995 to 1996

The greatest increase in sales in Oakdale occurred from 1994 to 1995.

Practice Problem 11

Based on the bar graph above, answer the following questions.

(a) What was the increase in homes sold in Riverside from 1994 to 1995?

(b) How many more homes were sold in Springfield in 1994 than in Riverside in 1994?

(c) Between what two years did Weston have the greatest increase in sales? ■

1.3 Exercises

Try to do problems 1–20 in 1 minute or less with no errors.

Subtract.

1. 8 − 2	**2.** 7 − 5	**3.** 13 − 5	**4.** 6 − 0	**5.** 6 − 5	**6.** 17 − 8	**7.** 15 − 6
8. 14 − 5	**9.** 16 − 0	**10.** 17 − 9	**11.** 18 − 9	**12.** 12 − 7	**13.** 11 − 4	**14.** 15 − 8
15. 13 − 7	**16.** 16 − 9	**17.** 11 − 8	**18.** 10 − 7	**19.** 15 − 6	**20.** 12 − 5	

Subtract. Check your answers by adding.

21. 98 − 37	**22.** 84 − 32	**23.** 85 − 73	**24.** 77 − 36	**25.** 126 − 95
26. 193 − 72	**27.** 768 − 143	**28.** 621 − 320	**29.** 1763 − 422	**30.** 1896 − 743
31. 24,396 − 13,205	**32.** 52,980 − 31,660	**33.** 986,302 − 433,201	**34.** 807,965 − 304,214	

Check each subtraction. If the problem has not been done correctly, find the correct answer.

35. 129 − 19 = 110	**36.** 186 − 45 = 141	**37.** 6180 − 1113 = 5067	**38.** 5976 − 3243 = 2733
39. 6030 − 5020 = 1020	**40.** 7890 − 3200 = 7670	**41.** 98,763 − 42,531 = 55,232	**42.** 47,969 − 33,846 = 14,223

Subtract. Use borrowing if necessary.

43. 93 − 47	**44.** 86 − 33	**45.** 125 − 88	**46.** 136 − 95	**47.** 451 − 376	**48.** 706 − 435
49. 905 − 324	**50.** 861 − 345	**51.** 10,000 − 6,704	**52.** 20,000 − 13,120	**53.** 152,000 − 117,908	
54. 361,000 − 121,520	**55.** 42,312 − 39,998	**56.** 54,913 − 29,997	**57.** 760,108 − 536,992	**58.** 580,092 − 349,905	

Solve.

59. $x + 14 = 19$

60. $x + 35 = 50$

61. $34 = x + 13$

62. $25 = x + 18$

63. $100 + x = 127$

64. $86 + x = 120$

Applications

Answer each question.

65. This year, for the college graduation ceremonies, the senior class voted for "Power of Example to Humanity." Martin Luther King received 765 votes, John F. Kennedy received 960 votes, and Mother Theresa received 778 votes. How many more votes did John F. Kennedy receive than Mother Theresa?

66. Three of the highest elevations on Earth are Mt. Everest, in Asia, which is 8846 meters high, Mt. Aconcagua, in Argentina, which is 6960 meters high, and Mt. McKinley in the United States (Alaska), which is 6194 meters high. How much higher is Mt. Everest than Mt. Aconcagua?

67. In 1994, the population of Ireland was approximately 3,539,296. In the same year, the population of Portugal was approximately 10,524,210. How much less than the population of Portugal was the population of Ireland in 1994?

68. The principality of Monaco, located on the Mediterranean coast of France, had a budget revenue of \$424,130,812 in 1994. The expenses for the year were \$376,211,007. How much was left over after expenses?

69. James earned \$475 painting a few rooms for his neighbor. He gave \$142 to his assistant and paid \$85 to the hardware store for materials. How much money did he actually receive after paying expenses?

70. The Ski Club raised \$1682 at a party for their favorite charity. The band received \$350 and they spent \$265 on refreshments. How much money was actually raised for the charity after paying expenses?

In answering problems 71–76, consider the following population table.

	1960	1970	1980	1990
Illinois	10,081,158	11,110,285	11,427,409	11,430,602
Michigan	7,823,194	8,881,826	9,262,044	9,295,297
Indiana	4,662,498	5,195,392	5,490,212	5,544,159
Minnesota	3,413,864	3,806,103	4,075,970	4,375,099

71. How much did the population of Minnesota increase from 1960 to 1990?

72. How much did the population of Michigan increase from 1960 to 1990?

73. In 1970, how much greater was the population of Michigan than the population of Indiana?

74. In 1980, how much greater was the population of Indiana than the population of Minnesota?

75. How much did the population of Illinois increase from 1970 to 1990?

76. How much did the population of Michigan increase from 1980 to 1990?

The number of real estate transfers in several towns during the years 1994 to 1996 is given in the following bar graph. Use the bar graph to answer questions 77–84. The figures in the bar graph reflect sales of single-family detached homes only.

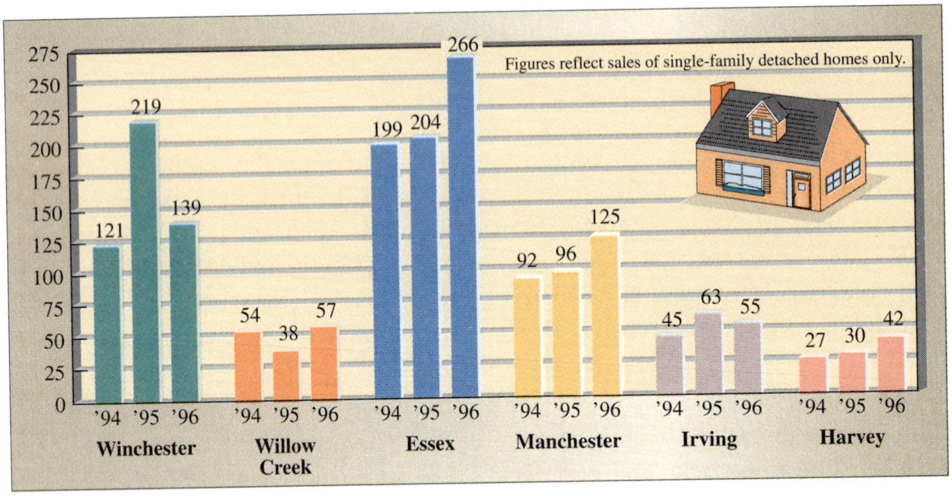

77. What was the increase in the number of homes sold in Manchester from 1995 to 1996?

78. What was the increase in the number of homes sold in Irving from 1994 to 1995?

79. What was the decrease in the number of homes sold in Winchester from 1995 to 1996?

80. What was the decrease in the number of homes sold in Willow Creek from 1994 to 1995?

81. Between what two years did the greatest change occur in the number of homes sold in Irving?

82. Between what two years did the greatest change occur in the number of homes sold in Winchester?

83. A realtor was trying to determine which two towns were closest to having the same number of sales of homes in 1994. What two towns should she select?

84. A realtor was trying to determine which two towns were closest to having the same number of sales of homes in 1995. What two towns should she select?

To Think About

85. In general, subtraction is not commutative. If a and b are whole numbers, $a - b \neq b - a$. For what types of numbers would it be true that $a - b = b - a$?

86. In general, subtraction is not associative. For example, $8 - (4 - 3) \neq (8 - 4) - 3$. In general, $a - (b - c) \neq (a - b) - c$. Can you find some numbers a, b, c for which $a - (b - c) = (a - b) - c$? (Remember, do operations inside the parentheses first.)

87. Subtract. 28,007,653,121,863 − 27,986,430,705,999

88. Subtract. 101,000,101,000,101 − 99,090,989,878,789

Cumulative Review Problems

89. Write in standard notation: eight million, four hundred sixty-six thousand, eighty-four

90. Write a word name for 296,308.

91. Add. 16 + 27 + 82 + 34 + 9

92. Add.
$$\begin{array}{r} 156{,}325 \\ +\ 963{,}209 \\ \hline \end{array}$$

Developing Your Study Skills

HOW TO DO HOMEWORK

Set aside time each day for your homework assignments. Do not attempt to do a whole week's worth on the weekend. Two hours spent studying outside of class for each hour in class is usual for college courses. You may need more than that for mathematics.

Before beginning to solve your homework problems, read your textbook very carefully. Expect to spend much more time reading a few pages of a mathematics textbook than several pages of another text. Read for complete understanding, not just for the general idea.

As you begin your homework assignments, read the directions carefully. You need to understand what is being asked. Concentrate on each exercise or problem, taking time to solve it accurately. Rushing through your work usually causes errors. Check your answers with those given in the back of the textbook. If your answer is incorrect, check to see that you are doing the right problem. Redo the problem, watching for little errors. If it is still wrong, check with a friend. Perhaps the two of you can figure out where you are going wrong.

Work on your assignments every day and do as many problems as it takes for you to know what you are doing. Begin by doing all the problems that have been assigned. If there are more available in that section of your text, then do more. When you think you have done enough problems to understand fully the topic at hand, do a few more to be sure. This may mean that you do many more problems than the instructor assigns, but you can never practice mathematics too much. Practice improves your skills and increases your accuracy, speed, competence, and confidence.

Also, check the examples in the textbook or in your notes for a similar exercise or problem. Can this one be solved in the same way? Give it some thought. You may want to leave it for a while by taking a break or doing a different problem. But come back later and try again. If you are still unable to figure it out, ask your instructor for help during office hours or in class.

1.4 Multiplication of Whole Numbers

MathPro Video 2 SSM

After studying this section, you will be able to:

1. *Master basic multiplication facts.*
2. *Multiply a single-digit number by a number with several digits.*
3. *Multiply a number by a number that is a power of 10.*
4. *Multiply a several-digit number by a several-digit number.*
5. *Use the properties of multiplication to perform efficient calculations.*
6. *Apply multiplication to real-life situations.*

1 *Mastering Basic Multiplication Facts*

Like subtraction, multiplication is related to addition. Suppose that the pastry chef at the Gourmet Restaurant bakes croissants on a sheet that holds four croissants across, with room for three rows. How many croissants does the sheet hold?

We can add 4 + 4 + 4 to get the total, or we can use a shortcut: 3 rows of 4 is the same as 3 times 4, which equals 12. This is **multiplication,** a shortcut for repeated addition.

The numbers that we multiply are called **factors.** The answer is called the **product.** For now, we will use × to show multiplication. 3 × 4, is read "3 times 4."

$$\underbrace{3}_{\text{factor}} \times \underbrace{4}_{\text{factor}} = \underbrace{12}_{\text{product}}$$

$$\begin{array}{rl} 3 & \text{factor} \\ \times 4 & \text{factor} \\ \hline 12 & \text{product} \end{array}$$

Your skill in multiplication depends on how well you know the basic multiplication facts. Look at the table on page 35. You should learn these facts well enough to quickly and correctly give the products of any two factors in the table. To check your knowledge, try Exercises 1.4, problems 3 and 4.

Study the table to see if you can discover any properties of multiplication. What do you see as results when you multiply zero by any number? When you multiply any number times zero, the result is zero. That is the **multiplication property of zero.**

$$2 \times 0 = 0 \qquad 5 \times 0 = 0 \qquad 0 \times 6 = 0 \qquad 0 \times 0 = 0$$

You may recall that zero plays a special role in addition. Zero is the *identity element* for addition. When we add any number to zero, that number does not change. Is there an identity element for multiplication? Look at the table. What is the identity element for multiplication?

What other properties of addition hold for multiplication? Is multiplication commutative? Does the order in which you multiply two numbers change the results? Find the product of 3×4. Then find the product of 4×3.

$$3 \times 4 = 12$$

$$4 \times 3 = 12$$

The **commutative property of multiplication** tells us that when we multiply two numbers, changing the order of the numbers gives the same result.

BASIC MULTIPLICATION FACTS

×	0	1	2	3	4	5	6	7	8	9
0	0	0	0	0	0	0	0	0	0	0
1	0	1	2	3	4	5	6	7	8	9
2	0	2	4	6	8	10	12	14	16	18
3	0	3	6	9	12	15	18	21	24	27
4	0	4	8	12	16	20	24	28	32	36
5	0	5	10	15	20	25	30	35	40	45
6	0	6	12	18	24	30	36	42	48	54
7	0	7	14	21	28	35	42	49	56	63
8	0	8	16	24	32	40	48	56	64	72
9	0	9	18	27	36	45	54	63	72	81

Quick Recall of Multiplication Facts

It is helpful if you can multiply quickly. See if you can do Example 1 correctly in 15 seconds or less. Repeat again with Practice Problem 1. Strive to obtain all answers correctly in 15 seconds or less.

EXAMPLE 1 Multiply.

(a) 5×7 **(b)** 8×9 **(c)** 6×8 **(d)** 9×3 **(e)** 7×8

(a) $\begin{array}{r} 7 \\ \times 5 \\ \hline 35 \end{array}$ **(b)** $\begin{array}{r} 9 \\ \times 8 \\ \hline 72 \end{array}$ **(c)** $\begin{array}{r} 8 \\ \times 6 \\ \hline 48 \end{array}$ **(d)** $\begin{array}{r} 3 \\ \times 9 \\ \hline 27 \end{array}$ **(e)** $\begin{array}{r} 8 \\ \times 7 \\ \hline 56 \end{array}$

Practice Problem 1 Multiply.

(a) $\begin{array}{r} 8 \\ \times 8 \\ \hline \end{array}$ **(b)** $\begin{array}{r} 7 \\ \times 6 \\ \hline \end{array}$ **(c)** $\begin{array}{r} 5 \\ \times 8 \\ \hline \end{array}$ **(d)** $\begin{array}{r} 9 \\ \times 7 \\ \hline \end{array}$ **(e)** $\begin{array}{r} 9 \\ \times 9 \\ \hline \end{array}$ ■

2 *Multiplying a Single-Digit Number by a Number with Several Digits*

EXAMPLE 2 Multiply. 4312×2

We first multiply the ones column, then the tens column, and so on, moving right to left.

$$\begin{array}{r} 4\,3\,1\,2 \\ \times \quad\quad 2 \\ \hline 8\,6\,2\,4 \end{array}$$

- $2 \times 2 = 4$ in the ones column
- $2 \times 1 = 2$ in the tens column
- $2 \times 3 = 6$ in the hundreds column
- $2 \times 4 = 8$ in the thousands column

Practice Problem 2 Multiply. 3021×3 ■

Usually, we will have to carry one digit of the result of some of the multiplication into the next left-hand column.

EXAMPLE 3 Multiply. 36×7

$$\begin{array}{r} {}^{4}\quad \\ 3\,6 \\ \times \quad 7 \\ \hline 2\,5\,2 \end{array}$$

carry number = 4

- $7 \times 6 = 42$. Leave the 2 in the ones column and carry the 4 to the next column.
- 7×3 tens $= 21$ tens. Add 21 tens + 4 tens to obtain 25 tens or 2 hundreds + 5 tens.

Practice Problem 3 Multiply. 43×8 ■

EXAMPLE 4 Multipy. 359×9

$$\begin{array}{r} {}^{5\;\,8}\quad \\ 3\,5\,9 \\ \times \quad\; 9 \\ \hline 3\,2\,3\,1 \end{array}$$

- $9 \times 9 = 81$. Leave the 1 in the ones column and carry the 8 to the top of the tens column.
- $9 \times 5 = 45$. Now add 45 tens + 8 tens = 53 tens or 5 hundreds + 3 tens. Leave the 3 in the tens column and carry the 5 to the top of the hundreds column.
- $9 \times 3 = 27$. Now add 27 hundreds + 5 hundreds to obtain 32 hundreds or 3 thousands + 2 hundreds.

Practice Problem 4 Multiply.

$$\begin{array}{r} 579 \\ \times \quad 7 \\ \hline \end{array}$$ ■

3 Multiplying by a Power of 10

Observe what happens when a number is multiplied by 10, 100, 1000, 10,000, and so on.

$$56 \times 10 = 560 \quad \text{(one zero)} \qquad 56 \times 100 = 5600 \quad \text{(two zeros)}$$

$$56 \times 1000 = 56{,}000 \quad \text{(three zeros)} \qquad 56 \times 10{,}000 = 560{,}000 \quad \text{(four zeros)}$$

A **power of 10** is a whole number that begins with 1 and ends in one or more zeros. The numbers 10, 100, 1000, 10,000, and so on are powers of 10.

> To multiply a whole number by a power of 10:
>
> **1.** Count the number of zeros in the power of 10.
>
> **2.** Attach that number of zeros to the right side of the other whole number to obtain the answer.

EXAMPLE 5 Multiply 358 by each number.

(a) 10 **(b)** 100 **(c)** 1000 **(d)** 100,000

(a) 358 × 10 = 3580 (*one zero*)

(b) 358 × 100 = 35,800 (*two zeros*)

(c) 358 × 1000 = 358,000 (*three zeros*)

(d) 358 × 100,000 = 35,800,000 (*five zeros*)

Practice Problem 5 Multiply 1267 by each number.

(a) 10 **(b)** 1000 **(c)** 10,000 **(d)** 1,000,000 ■

How can we handle zeros in multiplication involving a number that is not 10, 100, 1000, or any other power of 10? Consider 32 × 400. We can rewrite 400 as 4 × 100, which gives us 32 × 4 × 100. We can simply multiply 32 × 4 and then attach two zeros for the factor 100. We find that 32 × 4 = 128. Attaching two zeros gives us 12,800, or 32 × 400 = 12,800.

EXAMPLE 6 Multiply. **(a)** 12 × 3000 **(b)** 25 × 600 **(c)** 43 × 40,000

(a) 12 × 3000 = 12 × 3 × 1000 = 36 × 1000 = 36,000

(b) 25 × 600 = 25 × 6 × 100 = 150 × 100 = 15,000

(c) 43 × 40,000 = 43 × 4 × 10,000 = 172 × 10,000 = 1,720,000

Practice Problem 6 Multiply. **(a)** 9 × 60,000 **(b)** 15 × 400 **(c)** 21 × 8000 ■

4 Multiplying a Several-Digit Number by a Several-Digit Number

EXAMPLE 7 Multiply. 234 × 21

We can consider 21 as 2 tens (20) and 1 one (1). First we multiply 234 by 1.

$$\begin{array}{r} 234 \\ \times \quad 21 \\ \hline 234 \end{array}$$

The 234 is called a **partial product.** We next multiply 234 by 20.

$$\begin{array}{r} 234 \\ \times \quad 21 \\ \hline 234 \\ 468 \end{array}$$

Note that 468—also called a partial product—is placed so that its ones digit appears in the tens column. This placement is necessary because we are multiplying by the digit in the tens column of the factor 21. We may add a zero to the right of the 468 to make sure that the 8 of 468 lines up in the tens column.

$$\begin{array}{r} 234 \\ \times \quad 21 \\ \hline 234 \\ 4680 \end{array}$$

We now add the two partial products to reach the final product, which is the solution.

$$\begin{array}{rl} 234 & \\ \times \quad 21 & \\ \hline 234 & \leftarrow \text{Multiply } 234 \times 1. \\ 4680 & \leftarrow \text{Multiply } 234 \times 20. \text{ First put a zero in the ones column to line up the digits.} \\ \hline 4914 & \leftarrow \text{Add the two partial products.} \end{array}$$

Practice Problem 7 Multiply.

$$\begin{array}{r} 323 \\ \times \quad 32 \\ \hline \end{array}$$ ■

EXAMPLE 8 Multiply. 671 × 35

$$\begin{array}{rl} 6\ 7\ 1 & \\ \times \quad 3\ 5 & \\ \hline 3\ 3\ 5\ 5 & \leftarrow \text{First multiply } 671 \times 5. \\ 2\ 0\ 1\ 3\ \textcircled{0} & \leftarrow \text{Now multiply } 671 \times 30. \\ \hline 2\ 3\ 4\ 8\ 5 & \leftarrow \text{Now add the two partial products.} \end{array}$$

Note: We could omit zero on this line and leave the ones place blank.

Practice Problem 8 Multiply.

$$\begin{array}{r} 385 \\ \times \quad 69 \\ \hline \end{array}$$ ■

EXAMPLE 9 Multiply. 14×20

$$\begin{array}{r} 14 \\ \times\ 20 \\ \hline 0 \\ 280 \\ \hline 280 \end{array}$$

0 ← Multiply 14 by 0.

280 ← Multiply 14 by 2 tens. Place 28 with the 8 in the tens column. To line up the digits for adding, we can insert a 0.

Now add the partial products. → 280

Notice that you will also get this result if you multiply $14 \times 2 = 28$ *and then attach a zero to multiply it by* 10: 280.

Practice Problem 9 Multiply. 34×20 ■

EXAMPLE 10 Multiply. 120×40

$$\begin{array}{r} 120 \\ \times\ \ 40 \\ \hline 0 \\ 4800 \\ \hline 4800 \end{array}$$

0 ← Multiply 120×0.

4800 ← Multiply 120 by 4 tens. The answer is 480 tens. We place the 0 of the 480 in the tens column. To line up the digits for adding, we can insert a 0.

Now add the partial products. → 4800

Notice that this result is the same as $12 \times 4 = 48$ *with two zeros attached:* 4800.

Practice Problem 10 Multiply. 130×50 ■

EXAMPLE 11 Multiply. 684×763

$$\begin{array}{rrrrrr} & & & 6 & 8 & 4 \\ & & \times & 7 & 6 & 3 \\ \hline & & 2 & 0 & 5 & 2 \\ & 4 & 1 & 0 & 4 & \\ 4 & 7 & 8 & 8 & & \\ \hline 5 & 2 & 1 & 8 & 9 & 2 \end{array}$$

2052 ← Multiply 684×3.

4104 ← Multiply 684×60. Note that we omit the final zero.

4788 ← Multiply 684×700. Note we omit final two zeros.

Practice Problem 11 Multiply. 923×675 ■

5 *Using the Properties of Multiplication to Perform Efficient Calculations*

When we add three numbers, we use the associative property. Recall that the associative property allows us to group the three numbers in any way. Thus to add $9 + 7 + 3$, we can group the numbers as $9 + (7 + 3)$ because it is easier to find the sum. $9 + (7 + 3) = 9 + 10 = 19$. We can demonstrate that multiplication is also associative.

Is this true? $\quad 2 \times (5 \times 3) = (2 \times 5) \times 3$

$$2 \times (15) = (10) \times 3$$
$$30 = 30$$

The final product is the same in both cases.

The way we group numbers to be multiplied does not change the product. This property is called the **associative property of multiplication.**

EXAMPLE 12 Multiply. $14 \times 2 \times 5$

Since we can group any two numbers together, let's take advantage of multiplying by 10.

$$14 \times 2 \times 5 = 14 \times (2 \times 5) = 14 \times 10 = 140$$

Practice Problem 12 Multiply. $25 \times 4 \times 17$ ■

For convenience, we list the properties of multiplication that we have discussed in this section.

1. Associative Property of Multiplication. When we multiply three numbers, the multiplication can be grouped in any way.	$(7 \times 3) \times 2 = 7 \times (3 \times 2)$ $21 \times 2 = 7 \times 6$ $42 = 42$
2. Commutative Property of Multiplication. Two numbers can be multiplied in either order with the same result.	$9 \times 8 = 8 \times 9$ $72 = 72$
3. Identity Property of One. When one is multiplied by a number, the result is that number.	$7 \times 1 = 7$ $1 \times 15 = 15$
4. Multiplication Property of Zero. The product of any number and zero yields zero as a result.	$0 \times 14 = 0$ $2 \times 0 = 0$

Sometimes you can use several properties in one problem to make the calculation easier.

EXAMPLE 13 Multiply. $7 \times 20 \times 5 \times 6$

$$
\begin{aligned}
7 \times 20 \times 5 \times 6 &= 7 \times (20 \times 5) \times 6 && \textit{Associative property} \\
&= 7 \times 6 \times (20 \times 5) && \textit{Commutative property} \\
&= 42 \times 100 \\
&= 4200
\end{aligned}
$$

Practice Problem 13 Multiply. $8 \times 4 \times 3 \times 25$ ■

Thus far we have discussed the properties of addition and the properties of multiplication. There is one more property that links both operations.

Before we discuss that property, we will illustrate several different ways of showing multiplication. The following are all the ways to show "3 times 4."

3×4	$(3)(4)$	$3(4)$ $(3)4$	$3 \cdot 4$	$3 * 4$
with an $\times$	with two sets of parentheses	with a single set of parentheses	with a dot	with a star

We will use parentheses to mean multiplication when we use the **distributive property.**

SIDELIGHT Why does our method of multiplying several-digit numbers work? Why can we say that 234×21 is the same as $234 \times 1 + 234 \times 20$?

The *distributive property of multiplication over addition* allows us to distribute the multiplication and then add the results. To illustrate, 234×21 can be written as $234(20 + 1)$. By the distributive property

$$
\begin{aligned}
234(20 + 1) &= (234 \times 20) + (234 \times 1) \\
&= \quad 4680 \quad + \quad 234 \\
&= 4914
\end{aligned}
$$

This is what we actually do when we multiply.

$$
\begin{array}{r}
234 \\
\times \ \ 21 \\
\hline
234 \\
4680 \\
\hline
4914
\end{array}
$$

Distributive Property of Multiplication over Addition Multiplication can be distributed over addition without changing the result.

$$5 \times (10 + 2) = (5 \times 10) + (5 \times 2)$$

6 *Applying Multiplication to Real-Life Situations*

To use multiplication in word problems, the number of items or the value of each item must be the same. Recall that multiplication is a quick way to do repeated addition where *each addend is the same.* In the beginning of the section we showed 3 rows of 4 croissants to illustrate 3×4. The number of croissants in each row was the same, 4. Look at another example. If we had six nickels, we could use multiplication to find the total value of the coins because the value of each nickel is the same, 5¢. Since $6 \times 5 = 30$, six nickels are worth 30¢.

In the following example the word *average* is used. The word average has several different meanings. In this example, we are told that the *average annual salary* of an employee at Software Associates is \$21,342. This means that we can calculate the total payroll as if each employee made \$21,342 even though we know that the president probably makes more than any other employee.

EXAMPLE 14 The average annual salary of an employee at Software Associates is \$21,342. There are 38 employees. What is the annual payroll?

$$\begin{array}{r} \$21{,}342 \\ \times \quad 38 \\ \hline 170\ 736 \\ 640\ 26 \\ \hline 810{,}996 \end{array}$$

The total annual payroll is about \$810,966.

Practice Problem 14 The average cost of a new car sold last year at Westover Chevrolet was \$17,348. The dealership sold 378 cars. What were the total sales of cars at the dealership last year? ■

Another useful application of multiplication is area. The following example involves the area of a rectangle.

EXAMPLE 15 What is the area of a rectangular hallway that measures 4 feet wide and 8 feet long?

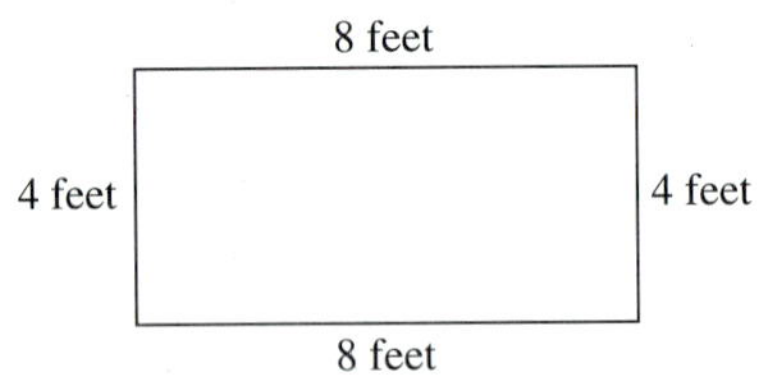

The **area** of a rectangle is the product of the length times the width. Thus for this hallway

$$\text{Area} = 8 \text{ feet} \times 4 \text{ feet} = 32 \text{ square feet}$$

The area of the hallway is 32 square feet.

Note: All measurements for area are given in square units such as square feet, square meters, square yards, etc.

Practice Problem 15 What is the area of a rectangular rug that measures 5 yards by 7 yards? ■

1.4 Exercises

Verbal and Writing Skills

1. Explain in your own words.

(a) the commutative property of multiplication

(b) the associative property of multiplication

2. How does the distributive property of multiplication over addition help us to multiply 4×13?

Complete the multiplication facts for each table. Strive for total accuracy, but work quickly. (For problems 3 and 4 allow a maximum of 5 minutes each.)

Multiply.

3.

$\times$	6	2	3	8	0	5	7	9	1	4
5										
7										
1										
0										
6										
2										
3										
8										
4										
9										

4.

$\times$	2	7	0	5	3	4	8	1	6	9
1										
6										
5										
3										
0										
9										
4										
7										
2										
8										

5. $\begin{array}{r} 32 \\ \times\ 3 \\ \hline \end{array}$ **6.** $\begin{array}{r} 21 \\ \times\ 4 \\ \hline \end{array}$ **7.** $\begin{array}{r} 104 \\ \times\ \ 2 \\ \hline \end{array}$ **8.** $\begin{array}{r} 302 \\ \times\ \ 3 \\ \hline \end{array}$ **9.** $\begin{array}{r} 6102 \\ \times\ \ \ 3 \\ \hline \end{array}$

10. $\begin{array}{r} 5203 \\ \times\ \ \ 2 \\ \hline \end{array}$ **11.** $\begin{array}{r} 101{,}204 \\ \times\ \ \ \ \ 3 \\ \hline \end{array}$ **12.** $\begin{array}{r} 301{,}200 \\ \times\ \ \ \ \ 4 \\ \hline \end{array}$ **13.** $\begin{array}{r} 14 \\ \times\ 5 \\ \hline \end{array}$ **14.** $\begin{array}{r} 12 \\ \times\ 4 \\ \hline \end{array}$

15. $\begin{array}{r} 87 \\ \times\ 6 \\ \hline \end{array}$ **16.** $\begin{array}{r} 95 \\ \times\ 7 \\ \hline \end{array}$ **17.** $\begin{array}{r} 326 \\ \times\ \ 5 \\ \hline \end{array}$ **18.** $\begin{array}{r} 732 \\ \times\ \ 6 \\ \hline \end{array}$ **19.** $\begin{array}{r} 1087 \\ \times\ \ \ 7 \\ \hline \end{array}$

20. $\begin{array}{r} 2605 \\ \times\ \ \ 9 \\ \hline \end{array}$ **21.** $\begin{array}{r} 12{,}526 \\ \times\ \ \ \ 8 \\ \hline \end{array}$ **22.** $\begin{array}{r} 48{,}761 \\ \times\ \ \ \ 7 \\ \hline \end{array}$ **23.** $\begin{array}{r} 235{,}702 \\ \times\ \ \ \ \ 4 \\ \hline \end{array}$ **24.** $\begin{array}{r} 127{,}054 \\ \times\ \ \ \ \ 6 \\ \hline \end{array}$

Multiply by powers of 10.

25. $\begin{array}{r} 156 \\ \times\ 10 \\ \hline \end{array}$

26. $\begin{array}{r} 278 \\ \times\ 10 \\ \hline \end{array}$

27. $\begin{array}{r} 27{,}158 \\ \times\ 100 \\ \hline \end{array}$

28. $\begin{array}{r} 89{,}361 \\ \times\ 100 \\ \hline \end{array}$

29. $\begin{array}{r} 482 \\ \times\ 1000 \\ \hline \end{array}$

30. $\begin{array}{r} 579 \\ \times\ 1000 \\ \hline \end{array}$

31. $\begin{array}{r} 316{,}250 \\ \times\ 10{,}000 \\ \hline \end{array}$

32. $\begin{array}{r} 25{,}346 \\ \times\ 10{,}000 \\ \hline \end{array}$

33. $\begin{array}{r} 423 \\ \times\ 20 \\ \hline \end{array}$

34. $\begin{array}{r} 134 \\ \times\ 20 \\ \hline \end{array}$

35. $\begin{array}{r} 2120 \\ \times\ 30 \\ \hline \end{array}$

36. $\begin{array}{r} 4230 \\ \times\ 20 \\ \hline \end{array}$

37. $\begin{array}{r} 14{,}000 \\ \times\ 4000 \\ \hline \end{array}$

38. $\begin{array}{r} 62{,}000 \\ \times\ 3000 \\ \hline \end{array}$

Multiply.

39. $\begin{array}{r} 1022 \\ \times\ 31 \\ \hline \end{array}$

40. $\begin{array}{r} 415 \\ \times\ 24 \\ \hline \end{array}$

41. $\begin{array}{r} 146 \\ \times\ 54 \\ \hline \end{array}$

42. $\begin{array}{r} 3021 \\ \times\ 12 \\ \hline \end{array}$

43. $\begin{array}{r} 89 \\ \times\ 64 \\ \hline \end{array}$

44. $\begin{array}{r} 68 \\ \times\ 49 \\ \hline \end{array}$

45. $\begin{array}{r} 607 \\ \times\ 25 \\ \hline \end{array}$

46. $\begin{array}{r} 817 \\ \times\ 84 \\ \hline \end{array}$

47. $\begin{array}{r} 569 \\ \times\ 73 \\ \hline \end{array}$

48. $\begin{array}{r} 915 \\ \times\ 47 \\ \hline \end{array}$

49. $\begin{array}{r} 912 \\ \times\ 76 \\ \hline \end{array}$

50. $\begin{array}{r} 498 \\ \times\ 39 \\ \hline \end{array}$

51. $\begin{array}{r} 5123 \\ \times\ 29 \\ \hline \end{array}$

52. $\begin{array}{r} 1268 \\ \times\ 38 \\ \hline \end{array}$

53. $\begin{array}{r} 9053 \\ \times\ 91 \\ \hline \end{array}$

54. $\begin{array}{r} 3078 \\ \times\ 72 \\ \hline \end{array}$

55. $\begin{array}{r} 178 \\ \times\ 235 \\ \hline \end{array}$

56. $\begin{array}{r} 493 \\ \times\ 416 \\ \hline \end{array}$

57. $\begin{array}{r} 678 \\ \times\ 132 \\ \hline \end{array}$

58. $\begin{array}{r} 392 \\ \times\ 187 \\ \hline \end{array}$

59. $\begin{array}{r} 2076 \\ \times\ 105 \\ \hline \end{array}$

60. $\begin{array}{r} 5092 \\ \times\ 302 \\ \hline \end{array}$

61. $\begin{array}{r} 3561 \\ \times\ 403 \\ \hline \end{array}$

62. $\begin{array}{r} 7023 \\ \times\ 2007 \\ \hline \end{array}$

63. $\begin{array}{r} 1023 \\ \times\ 4005 \\ \hline \end{array}$

64. $\begin{array}{r} 1298 \\ \times\ 304 \\ \hline \end{array}$

65. $\begin{array}{r} 260 \\ \times\ 40 \\ \hline \end{array}$

66. $\begin{array}{r} 307 \\ \times\ 300 \\ \hline \end{array}$

67. $\begin{array}{r} 307 \\ \times\ 30 \\ \hline \end{array}$

68. $\begin{array}{r} 150 \\ \times\ 200 \\ \hline \end{array}$

69. $7 \cdot 2 \cdot 5$

70. $8 \cdot 3 \cdot 2$

71. $11 \cdot 7 \cdot 4$

72. $15 \cdot 4 \cdot 4$

73. $10 \cdot 7 \cdot 10$

74. $20 \cdot 5 \cdot 3$

75. $12 \cdot 3 \cdot 5 \cdot 2$

76. $7 \cdot 6 \cdot 2 \cdot 5$

77. What is x if $x = 8 \cdot 7 \cdot 6 \cdot 0$?

78. What is x if $x = 3 \cdot 12 \cdot 0 \cdot 5$?

Applications

79. Find the area of a rectangular university fitness center if the fitness center floor is 48 feet long and 67 feet wide.

80. Find the area of a rectangular computer chip that is 17 millimeters long and 9 millimeters wide.

81. The Police Department is purchasing 35 bulletproof vests at $345 each. What is the total cost of purchasing these vests?

82. The Andersonville College football team is staying overnight at a local Holiday Inn. They have been quoted a rate of $43 per night per room. The team will need 27 rooms. What will be the team cost to stay overnight at this motel?

83. The student commons food supply needs to purchase espresso coffee. Find the cost of purchasing 240 pounds of espresso coffee beans at $5 per pound.

84. The music department of Wheaton College wishes to purchase 345 sets of headphones at the music supply store at a cost of $8 each. What will be the total amount of the purchase?

85. Helen pays $266 per month for her car payment on her new Honda Civic. What is her automobile payment cost for a one-year period?

86. A company rents a Ford Escort for a salesman at $276 per month for eight months. What is the cost for the car rental during this time?

87. Marcos has a Geo Prism that gets 34 miles per gallon during highway driving. Approximately how far can he travel if he has 18 gallons of gas in the tank?

88. Cheryl has a subcompact car that gets 48 miles per gallon on highway driving. Approximately how far can she travel if she has 12 gallons of gas in the tank?

89. Today, the Burger Basket has 12 employees who made 85 hamburgers each. How many burgers were made in total?

90. Fun 'n' Games Amusement Park has 102 employees on staff. They are all scheduled to work 38 hours per week. How many working hours should be scheduled in one week for all employees?

91. The country of Haiti has an average per capita (per person) income of $800. If the approximate population of Haiti is 6,530,000, what is the approximate total yearly income of the entire country.

92. The kingdom of the Netherlands (Holland) has an approximate population of 15,500,000. The average per capita (per person) income is $17,200. What is the approximate total yearly income of the entire country?

To Think About

93. Multiply.
$4 \times 9 \times 25 \times 12 \times 5$

Find the value of x in each equation.

94. $5(x) = 40$

95. $7(x) = 56$

96. $72 = 8(x)$

97. $63 = 9(x)$

98. Would the distributive property of multiplication be true for Roman numerals such as (XII) × (IV)? Why or why not?

99. We saw that multiplication is distributive over addition. Is it distributive over subtraction? Why or why not? Give examples.

Cumulative Review Problems

100. Subtract.
$$\begin{array}{r} 34{,}084 \\ -\ 27{,}328 \\ \hline \end{array}$$

101. Add.
$$\begin{array}{r} 263 \\ 27 \\ 891 \\ 5 \\ +\ 63 \\ \hline \end{array}$$

102. Albert Gachet earns $156 per week. He has withheld weekly $12 for taxes, $2 for dues, and $3 for retirement. How much is left?

103. Mrs. May Washington retired at $743 per month. Her cost-of-living raise increased her payment to $802 per month. What was the increase?

Developing Your Study Skills

WHY IS HOMEWORK NECESSARY?

Mathematics involves mastering a set of skills that you learn by practicing, not by watching someone else do it. Your instructor may make solving a mathematics problem look very easy, but for you to learn the necessary skills, you must practice them over and over again, just as your instructor once had to. There is no other way. Learning mathematics is like learning to play a musical instrument, to type, or to play a sport. No matter how much you watch someone else do it, no matter how many books you read on ''how to'' do it, no matter how easy it seems to be, the key to success is practice on a regular basis.

Homework provides this practice. The amount of practice needed varies for each individual, but usually students need to do most or all of the exercises provided at the end of each section in the text. The more problems you do, the better you get. Some problems in a set are more difficult than others, and some stress different concepts. Only by working all the problems will you cover the full range of difficulty.

1.5 Division of Whole Numbers

After studying this section, you will be able to:

1 *Perform simple division using basic division facts.*
2 *Perform long division by a one-digit number.*
3 *Perform long division by a two- or three-digit number.*
4 *Apply division to real-life situations.*

MathPro Video 2 SSM

1 *Mastering Basic Division Facts*

Suppose that we have eight quarters and want to divide them into two equal piles. We would discover that each pile contains four quarters.

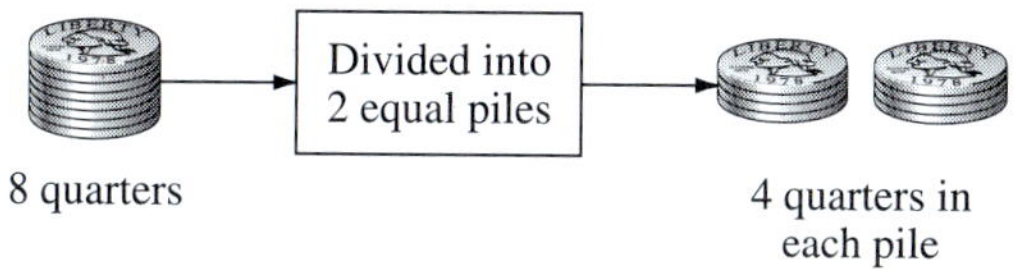

In mathematics we would express this thought by saying that

$$8 \div 2 = 4$$

We know that this answer is right because two piles of four quarters is the same dollar amount as eight quarters. In other words, we know that $8 \div 2 = 4$ because $2 \times 4 = 8$. These two mathematical sentences are called **related sentences.** The division sentence $8 \div 2 = 4$ is related to the multiplication sentence $2 \times 4 = 8$.

In fact, in mathematics we usually define division in terms of multiplication. The answer to the division problem $12 \div 3$ is that number which when multiplied by 3 yields 12. Thus

$$12 \div 3 = 4 \text{ because } 3 \times 4 = 12$$

Suppose that a surplus of \$30 in the French Club budget at the end of the year is to be equally divided among the five club members. We would want to divide the \$30 into five equal parts. We would write $30 \div 5 = 6$ because $5 \times 6 = 30$. Thus each of the five people would get \$6 in this situation.

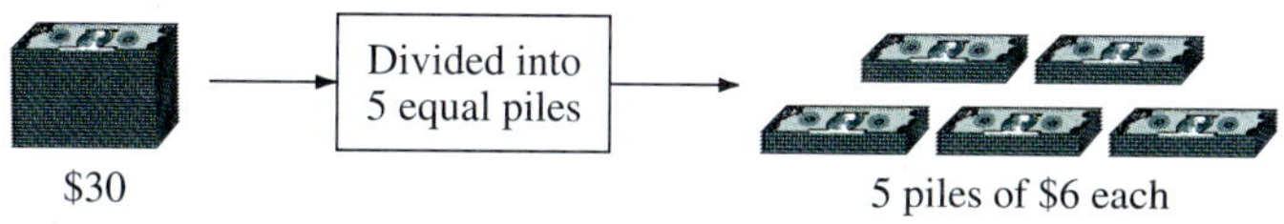

As a mathematical sentence, $30 \div 5 = 6$.

The division problem $30 \div 5 = 6$ could also be written $\frac{30}{5} = 6$ or $5\overline{)30}$ with quotient 6.

When referring to division, we sometimes use the words **divisor, dividend,** and **quotient** to identify the three parts.

$$\begin{array}{r} \text{quotient} \\ \text{divisor}\overline{)\text{dividend}} \end{array}$$

With $30 \div 5 = 6$, 30 is the dividend, 5 is the divisor, and 6 is the quotient.

$$\begin{array}{rl} 6 \longleftarrow & \text{quotient} \\ \text{divisor} \longrightarrow 5\overline{)30} \longleftarrow & \text{dividend} \end{array}$$

So the quotient is the answer to a division problem. It is important that you be able to do short problems involving basic division facts quickly.

EXAMPLE 1 Divide.

(a) $12 \div 4$ **(b)** $81 \div 9$ **(c)** $56 \div 8$ **(d)** $54 \div 6$

(a) $4\overline{)12}$ = 3 **(b)** $9\overline{)81}$ = 9 **(c)** $8\overline{)56}$ = 7 **(d)** $6\overline{)54}$ = 9

Practice Problem 1 Divide.

(a) $36 \div 4$ **(b)** $25 \div 5$ **(c)** $72 \div 9$ **(d)** $30 \div 6$ ■

Zero can be divided by any nonzero number, but division by zero is not possible. Why is this?

Suppose that we could divide by zero. Then $7 \div 0 =$ some number. Let us represent "some number" by the letter a.

$$\text{If } 7 \div 0 = a, \text{ then } 7 = 0 \times a$$

because every division problem has a related multiplication problem. But zero times any number is zero, $0 \times a = 0$. Thus

$$7 = 0 \times a = 0$$

That is, $7 = 0$, which we know is not true. Therefore, our assumption that $7 \div 0 = a$ is wrong. Thus we conclude that we cannot divide by zero.

It is helpful to remember the following basic concepts:

Division Problems Involving the Number 1 and the Number 0

1. Any nonzero number divided by itself is 1 $(7 \div 7 = 1)$.
2. Any number divided by 1 remains unchanged $(29 \div 1 = 29)$.
3. Zero may be divided by any nonzero number; the result is always zero $(0 \div 4 = 0)$.
4. Zero can never be the divisor in a division problem ($3 \div 0$ cannot be done; $0 \div 0$ cannot be done).

EXAMPLE 2 Divide if possible. If not possible, state why.

(a) $8 \div 8$ **(b)** $9 \div 1$ **(c)** $0 \div 6$ **(d)** $20 \div 0$ **(e)** $0 \div 0$

(a) $\frac{8}{8} = 1$ Any number divided by itself is 1.

(b) $\frac{9}{1} = 9$ Any number divided by 1 remains unchanged.

(c) $\frac{0}{6} = 0$ Zero divided by any nonzero number is zero.

(d) $\frac{20}{0}$ cannot be done Division by zero is impossible.

(e) $\frac{0}{0}$ cannot be done Division by zero is impossible.

Practice Problem 2 Divide, if possible.

(a) $7 \div 1$ **(b)** $\frac{8}{8}$ **(c)** $\frac{0}{5}$ **(d)** $12 \div 0$ **(e)** $\frac{0}{0}$ ■

2 *Performing Division by a One-Digit Number*

Our accuracy with division is improved if we have a checking procedure. For each division fact, there is a related multiplication fact.

$$\text{If } 20 \div 4 = 5, \text{ then } 20 = 4 \times 5$$

$$\text{If } 36 \div 9 = 4, \text{ then } 36 = 9 \times 4$$

We will often use multiplication to check our answers.

When two numbers do not divide exactly, a number called the **remainder** is left over. For example, 13 cannot be divided exactly by 2. The number 1 is left over. We call this 1 the *remainder.*

$$\begin{array}{r} 6 \\ 2\overline{)13} \\ \underline{12} \\ 1 \end{array} \longleftarrow \text{remainder}$$

Thus $13 \div 2 = 6$ with a remainder of 1. We can abbreviate this answer as

6 R 1

To check this division, we multiply $2 \times 6 = 12$ and add the remainder $12 + 1 = 13$. That is, $2 \times 6 + 1 = 13$. The result will be the dividend if the division was done correctly. The following fact shows you how to check a division that has a remainder.

> (divisor × quotient) + remainder = dividend

EXAMPLE 3 Divide. $33 \div 4$. Check your answer.

$$\begin{array}{rl} 8 & \longrightarrow \text{How many times can 4 be divided into 33? 8.} \\ 4\overline{)33} & \\ \underline{32} & \longleftarrow \text{What is } 8 \times 4\text{? 32.} \\ 1 & \longleftarrow \text{What is 33 subtract 32? 1.} \end{array}$$

The answer is 8 with a remainder of 1. We abbreviate 8 R 1.

Check.

$$\begin{array}{rl} 8 & \text{Multiply. } 8 \times 4 = 32. \\ \underline{\times\, 4} & \\ 32 & \\ \underline{1} & \text{Add the remainder. } 32 + 1 = 33. \\ 33 & \text{Because the dividend is 33, the answer is correct.} \end{array}$$

Practice Problem 3 Divide. $45 \div 6$. Check your answer. ■

EXAMPLE 4 Divide. 158 ÷ 5. Check your answer.

```
   31
5)158     5 divided into 15?  3.
  15  ←What is 3 × 5?  15.
  --
   08←15 subtract 15?  0. Bring down 8.
    5←5 divided into 8?  1. What is 1 × 5?  5.
    -
    3←8 subtract 5?  3.
```

The answer is 31 R 3. Check.

```
   31    Multiply. 31 × 5 = 155.
 ×  5
 ----
  155
    3    Add the remainder 3.
 ----
  158    Because the dividend is 158, the answer is correct.
```

Practice Problem 4 Divide. 129 ÷ 6. Check your answer. ■

EXAMPLE 5 Divide. 3672 ÷ 7

```
    524
 7)3672    How many times can 7 be divided into 36?  5.
   35   ←What is 5 × 7?  35.
   --
    17  ←36 subtract 35?  1. Bring down 7.
    14  ←7 divided into 17?  2. What is 2 × 7?  14.
    --
     32←17 subtract 14?  3. Bring down 2.
     28←7 divided into 32?  4. What is 4 × 7?  28.
     --
      4←32 subtract 28?  4.
```

The answer is 524 R 4.

Practice Problem 5 Divide. 4237 ÷ 8 ■

3 *Performing Division by a Two- or Three-Digit Number*

When the divisor has more than one digit, an estimation technique may help. Figure how many times the first digit of the divisor goes into the first two digits of the dividend. Try this answer as the first number in the quotient.

EXAMPLE 6 Divide. 283 ÷ 41

```
First guess:     7
             41)283
                287  too large
```

How many times can the first digit of the divisor (4) be divided into the first two digits of the dividend (28)? 7. We try the answer 7 as the first number of the quotient. We multiply 7 × 41 = 287. We see that 287 is larger than 283.

Second guess:

$$\begin{array}{r} 6 \\ 41\overline{)283} \\ \underline{246} \\ 37 \end{array}$$

Because 7 is slightly too large, we try 6.

246 ← 6 × 41? 246.

37 ← 283 subtract 246? 37.

The answer is 6 R 37. (Note that the remainder must always be less than the divisor.)

Practice Problem 6 Divide. 229 ÷ 32 ■

EXAMPLE 7 Divide. 33,897 ÷ 56

First guess:

$$\begin{array}{r} 60 \\ 56\overline{)33897} \\ \underline{336} \\ 29 \end{array}$$

How many times can 33 be divided by 5? 6.

336 ← What is 6 × 56? 336.

338 subtract 336? 2. Bring down 9.

56 cannot be divided into 29. Write 0 in quotient.

Second set of steps:

$$\begin{array}{r} 605 \\ 56\overline{)33897} \\ \underline{336} \\ 297 \\ \underline{280} \\ 17 \end{array}$$

Bring down 7.

How many times can 5 be divided into 29? 5.

280 ← What is 5 × 56? 280. Subtract 297 − 280.

Remainder is 17.

The answer is 605 R 17.

Practice Problem 7 Divide. 42,183 ÷ 33 ■

EXAMPLE 8 Divide. 5629 ÷ 134

$$\begin{array}{r} 42 \\ 134\overline{)5629} \\ \underline{536} \\ 269 \\ \underline{268} \\ 1 \end{array}$$

How many times does 134 divide into 562?
We guess by saying that 1 divides into 5 five times,
but this is too large. (5 × 134 = 670!)
So we try 4. What is 4 × 134? 536.
Subtract 562 − 536. We obtain 26. Bring down 9.
How many times does 134 divide into 269?
We guess by saying that 1 divided into 2 goes 2 times.
What is 2 × 134? 268. Subtract 269 − 268.
The remainder is 1.

The answer is 42 R 1.

Practice Problem 8 Divide. 3227 ÷ 128 ■

4 *Applying Division to Real-Life Situations*

When you solve a word problem that requires division, you will be given the total number and asked to calculate the number of items in each group or to calculate the number of groups. In the beginning of this section we showed 8 quarters (the total number) and we divided them into 2 equal piles (the number of groups). Division was used to find how many quarters were in each pile (the number in each group). That is, $8 \div 2 = 4$. There were 4 quarters in each pile.

Let's look at another example. Suppose that \$30 is to be divided equally among the members of a group. If each person receives \$6, how many people are in the group? We use division, $30 \div 6 = 5$, to find that there are 5 people in the group.

EXAMPLE 9 City Service Realty just purchased nine identical computers for the realtors in the office. The total cost for the nine computers was \$25,848. What was the cost of one computer? Check your answer.

To find the cost of one computer, we need to divide the total cost by 9. Thus we will calculate $\$25,848 \div 9$.

$$\begin{array}{r} 2872 \\ 9\overline{)25848} \\ \underline{18} \\ 78 \\ \underline{72} \\ 64 \\ \underline{63} \\ 18 \\ \underline{18} \\ 0 \end{array}$$

Therefore, the cost of one computer is \$2872. In order to check our work we will need to see if 9 computers each costing \$2872 will in fact result in a total of \$25,848. We use multiplication to check division.

$$\begin{array}{r} 2872 \\ \times \quad 9 \\ \hline 25848 \end{array} \checkmark$$

We did obtain 25,848. Our answer is correct.

Practice Problem 9 The Dallas police department purchased seven identical police cars at a total cost of \$117,964. Find the cost of one car. Check your answer. ■

In the following example you will see the word *average* used as it applies to division. The problem states that a car traveled 1144 miles in 22 hours. The problem asks you to find the average speed in miles per hour. This means that we will treat the problem as if the speed of the car was the same during each hour of the trip. We will use division to solve.

EXAMPLE 10 A car traveled 1144 miles in 22 hours. What was the average speed in miles per hour?

When doing distance problems, it is helpful to remember that distance ÷ time = rate. We need to divide 1144 miles by 22 hours to obtain the rate or speed in miles per hour.

$$\begin{array}{r} 52 \\ 22\overline{)1144} \\ \underline{110} \\ 44 \\ \underline{44} \\ 0 \end{array}$$

The car traveled an average of 52 miles per hour.

Practice Problem 10 An airplane traveled 5138 miles in 14 hours. What was the average speed in miles per hour? ■

Developing Your Study Skills

TAKING NOTES IN CLASS

An important part of mathematics studying is taking notes. In order to take meaningful notes, you must be an active listener. Keep your mind on what the instructor is saying, and be ready with questions whenever you do not understand something.

If you have previewed the lesson material, you will be prepared to take good notes. The important concepts will seem somewhat familiar. You will have a better idea of what needs to be written down. If you frantically try to write all that the instructor says or copy all the examples done in class, you may find your notes to be nearly worthless when you look at them at home. You may find that you are unable to make sense of what you have written.

Write down *important* ideas and examples as the instructor lectures, making sure that you are listening and following the logic. Include any helpful hints or suggestions that your instructor gives you or refers to in your text. You will be amazed at how easily you will forget these if you do not write them down. Try to review your notes the *same day* sometime after class. You will find the material in your notes easier to understand if you have attended class within the last few hours.

Successful note taking requires active listening and processing. Stay alert in class. You will realize the advantages of taking your own notes over copying those of someone else.

1.5 Exercises

Verbal and Writing Skills

1. Explain in your own words what happens when you
   (a) divide a nonzero number by itself **(b)** divide a number by 1

   (c) divide zero by a nonzero number **(d)** divide a number by 0

Divide. See if you can work problems 2–30 in 3 minutes or less.

2. $5\overline{)35}$ **3.** $6\overline{)42}$ **4.** $4\overline{)32}$ **5.** $8\overline{)24}$ **6.** $9\overline{)27}$ **7.** $8\overline{)40}$

8. $7\overline{)56}$ **9.** $9\overline{)36}$ **10.** $4\overline{)12}$ **11.** $7\overline{)21}$ **12.** $9\overline{)81}$ **13.** $8\overline{)56}$

14. $6\overline{)54}$ **15.** $7\overline{)63}$ **16.** $4\overline{)28}$ **17.** $8\overline{)72}$ **18.** $8\overline{)64}$ **19.** $9\overline{)63}$

20. $9\overline{)72}$ **21.** $6\overline{)24}$ **22.** $1\overline{)8}$ **23.** $1\overline{)7}$ **24.** $5\overline{)0}$ **25.** $6\overline{)0}$

26. $6\overline{)48}$ **27.** $54 \div 6$ **28.** $42 \div 7$ **29.** $6 \div 6$ **30.** $5 \div 5$

Divide. Check your answer on 31–42.

31. $32 \div 6$ **32.** $37 \div 5$ **33.** $76 \div 8$ **34.** $39 \div 9$ **35.** $128 \div 5$

36. $6\overline{)103}$ **37.** $7\overline{)165}$ **38.** $8\overline{)124}$ **39.** $9\overline{)288}$ **40.** $7\overline{)126}$

41. $5\overline{)185}$ **42.** $8\overline{)224}$ **43.** $4\overline{)1289}$ **44.** $3\overline{)758}$ **45.** $6\overline{)763}$

46. $7\overline{)403}$ **47.** $8\overline{)6024}$ **48.** $9\overline{)2286}$ **49.** $5\overline{)6789}$ **50.** $5\overline{)7324}$

51. $7\overline{)12304}$ **52.** $6\overline{)18127}$ **53.** $9\overline{)22305}$ **54.** $8\overline{)26037}$

Mixed Practice

Divide.

55. $185 \div 36$ **56.** $152 \div 48$ **57.** $267 \div 52$ **58.** $321 \div 53$ **59.** $427 \div 61$

60. $72\overline{)432}$ **61.** $12\overline{)1930}$ **62.** $13\overline{)6810}$ **63.** $30\overline{)1452}$ **64.** $40\overline{)1125}$

65. $15\overline{)9236}$ **66.** $16\overline{)8314}$ **67.** $36\overline{)7568}$ **68.** $32\overline{)3527}$ **69.** $18\overline{)3643}$

70. $19\overline{)2174}$ **71.** $124\overline{)500}$ **72.** $132\overline{)792}$ **73.** $322\overline{)2280}$ **74.** $41\overline{)205000}$

75. $3483 \div 129 = x$. What is the value of x?

76. $2214 \div 123 = x$. What is the value of x?

Applications

Solve each problem.

77. Western Saddle Stable uses 21,900 pounds of feed per year to feed its 30 horses. How much does each horse eat per year?

78. A *run* in skiing is going from the top of the ski lift to the bottom. If over seven days 431,851 runs were made, what was the average number of ski runs per day?

79. A telethon raised \$3,677,880 over 15 hours. What was the average amount of money raised per hour?

80. A factory that produces silver earrings made 864 pairs of hoops in 36 hours. What was the average amount of earrings produced per hour?

81. A group of 8 old friends invested the same amount each in a beach property that sold for \$369,432. How much did each friend pay?

82. A horse and carriage company in New York City bought 7 new carriages at exactly the same price each. The total bill was \$147,371. How much did each carriage cost?

83. A new sorting machine sorts 26 letters per minute. The machine is fed 884 letters. How many minutes will it take to sort the letters?

84. Chung wishes to pay off a loan of \$4104 in 24 months. How large will his monthly payments be?

To Think About

85. Division is not commutative. For example, $12 \div 4 \neq 4 \div 12$. If $a \div b = b \div a$, what must be true of the numbers a and b besides the fact that $b \neq 0$ and $a \neq 0$?

86. You can think of division as repeated subtraction. Show how $874 \div 138$ is related to repeated subtraction.

Cumulative Review Problems

Multiply.

87. $\begin{array}{r} 128 \\ \times\ 43 \\ \hline \end{array}$

88. $\begin{array}{r} 7162 \\ \times\ 145 \\ \hline \end{array}$

Add.

89. $316{,}214 + 89{,}981$

Subtract.

90. $1{,}360{,}000 - 1{,}293{,}156$

Putting Your Skills to Work

BUYING OR LEASING A CAR

Today over $\frac{1}{3}$ of all potential new car buyers decide to lease the car instead of purchasing it. What are the mathematics of buying versus leasing? How do we compare the costs?

Charlie and Sue Downey are considering leasing a new Dodge Stratus for \$199 per month or purchasing the car for \$14,996. See if you can determine the needed mathematical facts in order to make a well-informed decision.

PROBLEMS FOR INDIVIDUAL INVESTIGATION

1. The lease is for a rental period of 48 months and requires a down payment of \$1000. What is the basic minimum cost to lease the car for 4 years?

2. If the car is purchased with a down payment of \$1000 and the remainder is paid for with a new car loan for 48 months with monthly payments of \$370, what will it cost to purchase the car?

PROBLEMS FOR GROUP INVESTIGATION AND COOPERATIVE STUDY

Together with some members of your class, see if you can answer the following questions.

3. If the Downeys plan to keep the car at the end of the 4-year lease, they will need to purchase it from the dealer at a cost of \$9352. In that case, what is the total cost of obtaining the car?

4. If the Downeys plan to give the car back to the dealer after 4 years, they must pay a mileage surcharge of 15 cents per mile for all miles in excess of 12,000 miles per year. If they normally drive a car 64,000 miles over a 4-year period, what will be the total approximate cost for them to lease the car for 4 years?

5. If the Downeys are considering leasing and buying the car after 4 years (question 3) or purchasing the car (question 2), what would you recommend to them if they plan to drive the car about 64,000 miles during the 4-year period?

INTERNET CONNECTIONS: Go to ''http://www.prenhall.com/tobey'' to be connected

Site: Microsoft Carpoint or a related site

This site includes a loan calculator which can help you determine the cost of financing a car. Use the loan calculator to answer the following questions.

6. Kimberly is buying a car for \$13,000. She plans to make a 20% down payment, and she will finance the rest of the purchase price over a 3-year period. Her interest rate will be 9%. Find the amount of her monthly payment, and the total amount that she will pay for the car.

7. Jonathan wants to buy a car. He plans to make a 15% down payment and finance the rest over a 4-year period. His interest rate will be 7%. If Jonathan wants to keep his monthly car payment under \$225, what is the most expensive car he can afford to buy?

1.6 Exponents and Order of Operations

MathPro Video 2 SSM

After studying this section, you will be able to:

1 *Evaluate expressions with whole-number exponents.*
2 *Perform several arithmetic operations in the proper order.*

1 *Evaluating Whole-Number Exponents*

Sometimes a simple math idea comes "disguised" in technical language. For example, an **exponent** is just a "shorthand" number that saves writing multiplication of the same numbers.

10^3 The exponent 3 means $10 \times 10 \times 10$.
(which takes longer to write)

The product 5×5 can be written as 5^2. The small number 2 is called the *exponent.* The exponent tells us how many factors are in the multiplication. The number 5 is called the **base.** The base is the number that is multiplied.

$$3 \times 3 \times 3 \times 3 = 3^4$$
(4 ← exponent; 3 ← base)

In 3^4 the base is 3 and the exponent is 4. (The 4 is sometimes called the *superscript.*) 3^4 is read as "3 to the fourth power."

EXAMPLE 1 Write each product in exponent form.

(a) $15 \times 15 \times 15$ **(b)** $7 \times 7 \times 7 \times 7 \times 7$

(a) $15 \times 15 \times 15 = 15^3$ **(b)** $7 \times 7 \times 7 \times 7 \times 7 = 7^5$

Practice Problem 1 Write each product in exponent form.

(a) $12 \times 12 \times 12 \times 12$ **(b)** $2 \times 2 \times 2 \times 2 \times 2 \times 2$ ■

EXAMPLE 2 Find the value of each expression.

(a) 3^3 **(b)** 7^2 **(c)** 2^5 **(d)** 1^8

(a) To find the value of 3^3, multiply the base 3 by itself 3 times.
$3^3 = 3 \times 3 \times 3 = 27$

(b) To find the value of 7^2, multiply the base 7 by itself 2 times.
$7^2 = 7 \times 7 = 49$

(c) $2^5 = 2 \times 2 \times 2 \times 2 \times 2 = 32$

(d) $1^8 = 1 \times 1 \times 1 \times 1 \times 1 \times 1 \times 1 \times 1 = 1$

Practice Problem 2 Find the value of each expression.

(a) 12^2 **(b)** 6^3 **(c)** 2^6 **(d)** 1^{10} ■

If a whole number does not have a visible exponent, the exponent is understood to be 1. Thus

$$3 = 3^1 \quad \text{or} \quad 10 = 10^1$$

In today's world, large numbers are often expressed as a power of 10.

$10^1 = 10 = 1$ ten	$10^4 = 10{,}000 = 1$ ten thousand
$10^2 = 100 = 1$ hundred	$10^5 = 100{,}000 = 1$ hundred thousand
$10^3 = 1000 = 1$ thousand	$10^6 = 1{,}000{,}000 = 1$ million

What does it mean to have an exponent of zero? What is 10^0? Any whole number that is not zero can be raised to the zero power. The result is 1. Thus $10^0 = 1$, $3^0 = 1$, $5^0 = 1$, and so on. Why is this? Let's reexamine the powers of 10. As we go down one line at a time, notice the pattern that occurs.

As we move down one line, we decrease the exponent by 1.

$$10^5 = 100{,}000$$
$$10^4 = 10{,}000$$
$$10^3 = 1000$$
$$10^2 = 100$$
$$10^1 = 10$$
$$10^0 = 1$$

As we move down one line, we divide the previous number by 10.

Therefore, we present the following definition.

Definition

For any whole number a other than zero, $a^0 = 1$.

If numbers having exponents are added to other numbers, it is first necessary to **evaluate,** or find the value of, the number that is raised to a power. Then we may combine the results with another number.

EXAMPLE 3 Find the value of each expression.

(a) $3^4 + 2^3$ **(b)** $5^3 + 7^0$ **(c)** $6^3 + 6$

(a) $3^4 + 2^3 = (3)(3)(3)(3) + (2)(2)(2) = 81 + 8 = 89$

(b) $5^3 + 7^0 = (5)(5)(5) + 1 = 125 + 1 = 126$

(c) $6^3 + 6 = (6)(6)(6) + 6 = 216 + 6 = 222$

Practice Problem 3 Find the value of each expression.

(a) $7^3 + 8^2$ **(b)** $9^2 + 6^0$ **(c)** $5^4 + 5$ ■

2 *Performing Several Arithmetic Operations in the Proper Order*

Sometimes the order in which we do things is not important. The order in which chefs hang up their pots and pans probably does not matter. The order in which they add and mix the elements in preparing food, however, makes all the difference in the world! If various cooks follow a recipe, though, they will get similar results. The recipe assures that the results will be consistent. It shows the **order of operations.**

In mathematics the order of operations is a list of priorities for working with the numbers in computational problems. This mathematical "recipe" tells how to handle certain indefinite computations. For example, how does a person solve $5 + 3 \times 2$?

A problem such as $5 + 3 \times 2$ sometimes causes students difficulty. Some people think $(5 + 3) \times 2 = 8 \times 2 = 16$. Some people think $5 + (3 \times 2) = 5 + 6 = 11$. Only one answer is right, 11. To obtain the right answer follow the steps outlined below.

Order of Operations

In the absence of grouping symbols:

Do first **1.** Simplify any expressions with exponents.
↕ **2.** Multiply or divide from left to right.
Do last **3.** Add or subtract from left to right.

EXAMPLE 4 Evaluate. $3^2 + 5 - 4 \times 2$

$3^2 + 5 - 4 \times 2 = 9 + 5 - 4 \times 2$ *Evaluate the expression with exponents.*
$= 9 + 5 - 8$ *Multiply from left to right.*
$= 14 - 8$ *Add from left to right.*
$= 6$ *Subtract.*

Practice Problem 4 Evaluate. $7 + 4^3 \times 3$ ■

EXAMPLE 5 Evaluate. $5 + 3 \times 6 - 4 + 12 \div 3$

There are no numbers to raise to a power, so we first do any multiplication or division in order from *left to right.*

$5 + 3 \times 6 - 4 + 12 \div 3 = 5 + 18 - 4 + 12 \div 3$ *Multiply.*
$= 5 + 18 - 4 + 4$ *Divide.*
$= 23 - 4 + 4$ *Add.*
$= 19 + 4$ *Subtract.*
$= 23$ *Add.*

Practice Problem 5 Evaluate. $7 - 3 \times 4 \div 2 + 5$ ■

EXAMPLE 6 Evaluate. $2^3 + 3^2 - 7 \times 2$

$2^3 + 3^2 - 7 \times 2 = 8 + 9 - 7 \times 2$	*Evaluate exponent expressions* $2^3 = 8$ *and* $3^2 = 9$.
$= 8 + 9 - 14$	*Multiply.*
$= 17 - 14$	*Add.*
$= 3$	*Subtract.*

Practice Problem 6 Evaluate. $4^3 - 2 + 3^2$ ■

You can change the order in which you compute by using grouping symbols. Place the numbers you want to calculate first within parentheses. This tells you to do those calculations first.

Order of Operations

With grouping symbols:

Do first
1. Perform operations inside the parentheses.
2. Simplify any expressions with exponents.
3. Multiply or divide from left to right.
4. Add or subtract from left to right.

Do last

EXAMPLE 7 Evaluate. $2 \times (7 + 5) \div 4 + 3 - 6$

First, we combine numbers inside the parentheses by adding the 7 to the 5. Multiplication and division have equal priority, so we work from left to right doing whichever of these operations comes first.

$= 2 \times 12 \div 4 + 3 - 6$	
$= 24 \div 4 + 3 - 6$	*Multiply.*
$= 6 + 3 - 6$	*Divide.*
$= 9 - 6$	*Add.*
$= 3$	*Subtract.*

Practice Problem 7 Evaluate. $(17 + 7) \div 6 \times 2 + 7 \times 3 - 4$ ■

EXAMPLE 8 Evaluate. $4^3 + 18 \div 3 - 2^4 - 3 \times (8 - 6)$

$= 4^3 + 18 \div 3 - 2^4 - 3 \times 2$	*Work inside the parentheses.*
$= 64 + 18 \div 3 - 16 - 3 \times 2$	*Evaluate exponents.*
$= 64 + 6 - 16 - 3 \times 2$	*Divide.*
$= 64 + 6 - 16 - 6$	*Multiply.*
$= 70 - 16 - 6$	*Add.*
$= 54 - 6$	*Subtract.*
$= 48$	*Subtract.*

Practice Problem 8 Evaluate. $5^2 - 6 \div 2 + 3^4 + 7 \times (12 - 10)$ ■

1.6 Exercises

Verbal and Writing Skills

1. Explain what the expression 5^3 means. Evaluate 5^3.
2. The ______________________ tells how many times to multiply the base.
3. The ______________________ is the number that is multiplied.
4. 10^5 is read as ____.
5. Explain the order in which we perform mathematical operations to ensure consistency.
6. Use the order of operations to solve $12 \times 5 + 3 \times 5 + 7 \times 5$. Is this the same as $5(12 + 3 + 7)$? Why or why not?

Write each number in exponent form.

7. $6 \times 6 \times 6 \times 6$

8. $2 \times 2 \times 2 \times 2 \times 2$

9. $5 \times 5 \times 5$

10. $12 \times 12 \times 12$

11. $8 \times 8 \times 8 \times 8$

12. $1 \times 1 \times 1 \times 1 \times 1 \times 1 \times 1$

13. 9

14. 27

Find the value of each expression.

15. 2^4 **16.** 3^3 **17.** 4^2 **18.** 5^2 **19.** 6^3 **20.** 10^3 **21.** 10^4 **22.** 1^{20}

23. 1^{17} **24.** 2^5 **25.** 2^6 **26.** 4^3 **27.** 3^4 **28.** 13^2 **29.** 15^2 **30.** 8^3

31. 7^3 **32.** 5^4 **33.** 4^4 **34.** 2^7 **35.** 9^0 **36.** 8^0 **37.** 25^2 **38.** 20^3

39. 10^6 **40.** 8^1 **41.** 4^5 **42.** 10^5 **43.** 9^1 **44.** 14^2 **45.** 7^4 **46.** 4^5

Find the value of each expression.

47. $3^4 + 7^0$

48. $4^3 + 4^0$

49. $6^3 + 3^2$

50. $7^3 + 4^2$

51. $8^3 + 8$

52. $9^2 + 9$

Work each problem, using the correct order of operations.

53. $5 \times 6 - 3 + 10$

54. $8 \times 3 - 4 \times 2$

55. $2 + 6 \times 12 \div 2$

56. $6 \times 7 - 4 \div 2$

57. $100 \div 5^2 + 3$

58. $24 \div 2^2 + 4$

59. $7 \times 3^2 + 4 - 8$

60. $2^3 \times 4 + 6 - 9$

61. $4 \times 9^2 - 4 \times (7 - 2)$ **62.** $5 \times 2^5 - 4 \times (11 - 8)$ **63.** $(400 \div 20) \div 20$ **64.** $(600 \div 30) \div 20$

65. $950 \div (25 \div 5)$ **66.** $875 \div (35 \div 7)$ **67.** $(16)(4) - (16 + 4)$ **68.** $(8)(30) - (8 + 30)$

69. $3^2 + 4^2 \div 2^2$ **70.** $7^2 + 9^2 \div 3^2$ **71.** $(6)(7) - (12 - 8) \div 4$ **72.** $(8)(9) - (15 - 5) \div 5$

73. $100 - 3^2 \times 4$ **74.** $130 - 4^2 \times 5$ **75.** $5^2 + 2^2 + 3^3$ **76.** $2^3 + 3^2 + 4^3$

77. $100 \div 5 \times 3 \times 2 \div 2$ **78.** $36 \div 4 \times 5 \times 6 \div 6$ **79.** $12^2 - 6 \times 3 \times 4 \times 0$ **80.** $14^2 - 5 \times 2 \times 3 \times 0$

81. $7^2 \times 5 \div 5$ **82.** $5^2 \times 4 \div 10$ **83.** $5 + 6 \times 2 \div 4 - 1$ **84.** $7 + 5 \times 4 \div 10 - 2$

85. $3 + 3^2 \times 6 + 4$ **86.** $5 + 4^3 \times 2 + 7$ **87.** $12 \div 2 \times 3 - 2^4$ **88.** $30 \div 10 \times 5 - 3^2$

89. $3^2 \times 6 \div 9 + 4 \times 3$ **90.** $5^2 \times 3 \div 25 + 7 \times 6$ **91.** $4(10 - 6)^3 \div (2 + 6)$ **92.** $3(12 - 5)^2 \div (4 + 3)$

93. $1200 - 2^3(3) \div 6$ **94.** $2150 - 3^4(2) \div 9$ **95.** $300 \div 25 - 10 + 2^3$ **96.** $400 \div 16 - 8 + 3^4$

97. $20 \div 4 \times 3 - 6 + 2 \times (1 + 4)$ **98.** $50 \div 10 \times 2 - 8 + 3 \times (17 - 11)$

99. $3 \times 5 + 7 \times (8 - 5) - 5 \times 4$ **100.** $(5 + 3) \times 9 + 6 \times 5 - 6 \times 6$

Cumulative Review Problems

101. Write in expanded notation. 156,312

102. Write in standard notation. Two hundred million, seven hundred sixty-five thousand, nine hundred nine

103. Write in words. 261,763,002

104. Add.
$$\begin{array}{r} 1563 \\ 2736 \\ 381 \\ 25 \\ +\ 9823 \\ \hline \end{array}$$

105. Subtract.
$$\begin{array}{r} 16{,}093 \\ -\ 14{,}937 \\ \hline \end{array}$$

To Think About

106. The earth rotates once every 86,164 seconds. Round this figure to the nearest ten thousand seconds.

107. The planet Saturn rotates once every 10 hours, 12 minutes. How many minutes in total is this? How many seconds?

1.7 Rounding and Estimation

MathPro Video 3 SSM

After studying this section, you will be able to:

1. *Round whole numbers.*
2. *Estimate the answer to a problem involving calculation with whole numbers.*

1 *Rounding Whole Numbers*

Large numbers are often expressed to the nearest hundred or to the nearest thousand, because an approximate number is ''good enough'' for certain uses.

Distances from the earth to other galaxies are measured in light-years. Although light really travels at 5,865,696,000,000 miles a year, we usually **round** this number to the nearest trillion and say it travels at 6,000,000,000,000 miles a year. To round a number, we first determine the place we are rounding to. In this case: trillion. Then we find which value is closest to the number that we are rounding. In this case, the number we want to round is closer to 6 trillion than to 5 trillion. How do we know the number is closer to 6 trillion than to 5 trillion?

To see which is the closest value, we may picture a **number line,** where whole numbers are represented by points on a line. To show how to use the number line in rounding, we will round 368 to the nearest hundred. 368 is between 300 and 400. When we round off, we pick the hundred 368 is ''closest to.'' We draw a number line to show 300 and 400. We also show the point midway between 300 and 400 to help us to determine which hundred 368 is closest to.

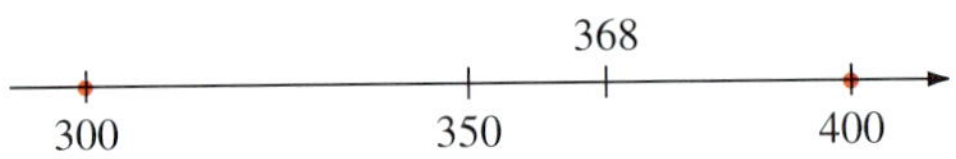

We find that the number 368 is closer to 400 than to 300, so we round 368 *up to* 400.

Let's look at another example. We will round 129 to the nearest hundred. 129 is between 100 and 200. We show this on the number line. We include the midpoint 150 as a guide.

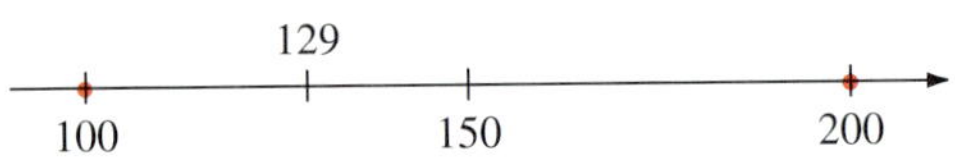

We find that the number 129 is closer to 100 than to 200, so we round 129 *down to* 100.

This leads us to this simple rule for rounding.

Rounding a Whole Number

1. If the first digit to the right of the round-off place is:
   - **(a)** *less than 5,* we make no change to the digit in the round-off place. (We know it is closer to the smaller number, so we round down.)
   - **(b)** *5 or more,* we increase the digit in the round-off place by 1. (We know it is closer to the larger number, so we round up.)
2. Then we replace the digits to the right of the round-off place by zeros.

EXAMPLE 1 Round 37,843 to the nearest thousand.

3 7,8 4 3 ↓	*According to the directions, the thousands will be the round-off place. We locate the thousands place.*
3 7(8)4 3 ↓ 3 8,0 0 0	*We see that the first digit to the right of the round-off place is 8, which is 5* or more. *We increase the thousands digit by 1, and replace all digits to the right by zero.*

We have rounded 37,843 to the nearest thousand: 38,000. This means that 37,843 is closer to 38,000 than to 37,000.

Practice Problem 1 Round 65,528 to the nearest thousand. ■

EXAMPLE 2 Round 2,445,360 to the nearest hundred thousand.

2,4 4 5,3 6 0 ↓	*Locate the hundred-thousands round-off place.*
2,4(4)5,3 6 0 ↓	*The first digit to the right of this is* less than 5, *so round down. Do not change the hundred-thousands digit.*
2,4 0 0,0 0 0	*Replace all digits to the right by zero.*

Practice Problem 2 Round 172,963 to the nearest ten thousand. ■

EXAMPLE 3 Round as indicated. **(a)** 561,328 to the nearest ten **(b)** 3,798,152 to the nearest hundred **(c)** 51,362,523 to the nearest million

(a) First locate the digit in the tens place.

561,3̓28	*The digit to the right of the tens is greater than 5.*
561,330	*Round up.*

561,328 rounded to the nearest ten is 561,330.

(b)	3,798,152	*The digit to the right of the hundreds is 5.*
	3,798,200	*Round up.*

3,798,152 rounded to the nearest hundred is 3,798,200.

(c)	51,362,523	*The digit to the right of the millions is less than 5.*
	51,000,000	*Round down.*

51,362,523 rounded to the nearest million is 51,000,000.

Practice Problem 3 Round as indicated. **(a)** 53,282 to nearest ten **(b)** 164,485 to nearest thousand **(c)** 1,365,273 to nearest hundred thousand ■

EXAMPLE 4 Round 763,571. **(a)** To the nearest thousand **(b)** To the nearest ten thousand **(c)** To the nearest million

(a) 763,571 = 764,000 to the nearest thousand.
The digit to the right of the thousands is 5. We rounded up.

(b) 763,571 = 760,000 to the nearest ten thousand.
The digit to the right of the ten thousands is less than 5. We rounded down.

(c) 763,571 does not have any digits for millions. If it helps, you can think of this number as 0,763,571. Since the digit to the right of the millions place is 7, we round up to obtain one million or 1,000,000.

Practice Problem 4 Round 935,682 as indicated. **(a)** To the nearest thousand **(b)** To the nearest hundred thousand **(c)** To the nearest million ■

EXAMPLE 5 Astronomers use the parsec as a measurement of distance. One **parsec** is approximately 30,900,000,000,000 kilometers. Round 1 parsec to the nearest trillion kilometers.

30,900,000,000,000 km is 31,000,000,000,000 km or 31 trillion km to the nearest trillion kilometers.

Practice Problem 5 One light-year is approximately 9,460,000,000,000,000 meters. Round to the nearest hundred trillion meters. ■

2 *Estimating the Answer to a Calculation with Whole Numbers*

Often we need to quickly check the answer of a calculation to be reasonably assured that the answer is correct. If you expected your bill to be "around $40" for the groceries you had selected and the cashier's total came to $41.89, you would probably be confident that the bill is correct and pay it. If, however, the cashier rang up a bill of $367, you would not just assume that it is correct. You would know an error had been made. If the cashier's total came to $60, you might not be certain, but you would probably suspect an error and check the calculation.

In mathematics we often **estimate,** or determine the approximate value of a calculation, if we need to do a quick check. There are many ways to estimate, but in this book we will use one simple principle of estimation.

Principle of Estimation

1. Round the numbers so that there is one nonzero digit in each number.
2. Perform the calculation with the rounded numbers.

EXAMPLE 6 Estimate the sum. 163 + 237 + 846 + 922

We first determine where to round each number in our problem to leave only one nonzero digit in each. In this case, we round all numbers to the nearest hundred. Then we perform the calculation with the rounded numbers.

Actual Sum	*Estimated Sum*
163	200
237	200
846	800
+ 922	+ 900
	2100

We estimate the answer to be 2100. If we calculate using the exact numbers, we obtain a sum of 2168, so our estimate is quite close to the actual sum.

Practice Problem 6 Estimate the sum. 3456 + 9876 + 5421 + 1278 ■

When we use the principle of estimation, we will not always round each number in a problem to the same place.

EXAMPLE 7 Charles and Melinda bought their first car last week. The selling price of this compact car was \$8980. The dealer preparation charge was \$289 and the sales tax was \$449. Estimate the total cost of the car that Charles and Melinda had to pay.

We round each number to have only one nonzero digit, and add the rounded numbers.

8980	9000
289	300
+ 449	+ 400
	9700

We estimate that the total cost is \$9700. (The exact answer is \$9718, so we see that our answer is quite close.)

Practice Problem 7 Jenna and Carlos purchased a new sofa for \$697, plus \$35 sales tax. The store also charged them \$19 to deliver the sofa. Estimate their total cost. ■

Now we turn to a case where an estimate can help us discover an error.

EXAMPLE 8 Roberto added together four numbers and obtained the following result. Estimate the sum and determine if the answer seems reasonable.

$$12{,}456 + 17{,}976 + 18{,}452 + 32{,}128 = 61{,}012$$

We round each number so that there is one nonzero digit. In this case, we round them all to the nearest ten thousand.

$$\begin{array}{r} 12{,}456 \\ 17{,}976 \\ 18{,}452 \\ +\ 32{,}128 \\ \hline \end{array} \qquad \begin{array}{r} 10{,}000 \\ 20{,}000 \\ 20{,}000 \\ +\ 30{,}000 \\ \hline 80{,}000 \end{array} \qquad \textit{Our estimate is } 80{,}000.$$

This is significantly different from 61,012, so we would suspect that an error has been made. In fact, Roberto did make an error. The exact sum is actually 81,012!

Practice Problem 8 Ming did the following calculation. Estimate to see if her sum appears to be correct or incorrect.

$$11{,}849 + 14{,}376 + 16{,}982 + 58{,}151 = 81{,}358$$ ■

Next we look at a subtraction example where estimation is used.

EXAMPLE 9 The profit from Techno Industries for the first quarter of the year was $642,987,000. The profit for the second quarter was $238,890,000. Estimate how much less the profit was for the second quarter than for the first quarter.

We round each number so that there is one nonzero digit. Then we subtract, using the two rounded numbers.

$$\begin{array}{r} 642{,}987{,}000 \\ -\ 238{,}890{,}000 \\ \hline \end{array} \qquad \begin{array}{r} 600{,}000{,}000 \\ -\ 200{,}000{,}000 \\ \hline 400{,}000{,}000 \end{array}$$

We estimate that the profit was $400,000,000 less for the second quarter.

Practice Problem 9 The 1990 population of Florida was 12,937,926. The 1990 population of California was 29,760,021. Estimate how many more people lived in California in 1990 than in Florida. ■

We also use this principle to estimate results of multiplication and division.

EXAMPLE 10 Estimate the product. 56,789 × 529

We round each number so that there is one nonzero digit. Then we multiply the rounded numbers to obtain our estimate.

$$\begin{array}{r} 56{,}789 \\ \times\quad 529 \\ \hline \end{array} \qquad \begin{array}{r} 60{,}000 \\ \times\quad 500 \\ \hline 30{,}000{,}000 \end{array}$$

We therefore estimate the product to be 30,000,000. (This is reasonably close to the exact answer of 30,041,381.)

Practice Problem 10 Estimate the product. 8945×7317 ■

EXAMPLE 11 Estimate the answer for the following division problem.

$$23\overline{)148{,}902}$$

We round each number to a number with one nonzero digit. Then we perform the division, using the two rounded numbers.

$$23\overline{)148{,}902} \qquad \begin{array}{r} 5000 \\ 20\overline{)100{,}000} \end{array}$$

Our estimate is 5000. (The exact answer is 6474. We see that our estimate is "in the ballpark" but is not very close to the exact answer. Remember, an estimate is just a rough approximation of the exact answer.)

Practice Problem 11 Estimate the answer for the following division problem.

$$39\overline{)75{,}342}$$ ■

Not all division estimates come out so easily. In some cases, you may need to carry out a long-division problem of several steps just to obtain the estimate.

EXAMPLE 12 Ralph drove his car a distance of 778 miles. He used 25 gallons of gas. Estimate how many miles he can travel on 1 gallon of gas.

In order to solve this problem, we need to divide 778 by 25 to obtain the number of miles Ralph gets with 1 gallon of gas. We round each number to a number with one nonzero digit and then perform the division, using the rounded numbers.

$$25\overline{)778} \qquad \begin{array}{r} 26 \\ 30\overline{)800} \\ \underline{60} \\ 200 \\ \underline{180} \\ 20 \end{array} \quad \textit{Remainder}$$

We obtain an answer of 26 with a remainder of 20. For our estimate we will use the whole number 27. Thus we estimate that the number of miles Ralph's car obtained on 1 gallon of gas was 27 miles. (This is reasonably close to the exact answer, which is just slightly more than 31 miles per gallon of gas.)

Practice Problem 12 The highway department purchased 58 identical trucks at a total cost of \$1,864,584. Estimate the cost for one truck. ■

1.7 Exercises

Verbal and Writing Skills

1. Explain the rule for rounding and provide examples.

2. What happens when you round 98 to the nearest ten?

Round to the nearest ten.

3. 83 **4.** 45 **5.** 65 **6.** 57 **7.** 92

8. 138 **9.** 526 **10.** 2834 **11.** 4235 **12.** 8963

Round to the nearest hundred.

13. 437 **14.** 735 **15.** 2781 **16.** 1258 **17.** 7692 **18.** 1643

Round to the nearest thousand.

19. 7621 **20.** 3754 **21.** 881 **22.** 679 **23.** 27,863 **24.** 94,671

Applications

Solve each problem.

25. The worst death rate from an earthquake was in Shaanxi, China, in 1556. That earthquake killed an estimated 832,400 people. Round this number to the nearest hundred thousand.

26. One light year (the distance light travels in 1 year) measures 5,878,612,843,000 miles. Round this figure to the nearest hundred million.

27. The Hubble Space Telescope's *Guide Star Catalogue* lists 15,169,873 stars. Round this figure to the nearest million.

28. In 1993, the countries included in the Far East and Oceanea utilized 15,817 barrels of petroleum per day. Round this value to the nearest thousand.

29. The total Native American population living on the Navajo and Trust Lands in the Arizona/New Mexico/Utah area numbered 143,405 in 1990. Round this figure to
(a) the nearest thousand
(b) the nearest hundred

30. A few years ago, there were 614,571 Catholic School pupils in the United States. Round this figure to
(a) the nearest thousand
(b) the nearest hundred

31. The total area of mainland China is 3,705,392 square miles, or, 9,596,960 square kilometers. For *both* square miles and square kilometers, round this figure to
(a) the nearest hundred thousand

(b) the nearest ten thousand

32. Recently, in the United States 10,957,442,856 pounds of apples were produced.
(a) Round this figure to the nearest hundred thousand.
(b) Round this figure to the nearest ten thousand.

Use the principle of estimation to find an estimate for each of the following calculations.

33. 456 + 234 + 875

34. 341 + 178 + 987

35. 34 + 78 + 59 + 31

36. 31 + 75 + 82 + 43

37. 146,270 + 47,566 + 96,112

38. 345,957 + 89,334 + 56,487

39. 567,984 − 129,562

40. 975,935 − 593,228

41. 78,945,000 − 61,076,500

42. 4,596,450 − 3,894,202

43. 33,261,378 − 18,199,276

44. 89,263,000 − 54,198,635

45. 47 × 62

46. 43 × 95

47. 1798 × 7

48. 3254 × 6

49. 631,540 × 312

50. 374,193 × 193

51. 24,318 ÷ 21

52. 61,986 ÷ 32

53. 156,721 ÷ 42

54. 581,361 ÷ 28

55. 6,787,129 ÷ 538

56. 7,863,127 ÷ 352

Estimate the result of each of the following calculations. Some results are correct and some are incorrect. Which results appear to be correct? Which results appear to be incorrect?

57.
$$\begin{array}{r} 361 \\ 522 \\ 873 \\ +\ 164 \\ \hline 1320 \end{array}$$

58.
$$\begin{array}{r} 476 \\ 124 \\ 516 \\ +\ 389 \\ \hline 1505 \end{array}$$

59.
$$\begin{array}{r} 97{,}635 \\ 52{,}123 \\ +\ 41{,}986 \\ \hline 291{,}744 \end{array}$$

60.
$$\begin{array}{r} 26{,}181 \\ 47{,}998 \\ +\ 63{,}271 \\ \hline 137{,}450 \end{array}$$

61.
$$\begin{array}{r} 257{,}163 \\ -\ \ 99{,}358 \\ \hline 157{,}805 \end{array}$$

62.
$$\begin{array}{r} 682{,}351 \\ -\ \ 96{,}891 \\ \hline 585{,}460 \end{array}$$

63.
$$\begin{array}{r} 78{,}126{,}345 \\ -\ 48{,}972{,}103 \\ \hline 19{,}154{,}242 \end{array}$$

64.
$$\begin{array}{r} 42{,}765{,}317 \\ -\ 29{,}318{,}274 \\ \hline 23{,}447{,}043 \end{array}$$

65. $\begin{array}{r} 216 \\ \times \ 24 \\ \hline 6184 \end{array}$

66. $\begin{array}{r} 578 \\ \times \ 32 \\ \hline 10{,}496 \end{array}$ ______

67. $\begin{array}{r} 5896 \\ \times \ 72 \\ \hline 424{,}512 \end{array}$

68. $\begin{array}{r} 8076 \\ \times \ 89 \\ \hline 718{,}764 \end{array}$ ______

69. $78\overline{)412{,}308}$ with quotient 5286

70. $58\overline{)277{,}994}$ with quotient 4793

71. $423\overline{)161{,}163}$ with quotient 381

72. $781\overline{)477{,}972}$ with quotient 612

Applications

73. A huge restaurant in New York City is 40 yards wide and 110 yards long. Estimate the number of square yards in the restaurant.

74. Larry and Nella are planning an outdoor wedding by the ocean. There is a beautiful meadow that is 35 feet wide and 62 feet long. Estimate the number of square feet in the meadow.

75. The last four times Mrs. Watkins went food shopping for her family of six kids, two adults, one cat, one dog, and a canary, she spent \$442, \$379, \$431, and \$333. Estimate what she spent to feed her family during this time period.

76. During the first week of October, the daily number of people buying football tickets for New England Patriots home games over four days was 975, 509, 784, and 881. Estimate the total number of people who bought football tickets for that 4-day period.

77. To get ready for bathing suit weather, Wendy did 85 sit-ups daily for 63 days. Estimate how many sit-ups she did during that period.

78. The local pizzeria makes 267 pizzas on an average day. Estimate how many pizzas were made in the last 134 days.

79. The price of a Superbowl ticket increased from \$30 in 1980 to \$175 in 1993. Estimate the increase. Determine the exact increase.

80. U.S. air travel increased from 457,000,000 passengers in 1990 to 509,000,000 in 1994. Round each figure to the nearest ten million. Then estimate the increase.

81. Tina determined that she spent \$780 last year renting videos. If she rented videos each week, estimate how much she spent weekly for videos last year.

82. Ted Kinney determined that he had earned \$5,125,780 as an attorney working 41 years in his law practice. Estimate his average annual salary over those 41 years.

To Think About

83. A space probe is sent at 23,560 miles per hour for a distance of 7,824,560,000 miles.

(a) How many *hours* will it take the space probe to travel that distance? (Estimate.)

(b) How many *days* will it take the space probe to travel that distance? (Estimate.)

84. A space probe is sent at 28,367 miles per hour for a distance of 9,348,487,000 miles.

(a) Estimate the number of *hours* it will take the space probe to travel that distance.

(b) Estimate the number of *days* it will take the space probe to travel that distance.

Cumulative Review Problems

Evaluate the following.

85. $26 \times 3 + 20 \div 4$

86. $5^2 + 3^2 - (17 - 10)$

87. $3 \times (16 \div 4) + 8 \times 2$

88. $126 + 4 - (20 \div 5)^3$

1.8 Applied Problems

After studying this section, you will be able to:

1. *Use the Mathematics Blueprint to solve problems involving one type of operation.*
2. *Use the Mathematics Blueprint to solve problems involving more than one type of operation.*

MathPro Video 3 SSM

1 *Solving Problems Involving One Type of Operation*

When a builder constructs a new home or office building, he often has a blueprint. This accurate drawing shows the basic structure of the building. It also shows the dimensions of the structure to be built. This blueprint serves as a useful reference throughout the construction process.

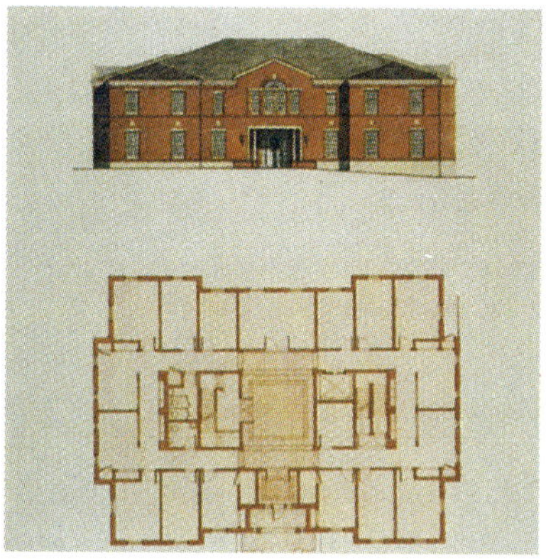

Similarly, when solving applied problems, it is helpful to have a "mathematics blueprint." This is a simple way to organize the information provided in the word problem. You record the facts you need to use and specify what you are solving for. You also record any other information that you feel will be helpful. We will use a Mathematics Blueprint for Problem Solving in the following situation.

EXAMPLE 1 Gerald made deposits of \$317, \$512, \$84, and \$161 into his checking account. He also made out checks for \$100 and \$125. What was the total of his deposits?

1. *Understand the problem.*

First we read over the problem carefully and we fill in the Mathematics Blueprint.

MATHEMATICS BLUEPRINT FOR PROBLEM SOLVING

Gather the Facts	What Am I Asked to Do?	How Do I Proceed?	Key Points to Remember
The **deposits** are \$317 \$512 \$84 \$161	Find the total of Gerald's four deposits.	I must add the four deposits to obtain the total.	Watch out! Don't use the **checks** of \$100 and \$125 in the calculation. We only want the total of the **deposits.**

2. *Solve and state the answer.*
We need to *add* to find the sum of the deposits.

$$\begin{array}{r} 317 \\ 512 \\ 84 \\ +\ 161 \\ \hline 1074 \end{array}$$

The total of the four deposits is \$1074.

3. *Check.*
Reread the problem. Be sure you have answered the question that was asked. Did they ask for the total of the deposits? Yes. ✓

Is the calculation correct? You can use estimation to check. Here we round each of the deposits so that we have one nonzero digit.

$$\begin{array}{rr} 317 & 300 \\ 512 & 500 \\ 84 & 80 \\ +\ 161 & +\ 200 \\ \hline & 1080 \end{array}$$

Our estimate is \$1080. \$1074 is close to our estimated answer of \$1080. Our answer is reasonable. ✓

Thus we conclude that the total of the four deposits is \$1074.

Practice Problem 1 Use the Mathematics Blueprint below to solve the following problem. Diane's paycheck shows deductions of \$135 for federal taxes, \$28 for state taxes, \$13 for FICA, and \$34 for health insurance. Her gross pay (amount before deductions) is \$1352. What is the total amount that is taken out of Diane's paycheck?

MATHEMATICS BLUEPRINT FOR PROBLEM SOLVING			
Gather the Facts	What Am I Asked to Do?	How Do I Proceed?	Key Points to Remember

Portland	Kansas
28353	30162

EXAMPLE 2 Theofilos looked at his odometer before he began his trip from Portland, Oregon, to Kansas City, Kansas. He checked his odometer again when he arrived in Kansas City. The two readings are shown on the right. How many miles did Theofilos travel?

1. *Understand the problem.*
   Determine what information is given.
   The mileage reading before the trip began and when the trip was over.
   What do you need to find?
   The number of miles traveled.

MATHEMATICS BLUEPRINT FOR PROBLEM SOLVING

Gather the Facts	What Am I Asked to Do?	How Do I Proceed?	Key Points to Remember
At start of trip odometer read 28,353 miles. At end of trip odometer read 30,162 miles.	Find out how many miles Theofilos traveled.	I must subtract the two mileage readings.	Subtract the mileage at the start of the trip from the mileage at the end of the trip.

2. *Solve and state the answer.*
   We need to subtract the two mileage readings to find the difference in the number of miles. This will give us the number of miles the car traveled on this trip alone.

   $30{,}162 - 28{,}353 = 1809$ The trip totaled 1809 miles.

3. *Check.*
   We estimate and compare the estimate with the answer above.

Kansas City	30,162→	→30,000	*We subtract*
Portland	28,353→	→28,000	*our rounded values.*
		2,000	

Our estimate is 2000 miles. We compare this estimate with our answer. Our answer is reasonable. ✓

Practice Problem 2 The table on the right shows the results of the 1992 presidential race in Ohio. By how many votes did the Democratic candidate beat the Republican candidate that year in Ohio?

1992 PRESIDENTIAL RACE, OHIO

Candidate	Number of Votes
Clinton	1,964,842
Bush	1,876,495
Perot	1,024,319

MATHEMATICS BLUEPRINT FOR PROBLEM SOLVING

Gather the Facts	What Am I Asked to Do?	How Do I Proceed?	Key Points to Remember

EXAMPLE 3 One horsepower is the power needed to lift 550 pounds a distance of 1 foot in 1 second. How many pounds can be lifted 1 foot in 1 second by 7 horsepower?

1. *Understand the problem.*
   Simplify the problem. If 1 horsepower can lift 550 pounds, how many pounds can be lifted by 7 horsepower? We draw and label a diagram.

We use the mathematics blueprint to organize the information.

MATHEMATICS BLUEPRINT FOR PROBLEM SOLVING			
Gather the Facts	**What Am I Asked to Do?**	**How Do I Proceed?**	**Key Points to Remember**
One horsepower will lift 550 pounds.	Find how many pounds can be lifted by 7 horsepower.	I need to multiply 550 by 7.	I do not use the information about moving one foot in one second.

2. *Solve and state the answer.*
   To solve the problem we multiply the 7 horsepower by 550 pounds for each horsepower.

$$\begin{array}{r} 550 \\ \times \quad 7 \\ \hline 3850 \end{array}$$

   We find that 7 horsepower moves 3850 pounds 1 foot in 1 second. We include 1 foot in 1 second in our answer because it is part of the unit of measure.

3. *Check.*
   We estimate our answer. We round 550 to 600 pounds.

$$600 \times 7 = 4200 \text{ pounds}$$

   Our estimate is 4200 pounds. Our calculations in step 2 gave us 3850. Is this reasonable? This answer is close to our estimate. Our answer is reasonable. ✓

Practice Problem 3 In a measure of liquid capacity, 1 gallon is 1024 fluid drams. How many fluid drams would be in 9 gallons?

MATHEMATICS BLUEPRINT FOR PROBLEM SOLVING			
Gather the Facts	**What Am I Asked to Do?**	**How Do I Proceed?**	**Key Points to Remember**

EXAMPLE 4 Laura can type 35 words per minute. She has to type an English theme that has 5180 words. How many minutes will it take her to type the theme? How many hours and how many minutes will it take her to type the theme?

1. *Understand the problem.*
We draw a picture. Each "package" of 1 minute is 35 words. We want to know how many packages make up 5180 words.

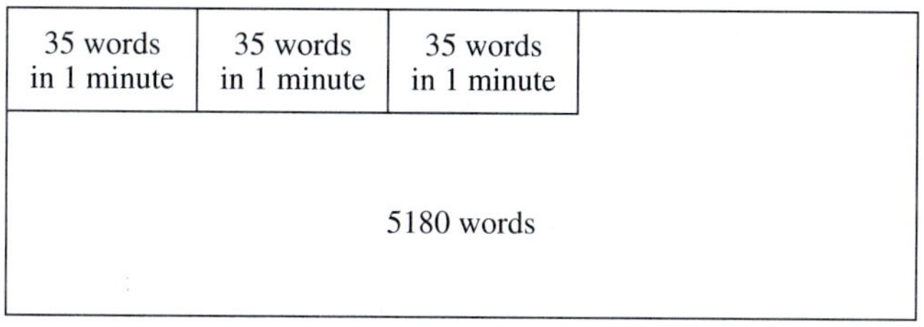

We use the mathematics blueprint to organize the information.

MATHEMATICS BLUEPRINT FOR PROBLEM SOLVING

Gather the Facts	What Am I Asked to Do?	How Do I Proceed?	Key Points to Remember
Laura can type 35 words per minute. She must type a paper with 5180 words.	Find out how many 35-word units are in 5180 words.	I need to divide 5180 by 35.	In converting minutes to hours, I will use the fact that 1 hour = 60 minutes.

2. *Solve and state the answer.*

Division

```
      148
 35)5180
    35
    ---
    168
    140
    ---
     280
     280
     ---
```

It will take 148 minutes.

We will change this answer to hours and minutes. Since 60 minutes = 1 hour, we divide 148 by 60. The quotient will tell us how many hours. The remainder will tell us how many minutes.

```
      2 R 28
 60)148
   -120
   ----
     28
```

Laura can type the theme in 148 minutes or 2 hours, 28 minutes.

3. *Check.*

The theme has *5180 words;* she can type 35 words per minute. 5180 words is approximately 5000 words.

5180 words →	5000 words rounded to nearest thousand.	$40\overline{)5000}$ = 125
35 words per minute →	40 words per minute rounded to nearest ten.	*We divide our estimated values.*

Our estimate is 125 minutes. This is close to our calculated answer. Our answer is reasonable. ✓

Practice Problem 4 Donna bought 45 shares of stock for \$1620. How much did the stock cost her per share?

MATHEMATICS BLUEPRINT FOR PROBLEM SOLVING			
Gather the Facts	What Am I Asked to Do?	How Do I Proceed?	Key Points to Remember

2 *Solving Problems Involving More Than One Type of Operation*

Sometimes a chart, table, or bill of sale can be used to help us organize the data in an applied problem.

EXAMPLE 5 Cleanway Rent-A-Car bought four luxury sedans at \$21,000 each, three compact sedans at \$14,000 each, and seven subcompact sedans at \$8000 each. What was the total cost of the purchase?

1. *Understand the problem.*
   We will make an imaginary bill of sale to help us to visualize the problem.
2. *Solve and state the answer.*
   We do the calculation and enter the results in the bill of sale.

CAR FLEET SALES, INC. HAMILTON, MASSACHUSETTS			
Customer: *Cleanway Rent-A-Car*			
Quantity	Type of Car	Cost per Car	Amount for This Type of Car
4	Luxury sedans	\$21,000	\$84,000 ($4 \times \$21{,}000 = \$84{,}000$)
3	Compact sedans	\$14,000	\$42,000 ($3 \times \$14{,}000 = \$42{,}000$)
7	Subcompact sedans	\$8,000	\$56,000 ($7 \times \$8{,}000 = \$56{,}000$)
		TOTAL	\$182,000 (Sum of the three amounts)

The total cost of all 14 cars is \$182,000.

3. *Check.*
   You may use estimation to check. The check is left to the student.

Practice Problem 5 Anderson Dining Commons purchased 50 tables at \$200 each, 180 chairs at \$40 each, and six moving carts at \$65 each. What was the cost of the total purchase? ■

EXAMPLE 6 Dawn had a balance of \$410 in her checking account last month. She made deposits of \$46, \$18, \$150, \$379, and \$22. She made out checks for \$316, \$400, and \$89. What is her balance?

1. *Understand the problem.*
   We want to *add* to get a total of all deposits and *add* to get a total of all checks.

   Old balance + total of deposits − total of checks = new balance

MATHEMATICS BLUEPRINT FOR PROBLEM SOLVING

Gather the Facts	What Am I Asked to Do?	How Do I Proceed?	Key Points to Remember
Old balance \$410 New deposits \$46 \$18 \$150 \$379 \$22 New checks \$316 \$400 \$89	Find the amount of money in checking account after deposits are made and checks are withdrawn.	**(a)** I need to calculate the total of the deposits and the total of the checks. **(b)** I add the total deposits to the old balance. **(c)** Then I subtract the total of the checks from that result.	Deposits are added to a checking account. Checks are subtracted from a checking account.

2. *Solve and state the answer.*

First we find

Total sum of deposits

$$\begin{array}{r} 46 \\ 18 \\ 150 \\ 379 \\ +\ 22 \\ \hline \$615 \end{array}$$

Then the

Total sum of checks

$$\begin{array}{r} 316 \\ 400 \\ +\ 89 \\ \hline \$805 \end{array}$$

Add the deposits to the old balance and subtract the amount of the checks.

$$\begin{array}{r|r} \text{Old balance} & 410 \\ +\ \text{total deposits} & +\ 615 \\ \cline{2-2} & 1025 \\ -\ \text{total checks} & -\ 805 \\ \cline{2-2} \text{New balance} & 220 \end{array}$$

The new balance of the checking account is \$220.00.

3. *Check.*

Work backward. You can add the total checks to the new balance and then subtract the total deposits. The result should be the old balance. Try it.

$$\begin{array}{r l} 410 & \text{Old balance } \checkmark \\ -\ 615 & \\ \hline 1025 & \uparrow \\ +\ 805 & \\ \hline 220 & \text{Work backward.} \end{array}$$

Practice Problem 6 Last month Bridget had \$498 in a savings account. She made two deposits: one for \$607 and one for \$163. The bank credited her with \$36 interest. Since last month, she has made four withdrawals: \$19, \$158, \$582, and \$74. What is her balance this month?

MATHEMATICS BLUEPRINT FOR PROBLEM SOLVING			
Gather the Facts	What Am I Asked to Do?	How Do I Proceed?	Key Points to Remember

■

EXAMPLE 7 When Lorenzo began his car trip, his gas tank was full and the odometer read 76,358 miles. He ended his trip at 76,668 miles and filled the gas tank with 10 gallons of gas. How many miles per gallon did he get with his car?

1. *Understand the problem.*

MATHEMATICS BLUEPRINT FOR PROBLEM SOLVING			
Gather the Facts	What Am I Asked to Do?	How Do I Proceed?	Key Points to Remember
End of trip, 76,668 miles Start of trip, 76,358 miles Used on trip, 10 gallons of gas	Find the number of miles per gallon that the car obtained on the trip.	**(a)** I need to subtract the two odometer readings to obtain the numbers of miles traveled. **(b)** I divide the number of miles driven by the number of gallons of gas used to get the number of miles obtained per gallon of gas.	The gas tank was full at the beginning of the trip. 10 gallons fills the tank at the end of the trip.

2. *Solve and state the answer.*
First we subtract the odometer readings to obtain the miles traveled.

$$\begin{array}{r} 76{,}668 \\ -\ 76{,}358 \\ \hline 310 \end{array}$$

The trip was 310 miles.
Next we divide the miles driven by the number of gallons.

$$\begin{array}{r} 31 \\ 10\overline{)310} \\ \underline{30} \\ 10 \\ \underline{10} \\ 0 \end{array}$$

Thus Lorenzo obtained 31 miles per gallon on the trip.

3. *Check.*
We do not want to round to one nonzero digit here, because, if we do, the result will be zero when we subtract. Thus we will round to the nearest hundred for the values of mileage.

76,668 becomes 76,700

76,358 becomes 76,400

Now we subtract the estimated values.

$$\begin{array}{r} 76{,}700 \\ -\ 76{,}400 \\ \hline 300 \end{array}$$

Thus we estimate the trip to be 300 miles.
Then we divide.

$$\begin{array}{r} 30 \\ 10\overline{)300} \end{array}$$

We obtain 30 miles per gallon for our estimate. This is very close to our calculated value of 31 miles per gallon. ✓

Practice Problem 7 Deidre took a car trip with a full tank of gas. Her trip began with the odometer at 50,698 and ended at 51,118 miles. She then filled the tank with 12 gallons of gas. How many miles per gallon did her car get on the trip?

MATHEMATICS BLUEPRINT FOR PROBLEM SOLVING			
Gather the Facts	What Am I Asked to Do?	How Do I Proceed?	Key Points to Remember

■

1.8 Exercises

Applications

Use the Mathematics Blueprint for Problem Solving to help you to solve the word problems in the exercise set.

1. The Federal Nigeria game preserve has 24,111 animals, 327 full time staff, and 793 volunteers. What is the total of these three groups?

2. The number of people who flew the shuttle between New York and Boston on the 3 p.m., 4 p.m., 5 p.m., and 6 p.m. flights today was 135, 167, 89, and 152, respectively. What was the total number of passengers today?

3. Wei Ling had a budget of \$37,650 for home renovations for Dr. Smith's residence. When all the work was done, she had actually spent \$42,318. How far over budget has she gone?

4. An airplane was flying at an altitude of 4760 feet. It had to clear a mountain of 4356 feet. By how many feet did it clear the mountain?

5. There are 250 hors d'oeuvres (canapés), in a box. The chef of Dexter's Hearthside ordered 15 boxes. How many hors d'oeuvres did she order?

6. There are 144 pencils in a gross. Mr. Jim Weston ordered 14 gross of pencils for the office. How many pencils did he order?

7. A 16-ounce can of beets cost 96¢. What is the unit cost of the beets? (How much did they cost per ounce?)

8. A 14-ounce can of chicken soup cost 98¢. What is the unit cost of the soup? (How much does the soup cost per ounce?)

9. Mike is a member of the Dartmouth College ski team. During the ski season, Mike had his skis tuned 14 times for a total of \$322. How much did it cost to tune his skis each time?

10. There are approximately 50,000 bison (American Buffalo) living in the United States. If Northwest Trek, the animal preserve located in Mt. Rainier National Park has 103 bison, how many bison are living elsewhere?

11. Salim can make 20 sandwiches in an hour. He has to make 180 sandwiches. How many hours will it take him to make the sandwiches? How many minutes?

12. If a new amusement park covers 43 acres and there are 44,010 square feet in 1 acre, how many square feet of land does the amusement park cover?

13. Every 60 minutes, the world population increases by 100,000 persons. How many people will be born during the next 480 minutes?

14. Roberto had \$2158 in his savings account six months ago. In the last six months he made four deposits: \$156, \$238, \$1119, and \$866. The bank deposited \$136 in interest over the six-month period. How much does he have in the savings account at present?

15. A games arcade has recently opened in a West Chicago neighborhood. The owners were nervous about whether it would be a success. Fortunately, the gross revenues over the last four weeks were \$7356, \$3257, \$4777, and \$4992. What was the gross income this month for the arcade?

16. The two largest cities in Saudi Arabia are Riyadh, the capital, with 1,250,000 people and Jeddah, with 900,000 people. What is the difference in population between these two cities?

Solve each applied problem. In each case, more than one type of operation is required.

17. Harry just got promoted to assistant store manager of a Wal-Mart store. He bought two suits at \$250 each, two shirts at \$35 each, two pairs of shoes at \$75 each, and three ties at \$22 each. What was the total cost of his new work wardrobe?

18. Hector bought three shirts at \$26 each, two pairs of pants at \$45 each, and three pairs of shoes at \$48 each. What was the total cost of his shopping spree?

19. Wei Mai Lee had a balance in her checking account of \$61. During the last few months, she has made deposits of \$385, \$945, \$732, and \$144. She wrote checks against her account for \$223, \$29, \$98, and \$435. When all the deposits are recorded and all the checks clear, what balance will she have in her checking account?

20. Wagner had a balance in his checking account of \$13. He made deposits of \$786, \$566, \$415, and \$50. He wrote out checks for \$554, \$351, \$14, and \$87. When all the deposits are recorded and all the checks clear, what balance will he have in his checking account?

21. Diana owns 85 acres of forest land in Oregon. She rents it to a timber grower for \$250 per acre per year. Her property taxes are \$57 per acre. How much profit does she make on the land each year?

22. Todd owns 13 acres of commercially zoned land in the city of Columbus, Ohio. He rents it to a construction company for \$12,350 per acre per year. His property taxes to the city are \$7362 per acre per year. How much profit does he make on the land each year?

23. Hanna wants to determine the miles-per-gallon rating of her Chevrolet Cavalier. She filled the tank when the odometer read 14,926 miles. She then drove her car on a trip. At the end of the trip the odometer read 15,276 miles. It took 14 gallons to fill the tank. How many miles per gallon does her car deliver?

24. Garcia wants to determine the miles-per-gallon rating of his Ford Contour. He filled the tank when the odometer read 36,339 miles. He drove the car for a week. The odometer then read 36,781 miles and the tank required 17 gallons to be filled. How many miles per gallon did Garcia's car achieve?

Cumulative Review Problems

25. Evaluate. 7^3

26. Perform in the proper order.
$3 \times 2^3 + 15 \div 3 - 4 \times 2$

27. Calculate. 126×38

28. Calculate. $12\overline{)3096}$

Chapter Organizer

Topic	Procedure	Examples
Place value of numbers, p. 4.	Each digit has a value depending on location. millions | hundred thousands | ten thousands | thousands | hundreds | tens | ones	In the number 2,896,341, what place value does 9 have? Ten thousands
Writing expanded notation, p. 5.	Take the number of each digit and multiply it by one, ten, hundred, thousand, . . . according to its place.	Write in expanded notation. 46,235 40,000 + 6000 + 200 + 30 + 5
Writing whole numbers in words, p. 6.	Take the number of each group of three digits and indicate if they are (millions) (thousands) (ones) xxx, xxx, xxx	Write in words. 134,718,216. One hundred thirty-four million, seven hundred eighteen thousand, two hundred sixteen
Adding whole numbers, p. 12.	Starting with right column, add each column separately. If a two-digit sum occurs, "carry" the first digit over to the next column to the left.	Add. (carries: 2 1) 258 367 291 + 453 1369
Subtracting whole numbers, p. 23.	Starting with right column, subtract each column separately. If necessary, borrow a unit from column to the left and bring it to right as a "10."	Subtract. (borrowing: 6 13 12) 16,742 − 12,395 4,347
Multiplying several factors, p. 40.	Keep multiplying from left to right. Take each product and multiply by next factor to right. Continue until all factors are used once. (Since multiplication is commutative and associative, the factors can be multiplied in any order.)	Multiply. $2 \times 9 \times 7 \times 6 \times 3$ $= 18 \times 7 \times 6 \times 3$ $= 126 \times 6 \times 3$ $= 756 \times 3$ $= 2268$
Multiplying several-digit numbers, p. 38.	Multiply top factor by ones digit, then by tens digit, then by hundreds digit. Add the partial products together.	Multiply. 567 × 238 4536 1701 1134 134,946
Dividing by a two- or three-digit number, p. 50.	Figure how many times the first digit of the divisor goes into the first two digits of the dividend. To try this answer, multiply it back to see if it is too large or small. Continue each step of long division until finished.	Divide. 589 238)140182 1190 2118 1904 2142 2142 0

Chapter Organizer

Topic	Procedure	Examples
Exponent form, p. 58.	To show in short form the repeated multiplication of the same number, write the number being multiplied. (This is the base.) Write in smaller print above the line the number of times it appears as a factor. (This is the exponent.) To evaluate the exponent form, write the factor the number of times shown in the exponent. Then multiply.	Write in exponent form. $10 \times 10 \times 10 \times 10 \times 10 \times 10 \times 10 \times 10$ 10^8 Evaluate. 6^3 $6 \times 6 \times 6 = 216$
Order of operations, p. 60.	**1.** Perform operations inside parentheses. **2.** First raise to a power. **3.** Then do multiplication and division in order from left to right. **4.** Then do addition and subtraction in order from left to right.	Evaluate. $2^3 + 16 \div 4^2 \times 5 - 3$ Raise to a power first. $8 + 16 \div 16 \times 5 - 3$ Then do multiplication or division from left to right. $8 + 1 \times 5 - 3$ $8 + 5 - 3$ Then do addition and subtraction. $13 - 3 = 10$
Rounding, p. 64.	**1.** If the first digit to right of round-off place is less than 5, the digit in round-off place is unchanged. **2.** If the first digit to right of round-off place is 5 or more, the digit in round-off place is increased by 1. **3.** Digits to right of round-off place are replaced by zeros.	Round to nearest hundred. ↓ 5 6,7 (4) 3 The digit 4 is less than 5. 56,700 Round to the nearest thousand. ↙ 1 2 8,5 1 7 The digit 5 is obviously 5 or greater. We increase the thousands digit by 1. 129,000
Estimating the answer to a calculation, p. 66.	**1.** Round each number so that there is one nonzero digit. **2.** Perform the calculation with the rounded numbers.	Estimate the answer. $45{,}780 \times 9453$ First we round. $50{,}000 \times 9000$ Then we multiply $\begin{array}{r} 50{,}000 \\ \times \quad 9{,}000 \\ \hline 450{,}000{,}000 \end{array}$ We estimate the answer to be 450,000,000.

Using the Mathematics Blueprint for Problem Solving, p. 73.

In solving an applied problem, students may find it helpful to complete the following steps. You will not use all the steps all the time. Choose the steps that best fit the conditions of the problem.

1. *Understand the problem.*
   (a) Read the problem carefully.
   (b) Draw a picture if this helps you to visualize the situation. Think about what facts you are given and what you are asked to find.
   (c) Use the Mathematics Blueprint for Problem Solving to organize your work. Follow these four parts.
      1. Gather the facts. (Write down specific values given in the problem.)
      2. What am I asked to do? (Identify what you must obtain for an answer.)
      3. How do I proceed? (What calculations need to be done.)
      4. Key points to remember. (Record any facts, warnings, formulas, or concepts you think will be important as you solve the problem.)
2. *Solve and state the answer.*
   (a) Perform the necessary calculations.
   (b) State the answer, including the unit of measure.
3. *Check.*
   (a) Estimate the answer to the problem. Compare this estimate to the calculated value. Is your answer reasonable?
   (b) Repeat your calculations.
   (c) Work backward from your answer. Do you arrive at the original conditions of the problem?

Example

The Manchester highway department has just purchased two pickup trucks and three dump trucks. The cost of a pickup truck is $17,920. The cost of a dump truck is $48,670. What was the cost to purchase these five trucks?

1. *Understand the problem.*

MATHEMATICS BLUEPRINT FOR PROBLEM SOLVING

Gather the Facts	What Am I Asked to Do?	How Do I Proceed?	Key Points to Remember
Buy 2 pickup trucks 3 dump trucks Cost Pickup: $17,920 Dump: $48,670	Find the total cost of the 5 trucks.	Find the cost of 2 pickup trucks. Find the cost of 3 dump trucks. Add to get final cost of all 5 trucks.	Multiply 2 times pickup truck cost. Multiply 3 times dump truck cost.

2. *Solve.*

Calculate cost of pickup trucks

$$\begin{array}{r} \$17,920 \\ \times\ 2 \\ \hline \$35,840 \end{array}$$

Calculate cost of dump trucks

$$\begin{array}{r} \$48,670 \\ \times\ 3 \\ \hline \$146,010 \end{array}$$

Find total cost. $35,840 + $146,010 = $181,850
The total cost of the five trucks is $181,850.

3. *Check.*

Estimate cost of pickup trucks $20,000 \times 2 = 40,000$

Estimate cost of dump trucks $50,000 \times 3 = 150,000$

Total estimate $40,000 + 150,000 = 190,000$

This is close to our calculated answer of $181,850. We determine that our answer is reasonable. ✓

Chapter 1 Review Problems

If you have trouble with a particular type of problem, review the examples in the section indicated for that group of problems. Answers to all exercises are located in the answer key.

1.1 *Write in words.*

1. 376

2. 5082

3. 109,276

4. 423,576,055

Write in expanded notation.

5. 4364

6. 27,986

7. 1,305,128

8. 42,166,037

Write in standard notation.

9. Nine hundred twenty-four

10. Six thousand ninety-five

11. One million, three hundred twenty-eight thousand, eight hundred twenty-eight

12. Forty-five million, ninety-two thousand, six hundred fifty-one

1.2 *Add.*

13. $\begin{array}{r} 36 \\ +\ 94 \\ \hline \end{array}$

14. $\begin{array}{r} 76 \\ +\ 39 \\ \hline \end{array}$

15. $\begin{array}{r} 127 \\ +\ 563 \\ \hline \end{array}$

16. $\begin{array}{r} 12 \\ 28 \\ 34 \\ +\ 76 \\ \hline \end{array}$

17. $\begin{array}{r} 123 \\ 61 \\ 9 \\ 84 \\ +\ 123 \\ \hline \end{array}$

18. $\begin{array}{r} 125 \\ 364 \\ +\ 980 \\ \hline \end{array}$

19. $\begin{array}{r} 937 \\ 405 \\ +\ 256 \\ \hline \end{array}$

20. $\begin{array}{r} 28{,}364 \\ +\ 97{,}059 \\ \hline \end{array}$

21. $\begin{array}{r} 1356 \\ 2892 \\ 561 \\ 89 \\ +\ 9805 \\ \hline \end{array}$

22. $\begin{array}{r} 26 \\ 503 \\ 935 \\ 1257 \\ +\ 7861 \\ \hline \end{array}$

1.3 *Subtract.*

23. $\begin{array}{r} 36 \\ -\ 19 \\ \hline \end{array}$

24. $\begin{array}{r} 54 \\ -\ 48 \\ \hline \end{array}$

25. $\begin{array}{r} 126 \\ -\ 99 \\ \hline \end{array}$

26. $\begin{array}{r} 543 \\ -\ 372 \\ \hline \end{array}$

27. $\begin{array}{r} 1296 \\ -\ 1137 \\ \hline \end{array}$

28. $\begin{array}{r} 9821 \\ -\ 4993 \\ \hline \end{array}$

29. $\begin{array}{r} 101{,}300 \\ -\ 98{,}274 \\ \hline \end{array}$

30. $\begin{array}{r} 201{,}010 \\ -\ 137{,}864 \\ \hline \end{array}$

31. $\begin{array}{r} 1{,}986{,}312 \\ -\ 1{,}761{,}555 \\ \hline \end{array}$

32. $\begin{array}{r} 7{,}216{,}003 \\ -\ 5{,}985{,}312 \\ \hline \end{array}$

1.4 *Multiply.*

33. $\begin{array}{r} 12 \\ \times\ 3 \\ \hline \end{array}$

34. $\begin{array}{r} 57 \\ \times\ 2 \\ \hline \end{array}$

35. $\begin{array}{r} 36 \\ \times\ 0 \\ \hline \end{array}$

36. $\begin{array}{r} 24 \\ \times\ 1 \\ \hline \end{array}$

37. $1 \times 3 \times 6$

38. $2 \times 4 \times 8$

39. $5 \times 7 \times 3$

40. $4 \times 6 \times 5$

41. $8 \times 1 \times 9 \times 2$

42. $7 \times 6 \times 0 \times 4$

43. $3 \cdot 4 \cdot 2 \cdot 2 \cdot 5$

44. $1 \cdot 2 \cdot 7 \cdot 3 \cdot 4$

45. $26{,}121 \times 100$

46. $84{,}312 \times 1000$

47. $832 \times 100{,}000$

48. $563 \times 1{,}000{,}000$

49. $\begin{array}{r} 36 \\ \times\ 24 \\ \hline \end{array}$

50. $\begin{array}{r} 58 \\ \times\ 32 \\ \hline \end{array}$

51. $\begin{array}{r} 150 \\ \times\ 27 \\ \hline \end{array}$

52. $\begin{array}{r} 360 \\ \times\ 38 \\ \hline \end{array}$

53. $\begin{array}{r} 709 \\ \times\ 36 \\ \hline \end{array}$

54. $\begin{array}{r} 502 \\ \times\ 48 \\ \hline \end{array}$

55. $\begin{array}{r} 123 \\ \times\ 714 \\ \hline \end{array}$

56. $\begin{array}{r} 431 \\ \times\ 623 \\ \hline \end{array}$

57. $\begin{array}{r} 1782 \\ \times\ 305 \\ \hline \end{array}$

58. $\begin{array}{r} 2057 \\ \times\ 124 \\ \hline \end{array}$

59. $\begin{array}{r} 300 \\ \times\ 500 \\ \hline \end{array}$

60. $\begin{array}{r} 400 \\ \times\ 600 \\ \hline \end{array}$

61. $\begin{array}{r} 1200 \\ \times\ 6000 \\ \hline \end{array}$

62. $\begin{array}{r} 2500 \\ \times\ 3000 \\ \hline \end{array}$

63. $\begin{array}{r} 100{,}000 \\ \times\ 20{,}000 \\ \hline \end{array}$

64. $\begin{array}{r} 300{,}000 \\ \times\ 40{,}000 \\ \hline \end{array}$

1.5 *Divide, if possible.*

65. $20 \div 10$

66. $40 \div 8$

67. $70 \div 5$

68. $36 \div 9$

69. $0 \div 8$

70. $12 \div 1$

71. $7 \div 1$

72. $0 \div 5$

73. $\frac{49}{7}$

74. $\frac{42}{6}$

75. $\frac{5}{0}$

76. $\frac{24}{6}$

77. $\frac{56}{8}$

78. $\frac{48}{8}$

79. $\frac{72}{9}$

80. $\frac{0}{0}$

Divide. Be sure to indicate the remainder, if one exists.

81. $7\overline{)875}$

82. $6\overline{)750}$

83. $5\overline{)1290}$

84. $4\overline{)1476}$

85. $3\overline{)77{,}622}$

86. $8\overline{)29{,}536}$

87. $6\overline{)221{,}748}$

88. $5\overline{)184{,}605}$

89. $8\overline{)127{,}890}$

90. $7\overline{)250{,}485}$

91. $67\overline{)490}$

92. $72\overline{)325}$

93. $21\overline{)666}$

94. $22\overline{)319}$

95. $68\overline{)2614}$

96. $76\overline{)4142}$

97. $35\overline{)9030}$

98. $45\overline{)4275}$

99. $132\overline{)7128}$

100. $204\overline{)3876}$

1.6 *Write in exponent form.*

101. 13×13

102. 24×24

103. $8 \times 8 \times 8 \times 8 \times 8$

104. $9 \times 9 \times 9 \times 9 \times 9$

Evaluate.

105. 2^6

106. 3^4

107. 5^3

108. 2^7

109. 7^2

110. 9^2

111. 6^3

112. 4^4

Perform each operation in proper order.

113. $6 \times 2 - 4 + 3$

114. $7 + 2 \times 3 - 5$

115. $2^5 + 4 - (5 + 3^2)$

116. $4^3 + 20 \div (2 + 2^3)$

117. $3^3 \times 4 - 6 \div 6$

118. $20 \div 20 + 5^2 \times 3$

119. $2^3 \times 5 \div 8 + 3 \times 4$

120. $3^2 \times 6 \div 3 + 5 \times 6$

121. $9 \times 2^2 + 3 \times 4 - 36 \div (4 + 5)$

122. $5 \times 3 + 5 \times 4^2 - 14 \div (6 + 1)$

Round to the nearest ten.

123. 5673

124. 1275

125. 15,305

126. 42,644

Round to the nearest thousand.

127. 12,350

128. 22,986

129. 675,800

130. 202,498

131. Round to the nearest hundred thousand. 5,668,243.

132. Round to the nearest ten thousand. 9,995,312.

Use the principle of estimation to find an estimate for each of the following calculations.

133. 589 + 622 + 933 + 864

134. 25,981 + 36,782 + 73,125

135. 29,378 − 17,924

136. 4,326,171 − 2,916,788

137. 1763 × 5782

138. 2,965,372 × 893

139. 83,421 ÷ 24

140. 7,963,127 ÷ 378

Estimate the result of each of the following calculations. Some results are correct and some are incorrect. Which results appear to be correct? Which results appear to be incorrect?

141. 87 + 36 + 94 + 55 = 272

142. 938,526 − 398,445 = 540,081

143.

$$\begin{array}{r} 176{,}394 \\ \times\ 5216 \\ \hline 92{,}007{,}114 \end{array}$$

144.

$$\begin{array}{r} 27{,}911 \\ 32\overline{)893{,}152} \end{array}$$

1.8 *Solve each applied problem.*

145. Ward types 25 words per minute. He typed for 7 minutes at that speed. How many words did he type?

146. The soft-drink cans come six to a package. There are 34 packages in the storage room. How many soft-drink cans are there?

147. Applepickers, Inc., bought a truck for $26,300, a car for $14,520, and a minivan for $18,650. What was the total purchase price?

148. Alfonso drove 1362 kilometers last summer, 562 km during Christmas break, and 473 km during spring break. How many total kilometers were driven?

149. Roberta was billed $11,658 for tuition. She received a $4630 grant. How much did she have to pay after the grant was deducted?

150. A plane was flying at 14,630 feet. It flew over a mountain 4329 feet high. How many feet was it from the plane to the top of the mountain?

151. Vincent bought 92 shares of stock for $5888. What was the cost per share?

152. The expedition cost a total of $32,544 for 24 paying passengers, who shared the cost equally. What was the cost per passenger?

153. Marcia's checking account balance last month was $436. She made deposits of $16, $98, $125, and $318. She made out checks of $29, $128, $100, and $402. What will be her balance this month?

154. Melissa's savings account balance last month was $810. The bank added $24 interest. Melissa deposited $105, $36, and $177. She made withdrawals of $18, $145, $250, and $461. What will be her balance this month?

155. Ali began a trip on a full tank of gas with the car odometer at 56,320 miles. He ended the trip at 56,720 miles and added 16 gallons of gas. How many miles per gallon did he get on the trip?

156. Amina began a trip on a full tank of gas with the car odometer at 24,396 miles. She ended the trip at 24,780 miles and added 16 gallons of gas. How many miles per gallon did she get on the trip?

157. The maintenance group bought three lawn mowers at $279, four power drills at $61, and two riding tractors at $1980. What was the total purchase price for these items?

158. The library bought 24 sets of shelves at $118 each, four desks at $120 each, and six chairs at $24 each. What was the total purchase price for these items?

Developing Your Study Skills

EXAM TIME: GETTING ORGANIZED

Studying adequately for an exam requires careful preparation. Begin early so that you will be able to spread your review over several days. Even though you may still be learning new material at this time, you can be reviewing concepts previously learned in the chapter. Giving yourself plenty of time for review will take the pressure off. You need this time to process what you have learned and to tie concepts together.

Adequate preparation enables you to feel confident and to think clearly with less tension and anxiety.

Chapter 1 Test

Do these problems simulating test conditions.

1. Write in words. 44,007,635

2. Write in expanded notation. 26,859

3. Write in standard notation: three million, five hundred eighty-one thousand, seventy-six.

Add.

4. $\begin{array}{r} 126 \\ 83 \\ 5 \\ 294 \\ +\ 192 \\ \hline \end{array}$

5. $\begin{array}{r} 470 \\ 386 \\ +\ 189 \\ \hline \end{array}$

6. $\begin{array}{r} 135{,}484 \\ 2{,}376 \\ 81{,}004 \\ +\ 100{,}113 \\ \hline \end{array}$

Subtract.

7. $\begin{array}{r} 7932 \\ -\ 513 \\ \hline \end{array}$

8. $\begin{array}{r} 300{,}523 \\ -\ 262{,}182 \\ \hline \end{array}$

9. $\begin{array}{r} 18{,}400{,}100 \\ -\ 13{,}174{,}332 \\ \hline \end{array}$

Multiply.

10. $1 \times 6 \times 9 \times 7$

11. $\begin{array}{r} 56 \\ \times\ 39 \\ \hline \end{array}$

12. $\begin{array}{r} 147 \\ \times\ 625 \\ \hline \end{array}$

13. $\begin{array}{r} 18{,}491 \\ \times \quad\quad 7 \\ \hline \end{array}$

1. ______
2. ______
3. ______
4. ______
5. ______
6. ______
7. ______
8. ______
9. ______
10. ______
11. ______
12. ______
13. ______

14. ______

15. ______

16. ______

17. ______

18. ______

19. ______

20. ______

21. ______

22. ______

23. ______

24. ______

25. ______

26. ______

27. ______

28. ______

29. ______

30. ______

Divide. If there is a remainder, be sure to state it as part of your answer.

14. $5\overline{)15{,}071}$

15. $7\overline{)15{,}323}$

16. $37\overline{)13{,}024}$

17. Write in exponent form.
$11 \times 11 \times 11$

18. Evaluate. 2^6

Perform each operation in proper order.

19. $5 + 6^2 - 2 \times (9 - 6)^2$

20. $2^3 + 4^3 + 18 \div 3$

21. $4 \times 6 + 3^3 \times 2 + 23 \div 23$

22. Round to the nearest ten. 26,453

23. Round to the nearest ten thousand. 6,462,431

24. Round to the nearest hundred thousand. 3,593,452

Estimate the answer to each of the following.

25. $4{,}867{,}010 \times 27{,}058$

26. $1423 + 4287 + 4103 + 8549$

Answer each question.

27. A cruise for 15 people cost \$32,220. If each person paid the same amount, how much will it cost each individual?

28. The river is 602 feet wide at Big Bend Corner. A boy is in the shallow water, 135 feet from the shore. How far is the boy from the other side of the river?

29. At the bookstore, Hector bought three notebooks at \$2 each, one textbook for \$45, two lamps at \$21 each, and two sweatshirts at \$17 each. What was his total bill?

30. Patricia is looking at her checkbook. She had a balance last month of \$31. She deposited \$902 and \$399. She made out checks for \$885, \$103, \$26, \$17, and \$9. What will be her new balance?

CHAPTER 2

Fractions

Crop profits, farm operations, equipment maintenance—it is easy to see that working on a farm involves a lot of calculations with whole numbers. However, did you know that many decisions on farms involving thousands of dollars are made based on the careful use of fractions? Are you confident enough in your knowledge of fractions that you could base the entire profit or loss of a farming operation on your accuracy in operations with fractions? Turn to the Putting Your Skills to Work Problems on page 153.

Pretest Chapter 2

1. __________
2. __________
3. __________
4. __________
5. __________
6. __________
7. __________
8. __________
9. __________
10. __________
11. __________
12. __________
13. __________
14. __________
15. __________
16. __________

If you are familiar with the topics in this chapter, take this test now. Check your answers with those in the back of the book. If an answer was wrong or you couldn't do a problem, study the appropriate section of the chapter.

If you are not familiar with the topics in this chapter, don't take this test now. Instead, study the examples, work the practice problems, and then take the test.

This test will help you identify which concepts you have mastered and which you need to study further.

Section 2.1

1. Use a fraction to represent the shaded part of the object.

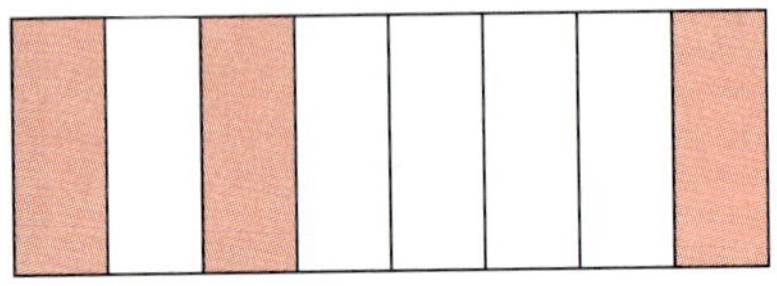

2. Draw a sketch to show $\frac{3}{6}$ of an object.

3. An inspector inspected 124 books. Of these, 17 were defective. Write a fraction that describes the part that was defective.

Section 2.2

Reduce each fraction.

4. $\dfrac{3}{18}$ **5.** $\dfrac{13}{39}$ **6.** $\dfrac{16}{112}$ **7.** $\dfrac{35}{49}$ **8.** $\dfrac{44}{121}$

Section 2.3

Change to an improper fraction.

9. $1\dfrac{2}{7}$ **10.** $6\dfrac{1}{9}$

Change to a mixed number.

11. $\dfrac{97}{4}$ **12.** $\dfrac{29}{5}$ **13.** $\dfrac{36}{17}$

Section 2.4

Multiply.

14. $\dfrac{5}{11} \times \dfrac{1}{4}$ **15.** $\dfrac{3}{7} \times \dfrac{14}{9}$ **16.** $12\dfrac{1}{3} \times 5\dfrac{1}{2}$

Section 2.5

Divide.

17. $\frac{3}{7} \div \frac{3}{7}$

18. $\frac{7}{16} \div \frac{7}{8}$

19. $6\frac{4}{7} \div 1\frac{5}{21}$

20. $8 \div \frac{12}{7}$

Section 2.6

Find the least common denominator.

21. $\frac{1}{8}, \frac{3}{4}, \frac{1}{2}$

22. $\frac{2}{9}, \frac{4}{45}$

23. $\frac{4}{11}, \frac{2}{55}$

24. $\frac{5}{24}, \frac{7}{36}$

Section 2.7

Add or subtract.

25. $\frac{7}{18} - \frac{3}{24}$

26. $\frac{5}{24} + \frac{4}{9} + \frac{1}{36}$

Section 2.8

Add or subtract.

27. $8 - 3\frac{2}{3}$

28. $1\frac{5}{6} + 2\frac{1}{7}$

29. $6\frac{4}{9} - 2\frac{3}{18}$

Section 2.9

Answer each question.

30. Miguel and Lee set out to hike $13\frac{1}{2}$ miles from Arlington to Concord. During the first 5 hours, they covered $6\frac{1}{3}$ miles going from Arlington to Bedford. How many miles are left to be covered from Bedford to Concord?

31. Robert harvested $4\frac{3}{12}$ tons of wheat. His helper harvested $3\frac{1}{18}$ tons of wheat. How much did they harvest together?

32. The students of Smith Dorm fifth floor contributed money to make a stock purchase of one share of stock each. They paid $776 in all to buy the shares of stock. The cost of buying one share was $\$43\frac{1}{9}$. How many students shared in the purchase?

17. ____________

18. ____________

19. ____________

20. ____________

21. ____________

22. ____________

23. ____________

24. ____________

25. ____________

26. ____________

27. ____________

28. ____________

29. ____________

30. ____________

31. ____________

32. ____________

2.1 Understanding Fractions

MathPro Video 3 SSM

After studying this section, you will be able to:

1. *Use a fraction to represent part of a whole.*
2. *Draw a sketch to illustrate a fraction.*
3. *Use fractions to represent the data from applied situations.*

1 *Using a Fraction to Represent Part of a Whole*

In Chapter 1 we studied whole numbers. In this chapter we will study a fractional part of a whole number. One way to represent parts of a whole is with fractions. The word *fraction* (like the word *fracture*) suggests that something is being broken. In mathematics, fractions represent the part that is "broken off" from a whole. The whole can be a single object (like a whole pie) or a group (the employees of a company). Here are some examples.

Single object

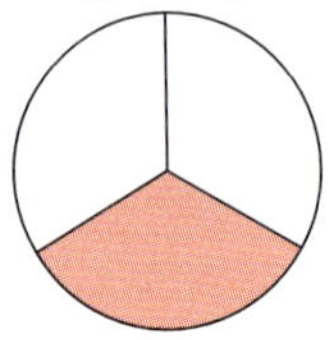

$\frac{1}{3}$

The whole is the pie on the left. The fraction $\frac{1}{3}$ represents the shaded part of the pie, 1 of 3 pieces.

A group: ACE company employs 150 men, 200 women.

$\frac{150}{350}$

The whole is the company of 350 people. The fraction $\frac{150}{350}$ represents that part of the company consisting of men. Note that the total number of employees is 150 men plus 200 women, or 350.

Recipe: Applesauce
4 apples
1/2 cup sugar
1 teaspoon cinnamon

The whole is 1 whole cup of sugar. This recipe calls for $\frac{1}{2}$ cup of sugar. Notice that in many real-life situations $\frac{1}{2}$ is written as 1/2.

When we say "$\frac{3}{8}$ of a pizza has been eaten," we mean 3 of 8 equal parts of a pizza have been eaten. (See the figure on the left.) When we write the fraction $\frac{3}{8}$, the number on the top, 3, is the **numerator,** and the number on the bottom, 8, is the **denominator.**

The numerator specifies how many parts ⟶ 3
The denominator specifies the total number of parts ⟶ 8

When we say, "$\frac{2}{3}$ of the marbles are red," we mean 2 marbles out of a total of 3 are red marbles.

Part we are interested in ⟶ 2 numerator
Total number in the group → 3 denominator

EXAMPLE 1 Use a fraction to represent the part of the whole shown.

(a) 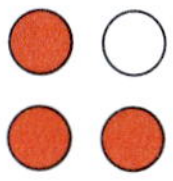

(b) 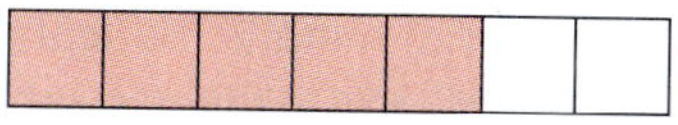

(c) 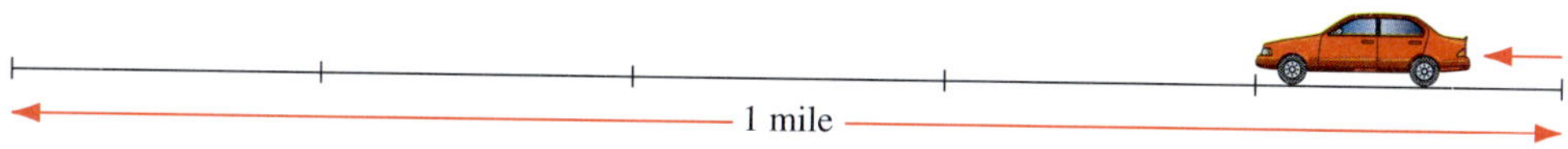

(a) Three out of four circles are red. The fraction is $\frac{3}{4}$.

(b) Five out of seven equal parts are shaded. The fraction is $\frac{5}{7}$.

(c) The mile is divided into five equal parts. The car has traveled 1 part out of 5 of the one mile distance. The fraction is $\frac{1}{5}$.

Practice Problem 1 Use a fraction to represent the shaded part of the whole.

(a) 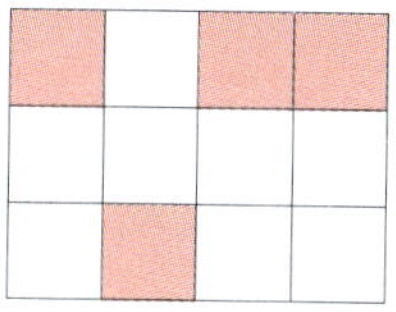

(b) 

(c) 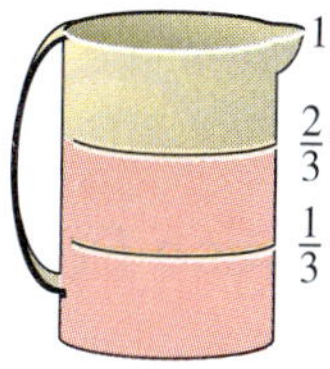

We can also think of a fraction as a division problem.

$$\frac{1}{3} = 1 \div 3 \qquad \text{and} \qquad 1 \div 3 = \frac{1}{3}$$

The division way of looking at fractions asks the questions:

What is the result of dividing one whole into three equal parts?

Thus we can say the fraction $\frac{a}{b}$ means the same as $a \div b$. However, special care must be taken with the number 0.

Suppose that we had four equal parts and we wanted to take none of them. We would want $\frac{0}{4}$ of the parts. Since $\frac{0}{4} = 0 \div 4 = 0$, we see that $\frac{0}{4} = 0$. Any fraction with a 0 numerator equals zero.

$$\frac{0}{8} = 0 \qquad \frac{0}{5} = 0 \qquad \frac{0}{13} = 0$$

What happens when zero is in the denominator? $\frac{4}{0}$ means 4 out of 0 parts. Taking 4 out of 0 does not make sense. We say $\frac{4}{0}$ is *not defined.*

$$\frac{3}{0}, \frac{7}{0}, \frac{4}{0} \qquad \textit{are not defined.}$$

We cannot have a fraction with 0 in the denominator. Since $\frac{4}{0} = 4 \div 0$, we say division by zero is *undefined.* We cannot divide by 0.

2 *Drawing a Sketch to Illustrate a Fraction*

Drawing a sketch of a mathematical situation is a powerful problem-solving technique. The picture often reveals information not always apparent in the words.

EXAMPLE 2 Draw a sketch to illustrate.

(a) $\frac{7}{11}$ of an object **(b)** $\frac{2}{9}$ of a group

(a) The easiest figure to draw is a rectangular bar. We call this a *fraction bar.*

We divide the bar into 11 equal parts. We then shade in 7 parts to show $\frac{7}{11}$.

(b) We draw 9 circles of equal size to represent a group of 9.

We shade in 2 of the 9 circles to show $\frac{2}{9}$.

Practice Problem 2 Draw a sketch to illustrate.

(a) $\frac{4}{5}$ of an object **(b)** $\frac{3}{7}$ of a group ■

Recall these facts about division problems involving the number 1 and the number 0.

Division Involving the Number 1 and the Number 0

1. Any nonzero number divided by itself is 1.

$$\frac{7}{7} = 1$$

2. Any number divided by 1 remains unchanged.

$$\frac{29}{1} = 29$$

3. Zero may be divided by any nonzero number; the result is always zero.

$$\frac{0}{4} = 0$$

4. Zero can never be the denominator in a division problem.

$\frac{0}{0}$ cannot be done $\frac{3}{0}$ cannot be done

3 *Using Fractions to Represent Data from Applied Situations*

Several real-life or applied situations can be described using fractions.

EXAMPLE 3 Use a fraction to describe each of the following situations.

(a) A baseball player gets a hit 5 out of 12 times at bat.

(b) There are 156 men and 185 women taking psychology this semester. Describe the part of the class that consists of women.

(c) APEX stock went up three-eighths of a point yesterday.

(a) The baseball player got a hit $\frac{5}{12}$ of his times at bat.

(b) The total class is $156 + 185 = 341$. The fractional part that is women is 185 out of 341. Thus $\frac{185}{341}$ of the class is women.

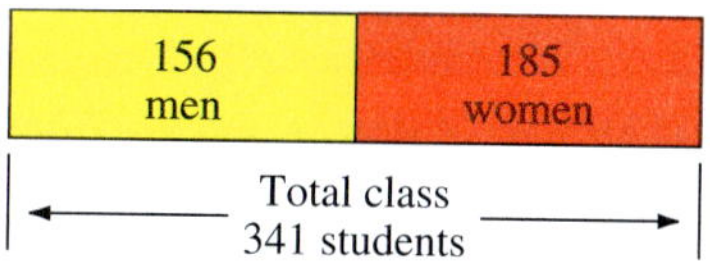

(c) The price of stock is calculated in dollars and part of a dollar. Three-eighths of a point is three-eighths of a dollar. The fraction is $\frac{3}{8}$.

Practice Problem 3 Use a fraction to describe each of the following situations.

(a) 9 out of the 17 players on the basketball team are on the Dean's List.

(b) The senior class has 382 men and 351 women. Describe the part of the class consisting of men.

(c) John needed seven-eighths of a yard of material. ■

EXAMPLE 4 Wanda made 13 calls, out of which she made five sales. Albert made 17 calls, out of which he made six sales. Write a fraction that describes for both people together the number of calls in which a sale was made compared with the total number of calls.

There are $5 + 6 = 11$ calls in which a sale was made.
There were $13 + 17 = 30$ total calls.

Thus $\dfrac{11}{30}$ of the calls resulted in a sale.

Practice Problem 4 An inspector found that one out of seven belts was defective. She also found that two out of nine shirts were defective. Write a fraction that describes what part of all the objects examined were defective. ■

2.1 Exercises

Verbal and Writing Skills

1. A __________ can be used to represent part of a whole or part of a group.
2. The ____________________ tells the number of parts we are interested in.
3. The __________ tells the total number of parts in the whole or in the group.
4. Describe a real-life situation that involves fractions.

Name the numerator and the denominator in each fraction.

5. $\frac{3}{5}$ 6. $\frac{9}{11}$ 7. $\frac{2}{3}$ 8. $\frac{3}{4}$ 9. $\frac{1}{17}$ 10. $\frac{1}{15}$

In problems 11–30, use a fraction to represent the shaded part of the object or the shaded portion of the set of objects.

11. 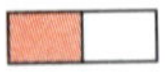

12. 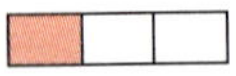

13. 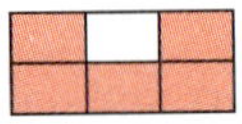

14. 

15. 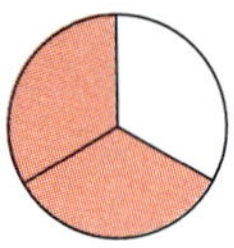

16. 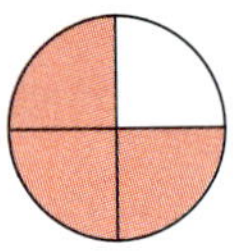

17. 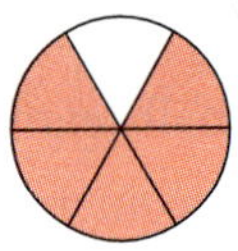

18. 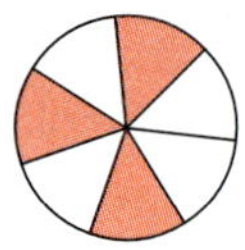

19. 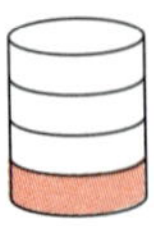

20. 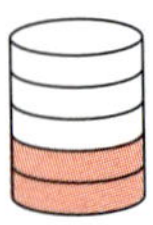

21. 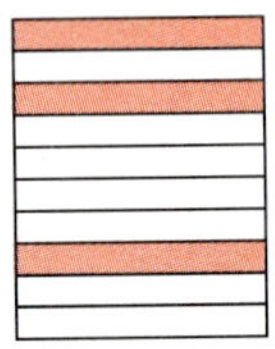

22. 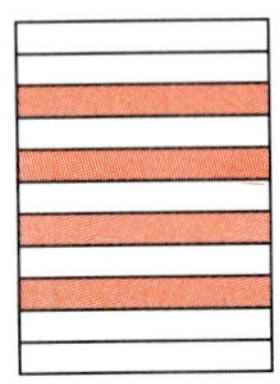

23. 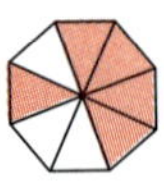

24. 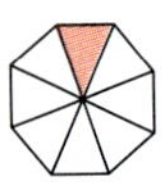

25. 

26. 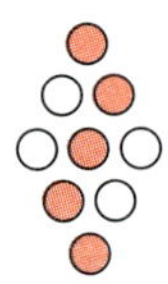

27. 

28. 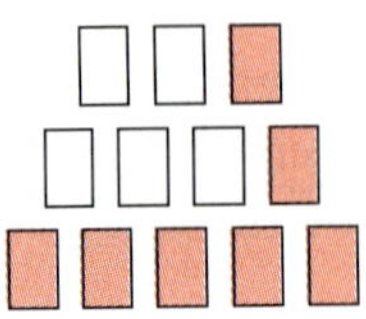

29. 

30. 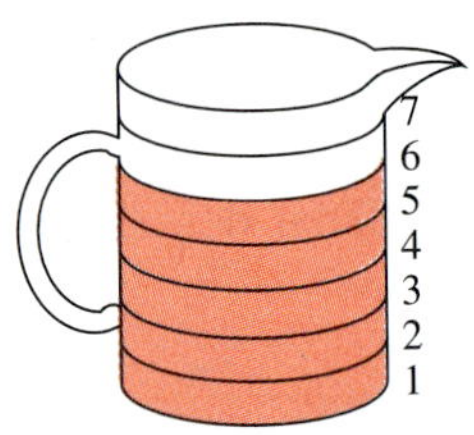

Draw a sketch to illustrate each fractional part.

31. $\frac{4}{5}$ of an object

32. $\frac{3}{7}$ of an object

33. $\frac{11}{13}$ of an object

34. $\frac{5}{12}$ of an object

35. $\frac{7}{10}$ of an object

36. $\frac{8}{15}$ of an object

Applications

37. The Crafts Fair had 165 exhibitors, 29 of whom were toy makers. What fractional part of the exhibitors at the Crafts Fair were toy makers?

38. The total purchase amount was 83¢, of which 5¢ was sales tax. What fractional part of the total purchase price was sales tax?

39. Lance bought a 100-CD jukebox for $750. Part of it was paid for with the $209 he earned parking cars for the valet service at a local wedding reception hall. What fractional part of the jukebox was paid for by his weekend earnings?

40. Charles drove for 47 minutes to go to the Laker game. 18 minutes of his trip were spent in bumper-to-bumper traffic. What fractional part of his time was spent in traffic?

41. The Democratic National Committee fund-raising event served 122 chicken dinners and 89 roast beef dinners to its contributors. What fractional part of the guests ate roast beef?

42. The East dormitory has 111 smokers and 180 nonsmokers. What fractional part of the dormitory has nonsmoking residents?

43. In the Golden West apartment building, 17 people own cats, 14 people own dogs, 4 own birds, and 15 own fish. What fractional part of the total pet owners have cats?

44. From Roger's paycheck $100 was withheld for federal income tax, $24 was withheld for state income tax, and $63 was withheld for social security. What fractional part of the withheld money is the federal income tax?

45. The picnic table held 2 bowls of corn, 3 bowls of potato salad, 4 bowls of baked beans, and 5 bowls of ribs. What fractional part of the bowls on the table contains either ribs or beans?

46. A box of compact discs contains 5 classical CDs, 6 jazz CDs, 4 sound tracks, and 24 blues CDs. What fractional part of the total CDs is either jazz or blues?

★**47.** The West Peabody Engine Company manufactured two items last week: 101 engines and 94 lawn mowers. It was discovered that 19 engines were defective and 3 lawn mowers were defective. Of the engines that were not defective, 40 were properly constructed but 42 were not of the highest quality. Of the lawn mowers that were not defective, 50 were properly constructed but 41 were not of the highest quality.

(a) What fractional part of all items manufactured was of the highest quality?

(b) What fractional part of all items manufactured was defective?

★**48.** A Chicago tour bus held 25 women and 33 men. 12 women wore jeans. 19 men wore jeans. In the group of 25 women, a subgroup of 8 women wore sandals. In the group of 19 men, a subgroup of 10 wore sandals.

(a) What fractional part of the people on the bus wore jeans?

(b) What fractional part of the women on the bus wore sandals?

To Think About

49. Illustrate a real-world example of the fraction $\frac{0}{6}$.

50. What happens when we try to illustrate a real-world example of the fraction $\frac{6}{0}$? Why?

Cumulative Review Problems

51. Add. $\begin{array}{r} 18 \\ 27 \\ 34 \\ 16 \\ 125 \\ +\ 21 \\ \hline \end{array}$

52. Subtract. $\begin{array}{r} 38{,}114 \\ -\ 27{,}008 \\ \hline \end{array}$

53. Multiply. $\begin{array}{r} 4136 \\ \times\ \ 29 \\ \hline \end{array}$

54. Divide. $12\overline{)2130}$

Developing Your Study Skills

PREVIEWING NEW MATERIAL

Part of your study time each day should consist of looking ahead to those sections in your text that are to be covered the following day. You do not necessarily have to study and learn the material on your own, but if you survey the concepts, terminology, diagrams, and examples, the new ideas will seem more familiar to you when the instructor presents them. You can take note of concepts that appear confusing or difficult and be ready to listen carefully for your instructor's explanations. You can be prepared to ask the questions that will increase your understanding. Previewing new material enables you to see what is coming and prepares you to be ready to absorb it.

2.2 Simplifying Fractions

After studying this section, you will be able to:

1 *Write a number as a product of prime factors.*
2 *Reduce a fraction.*
3 *Determine if two fractions are equal.*

MathPro Video 4 SSM

1 *Writing a Number as a Product of Prime Factors*

A **prime number** is a whole number greater than 1 that cannot be evenly divided except by itself and 1. If you examine all the whole numbers from 1 to 50, you will find 15 prime numbers.

The First 15 Prime Numbers

2, 3, 5, 7, 11, 13, 17, 19, 23, 29, 31, 37, 41, 43, 47

A **composite number** is a whole number greater than 1 that can be divided by whole numbers other than itself. The number 12 is a composite number.

$$12 = 2 \times 6 \qquad \text{and} \qquad 12 = 3 \times 4$$

The number 1 is neither a prime nor a composite number. The number 0 is neither a prime nor a composite number.

Recall that factors are numbers that are multiplied together. Prime factors are prime numbers. To check to see if a number is prime or composite, simply divide the smaller primes (such as 2, 3, 5, 7, 11, . . .) into the given number. If the number can be divided exactly without a remainder by one of the smaller primes, it is a composite and not a prime.

Some students find the following rules helpful when deciding if a number can be divided by 2, 3, or 5.

Divisibility Tests

1. A number is divisible by 2 if the last digit is 0, 2, 4, 6, or 8.
2. A number is divisible by 3 if the sum of the digits is divisible by 3.
3. A number is divisible by 5 if the last digit is 0 or 5.

To illustrate:

1. 478 is divisible by 2 since it ends in 8.
2. 531 is divisible by 3 since when we add the digits of 531 (5 + 3 + 1), we get 9, which is divisible by 3.
3. 985 is divisible by 5 since it ends in 5.

EXAMPLE 1 Write each whole number as the product of prime factors.

(a) 12 **(b)** 60 **(c)** 168

(a) To start, write 12 as the product of any two factors. We will write 12 as 4×3.

$12 = 4 \times 3$ *Now check whether the factors are prime. If not, factor these.*

$2 \times 2 \times 3$

$12 = 2 \times 2 \times 3$ *Now all factors are prime, so 12 is completely factored.*

Instead of writing $2 \times 2 \times 3$, we can write $2^2 \times 3$.

Note: To start, we could write 12 as 2×6. Begin this way and follow the steps above. Is the product of prime factors the same? Will this always be true?

(b) We follow the same steps as in (a).

$60 = 6 \times 10$

$3 \times 2 \times 2 \times 5$ *Check that all factors are prime.*

$60 = 2 \times 2 \times 3 \times 5$

Instead of writing $2 \times 2 \times 3 \times 5$, we can write $2^2 \times 3 \times 5$.

Note that in the final answer the prime factors are listed in order from least to greatest.

(c) $168 = 4 \times 42$

$= 2 \times 2 \times 2 \times 21$

$= 2 \times 2 \times 2 \times 3 \times 7$

$168 = 2 \times 2 \times 2 \times 3 \times 7$ *Check that all factors are prime.*

Instead of writing $2 \times 2 \times 2 \times 3 \times 7$, we can write $2^3 \times 3 \times 7$.

Practice Problem 1 Write each whole number as a product of primes.

(a) 18 **(b)** 72 **(c)** 400 ■

Suppose we started Example 1(c) by writing $168 = 14 \times 12$. Would we get the same answer? Would our answer be correct? Let's compare.

$168 = 4 \times 42$	$168 = 14 \times 12$
$= 2 \times 2 \times 2 \times 21$	$= 2 \times 7 \times 4 \times 3$
$168 = 2 \times 2 \times 2 \times 3 \times 7$	$= 2 \times 7 \times 2 \times 2 \times 3$
or $168 = 2^3 \times 3 \times 7$	$168 = 2 \times 2 \times 2 \times 3 \times 7$
	or $168 = 2^3 \times 3 \times 7$

The result is the same.

The order of prime factors is not important because multiplication is commutative. No matter how we start, when we factor a composite number, we always get exactly the same prime factors.

> **The Fundamental Theorem of Arithmetic**
>
> Every composite number can be written in exactly one way as a product of prime numbers.

You will be able to check this theorem out for yourself. See Exercises 2.2, problems 7 through 26. Writing a number as a product of prime factors is also called **prime factorization.**

2 *Reducing Fractions*

You know that $5 + 2$ and $3 + 4$ are two ways to write the same number. We say they are *equivalent* because they are *equal* to the same *value.* They are both ways of writing the value 7.

Like whole numbers, fractions can be written in more than one way. For example, $\frac{2}{4}$ and $\frac{1}{2}$ are two ways to write the same number. The value of the fractions is the same. When we use fractions, we often need to write them in another form. If we make the numerator and denominator smaller, we *simplify* the fractions.

Compare the two fractions in the drawings on the right. In each picture the shaded part is the same size. The fractions $\frac{3}{4}$ and $\frac{6}{8}$ are called **equivalent fractions.** The fraction $\frac{3}{4}$ is in simplest form. To see how we can change $\frac{6}{8}$ to $\frac{3}{4}$, we look at a property of the number 1.

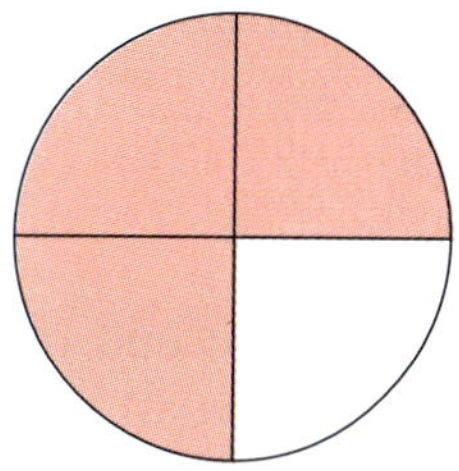

$\frac{3}{4}$ of the circle is shaded.

> Any nonzero number divided by itself is 1.
>
> $$\frac{5}{5} = \frac{17}{17} = \frac{c}{c} = 1.$$

Thus, if we multiply a fraction by $\frac{5}{5}$ or $\frac{17}{17}$ or $\frac{c}{c}$ (remember c cannot be zero), the value of the fraction is unchanged because we are multiplying by a form of 1. We can use this rule to show that $\frac{3}{4}$ and $\frac{6}{8}$ are equivalent.

$$\frac{3}{4} \times \frac{2}{2} = \frac{6}{8}$$

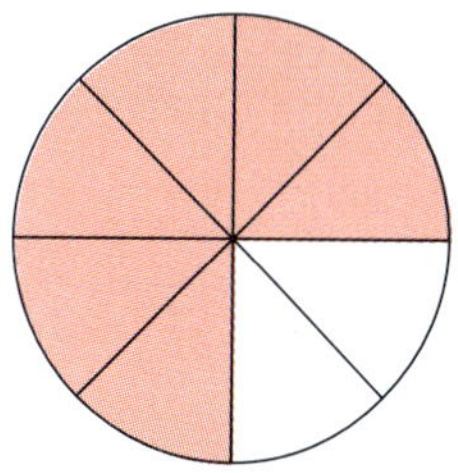

$\frac{6}{8}$ of the circle is shaded.

In general, if b and c are not zero

$$\frac{a}{b} = \frac{a \times c}{b \times c}$$

To reduce a fraction, we find a common factor in the numerator and in the denominator and divide it out. In the fraction $\frac{6}{8}$, the common factor is 2.

$$\frac{6}{8} = \frac{3 \times 2}{4 \times 2} = \frac{3}{4}$$

$$\frac{6}{8} = \frac{3}{4}$$

For all fractions (if a, b, and c are not zero), if c is a common factor,

$$\frac{a}{b} = \frac{a \div c}{b \div c}$$

A fraction is called **simplified** or **in lowest terms** if the numerator and the denominator have no common factor other than 1.

EXAMPLE 2 Simplify (write in lowest terms). **(a)** $\frac{15}{25}$ **(b)** $\frac{42}{56}$

(a) $\frac{15}{25} = \frac{15 \div 5}{25 \div 5} = \frac{3}{5}$ *The common factor is 5. Divide the numerator and the denominator by 5.*

(b) $\frac{42}{56} = \frac{42 \div 14}{56 \div 14} = \frac{3}{4}$ *The common factor is 14. Divide the numerator and the denominator by 14.*

Perhaps 14 was not the first common factor you thought of. Perhaps you did see the common factor 2. Divide out 2. Then look for another common factor, 7. Now divide out 7.

$$\frac{42}{56} = \frac{42 \div 2}{56 \div 2} = \frac{21}{28} = \frac{21 \div 7}{28 \div 7} = \frac{3}{4}$$

If we do not see large factors at first, sometimes simplifying a fraction by dividing both numerator and denominator by a smaller common factor can be done several times, until no common factors are left.

Practice Problem 2 Reduce (or simplify) by dividing out common factors.

(a) $\frac{30}{42}$ **(b)** $\frac{60}{132}$ ■

A second method to reduce or simplify fractions is called the *method of prime factors.* We factor the numerator and the denominator into prime numbers. We then divide the numerator and the denominator by any common prime factors.

EXAMPLE 3 Simplify the fractions by the method of prime factors.

(a) $\frac{35}{42}$ **(b)** $\frac{70}{110}$

(a) $\frac{35}{42} = \frac{5 \times 7}{2 \times 3 \times 7}$ *We factor 35 and 42 into prime factors. The common prime factor is 7.*

$= \frac{5 \times \overset{1}{\cancel{7}}}{2 \times 3 \times \underset{1}{\cancel{7}}}$ *Now we divide out 7.*

$= \frac{5 \times 1}{2 \times 3 \times 1} = \frac{5}{6}$ *We multiply the factors in the numerator and denominator to write the reduced or simplified form.*

Thus $\frac{35}{42} = \frac{5}{6}$, and $\frac{5}{6}$ is the simplified form.

(b) $\frac{70}{110} = \frac{2 \times 5 \times 7}{2 \times 5 \times 11} = \frac{\overset{1}{\cancel{2}} \times \overset{1}{\cancel{5}} \times 7}{\underset{1}{\cancel{2}} \times \underset{1}{\cancel{5}} \times 11} = \frac{7}{11}$

Practice Problem 3 Simplify the fractions by the method of prime factors.

(a) $\frac{120}{135}$ **(b)** $\frac{715}{880}$ ■

3 *Determining If Two Fractions Are Equal*

After we simplify, how can we check that a reduced fraction is *equivalent* to the original fraction? If two fractions are equal, their diagonal products are equal. This is called the *equality test for fractions.* If $\frac{3}{4} = \frac{6}{8}$, then

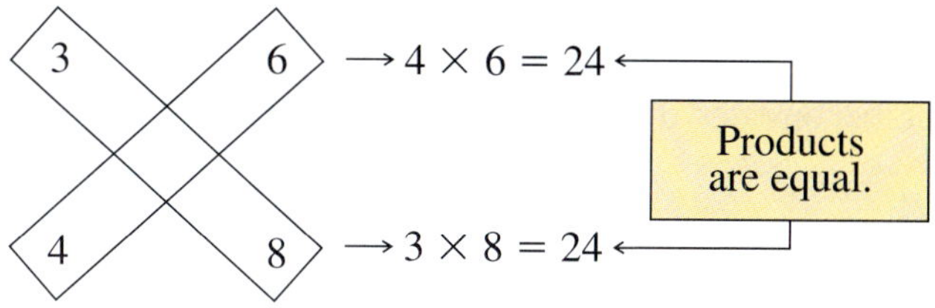

If two fractions are unequal (we use the symbol ≠), their diagonal products are unequal. If $\frac{5}{6} \neq \frac{6}{7}$, then

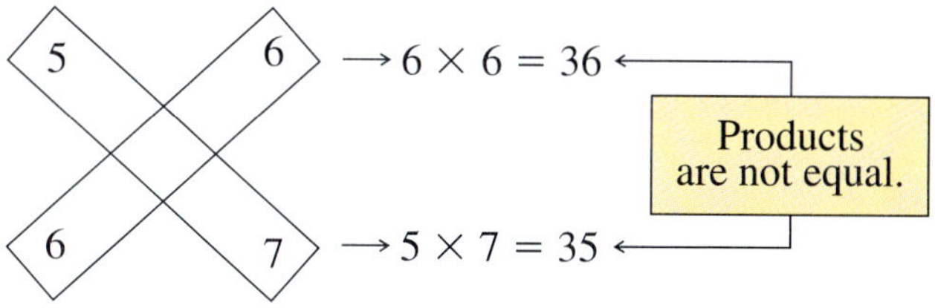

Since $36 \neq 35$, we know that $\frac{5}{6} \neq \frac{6}{7}$. The test can be described in this way.

> **Equality Test for Fractions**
>
> For any two fractions where a, b, c, and d are whole numbers and $b \neq 0$, $d \neq 0$, if $\frac{a}{b} = \frac{c}{d}$, then $a \times d = b \times c$.

EXAMPLE 4 Are these fractions equal? Use the equality test.

(a) $\frac{2}{11} \stackrel{?}{=} \frac{18}{99}$ **(b)** $\frac{3}{16} \stackrel{?}{=} \frac{12}{62}$

(a)

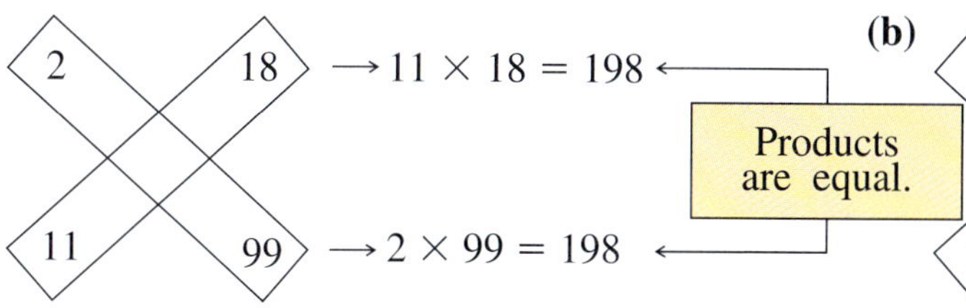

Since $198 = 198$, we know that $\frac{2}{11} = \frac{18}{99}$.

(b)

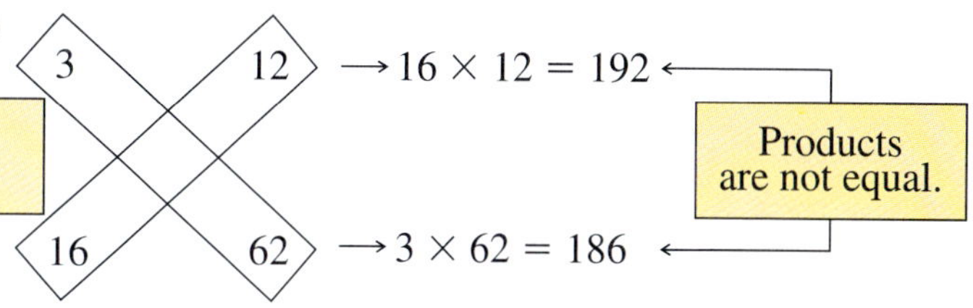

Since $192 \neq 186$, we know that $\frac{3}{16} \neq \frac{12}{62}$.

Practice Problem 4 Test whether the following fractions are equal.

(a) $\frac{84}{108} \stackrel{?}{=} \frac{7}{9}$ **(b)** $\frac{3}{7} \stackrel{?}{=} \frac{79}{182}$ ■

2.2 Exercises

Verbal and Writing Skills

1. Which of these whole numbers are prime?
   4, 12, 11, 15, 6, 19, 1, 41, 38, 24, 5, 46

2. A prime number is a whole number greater than 1 that cannot be evenly ______________ except by itself and 1.

3. A ______________ ______________ is a whole number greater than 1 that can be divided by whole numbers other than itself and 1.

4. Every composite number can be written in exactly one way as a ______________ of ______________ numbers.

5. Give an example of a composite number written as a product of primes.

6. Give an example of equivalent (equal) fractions.

Write each number as a product of prime factors.

7. 15 **8.** 9 **9.** 6 **10.** 8 **11.** 49 **12.** 25 **13.** 64

14. 81 **15.** 55 **16.** 42 **17.** 35 **18.** 36 **19.** 75 **20.** 125

21. 54 **22.** 90 **23.** 84 **24.** 56 **25.** 98 **26.** 65

Identify which of these whole numbers is prime. If a number is composite, write it as the product of prime factors.

27. 31 **28.** 47 **29.** 57 **30.** 51

31. 67 **32.** 71 **33.** 62 **34.** 91

35. 89 **36.** 97 **37.** 127 **38.** 137

39. 121 **40.** 49 **41.** 161 **42.** 169

Reduce each fraction by finding a common factor in the numerator and in the denominator and dividing by the common factor.

43. $\frac{18}{27}$ **44.** $\frac{16}{24}$ **45.** $\frac{32}{48}$ **46.** $\frac{30}{42}$

47. $\frac{30}{48}$ **48.** $\frac{48}{64}$ **49.** $\frac{42}{48}$ **50.** $\frac{35}{80}$

Reduce each fraction by the method of prime factors.

51. $\frac{3}{15}$ **52.** $\frac{7}{21}$ **53.** $\frac{39}{52}$ **54.** $\frac{63}{84}$

55. $\frac{18}{24}$ **56.** $\frac{65}{91}$ **57.** $\frac{27}{45}$ **58.** $\frac{28}{42}$

Mixed Practice

Reduce each fraction by any method.

59. $\frac{33}{36}$ **60.** $\frac{40}{96}$ **61.** $\frac{65}{169}$ **62.** $\frac{21}{98}$ **63.** $\frac{88}{121}$

64. $\frac{165}{180}$ **65.** $\frac{150}{200}$ **66.** $\frac{175}{300}$ **67.** $\frac{119}{210}$ **68.** $\frac{99}{189}$

Are these fractions equal? Why?

69. $\frac{3}{17} \stackrel{?}{=} \frac{15}{85}$ **70.** $\frac{12}{13} \stackrel{?}{=} \frac{36}{39}$ **71.** $\frac{12}{40} \stackrel{?}{=} \frac{9}{30}$ **72.** $\frac{24}{72} \stackrel{?}{=} \frac{15}{45}$

73. $\frac{23}{27} \stackrel{?}{=} \frac{92}{107}$ **74.** $\frac{70}{120} \stackrel{?}{=} \frac{41}{73}$ **75.** $\frac{35}{55} \stackrel{?}{=} \frac{69}{110}$ **76.** $\frac{18}{24} \stackrel{?}{=} \frac{23}{28}$

77. $\frac{65}{70} \stackrel{?}{=} \frac{13}{14}$ **78.** $\frac{98}{182} \stackrel{?}{=} \frac{7}{13}$

Applications

Reduce the fractions in your answer.

79. Every few weeks, Harold and Carolyn Bossél clean out their pantry to make sure that none of the food has gone bad. This week, they found that 4 out of the 6 boxes of cereal on the shelves were stale. What fractional part of the boxes of cereal were stale?

80. Medical students frequently work long hours. James worked a 14-hour shift, spending 10 hours in the emergency room and 4 hours in surgery. What fractional part of his shift was he in the emergency room? What fractional part of his shift was he in surgery?

81. At the local movie theatre, 65 out of 95 movies lasted only one week. What fractional part of the movies were more successful (lasted more than one week?)

82. William works for a wireless communications company that makes beepers and mobile phones. He inspected 315 beepers and found that 20 were defective. What fractional part of the beepers were not defective?

83. Rosie bought a used convertible for \$16,000. She then paid \$2000 to make it look and run like new. What fractional part of her total investment in the car went to make it look and run like new?

84. Monique's sister and her husband have been working two jobs each to put a down payment on a plot of land where they plan to build their house. The purchase price is \$42,500. They have saved \$5500. What fractional part of the cost of the land have they saved?

Cumulative Review Problems

85. Multiply. 386×425

86. Divide. $15{,}552 \div 12$

Estimate *the answer to the following.*

87. $176{,}328 + 43{,}926 + 786{,}035 + 92{,}196 + 543{,}076$

88. $58 \times 62 \times 384 \times 926 \times 18$

2.3 Improper Fractions and Mixed Numbers

After studying this section, you will be able to:

1. *Change a mixed number to an improper fraction.*
2. *Change an improper fraction to a mixed number.*
3. *Reduce a mixed number or an improper fraction.*

MathPro Video 4 SSM

1 *Changing a Mixed Number to an Improper Fraction*

Let's look at the value of a fraction. The value of any fraction may be

- Less than 1 or
- Equal to 1 or
- Greater than 1

We have names for different kinds of fractions. If the value of a fraction is less than 1, we call the fraction proper.

$$\frac{3}{5}, \frac{5}{7}, \frac{1}{8} \quad \textit{are called } \textbf{\textit{proper fractions.}}$$

Notice that the numerator is less than the denominator. If the numerator is less than the denominator, the fraction is a proper fraction.

If the value of a fraction is greater than 1, the quantity can be written as an improper fraction or as a mixed number.

Suppose that we have $1\frac{1}{6}$ of a pizza. We have 1 whole pizza and $\frac{1}{6}$ of a pizza. We could write this as $1\frac{1}{6}$. $1\frac{1}{6}$ is called a mixed number. A **mixed number** is the sum of a whole number greater than zero and a proper fraction. The notation $1\frac{1}{6}$ actually means $1 + \frac{1}{6}$. The plus sign is not usually shown.

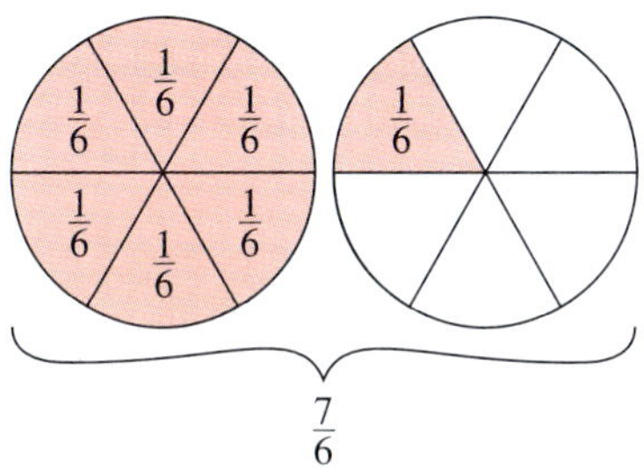

Another way of writing $1\frac{1}{6}$ pizza is to write $\frac{7}{6}$ pizza. $\frac{7}{6}$ is called an improper fraction. Notice the numerator is greater than the denominator. If the numerator is greater than or equal to the denominator, the fraction is an improper fraction.

$$\frac{7}{6}, \frac{6}{6}, \frac{5}{4}, \frac{8}{3}, \frac{2}{2} \quad \textit{are } \textbf{\textit{improper fractions.}}$$

The following chart will help you to visualize these different fractions and their names.

Value Less Than 1	Value Equal To 1	Value Greater Than 1
Proper Fraction	Improper Fraction	Improper Fraction or Mixed Number

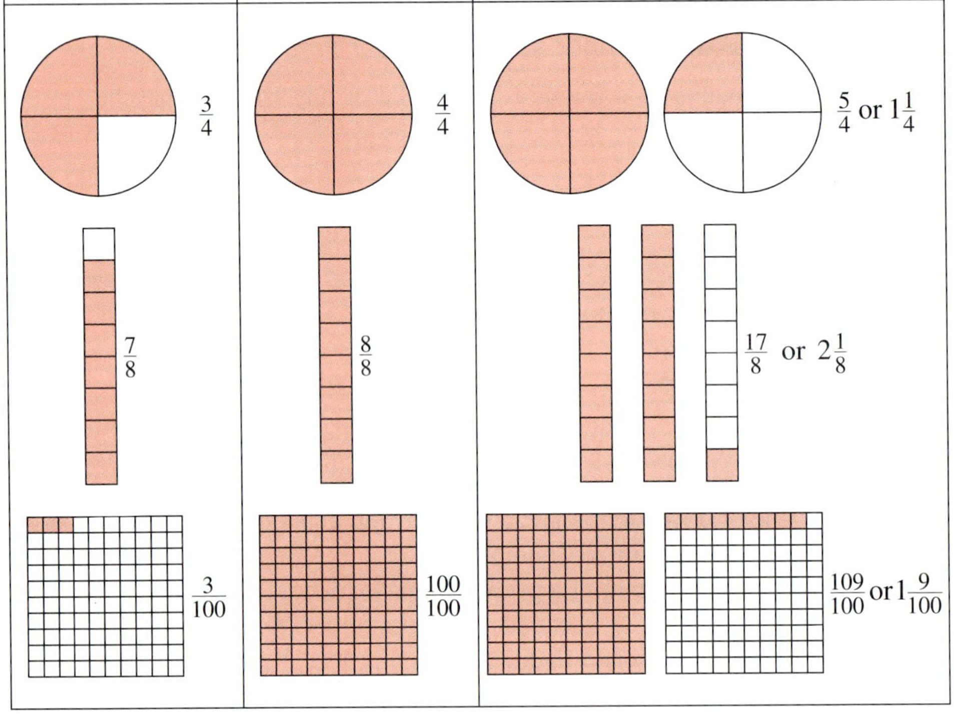

Because improper fractions are easier to add, subtract, multiply, and divide than mixed numbers, we change mixed numbers to improper fractions.

Changing a Mixed Number to an Improper Fraction

1. Multiply the whole number by the denominator of the fraction.
2. Add the numerator of the fraction to the product found in step 1.
3. Write the sum found in step 2 over the denominator of the fraction.

EXAMPLE 1 Change the mixed numbers to improper fractions.

(a) $3\frac{2}{5}$ **(b)** $5\frac{4}{9}$ **(c)** $18\frac{3}{5}$

Multiply the whole number by the denominator.

Add the numerator to the product.

(a) $3\frac{2}{5} \quad \frac{3 \times 5 + 2}{5} = \frac{15 + 2}{5} = \frac{17}{5} \longleftarrow$ Write the sum over the denominator.

(b) $5\frac{4}{9} = \frac{9 \times 5 + 4}{9} = \frac{45 + 4}{9} = \frac{49}{9}$

(c) $18\frac{3}{5} = \frac{5 \times 18 + 3}{5} = \frac{90 + 3}{5} = \frac{93}{5}$

Practice Problem 1 Change the mixed numbers to improper fractions.

(a) $4\frac{3}{7}$ **(b)** $6\frac{2}{3}$ **(c)** $19\frac{4}{7}$ ■

2 *Changing an Improper Fraction to a Mixed Number*

We often need to change an improper fraction to a mixed number.

> **Changing an Improper Fraction to a Mixed Number**
>
> **1.** Divide the numerator by the denominator.
>
> **2.** Write the quotient followed by the fraction with the remainder over the denominator.
>
> $$\text{quotient}\,\frac{\text{remainder}}{\text{denominator}}$$

EXAMPLE 2 Write as a mixed number.

(a) $\frac{13}{5}$ **(b)** $\frac{29}{7}$ **(c)** $\frac{105}{31}$ **(d)** $\frac{85}{17}$

(a) We divide the denominator 5 into 13.

$$\begin{array}{r} 2 \leftarrow \text{quotient} \\ 5\overline{)13} \\ \underline{10} \\ 3 \leftarrow \text{remainder} \end{array}$$

The answer is in the form $\text{quotient}\,\frac{\text{remainder}}{\text{denominator}}$

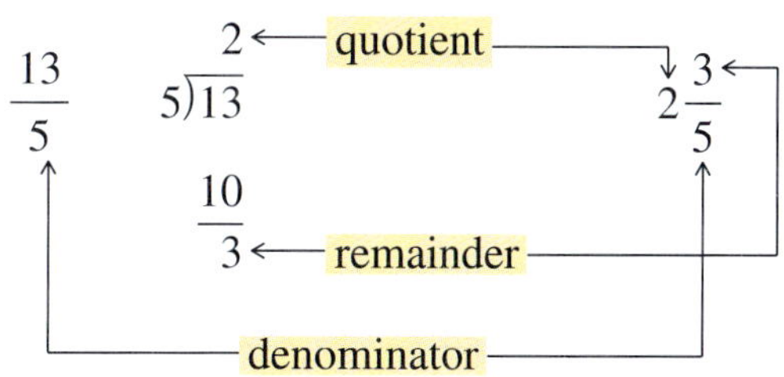

Thus $\frac{13}{5} = 2\frac{3}{5}$.

(b) $\begin{array}{r} 4 \\ 7\overline{)29} \\ \underline{28} \\ 1 \end{array}$ $\qquad \frac{29}{7} = 4\frac{1}{7}$

(c) $31\overline{)105}$ with quotient 3; 93; remainder 12 $\qquad \dfrac{105}{31} = 3\dfrac{12}{31}$

(d) $17\overline{)85}$ with quotient 5; 85; remainder 0 $\qquad$ The remainder is 0, so $\dfrac{85}{17} = 5$, a whole number.

Practice Problem 2 Write as a mixed number or a whole number.

(a) $\dfrac{17}{4}$ (b) $\dfrac{36}{5}$ (c) $\dfrac{116}{27}$ (d) $\dfrac{91}{13}$ ■

3 *Reducing a Mixed Number or an Improper Fraction*

Mixed numbers and improper fractions may need to be reduced if they are not in simplest form. Recall that we look for common factors in the numerator and the denominator of the fraction. Then we divide the numerator and the denominator by the common factor.

EXAMPLE 3 Reduce the improper fraction. $\dfrac{22}{8}$

$$\frac{22}{8} = \frac{\overset{1}{\cancel{2}} \times 11}{\underset{1}{\cancel{2}} \times 2 \times 2} = \frac{11}{4}$$

Practice Problem 3 Reduce the improper fraction. $\dfrac{51}{15}$ ■

EXAMPLE 4 Reduce the mixed number. $4\dfrac{21}{28}$

We do not need to reduce the whole number 4, only the fraction $\dfrac{21}{28}$.

$$\frac{21}{28} = \frac{3 \times \overset{1}{\cancel{7}}}{4 \times \underset{1}{\cancel{7}}} = \frac{3}{4}$$

Therefore, $4\dfrac{21}{28} = 4\dfrac{3}{4}$.

Practice Problem 4 Reduce the mixed number. $3\dfrac{16}{80}$ ■

If an improper fraction contains a very large numerator and denominator, it is best to change the fraction to a mixed number before reducing.

EXAMPLE 5 Reduce $\frac{945}{567}$ by first changing to a mixed number.

$$\begin{array}{r} 1 \\ 567\overline{)945} \\ \underline{567} \\ 378 \end{array} \qquad \text{so} \quad \frac{945}{567} = 1\frac{378}{567}$$

To reduce the fraction we write

$$\frac{378}{567} = \frac{3 \times 3 \times 3 \times 2 \times 7}{3 \times 3 \times 3 \times 3 \times 7} = \frac{\overset{1}{\cancel{3}} \times \overset{1}{\cancel{3}} \times \overset{1}{\cancel{3}} \times 2 \times \overset{1}{\cancel{7}}}{\underset{1}{\cancel{3}} \times \underset{1}{\cancel{3}} \times \underset{1}{\cancel{3}} \times 3 \times \underset{1}{\cancel{7}}} = \frac{2}{3}$$

So $\frac{945}{567} = 1\frac{378}{567} = 1\frac{2}{3}$.

Problems like Example 5 can be done in several different ways. It is not necessary to follow these exact steps when reducing this fraction.

Practice Problem 5 Reduce $\frac{1001}{572}$ by first changing to a mixed number. ■

To Think About A student concluded that just by looking at the denominator he could tell that the fraction $\frac{1655}{97}$ cannot be reduced unless $1655 \div 97$ is a whole number. How did he come to that conclusion? Note that 97 is a prime number. The only factors of 97 are 97 and 1. Therefore, *any* fraction with 97 in the denominator can be reduced only if 97 is a factor of the numerator. Since $1655 \div 97$ is not a whole number (see the division below), it is therefore impossible to reduce $\frac{1655}{97}$.

$$\begin{array}{r} 17 \\ 97\overline{)1655} \\ \underline{97} \\ 685 \\ \underline{679} \\ 6 \end{array}$$

You may explore this idea in Exercises 2.3, problems 81 and 82.

2.3 Exercises

Change each mixed number to an improper fraction.

1. $3\frac{1}{2}$ **2.** $5\frac{1}{4}$ **3.** $4\frac{2}{3}$ **4.** $3\frac{5}{6}$ **5.** $2\frac{3}{7}$ **6.** $3\frac{3}{8}$

7. $5\frac{3}{10}$ **8.** $4\frac{7}{10}$ **9.** $4\frac{5}{6}$ **10.** $9\frac{2}{3}$ **11.** $21\frac{2}{3}$ **12.** $13\frac{1}{3}$

13. $9\frac{1}{6}$ **14.** $56\frac{1}{2}$ **15.** $28\frac{1}{6}$ **16.** $6\frac{6}{7}$ **17.** $10\frac{11}{12}$ **18.** $13\frac{5}{7}$

19. $7\frac{9}{10}$ **20.** $4\frac{1}{50}$ **21.** $8\frac{1}{25}$ **22.** $66\frac{2}{3}$ **23.** $105\frac{1}{2}$ **24.** $207\frac{2}{3}$

25. $164\frac{2}{3}$ **26.** $33\frac{1}{3}$ **27.** $7\frac{14}{15}$ **28.** $4\frac{26}{27}$ **29.** $5\frac{13}{25}$ **30.** $6\frac{18}{19}$

Change each improper fraction to a mixed number or a whole number.

31. $\frac{7}{5}$ **32.** $\frac{10}{3}$ **33.** $\frac{11}{4}$ **34.** $\frac{9}{5}$ **35.** $\frac{15}{6}$ **36.** $\frac{23}{6}$ **37.** $\frac{27}{8}$

38. $\frac{48}{16}$ **39.** $\frac{65}{13}$ **40.** $\frac{56}{15}$ **41.** $\frac{86}{9}$ **42.** $\frac{47}{2}$ **43.** $\frac{28}{13}$ **44.** $\frac{54}{17}$

45. $\frac{51}{16}$ **46.** $\frac{19}{3}$ **47.** $\frac{28}{3}$ **48.** $\frac{100}{3}$ **49.** $\frac{35}{2}$ **50.** $\frac{132}{11}$ **51.** $\frac{91}{7}$

52. $\frac{183}{7}$ **53.** $\frac{200}{9}$ **54.** $\frac{156}{13}$ **55.** $\frac{102}{17}$ **56.** $\frac{105}{8}$ **57.** $\frac{403}{11}$ **58.** $\frac{212}{9}$

Reduce each mixed number.

59. $2\frac{9}{12}$ **60.** $2\frac{10}{15}$ **61.** $4\frac{11}{66}$ **62.** $3\frac{15}{90}$ **63.** $12\frac{15}{40}$ **64.** $15\frac{12}{36}$

Reduce each improper fraction.

65. $\frac{24}{6}$ **66.** $\frac{32}{6}$ **67.** $\frac{36}{15}$ **68.** $\frac{32}{18}$ **69.** $\frac{78}{9}$ **70.** $\frac{143}{22}$

Change to a mixed number and reduce.

71. $\frac{340}{126}$ **72.** $\frac{386}{226}$ **73.** $\frac{986}{424}$ **74.** $\frac{764}{328}$ **75.** $\frac{950}{350}$ **76.** $\frac{1008}{1000}$

Applications

77. The Science Museum is hanging banners all over the building to commemorate the Apollo astronauts. The art department is using $360\frac{2}{3}$ yards of starry sky parachute fabric. Change this number to an improper fraction.

78. For the Northwestern University alumni homecoming, the students studying sculpture have made a giant replica of the school using $244\frac{3}{4}$ pounds of clay. Change this number to an improper fraction.

79. A Cape Cod cranberry bog was contaminated by waste from an abandoned army base. Damage was done to $\frac{151}{3}$ acres of land. Write this as a mixed number.

80. We are helping to build housing for the homeless one weekend per month. There are plans for a porch which will use $\frac{267}{8}$ feet of lumber. Write this as a mixed number.

To Think About

81. Can $\frac{5687}{101}$ be reduced? Why or why not?

82. Can $\frac{9810}{157}$ be reduced? Why or why not?

Cumulative Review Problems

83. Add. 16,385 + 4126 + 8056

84. Subtract. 1,398,210 − 1,137,963

85. Estimate the answer.

78,964 × 229,350

86. Estimate the answer.

872,365 ÷ 286

2.4 Multiplication of Fractions and Mixed Numbers

MathPro Video 4 SSM

After studying this section, you will be able to:

1. *Multiply two fractions that are proper or improper.*
2. *Multiply a whole number by a fraction.*
3. *Multiply mixed numbers.*

1 *Multiplying Two Fractions That Are Proper or Improper*

Fudge Squares

Ingredients:

2 cups sugar	1/4 teaspoon salt
4 oz chocolate	1 teaspoon vanilla
1/2 cup butter	1 cup all-purpose flour
4 eggs	1 cup nutmeats

Suppose you want to cook an amount equal to half of what the recipe shown calls for. You would multiply the measure given for each ingredient by $\frac{1}{2}$.

$\frac{1}{2}$ of 2 cups sugar	$\frac{1}{2}$ of $\frac{1}{4}$ teaspoon salt
$\frac{1}{2}$ of 4 oz chocolate	$\frac{1}{2}$ of 1 teaspoon vanilla
$\frac{1}{2}$ of $\frac{1}{2}$ cup butter	$\frac{1}{2}$ of 1 cup all-purpose flour
$\frac{1}{2}$ of 4 eggs	$\frac{1}{2}$ of 1 cup nutmeats

We often use multiplication of fractions to describe taking a fractional part of something. To find $\frac{1}{2}$ of $\frac{3}{7}$, we multiply

$$\frac{1}{2} \times \frac{3}{7} = \frac{3}{14}$$

We begin with a bar that is $\frac{3}{7}$ shaded. To find $\frac{1}{2}$ of $\frac{3}{7}$ we divide the bar in half and take $\frac{1}{2}$ of the shaded section. $\frac{1}{2}$ of $\frac{3}{7}$ yields 3 out of 14 squares.

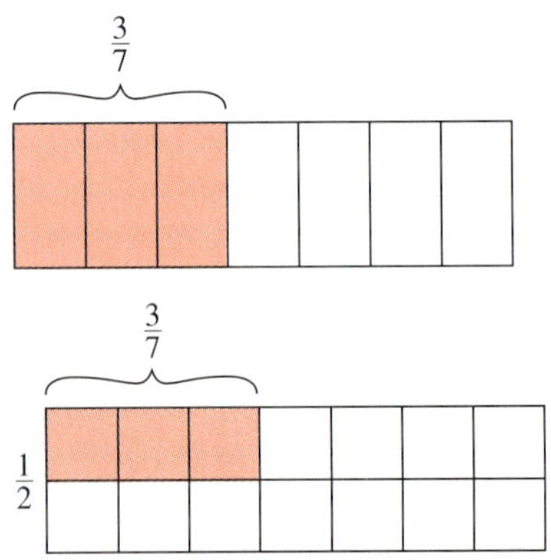

Thus $\frac{1}{2} \times \frac{3}{7} = \frac{3}{14}$

When you multiply two fractions together, you get a smaller fraction.

To multiply two fractions, we multiply the numerators and multiply the denominators.

$$\frac{2}{3} \times \frac{5}{7} = \frac{10}{21} \begin{array}{l} \longleftarrow 2 \times 5 = 10 \\ \longleftarrow 3 \times 7 = 21 \end{array}$$

In general, for all positive whole numbers a, b, c, and d,

$$\frac{a}{b} \times \frac{c}{d} = \frac{a \times c}{b \times d}$$

EXAMPLE 1 Multiply.

(a) $\frac{3}{8} \times \frac{5}{7}$ **(b)** $\frac{1}{11} \times \frac{2}{13}$

(a) $\frac{3}{8} \times \frac{5}{7} = \frac{3 \times 5}{8 \times 7} = \frac{15}{56}$ **(b)** $\frac{1}{11} \times \frac{2}{13} = \frac{1 \times 2}{11 \times 13} = \frac{2}{143}$

Practice Problem 1 Multiply.

(a) $\frac{6}{7} \times \frac{3}{13}$ **(b)** $\frac{1}{5} \times \frac{11}{12}$ ■

Some products may be reduced.

$$\frac{12}{35} \times \frac{25}{18} = \frac{300}{630} = \frac{10}{21}$$

By simplifying before multiplication, the reducing can be done more easily. In a multiplication problem, a factor in the numerator can be paired with a common factor in the denominator of the same or a different fraction. We then divide numerator and denominator by their common factor.

EXAMPLE 2 Simplify first and then multiply.

$$\frac{12}{35} \times \frac{25}{18}$$

$\frac{\overset{2}{\cancel{12}}}{35} \times \frac{25}{\underset{3}{\cancel{18}}}$ *First divide 12 by 6 and 18 by 6.*

$\frac{\overset{2}{\cancel{12}}}{\underset{7}{\cancel{35}}} \times \frac{\overset{5}{\cancel{25}}}{\underset{3}{\cancel{18}}}$ *Next divide 25 by 5 and 35 by 5.*

$\frac{2}{7} \times \frac{5}{3} = \frac{10}{21}$ *Then multiply the simplified fractions.*

The answer is $\frac{10}{21}$.

Practice Problem 2 Simplify first and then multiply.

$$\frac{55}{72} \times \frac{16}{33}$$ ■

2 *Multiplying a Whole Number by a Fraction*

When multiplying a fraction by a whole number, it is more convenient to express the whole number as a fraction with a denominator of 1. We know that $5 = \frac{5}{1}$, $7 = \frac{7}{1}$, and so on.

EXAMPLE 3 Multiply.

(a) $5 \times \frac{3}{8}$ **(b)** $\frac{22}{7} \times 14$

(a) $5 \times \frac{3}{8} = \frac{5}{1} \times \frac{3}{8} = \frac{15}{8}$ or $1\frac{7}{8}$

(b) $\frac{22}{7} \times 14 = \frac{22}{\cancel{7}_1} \times \frac{\cancel{14}^2}{1} = \frac{44}{1} = 44$

Practice Problem 3 Multiply.

(a) $7 \times \frac{5}{13}$ **(b)** $\frac{13}{16} \times 8$ ■

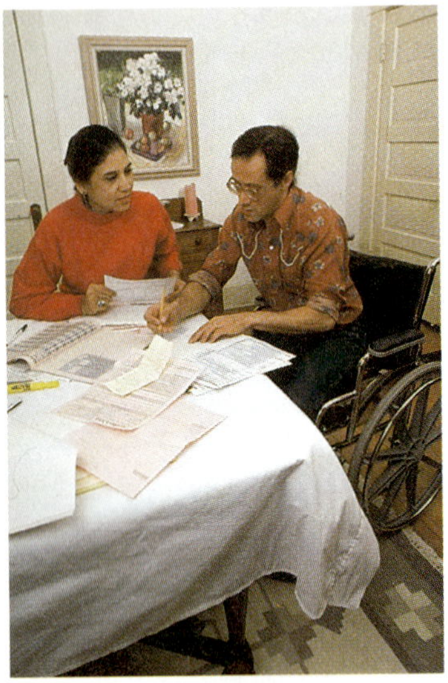

EXAMPLE 4 Mr. and Mrs. Jones found that $\frac{2}{7}$ of their income went to pay federal income taxes. Last year they earned \$37,100. How much did they pay in taxes?

We need to find $\frac{2}{7}$ of \$37,100. So we must multiply $\frac{2}{7} \times 37{,}100$.

$$\frac{2}{\cancel{7}_1} \times \cancel{37{,}100}^{5300} = \frac{2}{1} \times 5300 = 10{,}600$$

They paid \$10,600 in federal income taxes.

Practice Problem 4 Fred and Linda own 98,400 square feet of land. They found that $\frac{3}{8}$ of the land is in a wetland area and cannot be used for building. How many square feet of land are in the wetland area? ■

3 *Multiplying Mixed Numbers*

To multiply a fraction by a mixed number or to multiply two mixed numbers, first change each mixed number to an improper fraction.

EXAMPLE 5 Multiply.

(a) $\frac{5}{7} \times 3\frac{1}{4}$ **(b)** $20\frac{2}{5} \times 6\frac{2}{3}$ **(c)** $\frac{3}{4} \times 1\frac{1}{2} \times \frac{4}{7}$ **(d)** $4\frac{1}{3} \times 2\frac{1}{4}$

(a) $\frac{5}{7} \times 3\frac{1}{4} = \frac{5}{7} \times \frac{13}{4} = \frac{65}{28}$ or $2\frac{9}{28}$

(b) $20\frac{2}{5} \times 6\frac{2}{3} = \frac{\cancel{102}^{34}}{\cancel{5}_1} \times \frac{\cancel{20}^{4}}{\cancel{3}_1} = \frac{136}{1} = 136$

(c) $\frac{3}{4} \times 1\frac{1}{2} \times \frac{4}{7} = \frac{3}{\cancel{4}_1} \times \frac{3}{2} \times \frac{\cancel{4}^1}{7} = \frac{9}{14}$

(d) $4\frac{1}{3} \times 2\frac{1}{4} = \frac{13}{\cancel{3}_1} \times \frac{\cancel{9}^3}{4} = \frac{39}{4}$ or $9\frac{3}{4}$

Practice Problem 5 Multiply.

(a) $2\frac{1}{6} \times \frac{4}{7}$ (b) $10\frac{2}{3} \times 13\frac{1}{2}$ (c) $\frac{3}{5} \times 1\frac{1}{3} \times \frac{5}{8}$ (d) $3\frac{1}{5} \times 2\frac{1}{2}$ ■

EXAMPLE 6 Find the area in square miles of a rectangle with width $1\frac{1}{3}$ miles and length $12\frac{1}{4}$ miles.

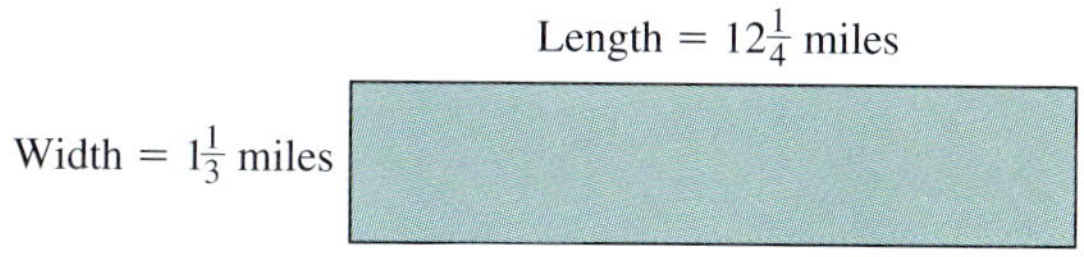

We find the area of a rectangle by multiplying the width times the length.

$$1\frac{1}{3} \times 12\frac{1}{4} = \frac{\cancel{4}^1}{3} \times \frac{49}{\cancel{4}_1} = \frac{49}{3} \text{ or } 16\frac{1}{3}$$

The area is $16\frac{1}{3}$ square miles.

Practice Problem 6 Find the area in square meters of a rectangle with width $1\frac{1}{5}$ meters and length $4\frac{5}{6}$ meters. ■

EXAMPLE 7 Find the value of x if

$$\frac{3}{7} \cdot x = \frac{15}{42}$$

The variable x represents a fraction. We know 3 times one number equals 15 and 7 times another equals 42.

Since $3 \cdot 5 = 15$
and $7 \cdot 6 = 42$ we know that $\frac{3}{7} \cdot \frac{5}{6} = \frac{15}{42}$

Therefore, $x = \frac{5}{6}$.

Practice Problem 7 Find the value of x if $\frac{8}{9} \cdot x = \frac{80}{81}$. ■

2.4 Exercises

Multiply.

1. $\frac{3}{5} \times \frac{7}{11}$
2. $\frac{1}{8} \times \frac{5}{11}$
3. $\frac{3}{4} \times \frac{5}{13}$
4. $\frac{4}{7} \times \frac{3}{5}$
5. $\frac{6}{5} \times \frac{10}{12}$
6. $\frac{7}{8} \times \frac{16}{21}$
7. $\frac{15}{7} \times \frac{8}{25}$
8. $\frac{7}{9} \times \frac{3}{21}$
9. $\frac{15}{28} \times \frac{7}{9}$
10. $\frac{5}{24} \times \frac{18}{15}$
11. $\frac{9}{10} \times \frac{35}{12}$
12. $\frac{12}{17} \times \frac{3}{24}$
13. $8 \times \frac{3}{7}$
14. $\frac{2}{11} \times 4$
15. $\frac{5}{16} \times 8$
16. $5 \times \frac{7}{25}$
17. $\frac{3}{7} \times \frac{2}{5} \times \frac{14}{9}$
18. $\frac{9}{5} \times \frac{3}{13} \times \frac{10}{27}$
19. $\frac{4}{5} \times \frac{1}{8} \times \frac{35}{7}$
20. $\frac{10}{13} \times \frac{26}{15} \times \frac{2}{3}$

Multiply. Change any mixed number to an improper fraction before multiplying.

21. $3\frac{1}{5} \times \frac{7}{8}$
22. $\frac{8}{11} \times 4\frac{3}{4}$
23. $1\frac{1}{4} \times 3\frac{2}{3}$
24. $2\frac{3}{5} \times 1\frac{4}{7}$
25. $2\frac{1}{2} \times 6$
26. $4\frac{1}{3} \times 9$
27. $2\frac{3}{10} \times \frac{3}{5}$
28. $4\frac{3}{5} \times \frac{1}{10}$
29. $1\frac{3}{16} \times 0$
30. $3\frac{7}{8} \times 1$
31. $4\frac{1}{5} \times 12\frac{2}{9}$
32. $5\frac{1}{4} \times 10\frac{3}{7}$
33. $8\frac{5}{6} \times \frac{2}{5}$
34. $\frac{3}{4} \times 9\frac{5}{7}$
35. $\frac{5}{5} \times 11\frac{5}{7}$
36. $0 \times 6\frac{2}{3}$

Solve for x.

37. $\frac{2}{7} \cdot x = \frac{18}{35}$
38. $\frac{5}{8} \cdot x = \frac{35}{88}$
39. $\frac{11}{12} \cdot x = \frac{121}{156}$
40. $\frac{12}{13} \cdot x = \frac{72}{91}$

Applications

41. A spy is running from his captors in a forest that is $8\frac{3}{4}$ miles long and $4\frac{1}{3}$ miles wide. Find the area of the forest where he is hiding. (*Hint:* The area of a rectangle is the product of the length times the width.)

42. An area in the Midwest is a designated tornado danger zone. The land is $22\frac{5}{6}$ miles long and $16\frac{1}{2}$ miles wide. Find the area of the tornado danger zone. (*Hint:* The area of a rectangle is the product of the length times the width.)

43. A jeep has $11\frac{1}{6}$ gallons of gas. The jeep averages 12 miles per gallon. How far will the jeep be able to go on what is in the tank?

44. A Lear Jet airplane has 360 gallons of fuel. The plane averages $4\frac{1}{3}$ miles per gallon. How far can the plane go?

45. We used $18\frac{1}{4}$ yards of wallpaper to decorate our TV room. If we were to wallpaper 8 rooms with the exact same measurements, how many yards of wallpaper would we need?

46. A recipe from Nanette's French cookbook for a scalloped potato tart requires $90\frac{1}{2}$ grams of grated cheese. How many grams of cheese would she need if she made one tart for each of her 18 cousins?

47. Value Hardware Store is making a custom color paint for Susan's kitchen walls. The paint requires $12\frac{1}{4}$ ounces of pigment. How many ounces of pigment would be necessary to make $\frac{3}{4}$ of the formula?

48. The propeller on the Ipswich River Cruise Boat turns 320 revolutions per minute. How fast would it turn at $\frac{3}{4}$ of that speed?

49. If you wanted to buy a share of Disney stock last year, you would have paid $\$72\frac{1}{4}$. How much money would you have needed to buy 85 shares?

50. If one share of Polaroid stock costs $\$45\frac{7}{8}$, how much money will be needed to buy 64 shares?

★ **51.** Of the students eligible to apply for grants, $\frac{3}{4}$ actually sent away for applications. Of the students who received applications, only $\frac{5}{6}$ sent their completed forms in.

(a) What fractional portion of the students actually sent in grant applications?

(b) If the school has 8600 eligible students, how many people actually sent in grant applications?

★ **52.** When Sue turned 13, her wealthy grandparents gave her an allowance and put $\frac{3}{5}$ of the money into a college fund. When she turned 18, her grandparents surprised her by telling her that $\frac{3}{8}$ of *that* money was put into a long-term fund.

(a) What portion of the original allowance went into the long-term fund?

(b) What amount was put into the long-term fund if the allowance is \$150 per month?

Cumulative Review Problems

53. A total of 16,399 cars used a toll bridge in January (31 days). What is the average number of cars using the bridge in one day?

54. The Office of Investors Services has 15,456 calls made per month by the sales personnel. There are 42 sales personnel in the office. What is the average number of calls made per month by one salesperson?

55. A computer printer can print 146 lines per minute. How many lines can it print in 12 minutes?

56. At cruising speed a new commercial jetplane uses 12,360 gallons of fuel per hour. How many gallons will be used in 14 hours of flying time?

2.5 Division of Fractions and Mixed Numbers

MathPro Video 5 SSM

After studying this section, you will be able to:

1 *Divide two proper or improper fractions.*
2 *Divide a whole number and a fraction.*
3 *Divide two mixed numbers or one mixed number and a fraction.*

1 *Dividing Two Proper or Improper Fractions*

Why would you divide fractions? Consider this problem.

- A copper pipe that is $\frac{3}{4}$ of a foot long is to be cut into $\frac{1}{4}$-foot pieces. How many pieces will there be?

To find how many $\frac{1}{4}$'s are in $\frac{3}{4}$, we divide $\frac{3}{4} \div \frac{1}{4}$. We draw a sketch.

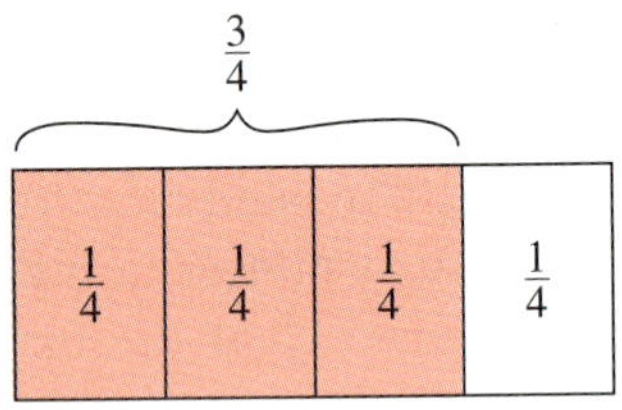

Notice there are three $\frac{1}{4}$'s in $\frac{3}{4}$.

How do we divide two fractions? We **invert** the second fraction and multiply.

$$\frac{3}{4} \div \frac{1}{4} = \frac{3}{\cancel{4}_1} \times \frac{\cancel{4}^1}{1} = \frac{3}{1} = 3$$

When we invert a fraction, we interchange the numerator and the denominator. If we invert $\frac{5}{9}$, we obtain $\frac{9}{5}$. If we invert $\frac{6}{1}$, we obtain $\frac{1}{6}$. Numbers such as $\frac{5}{9}$ and $\frac{9}{5}$ are called **reciprocals** of each other.

Rule for Division of Fractions

To divide two fractions, we invert the second fraction and multiply.

$$\frac{a}{b} \div \frac{c}{d} = \frac{a}{b} \times \frac{d}{c}$$

(when b, c, and d are not zero).

EXAMPLE 1 Divide. **(a)** $\frac{3}{11} \div \frac{2}{5}$ **(b)** $\frac{5}{8} \div \frac{25}{16}$

(a) $\frac{3}{11} \div \frac{2}{5} = \frac{3}{11} \times \frac{5}{2} = \frac{15}{22}$ **(b)** $\frac{5}{8} \div \frac{25}{16} = \frac{\cancel{5}^1}{\cancel{8}_1} \times \frac{\cancel{16}^2}{\cancel{25}_5} = \frac{2}{5}$

Practice Problem 1 Divide. **(a)** $\frac{7}{13} \div \frac{3}{4}$ **(b)** $\frac{16}{35} \div \frac{24}{25}$ ■

2 *Dividing a Whole Number and a Fraction*

When dividing with whole numbers, it is helpful to remember that for any whole number a, $a = \dfrac{a}{1}$.

EXAMPLE 2 Divide. **(a)** $\dfrac{3}{7} \div 2$ **(b)** $5 \div \dfrac{10}{13}$

(a) $\dfrac{3}{7} \div 2 = \dfrac{3}{7} \div \dfrac{2}{1} = \dfrac{3}{7} \times \dfrac{1}{2} = \dfrac{3}{14}$

(b) $5 \div \dfrac{10}{13} = \dfrac{5}{1} \div \dfrac{10}{13} = \dfrac{\overset{1}{\cancel{5}}}{1} \times \dfrac{13}{\underset{2}{\cancel{10}}} = \dfrac{13}{2}$ or $6\dfrac{1}{2}$

Practice Problem 2 Divide. **(a)** $\dfrac{3}{17} \div 6$ **(b)** $14 \div \dfrac{7}{15}$ ■

EXAMPLE 3 Divide, if possible.

(a) $\dfrac{23}{25} \div 1$ **(b)** $1 \div \dfrac{7}{5}$ **(c)** $0 \div \dfrac{4}{9}$ **(d)** $\dfrac{3}{17} \div 0$

(a) $\dfrac{23}{25} \div 1 = \dfrac{23}{25} \times \dfrac{1}{1} = \dfrac{23}{25}$ ← *Any fraction can be multiplied by 1 without changing the value of the fraction.*

(b) $1 \div \dfrac{7}{5} = \dfrac{1}{1} \times \dfrac{5}{7} = \dfrac{5}{7}$ ←

(c) $0 \div \dfrac{4}{9} = \dfrac{0}{1} \times \dfrac{9}{4} = \dfrac{0}{4} = 0$ *Zero divided by any nonzero number is zero.*

(d) $\dfrac{3}{17} \div 0$ *You can never divide by zero. This problem cannot be done.*

Practice Problem 3 Divide, if possible.

(a) $1 \div \dfrac{11}{13}$ **(b)** $\dfrac{14}{17} \div 1$ **(c)** $\dfrac{3}{11} \div 0$ **(d)** $0 \div \dfrac{9}{16}$ ■

SIDELIGHT Why do we divide by inverting the second fraction and multiplying? What is really going on when we do this? We are actually multiplying by 1. To see why, consider the following.

$$\frac{3}{7} \div \frac{2}{3} = \frac{\frac{3}{7}}{\frac{2}{3}}$$ *We write the division by using another fraction bar.*

$$= \frac{\frac{3}{7}}{\frac{2}{3}} \times 1$$ *Any fraction can be multiplied by 1 without changing the value of the fraction. This is the fundamental rule of fractions.*

$$= \frac{\frac{3}{7}}{\frac{2}{3}} \times \frac{\frac{3}{2}}{\frac{3}{2}}$$ *Any nonzero number divided by itself equals 1.*

$$= \frac{\frac{3}{7} \times \frac{3}{2}}{\frac{2}{3} \times \frac{3}{2}}$$ *Definition of multiplication of fractions.*

$$= \frac{\frac{3}{7} \times \frac{3}{2}}{1} = \frac{3}{7} \times \frac{3}{2}$$ *Any number can be written as a fraction with a denominator of 1 without changing its value.*

Thus

$$\frac{3}{7} \div \frac{2}{3} = \frac{3}{7} \times \frac{3}{2} = \frac{9}{14}$$

3 *Dividing Mixed Numbers*

If one or more mixed numbers are involved in the division, they should be converted to improper fractions first.

EXAMPLE 4 Divide. **(a)** $3\frac{7}{15} \div 1\frac{1}{25}$ **(b)** $\frac{3}{5} \div 2\frac{1}{7}$

(a) $3\frac{7}{15} \div 1\frac{1}{25} = \frac{52}{15} \div \frac{26}{25} = \frac{\overset{2}{\cancel{52}}}{\underset{3}{\cancel{15}}} \times \frac{\overset{5}{\cancel{25}}}{\underset{1}{\cancel{26}}} = \frac{10}{3}$ or $3\frac{1}{3}$

(b) $\frac{3}{5} \div 2\frac{1}{7} = \frac{3}{5} \div \frac{15}{7} = \frac{\overset{1}{\cancel{3}}}{5} \times \frac{7}{\underset{5}{\cancel{15}}} = \frac{7}{25}$

Practice Problem 4 Divide. **(a)** $1\frac{1}{5} \div \frac{7}{10}$ **(b)** $2\frac{1}{4} \div 1\frac{7}{8}$ ■

The division of two fractions may be indicated by a wide fraction bar.

EXAMPLE 5 Divide. **(a)** $\dfrac{10\frac{2}{9}}{2\frac{1}{3}}$ **(b)** $\dfrac{1\frac{1}{15}}{3\frac{1}{3}}$

(a) $$\dfrac{10\frac{2}{9}}{2\frac{1}{3}} = 10\frac{2}{9} \div 2\frac{1}{3} = \frac{92}{9} \div \frac{7}{3} = \frac{92}{\cancel{9}_{3}} \times \frac{\cancel{3}^{1}}{7} = \frac{92}{21} \text{ or } 4\frac{8}{21}$$

(b) $$\dfrac{1\frac{1}{15}}{3\frac{1}{3}} = 1\frac{1}{15} \div 3\frac{1}{3} = \frac{16}{15} \div \frac{10}{3} = \frac{\cancel{16}^{8}}{\cancel{15}_{5}} \times \frac{\cancel{3}^{1}}{\cancel{10}_{5}} = \frac{8}{25}$$

Practice Problem 5 Divide. **(a)** $\dfrac{5\frac{2}{3}}{7}$ **(b)** $\dfrac{1\frac{2}{5}}{2\frac{1}{3}}$ ■

EXAMPLE 6 Find the value of x if $x \div \frac{8}{7} = \frac{21}{40}$.

First we will change the division problem to an equivalent multiplication problem.

$$x \div \frac{8}{7} = \frac{21}{40}$$

$$x \cdot \frac{7}{8} = \frac{21}{40}$$

x represents a fraction.

In the numerator above, we want to know what times 7 equals 21. In the denominator, we want to know what times 8 equals 40.

$$\frac{3}{5} \cdot \frac{7}{8} = \frac{21}{40}$$

Thus $x = \frac{3}{5}$.

Practice Problem 6 Find the value of x if $x \div \frac{3}{2} = \frac{22}{36}$. ■

EXAMPLE 7 There are 117 milligrams of cholesterol in $4\frac{1}{3}$ cups of milk. How much cholesterol is contained in 1 cup of milk?

We want to divide the 117 by $4\frac{1}{3}$ to find out how much is in 1 cup.

$$117 \div 4\frac{1}{3} = 117 \div \frac{13}{3} = \frac{\overset{9}{\cancel{117}}}{1} \times \frac{3}{\underset{1}{\cancel{13}}} = \frac{27}{1} = 27$$

Thus there are 27 milligrams of cholesterol in 1 cup of milk.

Practice Problem 7 A copper pipe that is $19\frac{1}{4}$ feet long will be cut into 14 equal pieces. How long will each piece be? ■

Developing Your Study Skills

WHY IS REVIEW NECESSARY?

You master a course in mathematics by learning the concepts one step at a time. There are basic concepts like addition, subtraction, multiplication, and division of whole numbers that are considered the foundation upon which all of mathematics is built. These must be mastered first. Then the study of mathematics is built step by step upon this foundation, each step supporting the next. The process is a carefully designed procedure, and so no steps can be skipped. A student of mathematics needs to realize the importance of this building process to succeed.

Because learning new concepts depends on those previously learned, students often need to take time to review. The reviewing process will strengthen the understanding and application of concepts that are weak due to lack of mastery or passage of time. Review at the right time on the right concepts can strengthen previously learned skills and make progress possible.

Timely, periodic review of previously learned mathematical concepts is absolutely necessary in order to master new concepts. You may have forgotten a concept or grown a bit rusty in applying it. Reviewing is the answer. Make use of any review sections in your textbook, whether they are assigned or not. Look back to previous chapters whenever you have forgotten how to do something. Study the examples and practice some exercises to refresh your understanding.

Be sure that you understand and can perform the computations of each new concept. This will enable you to be able to move successfully on to the next ones.

Remember, mathematics is a step-by-step building process. Learn one concept at a time, skipping none, and reinforce and strengthen with review whenever necessary.

2.5 Exercises

Divide, if possible.

1. $\frac{2}{5} \div \frac{7}{2}$

2. $\frac{3}{4} \div \frac{5}{6}$

3. $\frac{3}{13} \div \frac{9}{26}$

4. $\frac{7}{8} \div \frac{2}{3}$

5. $\frac{5}{6} \div \frac{25}{27}$

6. $\frac{3}{14} \div \frac{6}{7}$

7. $\frac{5}{9} \div \frac{10}{27}$

8. $\frac{8}{15} \div \frac{24}{35}$

9. $\frac{4}{5} \div 1$

10. $1 \div \frac{3}{7}$

11. $2 \div \frac{7}{8}$

12. $\frac{3}{11} \div 4$

13. $\frac{2}{9} \div \frac{1}{6}$

14. $\frac{3}{4} \div \frac{2}{3}$

15. $\frac{4}{15} \div \frac{4}{15}$

16. $\frac{2}{7} \div \frac{2}{7}$

17. $\frac{3}{7} \div \frac{7}{3}$

18. $\frac{11}{12} \div \frac{1}{5}$

19. $\frac{4}{3} \div \frac{7}{27}$

20. $\frac{8}{9} \div \frac{5}{81}$

21. $0 \div \frac{3}{17}$

22. $0 \div \frac{5}{16}$

23. $\frac{18}{19} \div 0$

24. $\frac{24}{29} \div 0$

25. $\frac{9}{16} \div \frac{3}{4}$

26. $\frac{3}{4} \div \frac{9}{16}$

27. $\frac{3}{7} \div \frac{15}{28}$

28. $\frac{5}{6} \div \frac{15}{18}$

29. $\frac{10}{25} \div \frac{20}{50}$

30. $\frac{3}{25} \div \frac{24}{125}$

31. $12 \div \frac{3}{4}$

32. $18 \div \frac{3}{7}$

33. $\frac{7}{8} \div 4$

34. $\frac{5}{6} \div 12$

35. $6000 \div \frac{6}{5}$

36. $8000 \div \frac{4}{7}$

37. $\dfrac{\frac{2}{5}}{3}$

38. $\dfrac{\frac{3}{7}}{4}$

39. $\dfrac{\frac{5}{8}}{\frac{25}{7}}$

40. $\dfrac{\frac{3}{16}}{\frac{5}{8}}$

41. $3\frac{1}{5} \div \frac{3}{10}$

42. $2\frac{1}{8} \div \frac{1}{4}$

43. $2\frac{1}{3} \div 6$

44. $6\frac{1}{2} \div 3$

45. $1\frac{7}{9} \div 2\frac{2}{3}$

46. $2\frac{1}{15} \div 3\frac{1}{3}$

47. $5 \div 1\frac{1}{4}$

48. $7 \div 1\frac{2}{5}$

49. $12\frac{1}{2} \div 5\frac{5}{6}$

50. $14\frac{2}{3} \div 3\frac{1}{2}$

51. $8\frac{1}{4} \div 2\frac{3}{4}$

52. $2\frac{3}{8} \div 5\frac{3}{7}$

53. $3\frac{1}{2} \div 1\frac{7}{9}$

54. $1\frac{1}{8} \div 2\frac{2}{3}$

55. $2\frac{1}{2} \div 5$

56. $3\frac{1}{7} \div 10$

57. $\dfrac{4\frac{1}{2}}{\frac{3}{8}}$

58. $\dfrac{6\frac{1}{2}}{\frac{3}{4}}$

59. $\dfrac{1\frac{1}{4}}{1\frac{7}{8}}$

60. $\dfrac{2\frac{3}{5}}{1\frac{7}{10}}$

61. $\dfrac{\frac{5}{9}}{2\frac{1}{3}}$

62. $\dfrac{\frac{2}{10}}{3\frac{1}{5}}$

63. $\dfrac{5\frac{1}{3}}{2\frac{1}{2}}$

64. $\dfrac{6\frac{3}{4}}{3\frac{1}{2}}$

Find the value of x in each of the following.

65. $x \div \frac{4}{3} = \frac{21}{20}$

66. $x \div \frac{2}{5} = \frac{15}{16}$

67. $x \div \frac{9}{5} = \frac{20}{63}$

68. $x \div \frac{7}{3} = \frac{9}{28}$

Applications

Answer each question.

69. A leather factory in Morocco tans leather. In order to make the leather soft, it has to soak in a vat of uric acid and other ingredients. The main holding storage tank holds $20\frac{1}{4}$ gallons of the tanning mixture. If the mixture is distributed evenly into 9 vats of equal size for the different colored leathers, how much will each vat hold?

70. A specially protected stretch of beach bordering the Great Barrier Reef in Australia is used for marine biology and ecological research. The beach, which is $7\frac{1}{2}$ miles long, has been broken up into 20 equal segments for comparison purposes. How long is each segment of the beach?

71. Bruce drove in a snowstorm to get to his favorite mountain to do some snowboarding. He traveled 125 miles in $3\frac{1}{3}$ hours. What was his average speed (in miles per hour)?

72. Roberto drove his truck to Cedarville, a distance of 200 miles, in $4\frac{1}{6}$ hours. What was his average speed (in miles per hour)?

73. The Patriotic Flag Company is making flags which need $3\frac{1}{8}$ yards of fabric. The warehouse has $87\frac{1}{2}$ yards of flag fabric available. How many flags can the factory make?

74. The school cafeteria is making hamburgers for the annual Senior Day Festival. The cooks have decided that because hamburger shrinks on the grill, they will allow $\frac{2}{3}$ pound of meat for each student. If the kitchen has $38\frac{2}{3}$ pounds of meat, how many students will be fed?

To Think About

75. Division of fractions is not commutative. For example, $\frac{2}{3} \div \frac{5}{7} \neq \frac{5}{7} \div \frac{2}{3}$.

In general, $\frac{a}{b} \div \frac{c}{d} \neq \frac{c}{d} \div \frac{a}{b}$ $\quad (a, b, c, d \text{ all} \neq 0)$.

But sometimes there are exceptions. Can you think of any numbers a, b, c, d such that for fractions $\frac{a}{b}$ and $\frac{c}{d}$ it would be true that $\frac{a}{b} \div \frac{c}{d} = \frac{c}{d} \div \frac{a}{b}$?

76. Can you think of a way to divide $3\frac{7}{12} \div \frac{1}{4}$ *without* changing $3\frac{7}{12}$ to an improper fraction? Try your method to see if it works.

Cumulative Review Problems

77. Write in words. 39,576,304

78. Write in expanded form. 459,273

79. Add. $126 + 34 + 9 + 891 + 12 + 27$

80. Write in standard notation. Eighty-seven million, five hundred ninety-five thousand, six hundred thirty-one

2.6 The Least Common Denominator and Building Up Fractions

MathPro Video 5 SSM

After studying this section, you will be able to:

1. *Find the least common denominator given two or three fractions.*
2. *Build up a fraction so that it has as a denominator a given least common denominator.*

1 *Finding the Least Common Denominator*

We need some way to determine which of two fractions is larger. Suppose that Marcia and Melissa each have some pizza left.

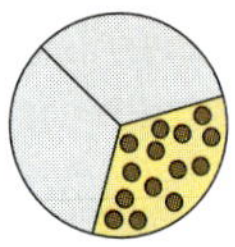

Marcia's Pizza

$\frac{1}{3}$ of a pizza left

Melissa's Pizza

$\frac{1}{4}$ of a pizza left

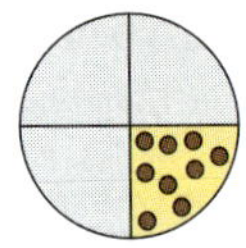

Who has more pizza left? How much more? Comparing the amounts of pizza left would be easy if each pizza had been cut into equal-sized pieces. If the original pizzas had each been cut into 12 pieces, we would be able to see that Marcia had $\frac{1}{12}$ of a pizza more than Melissa had.

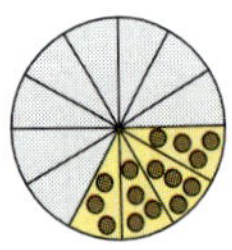

Marcia's Pizza

$\left(\text{We know that } \frac{4}{12} = \frac{1}{3} \text{ by reducing.}\right)$

Melissa's Pizza

$\left(\text{We know that } \frac{3}{12} = \frac{1}{4} \text{ by reducing.}\right)$

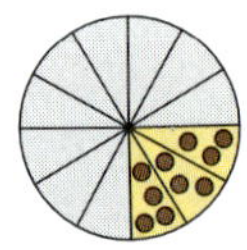

The denominator 12 is common to the fractions $\frac{4}{12}$ and $\frac{3}{12}$. We call a denominator that allows us to compare fractions the *least common denominator,* abbreviated LCD. The number 12 is the least common denominator for the fractions $\frac{1}{3}$ and $\frac{1}{4}$.

> **Definition**
>
> The least common denominator (LCD) of two or more fractions is the smallest number that can be divided evenly by each of the fractions' denominators.

Let's check. $12 \div 3 = 4$. $12 \div 4 = 3$. We have no remainder, so 12 is the LCD. Suppose we had thought the number 6 might be the LCD. $6 \div 3 = 2$ with no remainder. But we see that $6 \div 4 = 1$ with a remainder of 2, so 6 cannot be the LCD. In this case, 12 is the LCD of $\frac{1}{3}$ and $\frac{1}{4}$. The number 12 is the smallest number that can be divided evenly by 3 and by 4.

In some problems you may be able to guess the LCD quite quickly. With practice, you can often find the LCD mentally. For example, you now know that if the denominators of two fractions are 3 and 4, the LCD is 12. For the fractions $\frac{1}{2}$ and $\frac{1}{4}$, the LCD is 4; for the fractions $\frac{1}{3}$ and $\frac{1}{6}$, the LCD is 6. We can see that if the denominator of one fraction divides without remainder into the denominator of another, the LCD of the two fractions is the larger of the denominators.

EXAMPLE 1 Determine the LCD of the fractions.

(a) $\frac{7}{15}$ and $\frac{4}{5}$ **(b)** $\frac{2}{3}$ and $\frac{5}{27}$

(a) Since 5 can be divided into 15, the LCD of $\frac{7}{15}$ and $\frac{4}{5}$ is 15.

(b) Since 3 can be divided into 27, the LCD of $\frac{2}{3}$ and $\frac{5}{27}$ is 27.

Practice Problem 1 Determine the LCD of the fractions.

(a) $\frac{3}{4}$ and $\frac{11}{12}$ **(b)** $\frac{1}{7}$ and $\frac{8}{35}$ ■

In a few cases, the LCD is the product of the two denominators.

EXAMPLE 2 Find the LCD for $\frac{1}{4}$ and $\frac{3}{5}$.

We see that $4 \times 5 = 20$. Also, 20 is the *smallest* number that can be divided without remainder by 4 and by 5. So the LCD = 20.

Practice Problem 2 Find the LCD for $\frac{3}{7}$ and $\frac{5}{6}$. ■

Sometimes we need to follow steps to find the LCD.

Three-Step Procedure for Finding the Least Common Denominator

1. Write each denominator as the product of prime factors.
2. List all the prime factors that appear in either product.
3. Form a product of those prime factors, using each factor the greatest number of times it appears in any one denominator.

EXAMPLE 3 Find the LCD by the three-step procedure.

(a) $\frac{5}{6}$ and $\frac{4}{15}$ **(b)** $\frac{7}{18}$ and $\frac{7}{30}$ **(c)** $\frac{10}{27}$ and $\frac{5}{18}$

(a) Step 1 Write each denominator as a product of prime factors.

$$6 = 2 \times 3 \qquad 15 = 5 \times 3$$

Step 2 The LCD will contain the factors 2, 3, and 5.

$$6 = 2 \times 3 \qquad 15 = 5 \times 3$$

$$\text{LCD} = 2 \times 3 \times 5$$

$$= 30$$

Step 3 is not needed. No factor was repeated in any one denominator.

(b) Step 1 Write each denominator as a product of prime factors.

$$18 = 2 \times 9 = 2 \times 3 \times 3$$

$$30 = 3 \times 10 = 2 \times 3 \times 5$$

Step 2 The LCD will be a product containing 2, 3, and 5.
Step 3 The LCD will contain the factor 3 twice since it occurs twice in the denominator 18.

Factor 3 occurs twice in one denominator.

$$18 = 2 \times 3 \times 3$$

$$\text{LCD} = 2 \times 3 \times 3 \times 5 = 90$$

(c) Write each denominator as a product of prime factors.

$$27 = 3 \times 3 \times 3 \qquad 18 = 3 \times 3 \times 2$$

Factor 3 occurs three times

The LCD will contain the factor 2 once but the factor 3 three times.

$$\text{LCD} = 2 \times 3 \times 3 \times 3 = 54$$

Practice Problem 3 Find the LCDs.

(a) $\frac{3}{14}$ and $\frac{1}{10}$ **(b)** $\frac{1}{15}$ and $\frac{7}{50}$ **(c)** $\frac{3}{16}$ and $\frac{5}{12}$ ■

A similar procedure can be used for three fractions.

EXAMPLE 4 Find the LCD of $\frac{7}{12}$, $\frac{1}{15}$, and $\frac{11}{30}$.

$$12 = 2 \times 2 \times 3$$

$$15 = \qquad 3 \times 5$$

$$30 = \qquad 2 \times 3 \times 5$$

$$\text{LCD} = 2 \times 2 \times 3 \times 5$$

$$= 60$$

Practice Problem 4 Find the LCD of $\frac{3}{49}$, $\frac{5}{21}$, and $\frac{6}{7}$. ■

2 *Building Up Fractions*

We cannot add fractions with unlike denominators. To change denominators, we must (1) find the LCD and (2) build up the addends—the fractions being added—into equivalent fractions that have the LCD as the denominator. We know now how to find the LCD. Let's look at how we build up fractions. We know, for example, that

$$\frac{1}{2} = \frac{2}{4} = \frac{50}{100} \qquad \frac{1}{4} = \frac{25}{100} \qquad \text{and} \qquad \frac{3}{4} = \frac{75}{100}$$

In these cases, we have mentally multiplied the given fraction by 1, in the form of a certain number, c, in the numerator and that same number, c, in the denominator.

$$\frac{1}{2} \times \frac{c}{c} = \frac{2}{4} \qquad \text{Here } c = 2,\ \frac{2}{2} = 1$$

$$\frac{1}{2} \times \frac{c}{c} = \frac{50}{100} \qquad \text{Here } c = 50,\ \frac{50}{50} = 1$$

This property is called the *building fraction property*.

> **Building Fraction Property**
>
> For whole numbers a, b, and c where $b \neq 0$, $c \neq 0$,
>
> $$\frac{a}{b} = \frac{a}{b} \times 1 = \frac{a}{b} \times \frac{c}{c} = \frac{a \times c}{b \times c}$$

EXAMPLE 5 Build each fraction to an equivalent fraction with the LCD.

(a) $\frac{3}{4}$, LCD = 28 **(b)** $\frac{4}{5}$, LCD = 45 **(c)** $\frac{1}{3}$ and $\frac{4}{5}$, LCD = 15

(a) $\frac{3}{4} \times \frac{c}{c} = \frac{?}{28}$ *We know that* $4 \times 7 = 28$, *so the value* c *that we multiply numerator and denominator by is* 7.

$$\frac{3}{4} \times \frac{7}{7} = \frac{21}{28}$$

(b) $\frac{4}{5} \times \frac{c}{c} = \frac{?}{45}$ *We know that* $5 \times 9 = 45$, *so* $c = 9$.

$$\frac{4}{5} \times \frac{9}{9} = \frac{36}{45}$$

(c) $\frac{1}{3} = \frac{?}{15}$ *We know that* $3 \times 5 = 15$, *so we multiply numerator and denominator by* 5.

$$\frac{1}{3} \times \frac{5}{5} = \frac{5}{15}$$

$\frac{4}{5} = \frac{?}{15}$ *We know that* $5 \times 3 = 15$, *so we multiply numerator and denominator by* 3.

$$\frac{4}{5} \times \frac{3}{3} = \frac{12}{15}$$

Thus $\frac{1}{3} = \frac{5}{15}$ and $\frac{4}{5} = \frac{12}{15}$.

Practice Problem 5 Build each fraction to an equivalent fraction with the LCD.

(a) $\frac{3}{5}$, LCD = 40 **(b)** $\frac{7}{11}$, LCD = 44 **(c)** $\frac{2}{7}$ and $\frac{3}{4}$, LCD = 28 ■

EXAMPLE 6

(a) Find the LCD of $\frac{1}{32}$ and $\frac{7}{48}$.

(b) Build the fractions to equivalent fractions that have the LCD as their denominators.

(a) First we find the prime factors of 32 and 48.

$$32 = 2 \times 2 \times 2 \times 2 \times 2$$

$$48 = 2 \times 2 \times 2 \times 2 \times 3$$

Thus the LCD will require a factor of 2 five times and a factor of 3 one time.

$$\text{LCD} = 2 \times 2 \times 2 \times 2 \times 2 \times 3 = 96$$

(b) $\frac{1}{32} = \frac{?}{96}$ *Since* $32 \times 3 = 96$ *we multiply by* 3.

$$\frac{1}{32} = \frac{1}{32} \times \frac{3}{3} = \frac{3}{96}$$

$\frac{7}{48} = \frac{?}{96}$ *Since* $48 \times 2 = 96$, *we multiply by* 2.

$$\frac{7}{48} = \frac{7}{48} \times \frac{2}{2} = \frac{14}{96}$$

Practice Problem 6

(a) Find the LCD of $\frac{3}{20}$ and $\frac{11}{15}$.

(b) Build the fractions to equivalent fractions that have the LCD as their denominators. ■

ALTERNATIVE METHOD Is there another method that can be used to find the least common denominator for two or more fractions? Yes. You can find the least common multiple (LCM) of the two denominators. Suppose that we want to find the least common multiple of 10 and 12. List the multiples of 10 (multiply by 1, 2, 3, 4, 5, and so on).

10, 20, 30, 40, 50, 60, 70, 80, . . .

List the multiples of 12.

12, 24, 36, 48, 60, 72, 84, 96, . . .

The first multiple that appears on both lists is the least common multiple. It is also the LCD for two fractions that have 10 and 12 for denominators.

2.6 Exercises

Find the LCD for each pair of fractions.

1. $\frac{1}{5}$ and $\frac{3}{10}$ **2.** $\frac{3}{8}$ and $\frac{5}{16}$ **3.** $\frac{3}{7}$ and $\frac{1}{4}$ **4.** $\frac{5}{6}$ and $\frac{3}{5}$ **5.** $\frac{2}{5}$ and $\frac{3}{7}$

6. $\frac{1}{16}$ and $\frac{2}{3}$ **7.** $\frac{1}{6}$ and $\frac{5}{9}$ **8.** $\frac{1}{4}$ and $\frac{3}{14}$ **9.** $\frac{3}{16}$ and $\frac{11}{24}$ **10.** $\frac{3}{10}$ and $\frac{8}{25}$

11. $\frac{1}{16}$ and $\frac{3}{4}$ **12.** $\frac{2}{11}$ and $\frac{1}{44}$ **13.** $\frac{5}{10}$ and $\frac{11}{45}$ **14.** $\frac{13}{20}$ and $\frac{17}{30}$ **15.** $\frac{7}{12}$ and $\frac{7}{30}$

16. $\frac{5}{6}$ and $\frac{7}{15}$ **17.** $\frac{5}{21}$ and $\frac{8}{35}$ **18.** $\frac{1}{20}$ and $\frac{7}{8}$ **19.** $\frac{5}{18}$ and $\frac{11}{12}$ **20.** $\frac{1}{24}$ and $\frac{7}{40}$

Find the LCD for each set of three fractions.

21. $\frac{2}{3}, \frac{1}{2}, \frac{5}{6}$ **22.** $\frac{1}{5}, \frac{1}{3}, \frac{7}{10}$ **23.** $\frac{5}{24}, \frac{11}{15}, \frac{7}{30}$ **24.** $\frac{5}{18}, \frac{5}{12}, \frac{13}{20}$ **25.** $\frac{5}{16}, \frac{11}{18}, \frac{1}{24}$

26. $\frac{5}{6}, \frac{1}{2}, \frac{5}{22}$ **27.** $\frac{7}{12}, \frac{1}{21}, \frac{3}{14}$ **28.** $\frac{1}{30}, \frac{3}{40}, \frac{7}{8}$ **29.** $\frac{7}{15}, \frac{11}{12}, \frac{7}{8}$ **30.** $\frac{5}{36}, \frac{2}{48}, \frac{1}{24}$

Build each fraction to an equivalent fraction with the denominator specified. State the numerator.

31. $\frac{1}{6} = \frac{?}{18}$ **32.** $\frac{1}{5} = \frac{?}{35}$ **33.** $\frac{5}{6} = \frac{?}{54}$ **34.** $\frac{4}{7} = \frac{?}{28}$ **35.** $\frac{4}{11} = \frac{?}{55}$ **36.** $\frac{2}{13} = \frac{?}{39}$

37. $\frac{7}{24} = \frac{?}{48}$ **38.** $\frac{3}{50} = \frac{?}{100}$ **39.** $\frac{8}{9} = \frac{?}{108}$ **40.** $\frac{6}{7} = \frac{?}{147}$ **41.** $\frac{13}{42} = \frac{?}{126}$ **42.** $\frac{14}{15} = \frac{?}{180}$

The LCD of each pair of fractions is listed. Build each fraction to an equivalent fraction that has the LCD as the denominator.

43. LCD = 36, $\frac{7}{12}$ and $\frac{5}{9}$ **44.** LCD = 20, $\frac{9}{10}$ and $\frac{3}{4}$ **45.** LCD = 200, $\frac{3}{25}$ and $\frac{7}{40}$

46. LCD = 72, $\frac{5}{24}$ and $\frac{7}{36}$ **47.** LCD = 150, $\frac{19}{25}$ and $\frac{26}{75}$ **48.** LCD = 240, $\frac{57}{120}$ and $\frac{39}{80}$

Find the LCD. Build up the fractions to equivalent fractions having the LCD as the denominator.

49. $\frac{5}{7}$ and $\frac{7}{42}$ **50.** $\frac{3}{8}$ and $\frac{13}{40}$ **51.** $\frac{5}{12}$ and $\frac{1}{16}$ **52.** $\frac{7}{15}$ and $\frac{4}{25}$ **53.** $\frac{13}{20}$ and $\frac{11}{16}$

54. $\frac{9}{12}$ and $\frac{13}{18}$ **55.** $\frac{7}{12}$ and $\frac{23}{30}$ **56.** $\frac{3}{16}$ and $\frac{23}{24}$ **57.** $\frac{5}{24}, \frac{11}{36}, \frac{3}{72}$ **58.** $\frac{1}{30}, \frac{7}{15}, \frac{1}{45}$

59. $\frac{3}{56}, \frac{7}{8}, \frac{5}{7}$ **60.** $\frac{5}{9}, \frac{1}{6}, \frac{3}{54}$ **61.** $\frac{5}{63}, \frac{4}{21}, \frac{8}{9}$ **62.** $\frac{3}{8}, \frac{5}{14}, \frac{13}{16}$

Applications

63. Suppose that you wish to compare the lengths of the three portions of a stainless steel pin that came out of the door.
(a) What is the LCD for the three fractions?

(b) Build up each fraction to an equivalent fraction that has the LCD as a denominator.

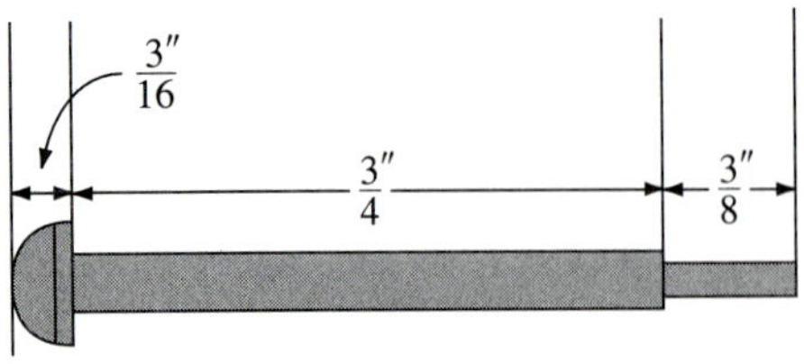

64. Suppose that you want to prepare a report on a plant that grew the following lengths during each week of a 3-week experiment.
(a) What is the LCD for the three fractions?

(b) Build up each fraction to an equivalent fraction that has the LCD for a denominator.

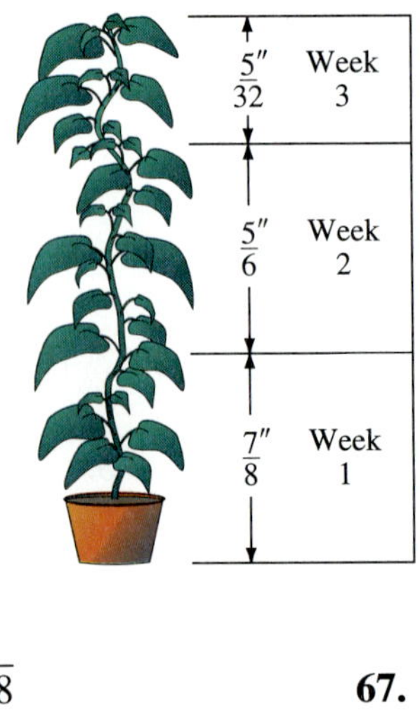

Cumulative Review Problems

65. Divide. $32\overline{)5699}$ **66.** Divide. $182\overline{)659{,}568}$ **67.** Multiply. 369×27

2.7 Addition and Subtraction of Fractions

After studying this section, you will be able to:

1 *Add or subtract fractions with a common denominator.*
2 *Add or subtract fractions without a common denominator.*

MathPro Video 5 SSM

1 *Adding and Subtracting Fractions with a Common Denominator*

You must have common denominators (denominators that are alike) to add or subtract fractions.

If your problem has fractions without a common denominator or if it has mixed numbers, you must use what you already know about changing the form of each fraction (how the fraction looks). Only after all the fractions have a common denominator can you add or subtract.

An important distinction: You must have common denominators to add or subtract fractions, but you need not have common denominators to multiply or divide fractions.

To add two fractions that have the same denominator, add the numerators and write the sum over the common denominator.

To illustrate we use $\frac{1}{5} + \frac{2}{5} = \frac{3}{5}$. The sketch shows that $\frac{1}{5} + \frac{2}{5} = \frac{3}{5}$.

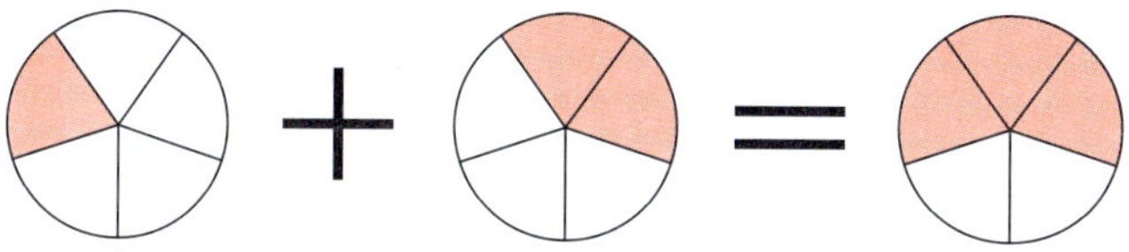

$\frac{1}{5}$ $\frac{2}{5}$ $\frac{3}{5}$

EXAMPLE 1 Add. $\frac{5}{13} + \frac{7}{13}$

$$\frac{5}{13} + \frac{7}{13} = \frac{12}{13}$$

Practice Problem 1 Add. $\frac{3}{17} + \frac{12}{17}$ ■

The answer may need to be reduced. Sometimes the answer may be written as a mixed number.

EXAMPLE 2 Add. **(a)** $\frac{4}{9} + \frac{2}{9}$ **(b)** $\frac{5}{7} + \frac{6}{7}$

(a) $\frac{4}{9} + \frac{2}{9} = \frac{6}{9} = \frac{2}{3}$ **(b)** $\frac{5}{7} + \frac{6}{7} = \frac{11}{7}$ or $1\frac{4}{7}$

Practice Problem 2 Add.

(a) $\frac{1}{12} + \frac{5}{12}$ **(b)** $\frac{13}{15} + \frac{7}{15}$ ■

A similar rule is followed for subtraction, except that the numerators are subtracted and the result placed over a common denominator. Be sure to reduce all answers when possible.

EXAMPLE 3 Subtract. **(a)** $\frac{5}{13} - \frac{4}{13}$ **(b)** $\frac{17}{20} - \frac{3}{20}$

(a) $\frac{5}{13} - \frac{4}{13} = \frac{1}{13}$ **(b)** $\frac{17}{20} - \frac{3}{20} = \frac{14}{20} = \frac{7}{10}$

Practice Problem 3 Subtract.

(a) $\frac{5}{19} - \frac{2}{19}$ **(b)** $\frac{21}{25} - \frac{6}{25}$ ■

2 *Adding or Subtracting Fractions without a Common Denominator*

If the two fractions do not have a common denominator, we follow the procedure in Section 2.6: Find the LCD and then build up each fraction so that its denominator is the LCD.

EXAMPLE 4 Add. $\frac{7}{12} + \frac{1}{4}$

The LCD is 12. $\frac{7}{12}$ already has the least common denominator.

$$\begin{array}{rcl} \frac{7}{12} & = & \frac{7}{12} \\ + \frac{1}{4} \times \frac{3}{3} & = & +\frac{3}{12} \\ \hline & & \frac{10}{12} = \frac{5}{6} \end{array}$$

Therefore we have found out that

$$\frac{7}{12} + \frac{1}{4} = \frac{7}{12} + \frac{3}{12} = \frac{10}{12} = \frac{5}{6}$$

It is very important to remember to reduce our final answer.

Practice Problem 4 Add. $\frac{2}{15} + \frac{2}{5}$ ■

EXAMPLE 5 Add. $\frac{7}{20} + \frac{4}{15}$

LCD = 60.

$$\frac{7}{20} \times \frac{3}{3} = \frac{21}{60} \qquad \frac{4}{15} \times \frac{4}{4} = \frac{16}{60}$$

Thus

$$\frac{7}{20} + \frac{4}{15} = \frac{21}{60} + \frac{16}{60} = \frac{37}{60}$$

Practice Problem 5 Add. $\frac{5}{12} + \frac{5}{16}$ ■

A similar procedure holds for the addition of three or more fractions.

EXAMPLE 6 Add. $\frac{3}{8} + \frac{5}{6} + \frac{1}{4}$

LCD = 24.

$$\frac{3}{8} \times \frac{3}{3} = \frac{9}{24} \qquad \frac{5}{6} \times \frac{4}{4} = \frac{20}{24} \qquad \frac{1}{4} \times \frac{6}{6} = \frac{6}{24}$$

$$\frac{3}{8} + \frac{5}{6} + \frac{1}{4} = \frac{9}{24} + \frac{20}{24} + \frac{6}{24} = \frac{35}{24} \quad \text{or} \quad 1\frac{11}{24}$$

Practice Problem 6 Add. $\frac{3}{16} + \frac{1}{8} + \frac{1}{12}$ ■

EXAMPLE 7 Subtract. $\frac{17}{25} - \frac{3}{35}$

LCD = 175.

$$\frac{17}{25} \times \frac{7}{7} = \frac{119}{175} \qquad \frac{3}{35} \times \frac{5}{5} = \frac{15}{175}$$

Thus

$$\frac{17}{25} - \frac{3}{35} = \frac{119}{175} - \frac{15}{175} = \frac{104}{175}$$

Practice Problem 7 Subtract. $\frac{9}{48} - \frac{5}{32}$ ■

EXAMPLE 8 John and Nancy have a house on $\frac{7}{8}$ acre of land. They have $\frac{1}{3}$ acre of land planted in grass. How much of the land is not planted in grass?

1. *Understand the problem.*
Draw a picture.

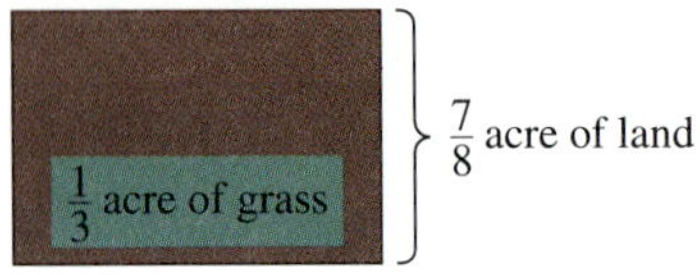

We need to subtract. $\frac{7}{8} - \frac{1}{3}$

2. *Solve and state the answer.*
The LCD is 24.

$$\frac{7}{8} \times \frac{3}{3} = \frac{21}{24} \qquad \frac{1}{3} \times \frac{8}{8} = \frac{8}{24}$$

$$\frac{7}{8} - \frac{1}{3} = \frac{21}{24} - \frac{8}{24} = \frac{13}{24}$$

We conclude that $\frac{13}{24}$ acre of the land is not planted in grass.

3. *Check.*
The check is up to the student.

Practice Problem 8 Leon had $\frac{9}{10}$ gallon of cleaning fluid in the garage. He used $\frac{1}{4}$ gallon to clean the garage floor. How much cleaning fluid is left? ■

EXAMPLE 9 Find the value of x in the equation $x + \frac{5}{6} = \frac{9}{10}$. Reduce your answer.

The LCD for the two fractions $\frac{5}{6}$ and $\frac{9}{10}$ is 30.

$$\frac{5}{6} \times \frac{5}{5} = \frac{25}{30} \qquad \frac{9}{10} \times \frac{3}{3} = \frac{27}{30}$$

Thus we can write the equation in the equivalent form.

$$x + \frac{25}{30} = \frac{27}{30}$$

The denominators are the same. Look at the numerators. We must add 2 to 25 to get 27.

$$\frac{2}{30} + \frac{25}{30} = \frac{27}{30}$$

So if $x = \frac{2}{30}$ we reduce the fraction to obtain $x = \frac{1}{15}$.

Practice Problem 9 Find the value of x in the equation $x + \frac{3}{10} = \frac{23}{25}$. ■

2.7 Exercises

Add or subtract. Simplify all answers.

1. $\frac{5}{9}+\frac{2}{9}$ **2.** $\frac{5}{8}+\frac{2}{8}$ **3.** $\frac{11}{30}+\frac{11}{30}$ **4.** $\frac{3}{28}+\frac{7}{28}$

5. $\frac{21}{23}-\frac{1}{23}$ **6.** $\frac{5}{24}-\frac{3}{24}$ **7.** $\frac{53}{88}-\frac{19}{88}$ **8.** $\frac{103}{110}-\frac{3}{110}$

Add or subtract. Simplify all answers.

9. $\frac{1}{4}+\frac{1}{3}$ **10.** $\frac{2}{7}+\frac{1}{2}$ **11.** $\frac{3}{10}+\frac{3}{20}$ **12.** $\frac{4}{9}+\frac{1}{6}$

13. $\frac{1}{8}+\frac{3}{4}$ **14.** $\frac{5}{16}+\frac{1}{2}$ **15.** $\frac{4}{5}+\frac{7}{10}$ **16.** $\frac{1}{3}+\frac{4}{5}$

17. $\frac{3}{10}+\frac{7}{100}$ **18.** $\frac{13}{100}+\frac{7}{10}$ **19.** $\frac{3}{25}+\frac{1}{35}$ **20.** $\frac{3}{15}+\frac{1}{25}$

21. $\frac{7}{8}+\frac{5}{12}$ **22.** $\frac{5}{6}+\frac{7}{8}$ **23.** $\frac{3}{8}+\frac{3}{10}$ **24.** $\frac{12}{35}+\frac{1}{10}$

25. $\frac{5}{12}-\frac{1}{6}$ **26.** $\frac{37}{20}-\frac{2}{5}$ **27.** $\frac{3}{7}-\frac{1}{5}$ **28.** $\frac{7}{8}-\frac{5}{6}$

29. $\frac{7}{8}-\frac{5}{16}$ **30.** $\frac{9}{50}-\frac{2}{25}$ **31.** $\frac{5}{12}-\frac{7}{30}$ **32.** $\frac{9}{24}-\frac{3}{8}$

33. $\frac{11}{12} - \frac{2}{3}$

34. $\frac{7}{10} - \frac{2}{5}$

35. $\frac{5}{6} - \frac{1}{3}$

36. $\frac{7}{9} - \frac{1}{2}$

37. $\frac{5}{12} - \frac{5}{16}$

38. $\frac{2}{5} - \frac{2}{15}$

39. $\frac{10}{16} - \frac{5}{8}$

40. $\frac{5}{6} - \frac{10}{12}$

41. $\frac{20}{35} - \frac{4}{7}$

42. $\frac{11}{20} - \frac{3}{8}$

43. $\frac{4}{5} + \frac{1}{20} + \frac{3}{4}$

44. $\frac{1}{3} + \frac{2}{24} + \frac{1}{6}$

45. $\frac{5}{30} + \frac{3}{40} + \frac{1}{8}$

46. $\frac{1}{12} + \frac{3}{14} + \frac{4}{21}$

47. $\frac{1}{5} + \frac{2}{3} + \frac{11}{15}$

48. $\frac{2}{3} + \frac{1}{6} + \frac{5}{12}$

Find the value of x in the equation.

49. $x + \frac{1}{7} = \frac{5}{14}$

50. $x + \frac{1}{8} = \frac{7}{16}$

51. $x + \frac{2}{3} = \frac{9}{11}$

52. $x + \frac{3}{4} = \frac{17}{18}$

53. $x - \frac{1}{5} = \frac{4}{12}$

54. $x - \frac{2}{3} = \frac{3}{11}$

Applications

55. This morning at breakfast, Keith used $\frac{4}{5}$ cup of corn flakes and Danny used $\frac{2}{3}$ cup. How many cups of cereal did Keith and Danny use?

56. Jackie started running on Wednesday. She ran $\frac{1}{2}$ mile. On Friday she ran $\frac{2}{3}$ mile. How many miles has she run so far this week?

57. Laurie purchased $\frac{2}{3}$ pound of bananas and $\frac{5}{6}$ pound of seedless grapes. How many pounds of fruit did she purchase?

58. Geoffrey is attaching solar energy panels to the roof of the science building. Each panel uses a bolt that goes through the panel to the frame of the roof ($4\frac{1}{3}$ inches), a washer $\frac{5}{16}$-inch thick, and a $\frac{3}{4}$-inch nut. How long a bolt does he need?

59. Consuela is a collage artist. She assembles different objects on canvases to present a message to the viewer. She is working on a collage that is attached to a dart board. She places a computer chip $\frac{1}{3}$ inch from the center bull's-eye. Then, continuing in a straight line from the bull's-eye, she places a tea bag $\frac{5}{9}$ inch from the computer chip. The last item she uses is a chewing gum wrapper placed $\frac{3}{4}$ inch from the tea bag. How far from the bull's-eye is the chewing gum wrapper?

60. An infant's father knows that straight apple juice is too strong for his daughter. Her bottle is $\frac{1}{2}$ full, and he adds $\frac{1}{3}$ of a bottle of water to dilute the apple juice.
(a) How much of the bottle is full?
(b) If she drinks $\frac{2}{5}$ of the bottle, how much is left?

61. The tomatoes in the Botany Department's organic garden needed a boost so Mary added $\frac{1}{3}$ cup of compost to every $\frac{3}{4}$ cup of soil. How many cups of the mixture did she end up with if she started with $5\frac{1}{4}$ cups of soil?

62. On Saturday morning the gasoline storage tank at Dusty's station was $\frac{11}{12}$ full. During Saturday and Sunday the attendants pumped out $\frac{2}{3}$ of the tank. How much remained in the tank when they opened Monday morning?

63. The manager at Fit Factory Health Club was going through his files for 1996 and discovered that only $\frac{9}{14}$ of the members actually used the club. When he checked the numbers from the previous year of 1995, he found that $\frac{5}{6}$ of the members had used the club. What fractional part of the membership represents the decrease in club usage?

64. Travis typed $\frac{11}{12}$ of his book report on his computer. Then he printed out $\frac{3}{5}$ of his book report on his computer printer. Suddenly, there was a power outage, and he discovered that he hadn't saved his book report before the power went off. What fractional part of the book report was lost when the power failed?

Cumulative Review Problems

65. Reduce to the lowest terms. $\dfrac{15}{85}$

66. Reduce to the lowest terms. $\dfrac{38}{57}$

67. Change to a mixed number. $\dfrac{123}{16}$

68. Change to an improper fraction. $14\dfrac{3}{7}$

2.8 Addition and Subtraction of Mixed Numbers

MathPro Video 6 SSM

After studying this section, you will be able to:

1 *Add two or more mixed numbers.*
2 *Subtract mixed numbers.*

1 *Adding Mixed Numbers*

When adding mixed numbers, it is best to add the fractions together and then add the whole numbers together.

EXAMPLE 1 Add. $3\frac{1}{8} + 2\frac{5}{8}$

$$\begin{array}{r} 3\ \boxed{\frac{1}{8}} \\ +2\ \boxed{\frac{5}{8}} \\ \hline 5\ \boxed{\frac{6}{8}} \end{array}$$

Add the whole numbers. $3 + 2 = 5$ → 5

← Add the fractions. $\frac{1}{8} + \frac{5}{8} = \frac{6}{8}$

$= 5\frac{3}{4}$ ← Reduce $\frac{6}{8} = \frac{3}{4}$

Practice Problem 1 Add. $5\frac{1}{12} + 9\frac{5}{12}$ ■

If the fraction portions of the mixed numbers do not have a common denominator, we must build up the fraction parts to obtain a common denominator before adding.

EXAMPLE 2 Add. $1\frac{2}{7} + 5\frac{1}{3}$

The LCD of $\frac{2}{7}$ and $\frac{1}{3}$ is 21.

$$\frac{2}{7} \times \frac{3}{3} = \frac{6}{21} \qquad \frac{1}{3} \times \frac{7}{7} = \frac{7}{21}$$

Thus $1\frac{2}{7} + 5\frac{1}{3} = 1\frac{6}{21} + 5\frac{7}{21}$.

$$\begin{array}{r} 1\ \boxed{\frac{6}{21}} \\ +5\ \boxed{\frac{7}{21}} \\ \hline 6\ \boxed{\frac{13}{21}} \end{array}$$

Add the whole numbers. $1 + 5$ → 6

← Add the fractions. $\frac{6}{21} + \frac{7}{21}$

Practice Problem 2 Add. $6\frac{1}{4} + 2\frac{2}{5}$ ■

If the sum of the fractions is an improper fraction, we convert it to a mixed number and add the whole numbers together.

EXAMPLE 3 Add. $6\frac{5}{6} + 4\frac{3}{8}$

The LCD of $\frac{5}{6}$ and $\frac{3}{8}$ is 24.

$$\begin{aligned} 6 \boxed{\frac{5}{6} \times \frac{4}{4}} &= 6 \boxed{\frac{20}{24}} \\ +\, 4 \boxed{\frac{3}{8} \times \frac{3}{3}} &= +\, 4 \boxed{\frac{9}{24}} \end{aligned}$$

Add the whole numbers. $\longrightarrow 10 \boxed{\frac{29}{24}} \longleftarrow$ *Add the fractions.*

$= 10 + \boxed{1\frac{5}{24}}$ *Since* $\frac{29}{24} = 1\frac{5}{24}$

$= 11\frac{5}{24}$ *We add the whole numbers* $10 + 1 = 11$.

Practice Problem 3 Add. $7\frac{1}{4} + 3\frac{5}{6}$ ■

2 *Subtracting Mixed Numbers*

Subtracting mixed numbers is like adding.

EXAMPLE 4 Subtract. $8\frac{5}{7} - 5\frac{5}{14}$

The LCD of $\frac{5}{7}$ and $\frac{5}{14}$ is 14.

$$\begin{aligned} 8 \boxed{\frac{5}{7} \times \frac{2}{2}} &= 8\frac{10}{14} \\ -\, 5\frac{5}{14} \quad &= 5\frac{5}{14} \end{aligned}$$

Subtract the whole numbers. $\longrightarrow 3\frac{5}{14} \longleftarrow$ Subtract the fractions.

Practice Problem 4 Subtract. $12\frac{5}{6} - 7\frac{5}{12}$ ■

Sometimes we must borrow before we can subtract.

EXAMPLE 5 Subtract. **(a)** $9\frac{1}{4} - 6\frac{5}{14}$ **(b)** $15 - 9\frac{3}{16}$

The following example is fairly challenging. Read through each step carefully. Be sure to have paper and pencil handy and see if you can verify each step.

(a) The LCD of $\frac{1}{4}$ and $\frac{5}{14}$ is 28.

$$9\boxed{\frac{1}{4} \times \frac{7}{7}} = 9\frac{7}{28}$$

$$-6\boxed{\frac{5}{14} \times \frac{2}{2}} = -6\frac{10}{28}$$

We cannot subtract $\frac{7}{28} - \frac{10}{28}$, so we will need to borrow.

We borrow 1 from 9 to obtain $9\frac{7}{28} = 8 + 1\frac{7}{28} = 8 + \frac{35}{28} = 8\frac{35}{28}$

We write this as

$$9\frac{7}{28} = 8\frac{35}{28}$$

$$-6\frac{10}{28} = -6\frac{10}{28}$$

$$\boxed{8 - 6 = 2} \rightarrow 2\frac{25}{28} \leftarrow \boxed{\frac{35}{28} - \frac{10}{28} = \frac{25}{28}}$$

(b) The LCD = 16.

$$15 = 14\frac{16}{16}$$

$$-9\frac{3}{16} = -9\frac{3}{16}$$

$$\boxed{14 - 9 = 5} \rightarrow 5\frac{13}{16} \leftarrow \boxed{\frac{16}{16} - \frac{3}{16} = \frac{13}{16}}$$

We borrow 1 from 15 to obtain $15 = 14 + 1 = 14 + \frac{16}{16} = 14\frac{16}{16}$

Practice Problem 5 Subtract. **(a)** $9\frac{1}{8} - 3\frac{2}{3}$ **(b)** $18 - 6\frac{7}{18}$ ■

EXAMPLE 6 A plumber had a pipe $5\frac{3}{16}$ inches long for a fitting under the sink. He needed a pipe that was $3\frac{7}{8}$ inches long, so he cut the pipe down. How much of the pipe did he cut off?

We will need to subtract $5\frac{3}{16} - 3\frac{7}{8}$ to find the length that was cut off.

$$\begin{array}{rcr} 5\frac{3}{16} & = & 5\frac{3}{16} \\ -\,3\frac{7}{8}\times\frac{2}{2} & = & -\,3\frac{14}{16} \\ \hline \end{array}$$

We borrow 1 from 5 to obtain $5\frac{3}{16} = 4 + 1\frac{3}{16} = 4 + \frac{19}{16}$

$$\begin{array}{r} 4\frac{19}{16} \\ -\,3\frac{14}{16} \\ \hline 1\frac{5}{16} \end{array}$$

$4 - 3 = 1$ → $1\frac{5}{16}$ ← $\frac{19}{16} - \frac{14}{16} = \frac{5}{16}$

The plumber had to cut off $1\frac{5}{16}$ inches of pipe.

Practice Problem 6 Hillary and Sam purchased $6\frac{1}{4}$ gallons of paint to paint the inside of their house. They used $4\frac{2}{3}$ gallons of paint. How much paint was left over? ■

ALTERNATIVE METHOD Can mixed numbers be added and subtracted as improper fractions? Yes. Recall Example 5(a).

$$9\frac{1}{4} - 6\frac{5}{14} = 2\frac{25}{28}$$

If we write $9\frac{1}{4} - 6\frac{5}{14}$ using improper fractions, we have $\frac{37}{4} - \frac{89}{14}$. Now we build up each of these improper fractions so that they both have the LCD for their denominators.

$$\begin{array}{rcl} \dfrac{37}{4} & \boxed{\times \dfrac{7}{7}} & = \dfrac{259}{28} \\ -\dfrac{89}{14} & \boxed{\times \dfrac{2}{2}} & = \dfrac{178}{28} \\ \hline & & \dfrac{81}{28} = 2\dfrac{25}{28} \end{array}$$

The same result is obtained as in Example 5(a). This method does not require borrowing. However, you do work with larger numbers. For more practice, see problems 47–50.

Developing Your Study Skills

PROBLEMS WITH ACCURACY

Strive for accuracy. Mistakes are often made because of human error rather than lack of understanding. Such mistakes are frustrating. A simple arithmetic or sign error can lead to an incorrect answer.

These five steps will help you cut down on errors.

1. Work carefully, and take your time. Do not rush through a problem just to get it done.
2. Concentrate on the problem. Sometimes problems become mechanical, and your mind begins to wander. You become careless and make a mistake.
3. Check your problem. Be sure that you copied it correctly from the book.
4. Check your computations from step to step. Check the solution to the problem. Does it work? Does it make sense?
5. Keep practicing new skills. Remember the old saying "Practice makes perfect." An increase in practice results in an increase in accuracy. Many errors are due simply to lack of practice.

There is no magic formula for eliminating all errors, but these five steps will be a tremendous help in reducing them.

2.8 Exercises

Add or subtract. Simplify all answers. Express the answer as a mixed number.

1. $7\frac{1}{8} + 2\frac{5}{8}$

2. $6\frac{3}{10} + 4\frac{1}{10}$

3. $15\frac{3}{14} - 11\frac{1}{14}$

4. $8\frac{3}{4} - 3\frac{1}{4}$

5. $12\frac{1}{3} + 5\frac{1}{6}$

6. $20\frac{1}{4} + 3\frac{1}{8}$

7. $5\frac{4}{5} + 10\frac{3}{10}$

8. $6\frac{3}{8} + 4\frac{1}{16}$

9. $1 - \frac{3}{7}$

10. $1 - \frac{9}{11}$

11. $1\frac{5}{6} + \frac{7}{8}$

12. $1\frac{2}{3} + \frac{5}{18}$

13. $7\frac{1}{2} + 8\frac{3}{4}$

14. $9\frac{1}{6} + 2\frac{2}{3}$

15. $8\frac{2}{3} - 5\frac{1}{6}$

16. $7\frac{4}{5} - 2\frac{1}{10}$

17. $18\frac{1}{6} - 10\frac{3}{4}$

18. $12\frac{4}{9} - 7\frac{5}{6}$

19. $30 - 15\frac{3}{7}$

20. $25 - 14\frac{2}{11}$

Add or subtract. Express the answer as a mixed number.

21. $15\frac{4}{15} + 26\frac{8}{15}$

22. $22\frac{1}{8} + 14\frac{3}{8}$

23. $4\frac{1}{3} + 2\frac{1}{4}$

24. $6\frac{1}{8} + 7\frac{3}{4}$

25. $2\frac{1}{15} + 14\frac{3}{5}$

26. $8\frac{1}{7} + 3\frac{11}{14}$

27. $47\frac{3}{10} + 26\frac{5}{8}$

28. $34\frac{1}{20} + 45\frac{8}{15}$

29. $19\frac{5}{6} - 14\frac{1}{3}$

30. $27\frac{11}{12} - 21\frac{1}{3}$

31. $3\frac{5}{6} - 3\frac{10}{12}$

32. $7\frac{1}{14} - 4\frac{3}{7}$

33. $12\frac{3}{20} - 7\frac{7}{15}$

34. $8\frac{5}{12} - 5\frac{9}{10}$

35. $12 - 3\frac{7}{15}$

36. $19 - 6\frac{3}{7}$

37. $\begin{array}{r} 120 \\ -\ 17\frac{3}{8} \\ \hline \end{array}$

38. $\begin{array}{r} 98 \\ -\ 89\frac{15}{17} \\ \hline \end{array}$

39. $\begin{array}{r} 3\frac{1}{8} \\ 2\frac{1}{3} \\ +7\frac{3}{4} \\ \hline \end{array}$

40. $\begin{array}{r} 4\frac{2}{3} \\ 3\frac{1}{5} \\ +6\frac{3}{4} \\ \hline \end{array}$

Applications

Solve each problem.

41. Lee Hong rode his mountain bike through part of the Sangre de Cristo Mountains in New Mexico. On Wednesday he rode $20\frac{3}{4}$ miles. On Thursday he rode $22\frac{3}{8}$ miles. What was his total biking distance during those two days?

42. Kathryn traveled on a flight that took $7\frac{5}{6}$ hours. She connected with another flight that took $5\frac{4}{5}$ hours. What was her total time in the air?

43. Renaldo bought $6\frac{3}{8}$ pounds of cheese for a party. The guests ate $4\frac{1}{3}$ pounds of the cheese. How much cheese did Renaldo have left over?

44. Shanna purchased stock at $\$21\frac{3}{8}$ per share. When her son was ready for college, she sold the stock at $\$93\frac{5}{8}$ per share. How much did she make per share for her son's tuition?

45. Jack's dog is too heavy. According to the veterinarian, the dog needs to lose 16 pounds over the next three months. During the first month he lost $5\frac{1}{8}$ pounds. The second month he lost $4\frac{2}{3}$ pounds.

(a) How much weight did Jack's dog lose during the first two months?

(b) How much weight does his dog need to lose in the third month to reach the goal advised by the veterinarian?

46. A young man has been under a doctor's care to lose weight. His doctor wanted him to lose 46 pounds in the first three months. He lost $17\frac{5}{8}$ pounds the first month and $13\frac{1}{2}$ pounds the second month.

(a) How much did he lose during the first two months?

(b) How much would he need to lose in the third month to reach the goal?

To Think About

Use improper fractions as discussed in the text to perform the following calculations.

47. $\frac{379}{8} + \frac{89}{5}$

48. $\frac{151}{6} - \frac{130}{7}$

49. $\frac{200}{3} - \frac{153}{7}$

50. $\frac{400}{9} + \frac{126}{6}$

Cumulative Review Problems

Multiply.

51. $\begin{array}{r} 12{,}367 \\ \times\ \ 9 \\ \hline \end{array}$

52. $\begin{array}{r} 304 \\ \times\ 128 \\ \hline \end{array}$

53. $\begin{array}{r} 6737 \\ \times\ \ 76 \\ \hline \end{array}$

54. $\begin{array}{r} 4050 \\ \times\ 2106 \\ \hline \end{array}$

Putting Your Skills to Work

FARMING ON THE HALVES

An interesting use of fractions is found among the rural farm areas of the Maritime Provinces (New Brunswick, Nova Scotia, and Prince Edward Island) of Canada. They have a custom of "farming on the halves" and farming on other fractional values as well.

PROBLEMS FOR INDIVIDUAL INVESTIGATION AND ANALYSIS

1. Everett Hatfield raises a blueberry crop in Nova Scotia on 40 acres of land. He has arranged with the local farmers' cooperative to "farm on the halves." He will share one half of the expenses of raising the crop and one half of the profits with the local cooperative. It cost \$1300 to burn a portion of the fields each year with a portable propane burner to maximize the yield of the blueberry crop. If his fields bring in 23 tons of blueberries, the price of blueberries is $\$2\frac{1}{2}$ per pound, and the workers are paid $\$1\frac{3}{4}$ per pound to pick the blueberries, how much profit will he take in?

2. Dick Wightman owns 100 acres of potato fields in Prince Edward Island. He will ask the farming cooperative to take three fifths of the expenses of raising the crop, and they will therefore also take three fifths of the profits from raising the crop. It cost $\$1\frac{1}{2}$ per acre to till the soil and plant and fertilize the potatoes. It cost $\$3\frac{1}{4}$ per acre to water and weed the potato fields. Dick sold $26\frac{1}{4}$ tons of potatoes from his fields. The cost of the laborers who harvested the potatoes was $\$\frac{1}{2}$ per pound. The price that he sold his potatoes for this year was $\$\frac{3}{4}$ per pound. How much profit will Dick take in?

PROBLEMS FOR GROUP INVESTIGATION AND COOPERATIVE PROBLEM SOLVING

Together with some other members of your class determine the following.

3. Loring Kerr needs to raise \$24,750 from farming wheat. He hopes to produce 300 bushels of wheat per acre on his 60-acre wheat fields. The expected price of wheat this year is $\$2\frac{1}{2}$ per bushel. The cost to harvest the wheat is \$200 per acre. He does the rest of the labor himself. The local cooperative has offered to pay for $\frac{3}{8}$ of the costs and take $\frac{3}{8}$ of the profits. Can he afford to take their offer? If not what fractional part of the costs and the profits should he have as a goal?

4. Betty and Walter Ames want to raise corn on their farm this year. They have reached an agreement with the local farming cooperative to pay for $\frac{2}{3}$ of the expenses and also to receive $\frac{2}{3}$ of the profits. The hope to produce 210 bushels of corn per acre from their farm. The expected price of corn this year is $\$1\frac{1}{2}$ per bushel. The cost to harvest corn is \$150 per acre. Betty and Walter do the rest of the labor themselves. They want to raise \$19,250 from the corn crop this year. They own 1000 acres of corn fields but only want to plant the minimum number of acres to achieve their financial goal. How many acres of corn should they plant?

INTERNET CONNECTIONS: Go to "http://www.prenhall.com/tobey" to be connected

Site: Farm Workers, by Type, 1910–1995 (United States Department of Agriculture) or a related site

This site shows a graph of the number of family workers and hired workers on American farms. Use the graph to answer the following questions.

5. Estimate the number of family workers in 1950.

6. Estimate the number of hired workers in 1925.

7. Estimate the total number of farm workers in 1985.

8. Tell what patterns you see in the graph.

2.9 Applied Problems Involving Fractions

MathPro Video 6 SSM

After studying this section, you will be able to:

1 *Solve applied problems with fractions.*

1 *Solving Applied Problems with Fractions*

All problem solving requires the same kind of thinking. In this section we will combine problem-solving skills with our new computational skills with fractions. Sometimes the difficulty is in figuring out what must be done. Sometimes it is in doing the computation. Remember that *estimating* is important in problem solving. We may use the following steps.

1. *Understand the problem.*
 (a) Read the problem carefully.
 (b) Draw a picture if this helps you.
 (c) Fill in the mathematics blueprint.

2. *Solve.*
 (a) Perform the calculations.
 (b) State the answer, including the unit of measure.

3. *Check.*
 (a) Estimate the answer.
 (b) Compare the exact answer with the estimate to see if your answer is reasonable.

EXAMPLE 1 Maria bought Digital Equipment Corporation stock at $\$26\frac{1}{8}$. The stock's value went up $\$2\frac{5}{8}$ the first week. The second week it went up $\$3\frac{3}{4}$. What was its value after these two increases?

1. *Understand the problem.*
We draw a picture to help us.

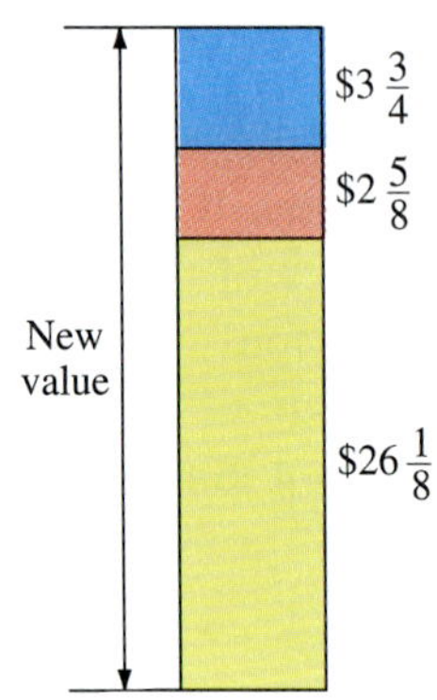

Then we fill in the Mathematics Blueprint.

MATHEMATICS BLUEPRINT FOR PROBLEM SOLVING

Gather the Facts	What Am I Asked to Do?	How Do I Proceed?	Key Points to Remember
Purchase Price: $\$26\frac{1}{8}$ Week One: up $\$2\frac{5}{8}$ Week Two: up $\$3\frac{3}{4}$	Find the value of the stock after the two increases.	Add the two increases to the purchase price.	When adding mixed numbers, add the whole numbers first and then add the fractions.

2. *Solve and state the answer.*

Add the three amounts. $26\frac{1}{8} + 2\frac{5}{8} + 3\frac{3}{4}$

$$\text{LCD} = 8 \qquad \begin{aligned} 26\frac{1}{8} &= 26\frac{1}{8} \\ 2\frac{5}{8} &= 2\frac{5}{8} \\ +\ 3\boxed{\frac{3}{4} \times \frac{2}{2}} &= +\ 3\frac{6}{8} \\ \hline & \quad 31\frac{12}{8} = 32\frac{4}{8} = 32\frac{1}{2} \end{aligned}$$

The value of the stock is now $\$32\frac{1}{2}$.

3. *Check.*

Estimate the sum by rounding each fraction to one nonzero digit.

Thus $\quad 26\frac{1}{8} + 2\frac{5}{8} + 3\frac{3}{4}$

becomes $\quad 30 + 3 + 4 = 37$

This is close to our answer, $32\frac{1}{2}$. Our answer seems reasonable.

One of the most important uses of estimation in mathematics is in the calculation of problems involving fractions. People find it easier to detect significant errors when working with whole numbers. However, the extra steps involved in the calculation with fractions and mixed numbers often distract our attention from an error that we should have detected.

Thus it is particularly critical to take the time to check your answer by estimating the results of the calculation by using whole numbers. Be sure to ask yourself, is this answer reasonable? Does this answer seem realistic? Only by estimating our results by using whole numbers will we be able to answer that question. It is this estimating skill that you will find more useful in your own life as a consumer and as a citizen.

Practice Problem 1 Nicole purchased the following amounts of gas for her car in the last three fill-ups: $18\frac{7}{10}$ gallons, $15\frac{2}{5}$ gallons, and $14\frac{1}{2}$ gallons. How many gallons did she buy altogether? ■

The word *diameter* has two common meanings. First, it means a line segment that passes through the center of and intersects a circle twice. Second, it means the *length* of this segment.

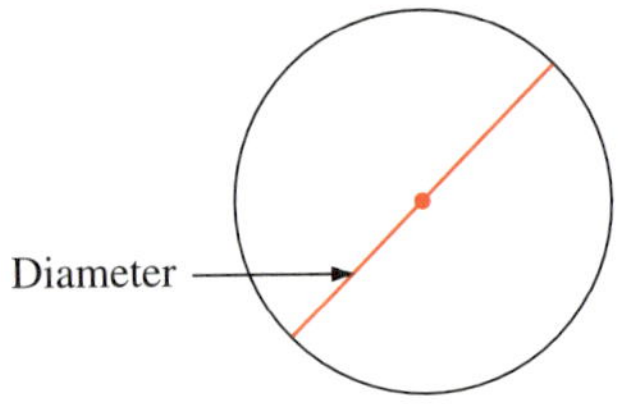

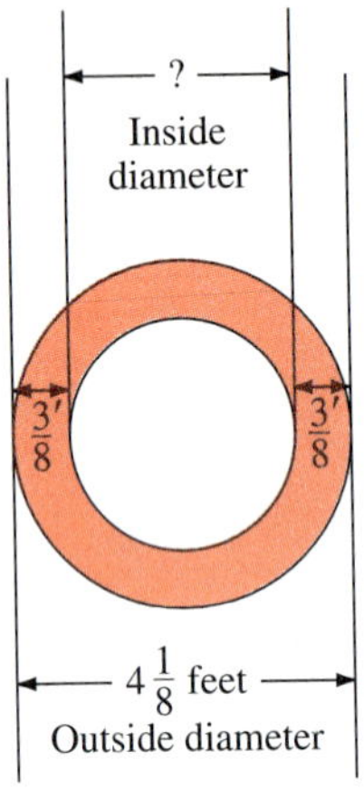

EXAMPLE 2 What is the inside diameter (distance across) of a cement storm drain pipe that has an outside diameter of $4\frac{1}{8}$ feet and is $\frac{3}{8}$ feet thick?

1. *Understand the problem.*
   Read the problem carefully. Draw a picture. The picture is in the margin on the left. Now fill in the Mathematics Blueprint.

MATHEMATICS BLUEPRINT FOR PROBLEM SOLVING

Gather the Facts	What Am I Asked to Do?	How do I Proceed?	Key Points to Remember
Outside diameter is $4\frac{1}{8}$ feet. Thickness is 3/8 foot on both ends of the diameter.	Find the *inside* diameter of the pipe.	Add the two measures of thickness. Then subtract this total from the outside diameter.	Since the LCD = 8, all fractions must have this denominator.

2. *Solve and state the answer.*
   Add the two thickness measurements together.
   Adding $\frac{3}{8} + \frac{3}{8} = \frac{6}{8}$ gives the total thickness of the pipe, $\frac{6}{8}$ foot. We will not reduce $\frac{6}{8}$ since the LCD is 8.
   We subtract the total of the two thickness measurements from the outside diameter.

$$\begin{array}{rcr} 4\frac{1}{8} & = & 3\frac{9}{8} \\ -\ \frac{6}{8} & = & -\ \frac{6}{8} \\ \hline & & 3\frac{3}{8} \end{array}$$

We borrow 1 from 4 to get $3 + 1\frac{1}{8}$ or $3\frac{9}{8}$.

The inside diameter is $3\frac{3}{8}$ feet.

3. *Check.*

We will work backward to check. We will use the exact values.

If we have done our work correctly, $\frac{3}{8}$ foot + $3\frac{3}{8}$ feet + $\frac{3}{8}$ foot should add up to the outside diameter, $4\frac{1}{8}$ feet.

$$\frac{3}{8} + 3\frac{3}{8} + \frac{3}{8} \stackrel{?}{=} 4\frac{1}{8}$$

$$3\frac{9}{8} \stackrel{?}{=} 4\frac{1}{8}$$

$$4\frac{1}{8} = 4\frac{1}{8} \quad \checkmark$$

Our answer of $3\frac{3}{8}$ feet is correct.

Practice Problem 2 A poster is $12\frac{1}{4}$ inches long. We want a $1\frac{3}{8}$-inch border on the top and a 2-inch border on the bottom. What is the length of the inside portion of the poster? ■

EXAMPLE 3 On Tuesday Michael earned $\$8\frac{1}{4}$ per hour working for 8 hours. He also earned overtime pay, which is $1\frac{1}{2}$ times his regular rate of $\$8\frac{1}{4}$, for 4 hours on Tuesday. How much pay did he earn altogether on Tuesday?

1. *Understand the problem.*

We draw a picture of the parts of Michael's pay on Tuesday.

Michael's earnings on Tuesday are the sum of two parts:

Pay at regular pay rate	+	Pay at overtime pay rate	=	Total pay for the day

Now fill in the Mathematics Blueprint.

MATHEMATICS BLUEPRINT FOR PROBLEM SOLVING

Gather the Facts	What Am I Asked to Do?	How Do I Proceed?	Key Points to Remember
He works 8 hours at $\$8\frac{1}{4}$ per hour. He works 4 hours at the overtime rate, $1\frac{1}{2}$ times the regular rate.	Find his total pay for Tuesday.	Find out how much he is paid for regular time. Find out how much he is paid for overtime. Then add the two.	The overtime rate is $1\frac{1}{2}$ multiplied by the regular rate.

2. *Solve and state the answer.*

Find his overtime pay rate.

$$1\frac{1}{2} \times 8\frac{1}{4} = \frac{3}{2} \times \frac{33}{4} = \frac{\$99}{8} \text{ per hour}$$

We leave our answer as an improper fraction because we will need to multiply it by another fraction.

How much was he paid for regular time? For overtime?

For 8 regular hours, he earned $8 \times 8\frac{1}{4} = \overset{2}{\cancel{8}} \times \frac{33}{\underset{1}{\cancel{4}}} = \$66.$

For 4 overtime hours, he earned $\overset{1}{\cancel{4}} \times \frac{99}{\underset{2}{\cancel{8}}} = \frac{99}{2} = \$49\frac{1}{2}.$

Now we add to find the total pay.

\$66	Pay at regular pay rate
$\$49\frac{1}{2}$	Pay at overtime rate

Michael earned $\$115\frac{1}{2}$ working on Tuesday.

3. *Check.*

We estimate his regular pay rate at \$8 per hour.

We estimate his overtime pay rate at $1\frac{1}{2} \times 8 = \frac{3}{2} \times 8 = 12$ or \$12 per hour.

$$8 \text{ hours} \times \$8 \text{ per hour} = \$64 \text{ regular pay}$$

$$4 \text{ hours} \times \$12 \text{ per hour} = \$48 \text{ overtime pay}$$

Estimated sum. $\$64 + \$48 \approx 60 + 50 = 110$

\$110 is close to our calculated value, $\$115\frac{1}{2}$, so our answer is reasonable. ✓

Practice Problem 3 A tent manufacturer uses $8\frac{1}{4}$ yards of waterproof duck cloth to make a regular tent. She uses $1\frac{1}{2}$ times that amount to make a large tent. How many yards of cloth will she need to make six regular tents and 16 large tents? ■

EXAMPLE 4 Alicia is buying some 8-foot boards for shelving. She wishes to make two bookcases, each with three shelves. Each shelf will be $3\frac{1}{4}$ feet long.

(a) How many boards does she need to buy?

(b) How many linear feet of shelving are actually needed to build the bookcases?

(c) How many linear feet of shelving are left over?

1. *Understand the problem.*

Draw a sketch of a bookcase. Each bookcase will have 3 shelves. Alicia is making 2 such bookcases. (Alicia's boards are for the shelves, not the sides.)

Now fill in the Mathematics Blueprint.

MATHEMATICS BLUEPRINT FOR PROBLEM SOLVING

Gather the Facts	What Am I Asked to Do?	How Do I Proceed?	Key Points to Remember
She needs three shelves for each bookcase. Each shelf is $3\frac{1}{4}$ feet long. She will make two bookcases. Shelves are cut from 8-foot boards.	Find out how many boards to buy. Find out how many feet of board is needed for shelves and how many feet will be left over.	First find out how many $3\frac{1}{4}$-foot shelves she can get from one board. Then see how many boards she needs to make all six shelves.	Each time she cuts up an 8-foot board, she will get some shelves and some leftover wood.

2. *Solve and state the answer.*

We want to know how many $3\frac{1}{4}$-foot boards are in an 8-foot board. By drawing a rough sketch, we would probably guess the answer is 2. To find exactly how many $3\frac{1}{4}$-foot long pieces are in 8 feet, we will use division.

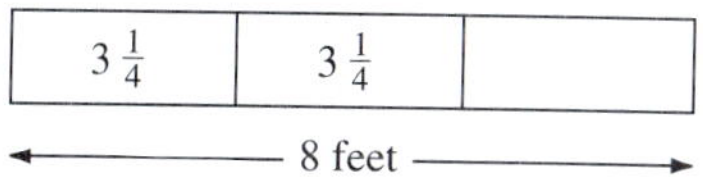

$$8 \div 3\frac{1}{4} = \frac{8}{1} \div \frac{13}{4} = \frac{8}{1} \times \frac{4}{13} = \frac{32}{13} = 2\frac{6}{13} \text{ boards}$$

She will get 2 shelves from each board, and some wood will be left over.

(a) How many boards does Alicia need to build two bookcases?

For 2 bookcases, she needs 6 shelves. This will provide 2 shelves out of each board. $6 \div 2 = 3$. She will need three of the 8-foot boards.

(b) How many linear feet of shelving are actually needed to build the bookcases?

She needs 6 shelves at $3\frac{1}{4}$ feet $= 6 \times 3\frac{1}{4} = \overset{3}{\cancel{6}} \times \frac{13}{\underset{2}{\cancel{4}}} = \frac{39}{2} = 19\frac{1}{2}$ linear feet. A total of $19\frac{1}{2}$ linear feet of shelving is needed.

(c) How many linear feet of shelving are left over?

Each time she uses one board she has $8 - 3\frac{1}{4} - 3\frac{1}{4} = 8 - 6\frac{1}{2} = 1\frac{1}{2}$ feet left over. Each of the three boards has $1\frac{1}{2}$ feet left over.

$3 \times 1\frac{1}{2} = 3 \times \frac{3}{2} = \frac{9}{2} = 4\frac{1}{2}$ linear feet. A total of $4\frac{1}{2}$ linear feet of shelving is left over.

3. *Check.*

Work backward. See if you can check that with three 8-foot boards you

(a) can make the 6 shelves for the two bookcases

(b) will use exactly $19\frac{1}{2}$ linear feet to make the shelves

(c) will have exactly $4\frac{1}{2}$ linear feet left over

The check is left to you.

Practice Problem 4 Michael is purchasing 12-foot boards for shelving. He wishes to make two bookcases, each with four shelves. Each shelf will be $2\frac{3}{4}$ feet long.

(a) How many boards does he need to buy?

(b) How many linear feet of shelving are actually needed to build the bookcases?

(c) How many linear feet of shelving are left over? ■

Another useful method for solving applied problems is called "Do a similar, simpler problem." When a problem seems difficult to understand because of the fractions, change the problem to an easier but similar problem. Then decide how to solve the simpler problem and use the same steps to solve the original problem. For example:

> How many gallons of water can a tank hold if its volume is $58\frac{2}{3}$ cubic feet? (1 cubic foot holds about $7\frac{1}{2}$ gallons.)

A similar, easier problem would be: "If 1 cubic foot holds about 8 gallons and a tank holds 60 cubic feet, how many gallons of water does the tank hold?"

The easier problem can be read more quickly and seems to make more sense. Probably we will see how to solve the easier problem right away: "I can find the number of gallons by multiplying 8×60." Therefore we can solve the first problem by multiplying $7\frac{1}{2} \times 58\frac{2}{3}$ to obtain the number of gallons of water. See the next example.

EXAMPLE 5 A fishing boat traveled $69\frac{3}{8}$ nautical miles in $3\frac{3}{4}$ hours. How many knots (nautical miles per hour) did the fishing boat average?

1. *Understand the problem.*

Let us think of a simpler problem. If a boat traveled 70 nautical miles in 4 hours, how many knots did it average? We would divide distance by time.

$$70 \div 4 = \text{average speed}$$

Likewise in our original problem we need to divide distance by time.

$$69\frac{3}{8} \div 3\frac{3}{4} = \text{average speed}$$

Now fill in the Mathematics Blueprint.

MATHEMATICS BLUEPRINT FOR PROBLEM SOLVING

Gather the Facts	What Am I Asked to Do?	How Do I Proceed?	Key Points to Remember
Distance is $69\frac{3}{8}$ nautical miles. Time is $3\frac{3}{4}$ hours.	Find the average speed of the boat.	Divide the distance in nautical miles by the time in hours.	You must change the mixed numbers to improper fractions before dividing.

2. *Solve and state the answer.*

Divide distance by time to get speed in knots.

$$69\frac{3}{8} \div 3\frac{3}{4} = \frac{555}{8} \div \frac{15}{4} = \frac{\overset{37}{\cancel{555}}}{\underset{2}{\cancel{8}}} \cdot \frac{\overset{1}{\cancel{4}}}{\underset{1}{\cancel{15}}}$$

$$= \frac{37}{2} \cdot \frac{1}{1} = \frac{37}{2} = 18\frac{1}{2} \text{ knots}$$

The speed of the boat was $18\frac{1}{2}$ knots.

3. *Check.*

We estimate $69\frac{3}{8} \div 3\frac{3}{4}$.

$$\text{Use } 70 \div 4 = 17\frac{1}{2} \text{ knots}$$

Our estimate is close to the calculated value.
Our answer is reasonable. ✓

Practice Problem 5 Alfonso traveled $199\frac{3}{4}$ miles in his car and used $8\frac{1}{2}$ gallons of gas. How many miles per gallon did he get? ■

2.9 Exercises

You may use the Mathematics Blueprint for problem solving to help you solve the following problems.

Applications

1. A truck hauled $5\frac{1}{4}$ tons of gravel on Monday, $4\frac{7}{8}$ tons on Tuesday, and $6\frac{1}{6}$ tons on Wednesday. How many tons were hauled over the three days?

2. A deli used $8\frac{1}{4}$ pounds of roast beef on Monday, $7\frac{1}{2}$ pounds on Tuesday, and $6\frac{7}{10}$ pounds on Wednesday. How many pounds were used over the three days?

3. The coastal area of North Carolina experienced devastating erosion caused by fierce storms. The National Guard assisted by hauling sand to try and rebuild the dunes. During three days in September 1996, the National Guard hauled $7\frac{5}{6}$ tons of sand on the first day, $8\frac{1}{8}$ tons on the second day, and $9\frac{1}{2}$ tons on the third. How many tons of sand were hauled?

4. Nancy wonders if a tree in her front yard will grow to be 8 feet tall. When the tree was planted, it was $2\frac{1}{6}$ feet above the ground in height. It grew $1\frac{1}{4}$ feet the first year after being planted. How many more feet does it need to grow to reach a height of 8 feet?

5. A bolt extends through $\frac{3}{4}$-inch-thick plywood, two washers that are each $\frac{1}{16}$ inch thick, and a nut that is $\frac{3}{16}$ inch thick. The bolt must be $\frac{1}{2}$ inch longer than the sum of the thicknesses of plywood, washers, and nut. What is the minimum length of the bolt?

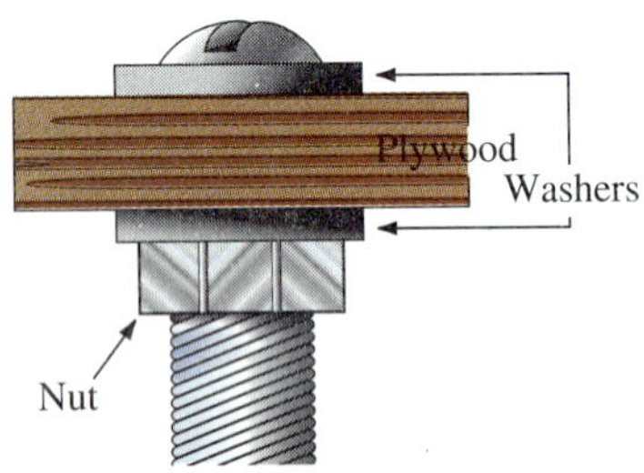

6. Repeat problem 5 with $\frac{11}{16}$-inch-thick plywood, two washers that are each $\frac{1}{8}$ inch thick, and a nut that is $\frac{1}{4}$ inch thick.

7. A cable wire is $65\frac{1}{6}$ inches long. It is to be cut into pieces that are $3\frac{5}{6}$ inch long to be packaged and sold separately in electronics stores. How many pieces will be made?

8. Hank is practicing to run in the Boston Marathon. His goal this week is to run 55 miles. Today he ran $15\frac{1}{4}$ miles. Yesterday he ran $14\frac{3}{8}$ miles. How many more miles must he run this week to meet his goal?

9. For a party of the British Literature Club using all "English foods," Nancy bought a $10\frac{2}{3}$-pound wheel of Stilton cheese, to go with the pears and the apples, at $\$8\frac{3}{4}$ per pound. How much did the wheel of Stilton cheese cost?

10. The following exaggerated story was repeated by certain basketball fans: Michael Jordan wasn't always 7 feet tall. When he was a little boy, at the height of $3\frac{1}{2}$ feet, no one knew that he would grow so quickly. The following year he grew $1\frac{5}{6}$ feet, and the year after that he grew $1\frac{1}{5}$ feet. How many more feet did he grow to reach 7 feet tall?

11. How many gallons can a tank hold that has a volume of $36\frac{3}{4}$ cubic feet? (Assume that 1 cubic foot holds about $7\frac{1}{2}$ gallons.)

12. A tank can hold a volume of $7\frac{1}{4}$ cubic feet. If it is filled with water, how much does the water weigh? (Assume that 1 cubic foot of water weighs $62\frac{1}{2}$ pounds.)

13. The night of the *Titanic* cruise ship disaster, the captain decided to run his ship at $22\frac{1}{2}$ knots, (nautical miles per hour). The *Titanic* traveled at that speed for $4\frac{3}{4}$ hours before it met its tragic demise. How far did the *Titanic* travel at this excessive speed before the disaster?

14. Barbara earns \$450 per week. She has $\frac{1}{5}$ of her income withheld for federal taxes, $\frac{1}{15}$ of her income withheld for state taxes, and $\frac{1}{25}$ of her income withheld for medical coverage. How much per week is left for Barbara after those three deductions?

15. Noriko earns \$660 per week. She has $\frac{1}{5}$ of her income withheld for federal taxes, $\frac{1}{15}$ of her income withheld for state taxes, and $\frac{1}{20}$ of her income withheld for medical coverage. How much per week is left for Noriko after these three deductions?

16. When it was new, Ginny Sue's car got $28\frac{1}{6}$ miles per gallon. It now gets $1\frac{5}{6}$ miles per gallon less. How far can she drive now if the car has $10\frac{3}{4}$ gallons in the tank?

17. The crafts store bought $36\frac{3}{4}$ ounces of potpourri (a type of air freshener) for \$2. The shop divides the potpourri into $\frac{7}{8}$-ounce bags to put into drawers and closets, and sells each bag for $\$3\frac{1}{2}$.

(a) How many $\frac{7}{8}$-ounce bags of potpourri can the shop make?

(b) If all of the bags are sold, how much money will the store take in?

(c) How much profit will it make?

18. A candle company purchased $48\frac{1}{8}$ pounds of wax to make specialty candles. It takes $\frac{5}{8}$ pound of wax to make one candle. The owners of the business plan to sell the candles for \$12 each. The specialty wax cost them \$2 per pound.

(a) How many candles can they make?

(b) How much does it cost to make one candle?

(c) How much profit will they make if they sell all of the candles?

19. Cecilia bought a loaf of sourdough bread that was made by a local gourmet bakery. The label said that the bread, plus its fancy box, weighed $18\frac{1}{2}$ ounces in total. Of this, $1\frac{1}{4}$ ounces turned out to be the weight of the ribbon. The box weighed $3\frac{1}{8}$ ounces.

(a) How many ounces of bread did she actually buy?

(b) The box stated its net weight as $14\frac{3}{4}$ ounces. (This means that she should have found $14\frac{3}{4}$ ounces of gourmet sourdough bread in the box.) How much in error was this measurement?

20. Tony was making brownies one night. The recipe calls for $3\frac{3}{4}$ cups of chocolate chips. He had started to put in the chips (he put in $1\frac{1}{8}$ cups) when his little brother came into the kitchen and threw in $1\frac{3}{4}$ cups of chocolate chips.

(a) How many more cups of chocolate chips does Tony need to add according to the recipe?

(b) He has $\dfrac{3}{4}$ cup of chocolate chips left in the bowl. Does he have enough to add the amount needed in (a)?

21. The largest Coast Guard boat stationed at San Diego can travel $160\frac{1}{8}$ nautical miles in $5\frac{1}{4}$ hours.
   (a) At how many knots is the boat traveling?

   (b) At this speed, how long would it take the Coast Guard boat to travel $213\frac{1}{2}$ nautical miles?

22. Russ and Norma's Mariah water ski boat can travel $72\frac{7}{8}$ nautical miles in $2\frac{3}{4}$ hours.
   (a) At how many knots is the boat traveling?

   (b) At this speed, how long would it take their water ski boat to travel $92\frac{3}{4}$ nautical miles?

★ 23. A Kansas wheat farmer has a storage bin with a capacity of $6856\frac{1}{4}$ cubic feet.
   (a) If a bushel of wheat is $1\frac{1}{4}$ cubic feet, how many bushels can it hold?
   (b) If a farmer wants to make a new storage bin $1\frac{3}{4}$ times larger, how many cubic feet will it hold?
   (c) How many bushels will the new bin hold?

★ 24. A Greyhound bus made a long-distance run in upstate New York, stopping several times. The entire trip took $24\frac{1}{6}$ hours. The bus stopped a total of $2\frac{1}{3}$ hours for restroom and meal stops. The bus also stopped $1\frac{1}{2}$ hours for repairs. The bus traveled 854 miles. What was its average speed in miles per hour when it was moving? (Do not count stop times.)

Cumulative Review Problems

25. Add. $\begin{array}{r} 16{,}846 \\ 19{,}321 \\ +\ 8{,}078 \\ \hline \end{array}$

26. Subtract. $\begin{array}{r} 193{,}705 \\ -\ 165{,}891 \\ \hline \end{array}$

27. Multiply. $\begin{array}{r} 1683 \\ \times\ \ 27 \\ \hline \end{array}$

28. Divide. $19\overline{)5285}$

Developing Your Study Skills

WHY STUDY MATHEMATICS?

Students often question the value of mathematics. They see little real use for it in their everyday lives. However, mathematics is often the key that opens the door to a better paying job.

In our present-day technological world, many people use mathematics daily. Many vocational and professional areas—such as the fields of business, statistics, economics, psychology, finance, computer science, chemistry, physics, engineering, electronics, nuclear energy, banking, quality control, and teaching—require a certain level of expertise in mathematics. Those who want to work in these fields must be able to function at a given mathematical level. Those who cannot will not be able to enter this job area.

So, whatever your field, be sure to realize the importance of mastering the basics of this course. It is very likely to help you advance to the career of your choice.

Chapter Organizer

Topic	Procedure	Examples
Concept of a fractional part, p. 96.	The numerator is the number of parts selected. The denominator is the number of total parts.	What part of this sketch is shaded? 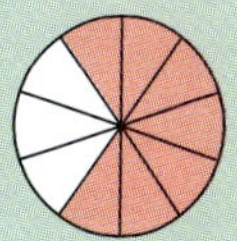 $\frac{7}{10}$
Prime factorization, p. 103.	Prime factorization is the writing of a number as the product of prime numbers.	Write the prime factorization of 36. $36 = 4 \times 9$ $2 \times 2 \quad 3 \times 3$ $= 2 \times 2 \times 3 \times 3$
Reducing fractions, p. 105.	**1.** Factor numerator and denominator into prime factors. **2.** Divide out factors common to numerator and denominator.	Reduce. $\frac{54}{90}$ $\frac{54}{90} = \frac{\overset{1}{\cancel{3}} \times \overset{1}{\cancel{3}} \times 3 \times \overset{1}{\cancel{2}}}{\underset{1}{\cancel{3}} \times \underset{1}{\cancel{3}} \times \underset{1}{\cancel{2}} \times 5} = \frac{3}{5}$
Changing mixed numbers to improper fractions, p. 111.	**1.** Multiply whole number by denominator. **2.** Add product to numerator. **3.** Place sum over denominator.	Write as an improper fraction. $7\frac{3}{4} = \frac{4 \times 7 + 3}{4} = \frac{28 + 3}{4} = \frac{31}{4}$
Changing an improper fraction to a mixed number, p. 113.	**1.** Divide denominator into numerator. **2.** The quotient is the whole number. **3.** The fraction is the remainder over the divisor.	Change to a mixed number. $\frac{32}{5}$ $5\overline{)32}$ quotient 6, $\frac{30}{2}$; $= 6\frac{2}{5}$
Multiplying fractions, p. 118.	**1.** Divide out common factors from the numerators and denominators whenever possible. **2.** Multiply numerators. **3.** Multiply denominators.	Multiply. $\frac{3}{7} \times \frac{5}{13} = \frac{15}{91}$ Multiply. $\frac{\overset{1}{\cancel{5}}}{\underset{1}{\cancel{8}}} \times \frac{\overset{2}{\cancel{16}}}{\underset{3}{\cancel{15}}} = \frac{2}{3}$
Multiplying mixed and/or whole numbers, p. 120.	**1.** Change any whole numbers to a fraction with a denominator of 1. **2.** Change any mixed numbers to improper fractions. **3.** Use multiplication rule for fractions.	Multiply. $7 \times 3\frac{1}{4}$ $\frac{7}{1} \times \frac{13}{4} = \frac{91}{4}$ $= 22\frac{3}{4}$
Dividing fractions, p. 124.	To divide two fractions, we invert the second fraction and multiply.	Divide. $\frac{3}{7} \div \frac{2}{9} = \frac{3}{7} \times \frac{9}{2} = \frac{27}{14}$ *or* $1\frac{13}{14}$
Dividing mixed numbers and/or whole numbers, p. 126.	**1.** Change any whole numbers to a fraction with a denominator of 1. **2.** Change any mixed numbers to improper fractions. **3.** Use rule for division of fractions.	Divide. $8\frac{1}{3} \div 5\frac{5}{9} = \frac{25}{3} \div \frac{50}{9}$ $= \frac{\overset{1}{\cancel{25}}}{\underset{1}{\cancel{3}}} \times \frac{\overset{3}{\cancel{9}}}{\underset{2}{\cancel{50}}} = \frac{3}{2}$ *or* $1\frac{1}{2}$

Chapter Organizer

Topic	Procedure	Examples
Finding the least common denominator, p. 132.	**1.** Write each denominator as the product of prime factors. **2.** List all the prime factors that appear in both products. **3.** Form a product of those factors, using each factor the greatest number of times it appears in any denomination.	Find LCD of $\frac{1}{10}$, $\frac{3}{8}$, and $\frac{7}{25}$. $10 = 2 \times 5$ $8 = 2 \times 2 \times 2$ $25 = 5 \times 5$ $\text{LCD} = 2 \times 2 \times 2 \times 5 \times 5$ $= 200$
Building fractions, p. 134.	**1.** Find how many times the original denominator can be divided into the new denominator. **2.** Multiply that value by numerator and denominator of original fraction.	Build $\frac{5}{7}$ to an equivalent fraction with a denominator of 42. First we find $7\overline{)42}$ = 6. Then we multiply by 6 for numerator and denominator. $\frac{5}{7} \times \frac{6}{6} = \frac{30}{42}$
Adding or subtracting fractions with a common denominator, p. 139.	**1.** Add or subtract the numerators. **2.** Keep the common denominator.	Add. $\frac{3}{13} + \frac{5}{13} = \frac{8}{13}$ Subtract. $\frac{15}{17} - \frac{12}{17} = \frac{3}{17}$
Adding or subtracting fractions without a common denominator, p. 140.	**1.** Find the LCD of the fractions. **2.** Build up each fraction, if needed, to obtain the LCD in the denominator. **3.** Follow the steps for adding and subtracting fractions with the same denominator.	Add. $\frac{1}{4} + \frac{3}{7} + \frac{5}{8}$ LCD = 56 $\quad \frac{1 \times 14}{4 \times 14} + \frac{3 \times 8}{7 \times 8} + \frac{5 \times 7}{8 \times 7}$ $= \frac{14}{56} + \frac{24}{56} + \frac{35}{56} = \frac{73}{56} = 1\frac{17}{56}$
Adding mixed numbers, p. 146.	**1.** Change fractional parts to equivalent fractions with LCD as a denominator, if needed. **2.** Add whole numbers and fractions separately. **3.** If improper fractions occur, change to mixed numbers and simplify.	Add. $6\frac{3}{4} + 2\frac{5}{8}$ $6\boxed{\frac{3}{4} \times \frac{2}{2}} = 6\frac{6}{8}$ $+2\frac{5}{8} \quad = 2\frac{5}{8}$ $8\frac{11}{8} = 9\frac{3}{8}$
Subtracting mixed numbers, p. 147.	**1.** Change fractional parts to equivalent fractions with LCD as a denominator, if needed. **2.** If necessary, borrow from whole number to subtract fractions. **3.** Subtract whole numbers and fractions separately.	Subtract. $8\frac{1}{5} - 4\frac{2}{3}$ $8\boxed{\frac{1}{5} \times \frac{3}{3}} = 8\frac{3}{15}$ $-4\boxed{\frac{2}{3} \times \frac{5}{5}} = -4\frac{10}{15}$ $= 7\frac{18}{15}$ $-4\frac{10}{15}$ $3\frac{8}{15}$

Using the Mathematics Blueprint for Problem Solving, p. 154.

In solving an applied problem with fractions, students may find it helpful to complete the following steps. You will not use the steps all of the time. Choose the steps that best fit the conditions of the problem.

1. *Understand the problem.*
   (a) Read the problem carefully.
   (b) Draw a picture if this helps you to visualize the situation. Think about what facts you are given and what you are asked to find.
   (c) It may help to write a similar, simpler problem to get started and to determine what operation to use.
   (d) Use the Mathematics Blueprint for Problem Solving to organize your work. Follow these four parts.
      1. Gather the facts. (Write down specific values given in the problem.)
      2. What am I asked to do? (Identify what you must obtain for an answer.)
      3. How do I proceed? (What calculations need to be done.)
      4. Key points to remember. (Record any facts, warnings, formulas, or concepts you think will be important as you solve the problem.)
2. *Solve and state the answer.*
   (a) Perform the necessary calculations.
   (b) State the answer, including the unit of measure.
3. *Check.*
   (a) Estimate the answer to the problem. Compare this estimate to the calculated value. Is your answer reasonable?
   (b) Repeat your calculations.
   (c) Work backward from your answer. Do you arrive at the original conditions of the problem?

EXAMPLE

A wire is $95\frac{1}{3}$ feet long. It is cut up into smaller, equal-sized pieces, each $4\frac{1}{3}$ feet long. How many pieces will there be?

1. *Understand the problem.*
   Let us draw a picture of the situation.
   How will we find the number of pieces?

Now we will use a simpler problem to clarify the idea. A wire 100 feet long is cut up into smaller pieces each 4 feet long. How many pieces will there be? We readily see that we would divide 100 by 4. Thus in our original problem we should divide $95\frac{1}{3}$ feet by $4\frac{1}{3}$ feet. This will tell us the number of pieces. Now we fill in the Mathematics Blueprint.

MATHEMATICS BLUEPRINT FOR PROBLEM SOLVING

Gather the Facts	What Am I Asked to Do?	How Do I Proceed?	Key Points to Remember
Wire is $95\frac{1}{3}$ feet. It is cut into equal pieces $4\frac{1}{3}$ feet long.	Determine how many pieces of wire there will be.	Divide $95\frac{1}{3}$ by $4\frac{1}{3}$.	Change mixed numbers to improper fractions before carrying out the division.

2. *Solve and state the answer.*
   We need to divide $95\frac{1}{3} \div 4\frac{1}{3}$

$$\frac{286}{3} \div \frac{13}{3} = \frac{\overset{22}{\cancel{286}}}{\underset{1}{\cancel{3}}} \times \frac{\overset{1}{\cancel{3}}}{\underset{1}{\cancel{13}}} = \frac{22}{1} = 22$$

   There will be 22 pieces of wire.

3. *Check.*
   Estimate. Round $95\frac{1}{3}$ to nearest ten = 100.
   Round $4\frac{1}{3}$ to nearest integer = 4.
   $100 \div 4 = 25$
   This is close to our estimate. Our answer is reasonable. ✓

Chapter 2 Review Problems

2.1 *In each problem, use a fraction to represent the shaded part of the object.*

1. 

2. 

Draw a sketch to illustrate the fraction.

3. $\frac{4}{7}$ of an object

4. $\frac{7}{9}$ of an object

5. An inspector looked at 31 parts and found 6 of them to be defective. What fractional part of these items was defective?

6. The dean asked 100 of the freshman if they would be staying in the dorm over the holidays. A total of 87 said they would not. What fractional part of the freshmen said they would not?

2.2 *Express each number as a product of prime factors.*

7. 42

8. 54

9. 168

Identify which of the following numbers is prime. If it is composite, express it as the product of prime factors.

10. 59

11. 78

12. 167

Reduce the fraction.

13. $\frac{13}{52}$

14. $\frac{12}{42}$

15. $\frac{21}{36}$

16. $\frac{26}{34}$

17. $\frac{168}{192}$

18. $\frac{51}{105}$

2.3 *Change each mixed number to an improper fraction.*

19. $4\frac{3}{8}$

20. $2\frac{19}{23}$

Change each improper fraction to a mixed number.

21. $\frac{19}{7}$

22. $\frac{63}{13}$

23. Reduce and leave your answer as a mixed number.
$3\frac{15}{55}$

24. Reduce and leave your answer as an improper fraction.
$\frac{234}{16}$

25. Change to a mixed number and then reduce.
$\frac{385}{240}$

2.4 *Multiply each fraction.*

26. $\frac{4}{7} \times \frac{5}{11}$

27. $\frac{7}{9} \times \frac{21}{35}$

28. $12 \times \frac{3}{7} \times 0$

29. $\frac{3}{5} \times \frac{2}{7} \times \frac{10}{27}$

30. $12 \times 8\frac{1}{5}$

31. $5\frac{3}{8} \times 3\frac{4}{5}$

32. $5\frac{1}{4} \times 4\frac{6}{7}$

33. $35 \times \frac{7}{10}$

34. One share of stock costs $\$37\frac{5}{8}$. How much money will 18 shares cost?

35. Find the area of a rectangle that is $18\frac{1}{5}$ inches wide and $26\frac{3}{4}$ inches long.

2.5 *Divide, if possible.*

36. $\frac{3}{7} \div \frac{2}{5}$

37. $\frac{9}{17} \div \frac{18}{5}$

38. $1200 \div \frac{5}{8}$

39. $\dfrac{2\frac{1}{4}}{3\frac{1}{3}}$

40. $\dfrac{20}{4\frac{4}{5}}$

41. $2\frac{1}{8} \div 20\frac{1}{2}$

42. $0 \div 3\frac{7}{5}$

43. $4\frac{2}{11} \div 3$

44. There are 420 calories in $2\frac{1}{4}$ cans of grape soda. How many calories are contained in 1 can of soda?

45. A dress requires $3\frac{1}{8}$ yards of fabric. Amy has $21\frac{7}{8}$ yards available. How many dresses can she make?

2.6 *Find the LCD for each set of fractions.*

46. $\frac{7}{14}$ and $\frac{3}{49}$

47. $\frac{7}{40}$ and $\frac{11}{30}$

48. $\frac{5}{18}, \frac{1}{6}, \frac{7}{45}$

Build each fraction to an equivalent fraction with the specified denominator.

49. $\frac{3}{7} = \frac{?}{56}$

50. $\frac{11}{24} = \frac{?}{72}$

51. $\frac{9}{43} = \frac{?}{172}$

52. $\frac{17}{18} = \frac{?}{198}$

2.7 *Add or subtract.*

53. $\frac{3}{7} - \frac{5}{14}$

54. $\frac{1}{2} + \frac{1}{3} + \frac{1}{4}$

55. $\frac{4}{7} + \frac{7}{9}$

56. $\frac{7}{8} - \frac{3}{5}$

57. $\frac{7}{30} + \frac{2}{21}$

58. $\frac{5}{18} + \frac{5}{12}$

59. $\frac{14}{15} - \frac{3}{25}$

60. $\frac{15}{16} - \frac{13}{24}$

2.8 *Add or subtract.*

61. $1 - \frac{17}{23}$

62. $6 - \frac{5}{9}$

63. $3 + 5\frac{2}{3}$

64. $8 + 12\frac{5}{7}$

65. $3\frac{1}{4} + 1\frac{5}{8}$

66. $7\frac{3}{16} - 2\frac{5}{6}$

67. $120 - 16\frac{2}{3}$

68. $22\frac{2}{3} + 48\frac{3}{4}$

69. Bob jogged $1\frac{7}{8}$ miles on Monday, $2\frac{3}{4}$ miles on Tuesday, and $4\frac{1}{10}$ miles on Wednesday. How many miles did he jog on these three days?

70. On a recent day, Coca-Cola opened at $\$36\frac{5}{8}$ a share and closed at $\$38\frac{1}{4}$. How much did a share of Coca-Cola gain that day?

71. A recipe calls for $3\frac{1}{3}$ cups of sugar and $4\frac{1}{4}$ cups of flour. How much sugar and how much flour would be needed for $\frac{1}{2}$ of that recipe?

72. Rafael traveled in a car that gets $24\frac{1}{4}$ miles per gallon. He had $8\frac{1}{2}$ gallons of gas in the gas tank. Approximately how far could he drive?

73. How many lengths (pieces) of pipe $3\frac{1}{5}$ inches long can be cut from a pipe 48 inches long?

74. A car radiator holds $15\frac{3}{4}$ liters. If it contains $6\frac{1}{8}$ liters of antifreeze and the rest is water, how much is water?

75. Delbert types 366 words in $12\frac{1}{5}$ minutes. How many words per minute did he type?

76. Alicia earns $\$4\frac{1}{2}$ dollars per hour for regular pay and $1\frac{1}{2}$ times that rate of pay for overtime. On Monday she worked 8 hours at regular pay and 3 hours at overtime. How much did she earn on Monday?

77. George purchased stock at $\$88\frac{3}{8}$ a share. He sold it in four months at $\$79\frac{5}{8}$ a share. How much did the value of the stock decrease during that time period?

78. A 3-inch bolt passes through $1\frac{1}{2}$ inches of pine board, a $\frac{1}{16}$-inch washer, and a $\frac{1}{8}$-inch nut. How many inches extend beyond the board, washer, and nut if the head of the bolt is $\frac{1}{4}$ inch long.

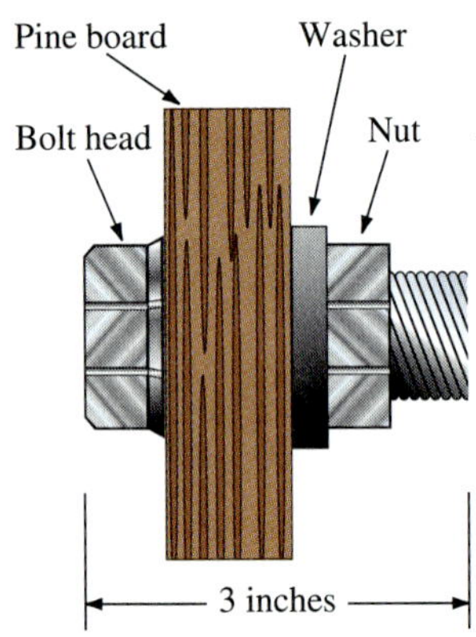

79. Francine has a take-home pay of \$880 per month. She gives $\frac{1}{10}$ of it to her church, spends $\frac{1}{2}$ of it for rent and food, and spends $\frac{1}{8}$ of it on electricity, heat, and telephone. How many dollars per month does she have left for other things?

80. Manuel's new car used $18\frac{2}{5}$ gallons of gas on a 460-mile trip.

(a) How many miles can his car travel on 1 gallon of gas?

(b) How much did his trip cost him in gasoline expense if the average cost of gasoline was $\$1\frac{1}{5}$ per gallon?

Chapter 2 Test

1. Use a fraction to represent the shaded part of the object.

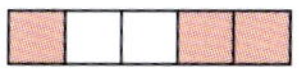

2. A basketball star shot at the hoop 388 times. The ball went in 311 times. Write a fraction that describes the part of the time that his shot went in.

Reduce each fraction.

3. $\dfrac{15}{70}$

4. $\dfrac{18}{42}$

5. $\dfrac{225}{50}$

6. Change to an improper fraction.
$6\dfrac{4}{5}$

7. Change to a mixed number.
$\dfrac{114}{14}$

Multiply.

8. $42 \times \dfrac{2}{7}$

9. $\dfrac{7}{9} \times \dfrac{2}{5}$

10. $4\dfrac{1}{3} \times 7\dfrac{1}{5}$

Divide.

11. $\dfrac{7}{8} \div \dfrac{5}{11}$

12. $5\dfrac{1}{7} \div 3$

13. $7\dfrac{1}{5} \div 1\dfrac{1}{25}$

14. $\dfrac{12}{31} \div \dfrac{8}{13}$

Find the least common denominator of each set of fractions.

15. $\dfrac{5}{24}$ and $\dfrac{7}{18}$

16. $\dfrac{3}{16}$ and $\dfrac{1}{24}$

17. $\dfrac{1}{4}, \dfrac{3}{8}, \dfrac{5}{6}$

18. Build the fraction to an equivalent fraction with the specified denominator. $\dfrac{5}{12} = \dfrac{?}{72}$

1. ____________________

2. ____________________

3. ____________________

4. ____________________

5. ____________________

6. ____________________

7. ____________________

8. ____________________

9. ____________________

10. ____________________

11. ____________________

12. ____________________

13. ____________________

14. ____________________

15. ____________________

16. ____________________

17. ____________________

18. ____________________

19. ____

20. ____

21. ____

22. ____

23. ____

24. ____

25. ____

26. ____

27. ____

28. ____

29. ____

30. (a) ____

(b) ____

Add or subtract.

19. $\frac{7}{9} - \frac{5}{12}$

20. $\frac{2}{15} + \frac{5}{12}$

21. $\frac{1}{4} + \frac{3}{7} + \frac{3}{14}$

22. $8\frac{3}{5} + 5\frac{4}{7}$

23. $18\frac{6}{7} - 13\frac{13}{14}$

24. $6 - 3\frac{5}{9}$

25. $30 - 8\frac{3}{4}$

Answer each question.

26. A hallway measures $8\frac{1}{6}$ yards by $5\frac{1}{7}$ yards. How many square yards is the area of the hallway?

27. A butcher has $18\frac{2}{3}$ pounds of steak that he wishes to place into packages that average $2\frac{1}{3}$ pounds each. How many packages can he make?

28. From central parking it is $\frac{9}{10}$ of a mile to the science building. Bob started at central parking and walked $\frac{1}{5}$ of a mile toward the science building. He stopped for coffee. When he finishes, how far does he have to walk?

29. Robin jogged $4\frac{1}{8}$ miles on Monday, $3\frac{1}{6}$ miles on Tuesday, and $6\frac{3}{4}$ miles on Wednesday. How far did she jog on those three days?

30. Mr. and Mrs. Samuel visited Florida and purchased 120 oranges. They gave $\frac{1}{4}$ to relatives, ate $\frac{1}{12}$ of them in the hotel, and gave $\frac{1}{3}$ of them to friends. They shipped the rest home to Illinois.

(a) How many oranges did they ship?

(b) If it costs 24¢ for each orange to be shipped to Illinois, what was the total of the shipping bill?

Cumulative Test for Chapters 1–2

One-half of this test is based on Chapter 1 material. The remainder is based on material covered in Chapter 2.

1. Write in words. 84,361,208

2. Add.
$$\begin{array}{r} 128 \\ 452 \\ 178 \\ 34 \\ +\ 77 \\ \hline \end{array}$$

3. Add.
$$\begin{array}{r} 156{,}200 \\ 364{,}700 \\ +\ 198{,}320 \\ \hline \end{array}$$

4. Subtract.
$$\begin{array}{r} 5718 \\ -\ 3643 \\ \hline \end{array}$$

5. Subtract.
$$\begin{array}{r} 1{,}000{,}361 \\ -\ \ 983{,}145 \\ \hline \end{array}$$

6. Multiply.
$$\begin{array}{r} 126 \\ \times\ 38 \\ \hline \end{array}$$

7. Multiply.
$$\begin{array}{r} 16{,}908 \\ \times\ \ 12 \\ \hline \end{array}$$

8. Divide. $7\overline{)30{,}150}$

9. Divide. $18\overline{)6642}$

10. Evaluate. 7^2

11. Round to the nearest thousand. 6,037,452

12. Perform the operations in their proper order.

$$4 \times 3^2 + 12 \div 6$$

13. Roone bought three shirts at \$26 and two pairs of pants at \$48. What was his total bill?

14. Leslie had a balance of \$64 in her checking account. She deposited \$1160. She made checks out for \$516, \$199, and \$203. What will be her new balance?

15. Eighty-four students enrolled in psychology. Fifty-five were women. Write a fraction that describes the part of the class that was made up of women.

1. ______
2. ______
3. ______
4. ______
5. ______
6. ______
7. ______
8. ______
9. ______
10. ______
11. ______
12. ______
13. ______
14. ______
15. ______

16. ______

17. ______

18. ______

19. ______

20. ______

21. ______

22. ______

23. ______

24. ______

25. ______

26. ______

27. ______

28. ______

29. ______

16. Reduce. $\frac{28}{52}$

17. Write as an improper fraction. $18\frac{3}{4}$

18. Write as a mixed number. $\frac{100}{7}$

19. Multiply. $3\frac{7}{8} \times 2\frac{5}{6}$

20. Divide. $\frac{44}{49} \div 2\frac{13}{21}$

21. Find the least common denominator of $\frac{6}{13}$ and $\frac{5}{39}$.

Add or subtract.

22. $\frac{7}{18} + \frac{5}{27}$

23. $2\frac{1}{8} + 6\frac{3}{4}$

24. $12\frac{1}{5} - 4\frac{2}{3}$

25. $\frac{11}{14} - \frac{9}{28}$

26. A truck hauled $9\frac{1}{2}$ tons of gravel on Monday. On Tuesday, it hauled $6\frac{3}{8}$ tons, and on Wednesday, $7\frac{1}{4}$ tons. How many tons were hauled on the three days?

27. Melinda traveled $221\frac{2}{5}$ miles on 9 gallons of gas. How many miles per gallon did her car get?

28. A recipe requires $3\frac{1}{4}$ cups of sugar and $2\frac{1}{3}$ cups of flour. Marcia wants to make $2\frac{1}{2}$ times that recipe. How many cups of each will she need?

29. A space probe travels at 28,356 miles per hour for 2142 hours. *Estimate* how many miles it travels.

CHAPTER 3
Decimals

There is a major problem that affects almost half of the college students of the United States during their first year of school. This problem costs them extra money and sometimes has serious financial consequences. You probably cannot guess what it is! Believe it or not—almost half of the first-year college students in the United States cannot balance their checkbook!

If you think you are above problems like that, please turn to the Putting Your Skills To Work on page 198.

Pretest Chapter 3

If you are familiar with the topics in this chapter, take this test now. Check your answers with those in the back of the book. If an answer was wrong or you couldn't do a problem, study the appropriate section of the chapter.

If you are not familiar with the topics in this chapter, don't take this test now. Instead, study the examples, work the practice problems, and then take the test.

This test will help you identify which concepts you have mastered and which you need to study further.

Section 3.1

1. Write a word name for the decimal. 47.813

2. Express this fraction as a decimal. $\frac{567}{10,000}$

Write as a fraction or a mixed number. Reduce whenever possible.

3. 2.11 **4.** 0.525

Section 3.2

5. Place the set of numbers in the proper order from smallest to largest. 1.6, 1.59, 1.61, 1.601

6. Round to the nearest tenth. 10.472935

7. Round to the nearest thousandth. 1.053458

Section 3.3

Add.

8.
$$\begin{array}{r} 5.12 \\ 4.7 \\ 8.03 \\ +\ 1.6 \\ \hline \end{array}$$

9. $24.613 + 0.273 + 2.305$

Subtract.

10.
$$\begin{array}{r} 42.16 \\ -\ 31.57 \\ \hline \end{array}$$

11. $13 - 4.108$

1. ______
2. ______
3. ______
4. ______
5. ______
6. ______
7. ______
8. ______
9. ______
10. ______
11. ______

Section 3.4

Multiply.

12. $\begin{array}{r} 11.67 \\ \times\ 0.03 \\ \hline \end{array}$ **13.** 4.7805×1000

14. 0.0005129×10^4

Section 3.5

Divide.

15. $0.09\overline{)0.03186}$

16. $0.3328 \div 2.6$

Section 3.6

17. Write as a decimal. $\dfrac{7}{16}$

18. Write as repeating decimal. $\dfrac{5}{22}$

19. Perform the operations in the correct order. $(0.2)^2 + 8.7 \times 0.3 - 1.68$

Section 3.7

20. When Marcia began her trip, the odometer on her car read 57,124.8. When she ended the trip, it read 57,312.8. She used 10.5 gallons of gas. How many miles per gallon did her car get? Round your answer to the nearest tenth.

21. A rectangular room is 10.5 yards long and 3.6 yards wide. How much will it cost to install carpeting for the entire room at $12.95 per square yard?

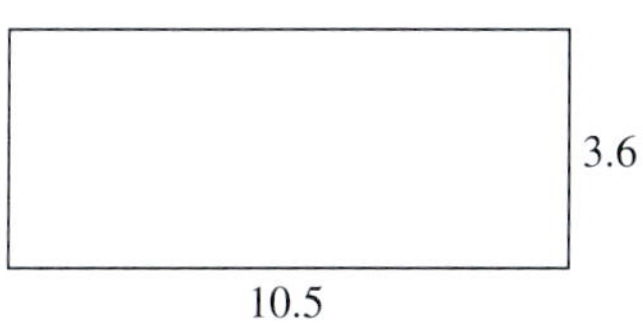

22. Tony worked 35 hours last week. He was paid $298.55. How much was he paid per hour?

12. ______

13. ______

14. ______

15. ______

16. ______

17. ______

18. ______

19. ______

20. ______

21. ______

22. ______

3.1 Decimal Notation

MathPro Video 6 SSM

After studying this section, you will be able to:

1 *Write a word name for a decimal fraction.*
2 *Change from fractional notation to decimal notation.*
3 *Change from decimal notation to fractional notation.*

1 *Writing a Word Name for a Decimal Fraction*

In Chapter 2 we discussed *fractions*—the set of numbers such as $\frac{1}{2}$, $\frac{2}{3}$, $\frac{1}{10}$, $\frac{6}{7}$, $\frac{18}{100}$, and so on. Now we will take a closer look at **decimal fractions**—that is, fractions with 10, 100, 1000, and so on, in the denominator, such as $\frac{1}{10}$, $\frac{18}{100}$, and $\frac{43}{1000}$.

Why, of all fractions, do we take special notice of these? Our hands have *ten* digits. Our U.S. money system is based on the dollar, which has 100 equal parts, or cents. And the international system of measurement called the *metric system* is based on 10 and powers of 10.

As with other numbers, these decimal fractions can be written in different ways (forms). For example, the shaded part of the whole in the drawing can be written

In words (one-tenth)
In fraction form ($\frac{1}{10}$)
In decimal form (0.1)

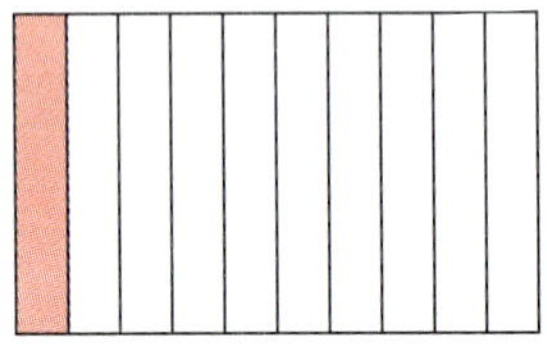

All mean the same quantity, namely 1 out of 10 equal parts of the whole. We'll see that when we use decimal notation, computations can be easily done based on the old rules for whole numbers and a few new rules about where to place the decimal point. In a world where calculators and computers are commonplace, many of the fractions we encounter are decimal fractions. A decimal fraction is a fraction whose denominator is a power of 10.

$\frac{7}{10}$ is a decimal fraction. $\frac{89}{10^2} = \frac{89}{100}$ is a decimal fraction.

Decimal fractions can be written with numerals in two ways: fractional form or decimal form.

FRACTIONAL FORM		DECIMAL FORM
$\frac{3}{10}$	=	0.3
$\frac{59}{100}$	=	0.59
$\frac{171}{1000}$	=	0.171

The zero in front of the decimal point is not actually required. We place it there simply to make sure that we don't miss seeing the decimal point. When a number is written in decimal form, the first digit to the right of the decimal point represents tenths, the next digit hundredths, the next digit thousandths, and so on. 0.9 means nine tenths and is equivalent to $\frac{9}{10}$. 0.51 means fifty-one hundredths and is equivalent to $\frac{51}{100}$. Some decimals are larger than 1. For example, 1.683 means one and six hundred eighty-three thousandths. It is equivalent to $1\frac{683}{1000}$. Note that the word *and* is used to indicate the decimal point. A place-value chart is helpful.

Decimal Place Values

Hundreds	Tens	Ones	Decimal Point	Tenths	Hundredths	Thousandths	Ten Thousandths
1	5	6	.	2	8	7	4
100	10	1	"and"	$\frac{1}{10}$	$\frac{1}{100}$	$\frac{1}{1,000}$	$\frac{1}{10,000}$

So, for 156.2874, we can write in words one hundred fifty-six and two thousand eight hundred seventy-four ten thousandths. We say ten thousandths because it is the name of the last decimal place on the right.

EXAMPLE 1 Write a word name for each decimal.

(a) 0.79 **(b)** 0.5308 **(c)** 1.6 **(d)** 23.765

(a) 0.79 = Seventy-nine hundredths

(b) 0.5308 = Five thousand three hundred eight ten thousandths

(c) 1.6 = One and six tenths

(d) 23.765 = Twenty-three and seven hundred sixty-five thousandths

Practice Problem 1 Write a word name for each decimal.

(a) 0.073 **(b)** 4.68 **(c)** 0.0017 **(d)** 561.78 ■

Sometimes, decimals are used where we would not expect them. For example, we commonly say that there are 365 days in a year with 366 days in every fourth year (or leap year). However, this is not quite correct. In fact from time to time further adjustments need to be made to the calendar to adjust for these inconsistencies. Astronomers know that a more accurate measure of a year is called a **tropical year** (measured from one equinox to the next). Rounded to the nearest hundred thousandth, 1 tropical year = 365.24122 days. This is read three hundred sixty-five and twenty-four thousand, one hundred twenty-two hundred thousandths. This approximate value is a more accurate measurement of the amount of time it takes the earth to complete one orbit around the sun.

Decimal notation is commonly used with money. When writing a check, we often write the amount that is less than 1 dollar, such as 23¢, as $\frac{23}{100}$ dollar.

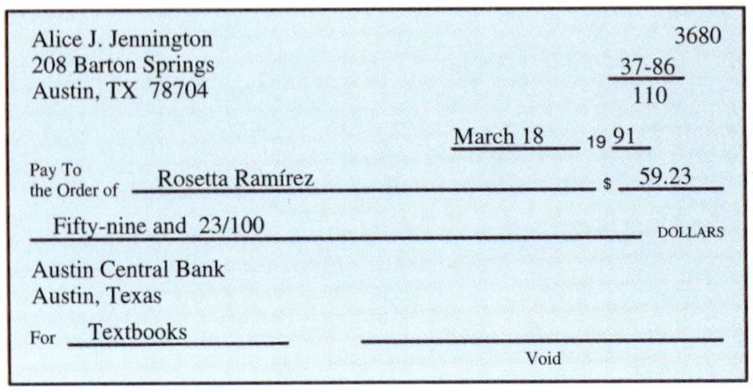

EXAMPLE 2 Write a word name for the amount on a check made out for \$672.89.

$$\text{Six hundred seventy-two and } \frac{89}{100} \text{ dollars}$$

Practice Problem 2 Write a word name for the amount of a check made out for \$7863.04. ■

2 Changing from Fractional Notation to Decimal Notation

It is helpful to be able to write decimals in both decimal notation and fractional notation. First we illustrate changing a fraction with a denominator of 10, 100, or 1000 into decimal form.

EXAMPLE 3 Write each fraction as a decimal.

(a) $\frac{8}{10}$ **(b)** $\frac{74}{100}$ **(c)** $1\frac{3}{10}$ **(d)** $2\frac{56}{1000}$

(a) $\frac{8}{10} = 0.8$ **(b)** $\frac{74}{100} = 0.74$

(c) $1\frac{3}{10} = 1.3$ **(d)** $2\frac{56}{1000} = 2.056$

Practice Problem 3 Write each fraction as a decimal.

(a) $\frac{9}{10}$ **(b)** $\frac{136}{1000}$ **(c)** $2\frac{56}{100}$ **(d)** $34\frac{86}{1000}$ ■

3 Changing from Decimal Notation to Fractional Notation

EXAMPLE 4 Write in fractional notation.

(a) 0.51 **(b)** 18.1 **(c)** 0.7611 **(d)** 1.363

(a) $0.51 = \frac{51}{100}$ **(b)** $18.1 = 18\frac{1}{10}$

(c) $0.7611 = \frac{7611}{10{,}000}$ **(d)** $1.363 = 1\frac{363}{1000}$

Practice Problem 4 Write in fractional notation.

(a) 0.37 **(b)** 182.3 **(c)** 0.7131 **(d)** 42.019 ■

When we convert from decimal form to fractional form, we reduce whenever possible.

EXAMPLE 5 Write in fractional form. Reduce whenever possible.

(a) 2.6 **(b)** 0.38 **(c)** 0.525 **(d)** 361.007

(a) $2.6 = 2\frac{6}{10} = 2\frac{3}{5}$ **(b)** $0.38 = \frac{38}{100} = \frac{19}{50}$

(c) $0.525 = \frac{525}{1000} = \frac{105}{200} = \frac{21}{40}$

(d) $361.007 = 361\frac{7}{1000}$ (cannot be reduced)

Practice Problem 5 Write in fraction form. Reduce whenever possible.

(a) 8.5 **(b)** 0.58 **(c)** 36.25 **(d)** 106.013 ■

EXAMPLE 6 A chemist found that the concentration of lead in a water sample was 5 parts per million. What fraction would represent the concentration of lead?

Five parts per million means 5 parts out of 1,000,000. As a fraction, this is $\frac{5}{1{,}000{,}000}$. We can reduce this by dividing numerator and denominator by five. Thus

$$\frac{5}{1{,}000{,}000} = \frac{1}{200{,}000}$$

The concentration of lead in the water sample is $\frac{1}{200{,}000}$.

Practice Problem 6 A chemist found that the concentration of PCBs in a water sample was 2 parts per billion. What fraction would represent the concentration of PCBs? ■

Developing Your Study Skills

STEPS TOWARD SUCCESS IN MATHEMATICS

Mathematics is a building process, mastered one step at a time. The foundation of this process is built on a few basic requirements. Those who are successful in mathematics realize the absolute necessity for building a study of mathematics on the firm foundation of these six minimum requirements.

1. Attend class every day.
2. Read the textbook.
3. Take notes in class.
4. Do assigned homework every day.
5. Get help immediately when needed.
6. Review regularly.

3.1 Exercises

Verbal and Writing Skills

1. Describe a decimal fraction and provide examples.

2. What word is used to describe the decimal point when writing the word name for a decimal that is greater than one?

3. What is the name of the last decimal place on the right for the decimal 132.45678?

4. When writing \$82.75 on a check, we write 75¢ as ______.

Write a word name for each decimal.

5. 0.57
6. 0.78
7. 3.8
8. 5.2
9. 0.124
10. 0.367
11. 28.0007
12. 54.0003

Write a word name as you would on a check.

13. \$36.18
14. \$25.99
15. \$1236.08
16. \$7652.02
17. \$10,000.76
18. \$20,000.67

Write each of the following in decimal notation.

19. seven tenths
20. twelve hundredths
21. thirty-nine thousandths
22. seventy-five thousandths
23. sixty-five ten thousandths
24. twenty-nine ten thousandths
25. two hundred eighty-six millionths
26. seven hundred sixteen millionths

Write each fraction as a decimal.

27. $\frac{3}{10}$ 28. $\frac{7}{10}$ 29. $\frac{76}{100}$ 30. $\frac{84}{100}$ 31. $\frac{771}{1000}$ 32. $\frac{652}{1000}$

33. $\frac{53}{1000}$ 34. $\frac{42}{1000}$ 35. $\frac{26}{10{,}000}$ 36. $\frac{36}{10{,}000}$ 37. $8\frac{3}{10}$ 38. $3\frac{1}{10}$

39. $84\frac{13}{100}$ 40. $52\frac{77}{100}$ 41. $1\frac{19}{1000}$ 42. $2\frac{23}{1000}$ 43. $126\frac{571}{10{,}000}$ 44. $198\frac{333}{10{,}000}$

Write in fractional notation. Reduce whenever possible.

45. 0.02 **46.** 0.05 **47.** 3.6 **48.** 7.4

49. 0.121 **50.** 0.143 **51.** 12.625 **52.** 29.875

53. 7.0015 **54.** 4.0016 **55.** 235.1254 **56.** 581.2406

57. 0.0187 **58.** 0.0209 **59.** 8.0108 **60.** 7.0605

61. 289.376 **62.** 423.814 **63.** 0.9889 **64.** 0.3211

Applications

65. Every year turtles lay eggs on the islands of South Carolina. Unfortunately, due to illegal polluting, a lot of the eggs are contaminated. If the turtle eggs contain more than 2 parts per one hundred million of chemical pollutants they will not hatch and the population will continue to head toward extinction. Write the above as a fraction in the lowest terms.

66. American Bald Eagles have been fighting extinction due to environmental hazards such as DDT, PCBs, and dioxin. The problem is with the food chain. Fish or rodents eat or drink contaminated food and/or water. Then, the eagles ingest the poison, which in turn affects the durability of the eagles' eggs. If it takes only 4 parts per million of certain chemicals to ruin an eagle egg, write this number as a fraction in lowest terms. In 1994 the Bald Eagle was removed from the endangered species list.

Cumulative Review Problems

67. Add.
$$\begin{array}{r} 156 \\ 84 \\ 39 \\ 463 \\ +\ 76 \\ \hline \end{array}$$

68. Subtract.
$$\begin{array}{r} 12{,}843 \\ -\ 11{,}905 \\ \hline \end{array}$$

69. Round to the nearest *hundred.* 56,758

70. Round to the nearest *thousand.* 8,069,482

3.2 Compare, Order, and Round Decimals

MathPro Video 7 SSM

After studying this section, you will be able to:

1 *Compare decimals.*
2 *Place decimals in order from smallest to largest.*
3 *Round decimals to a specified decimal place.*

1 *Comparing Decimals*

Number Line

All of the numbers we have studied have a specific order. To illustrate this order, we can place the numbers on a number line. Look at the number line on the left. Each number has a specific place on it. The arrow points in the direction of increasing value. Thus, if one number is to the right of a second number, it is larger, or greater, than that number. Since 5 is to the right of 2 on the number line, we say that 5 is greater than 2. We write $5 > 2$.

Since 4 is to the left of 6 on the number line, we say that 4 is less than 6. We write $4 < 6$.

The symbols "$>$" and "$<$" are called **inequality symbols.**

$a < b$ is read "a is less than b."

$a > b$ is read "a is greater than b."

We can assign exactly one point on the number line to each decimal number. When two decimal numbers are placed on a number line, the one farther to the right is the larger. Thus we can say that $3.4 > 2.7$ and $4.3 > 4.0$. We can also say that $0.5 < 1.0$ and $1.8 < 2.2$. Why?

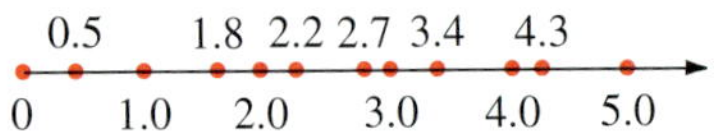

To compare or order decimals without using a number line, we compare each digit.

> **Comparing Two Numbers in Decimal Notation**
>
> **1.** Start at the left and compare corresponding digits. If the digits are the same, move one place to the right.
>
> **2.** When two digits are different, the larger number is the one with the larger digit.

EXAMPLE 1 Write an inequality statement with 0.167 and 0.166.

The number 1 in the tenths place is the same.

0.1 6 7   0.1 6 6

The number 6 in the hundredths place is the same.

0.1 6 7   0.1 6 6

The numbers in the thousandths place differ.

0. 1 6 7   0.1. 6 6

Since $7 > 6$, we know that $0.167 > 0.166$.

Practice Problem 1 Write an inequality statement with 5.74 and 5.75. ■

Whenever necessary, extra zeros can be written to the right of the last digit—that is, to the right of the decimal point—without changing the value of the decimal. Thus

$$0.56 = 0.56000 \qquad \text{and} \qquad 0.7768 = 0.77680$$

The zero to the left of the decimal point is optional. Thus $0.56 = .56$. Both notations are used. You are encouraged to place a zero to the left of the decimal point so that you don't miss the decimal point when you work with decimals.

EXAMPLE 2 Fill in the blank with one of the symbols $<$, $=$, or $>$.

$$0.77 \underline{\qquad} 0.777$$

If we add a zero to the first decimal

$$0.77\underline{0} \qquad\qquad 0.77\underline{7}$$

we see that the tenths and hundredths digits are equal. But the thousandths digits differ. Since $0 < 7$, we have $0.770 < 0.777$.

Practice Problem 2 Fill in the blank with one of the symbols $<$, $=$, or $>$. ■

$$0.894 \underline{\qquad} 0.89$$

2 *Placing Decimals in Order*

You can place three or more decimals in order. If you are asked to order the decimals from smallest to largest, look for the smallest decimal and place it first.

EXAMPLE 3 Place the following five decimal numbers in order from smallest to largest.

1.834, 1.83, 1.381, 1.38, 1.8

First we add zeros to make the comparison easier.

1.834, 1.830, 1.381, 1.380, 1.800

Now we rearrange with smallest first.

1.380, 1.381, 1.800, 1.830, 1.834

Practice Problem 3 Place the following five decimal numbers in order from smallest to largest.

2.45, 2.543, 2.46, 2.54, 2.5 ■

3 *Rounding Decimals*

Sometimes in calculations involving money we see numbers like \$386.432 and \$29.5986. To make these useful, we usually round them to the nearest cent. \$386.432 is rounded to \$386.43. \$29.5986 is rounded to \$29.60. A general rule for rounding decimals follows.

> **Rounding Decimals**
>
> **1.** Find the decimal place (units, tenths, hundredths, and so on) for which rounding off is required.
>
> **2.** If the first digit to the right of this place is less than 5, drop it and all digits to the right of it.
>
> **3.** If the first digit to the right of this place is 5 or greater, increase the digit by one. Drop all digits to the right of this place.

EXAMPLE 4 Round 156.37 to the nearest tenth.

156.3 7

↑ We find the tenths place.

Note that 7, the next place to the right, is greater than 5. We round up to 156.4 and drop the digits to the right. The answer is 156.4.

Practice Problem 4 Round 723.88 to the nearest tenth. ■

EXAMPLE 5 Round to the nearest thousandth.

(a) 0.06358 **(b)** 128.37448

(a) 0.06358

↑ We locate the thousandths place.

Note that the digit to the right of the thousandths place is 5. We round up to 0.064 and drop all the digits to the right.

(b) 128.37448

↑ We locate the thousandths place.

Note that the digit to the right of thousandths place is less than 5. We round to 128.374 and drop all the digits to the right.

Practice Problem 5 Round to the nearest thousandth.

(a) 12.92647 **(b)** 0.007892 ■

Remember that rounding up to the next digit in a position may result in several digits being changed.

EXAMPLE 6 Round to the nearest hundredth. Fred and Linda used 203.9964 kilowatt hours of electricity in their house in May.

We locate the hundredths place

203.9**9**64

Since the digit to the right of hundredths is greater than 5, we round up. This affects the next two positions. Do you see why? The result is 204.00 kilowatt hours.

Practice Problem 6 Round to the nearest tenth. Last month the college gymnasium used 15,699.953 kilowatt hours of electricity. ■

Sometimes we round a decimal to the nearest whole number. For example, when writing figures on income tax forms, a taxpayer may round all figures to the nearest dollar.

EXAMPLE 7 To complete her income tax return, Marge needs to round these figures to the nearest whole dollar.

Medical bills	$779.86	Taxes	$563.49
Retirement contributions	$674.38	Contributions to charity	$534.77

Round the amounts.

	ORIGINAL FIGURE	ROUNDED TO NEAREST DOLLAR
Medical bills	779.86	780
Taxes	563.49	563
Retirement	674.38	674
Charity	534.77	535

Practice Problem 7 Round to the nearest whole dollar the following figures.

Medical bills	$375.50	Taxes	$981.39
Retirement contributions	$980.49	Contributions to charity	$817.65 ■

WARNING Why is it so important to consider only *one* digit to the right of the desired round-off position? What is wrong with rounding in steps? Suppose that Mark rounds 1.349 to the nearest tenth in steps. First he rounds 1.349 to 1.35 (nearest hundredth). Then he rounds 1.35 to 1.4 (nearest tenth). What is wrong with this reasoning?

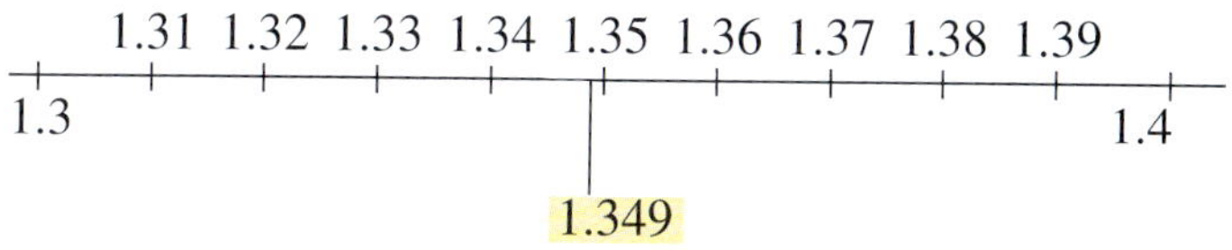

To round 1.349 to the nearest tenth, we ask if 1.349 is closer to 1.3 or to 1.4. It is closer to 1.3. Mark got 1.4, so he is not correct. He "rounded in steps" by first moving to 1.35, thus increasing the error and moving in the wrong direction. To control rounding errors, we consider *only* the first digit to the right of the decimal place to which we are rounding.

3.2 Exercises

Fill in the blank with one of the symbols <, =, or >.

1. 1.3 ____ 1.29 **2.** 2.6 ____ 2.58 **3.** 0.68 ____ 0.681 **4.** 72.54 ____ 72.56

5. 18.92 ____ 18.93 **6.** 0.460 ____ 0.46 **7.** 0.0006 ____ 0.0005 **8.** 0.0037 ____ 0.003

9. 1.0024 ____ 1.003 **10.** 1.003 ____ 1.004 **11.** 126.34 ____ 125.35 **12.** 406.78 ____ 407.75

13. 16.0572 ____ 16.0574 **14.** 18.00039 ____ 18.00038 **15.** $\frac{8}{10}$ ____ 0.08 **16.** $\frac{5}{100}$ ____ 0.005

Arrange each set of decimals from smallest to largest.

17. 12.6, 12.8, 12.65 **18.** 18.32, 18.038, 18.04 **19.** 0.0071, 0.05, 0.007

20. 0.0025, 0.0052, 0.002 **21.** 1.8, 1.1, 1.81, 1.79 **22.** 2.7, 2.5, 2.53, 2.48

23. 26.034, 26.003, 26.04, 26.033 **24.** 33.082, 33.02, 33.088, 33.079

25. 18.006, 18.060, 18.066, 18.606, 18.065 **26.** 15.020, 15.002, 15.001, 15.018, 15.0019

Round each decimal to the nearest tenth.

27. 5.67 **28.** 8.35 **29.** 29.49 **30.** 38.48 **31.** 578.064 **32.** 311.091 **33.** 2176.83 **34.** 4082.74

Round each decimal to the nearest hundredth.

35. 26.032 **36.** 47.071 **37.** 5.76582 **38.** 2.98613

39. 156.1197 **40.** 283.2168 **41.** 2786.706 **42.** 4609.285

Round to the nearest given place.

43. 1.06132 thousandths **44.** 8.10263 thousandths

45. 0.047357 ten thousandths **46.** 0.063148 ten thousandths

47. 5.00761238 hundred thousandths **48.** 4.01062378 hundred thousandths

49. 0.00753682 millionths **50.** 0.00964983 millionths

51. 129.08939 nearest whole number **52.** 208.4372 nearest whole number

Round each amount to the nearest dollar.

53. \$2536.85

54. \$5319.62

55. \$10,098.47

56. \$20,159.48

Round each amount to the nearest cent.

57. \$56.9832

58. \$28.7619

59. \$5783.716

60. \$3928.649

Applications

61. If an average is taken of all the men and women on the planet, the average man consumes 0.095 kilogram of protein per day, while the average woman consumes 0.066 kilogram of protein per day. Round these values to the nearest hundredth.

62. In the 1960s, the average person in the United States ate 10.62 pounds of chocolate per year. In the 1990s, the average person eats 21.78 pounds of chocolate per year. Round these values to the nearest tenth.

To Think About

63. Arrange in order from smallest to largest.

$$0.61,\ 0.062,\ \frac{6}{10},\ 0.006,\ 0.0059,\ \frac{6}{100},\ 0.0601,\ 0.0519,\ 0.0612$$

64. Arrange in order from smallest to largest.

$$1.05,\ 1.512,\ \frac{15}{10},\ 1.0513,\ 0.049,\ \frac{151}{100},\ 0.0515,\ 0.052,\ 1.051$$

65. A person wants to round 86.23498 to the nearest hundredth. He first rounds 86.23498 to 86.2350. He then rounds to 86.235. Finally, he rounds to 86.24. What is wrong with his reasoning?

66. Fred is checking the calculations on his monthly bank statement. An interest charge of \$16.3724 was rounded to \$16.38. An interest charge of \$43.7214 was rounded to \$43.73. What rule does the bank use for rounding off to the nearest cent?

Cumulative Review Problems

67. Add. $3\frac{1}{4} + 2\frac{1}{2} + 6\frac{3}{8}$

68. Subtract. $27\frac{1}{5} - 16\frac{3}{4}$

69. Mary drove her Dodge Caravan on a trip. At the start of the trip the odometer (which measures distance) read 46,381. At the end of the trip it read 47,073. How many miles long was the trip?

70. The cost of four pieces of equipment for the college computer center was \$1736, \$2714, \$892, and \$4316. What was their total cost?

3.3 Addition and Subtraction of Decimals

MathPro Video 7 SSM

After studying this section, you will be able to:

1 *Add two or more decimals.*
2 *Subtract two decimals.*

The Apple Crate Restaurant

2 apple pie specials	$4.95
2 cups coffee	$1.85
	$6.80
tax	.34
total	$7.14

Please pay cashier

1 *Adding Decimals*

We often add decimals when we check the addition of our bill at a restaurant or at a store. We can relate addition of decimals to addition of fractions. For example,

$$\frac{3}{10} + \frac{6}{10} = \frac{9}{10} \qquad \text{and} \qquad 1\frac{1}{10} + 2\frac{8}{10} = 3\frac{9}{10}$$

These same problems can be written more efficiently as decimals.

$$\begin{array}{r} 0.3 \\ +\ 0.6 \\ \hline 0.9 \end{array} \qquad \begin{array}{r} 1.1 \\ +\ 2.8 \\ \hline 3.9 \end{array}$$

The steps to follow when adding decimals are listed below.

Adding Decimals

1. Write the numbers to be added vertically and line up the decimal points. Extra zeros may be placed to right of the decimal points if needed.
2. Add all the digits with the same place value, starting with the right column and moving to the left.
3. Place the decimal point of the sum in line with the decimal points of the numbers added.

EXAMPLE 1 Add.

(a) 2.8 + 5.6 + 3.2 **(b)** 158.26 + 200.07 + 315.98

(c) 5.3 + 26.182 + 0.0007 + 6.24

(a)
$$\begin{array}{r} {}^{1} \\ 2.8 \\ 5.6 \\ +\ 3.2 \\ \hline 11.6 \end{array}$$

(b)
$$\begin{array}{r} {}^{1\,1}{}^{2} \\ 158.26 \\ 200.07 \\ +\ 315.98 \\ \hline 674.31 \end{array}$$

(c)
$$\begin{array}{r} {}^{1}{}^{1} \\ 5.3000 \\ 26.1820 \\ 0.0007 \\ +\ \ 6.2400 \\ \hline 37.7227 \end{array}$$

Extra zeros have been added to make the problem easier.

Practice Problem 1 Add.

(a)
$$\begin{array}{r} 9.8 \\ 3.6 \\ +\ 5.4 \\ \hline \end{array}$$

(b)
$$\begin{array}{r} 300.72 \\ 163.75 \\ +\ 291.08 \\ \hline \end{array}$$

(c) 8.9 + 37.056 + 0.0023 + 9.45 ■

SIDELIGHT When we add decimals like 3.1 + 2.16 + 4.007, we may write in zeros as below.

$$\begin{array}{r} 3.100 \\ 2.160 \\ +\ 4.007 \\ \hline 9.267 \end{array}$$

What are we really doing here? What is the advantage of adding these extra zeros?

"Decimals" means "decimal fractions." If we look at the number as fractions, we see that we are actually using the property of multiplying a fraction by 1 in order to obtain common denominators. Look at the problem this way:

$$\left.\begin{array}{l} 3.1 = 3\frac{1}{10} \\ 2.16 = 2\frac{16}{100} \\ 4.007 = 4\frac{7}{1000} \end{array}\right\}$$

The least common denominator is 1000. *To obtain the common denominator for the first two fractions, we multiply.*

$$\left.\begin{array}{lll} 3 & \frac{1}{10} \times \frac{100}{100} & = 3\frac{100}{1000} \\ 2 & \frac{16}{100} \times \frac{10}{10} & = 2\frac{160}{1000} \\ +\ 4 & \frac{7}{1000} & = 4\frac{7}{1000} \end{array}\right\}$$

$$9\frac{267}{1000} = 9.267$$

Once we obtain a common denominator, we can add the three fractions.

This is the answer we arrived at above using the decimal form for each number. Thus writing in zeros in a decimal fraction is really an easy way to transform fractions to equivalent fractions with a common denominator. Working with decimal fractions is easier than working with other fractions.

CALCULATOR

Adding Decimals

The calculator can be used to verify your work. You can use your calculator to add decimals. To find 23.08 + 8.53 + 9.31 enter:

23.08 [+] 8.53 [+] 9.31 [=]

Display: [40.92]

The final digit of an odometer measures tenths of a mile. The odometer reading shown in the odometer on the right is 38,516.2 miles.

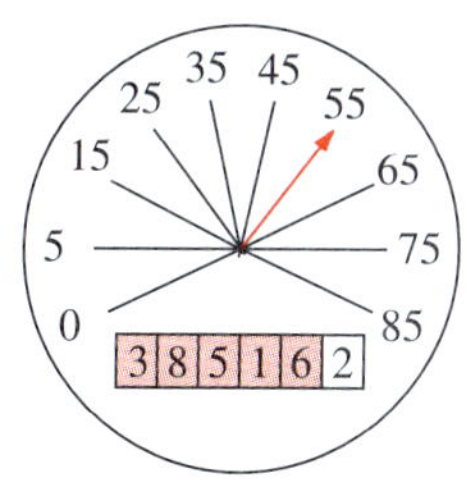

EXAMPLE 2 Barbara checked her odometer before the summer began. It read 49,645.8 miles. She traveled 3852.6 miles that summer in her car. What was the odometer reading at the end of the summer?

$$\begin{array}{r} \overset{1\,1}{4}\overset{}{9},6\overset{1}{4}5.8 \\ +\ \ 3,852.6 \\ \hline 53,498.4 \end{array}$$

The odometer read 53,498.4 miles.

Practice Problem 2 A car odometer read 93,521.8 miles before a trip of 1634.8 miles. What was the final odometer reading? ■

EXAMPLE 3 During his first semester at Tarrant County Community College, Kelvey deposited checks into his checking account in the amounts of $98.64, $157.32, $204.81, $36.07, and $229.89. What was the sum of his five checks?

$$\begin{array}{r} {\scriptstyle 2\,3\,2\;\;2} \\ \$\;\;98.64 \\ 157.32 \\ 204.81 \\ 36.07 \\ +\;\;229.89 \\ \hline \$726.73 \end{array}$$

Practice Problem 3 Beforing buying textbooks for the fall semester, Will deposited the following checks into his account: $80.95, $133.91, $256.47, $53.08, and $381.32. What was the sum of his five checks? ■

2 *Subtracting Decimals*

It is important to see the relationship between the decimal form of a mixed number and the fractional form of a mixed number. This relationship helps us to understand why calculations with decimals are done the way they are. When we subtract mixed numbers with common denominators, sometimes we must borrow from the whole number.

$$\begin{array}{rcr} 5\frac{4}{10} & = & 4\frac{14}{10} \\ -\;2\frac{7}{10} & = & -\;2\frac{7}{10} \\ \hline & & 2\frac{7}{10} \end{array}$$

We could write the same problem in decimal form:

$$\begin{array}{r} {\scriptstyle 4\;14} \\ \not{5}.\not{4} \\ -\;2.7 \\ \hline 2.7 \end{array}$$

Subtraction of decimals is thus similar to subtraction of fractions (we get the same result), but it's usually easier to subtract with decimals than to subtract with fractions.

Subtracting Decimals

1. Write the decimals to be subtracted vertically and line up the decimal points. Additional zeros may be placed to the right if not all numbers have the same number of decimal places.
2. Subtract all digits with the same place value, starting with the right column and moving to the left. Borrow when necessary.
3. Place the decimal point of the difference in line with the decimal point of the two numbers being subtracted.

EXAMPLE 4 Subtract.

(a)
$$\begin{array}{r} 84.8 \\ -\ 27.3 \\ \hline \end{array}$$

(b)
$$\begin{array}{r} 1076.320 \\ -\ \ 983.518 \\ \hline \end{array}$$

(a)
$$\begin{array}{r} \overset{7}{\cancel{8}}\,\overset{14}{\cancel{4}}\,.8 \\ -\ 2\ 7\ .3 \\ \hline 5\ 7\ .5 \end{array}$$

(b)
$$\begin{array}{r} \cancel{1}\,\overset{9}{\overset{\cancel{10}}{\cancel{0}}}\,\overset{17}{\cancel{7}}\,\overset{5}{\cancel{6}}.\,\overset{13}{\cancel{3}}\,\overset{1}{\cancel{2}}\,\overset{10}{\cancel{0}} \\ -\ \ 9\ 8\ 3.\ 5\ 1\ 8 \\ \hline 9\ 2.\ 8\ 0\ 2 \end{array}$$

Practice Problem 4 Subtract.

(a)
$$\begin{array}{r} 38.8 \\ -\ 26.9 \\ \hline \end{array}$$

(b)
$$\begin{array}{r} 2034.908 \\ -\ 1986.325 \\ \hline \end{array}$$ ■

When the two numbers being subtracted do not have the same number of decimal places, write in zeros as needed.

EXAMPLE 5 Subtract. **(a)** 12 − 8.362 **(b)** 156.381 − 99.82

(a)
$$\begin{array}{r} \cancel{1}\,\overset{11}{\cancel{2}}\,.\overset{9}{\overset{\cancel{10}}{\cancel{0}}}\,\overset{9}{\overset{\cancel{10}}{\cancel{0}}}\,\overset{10}{\cancel{0}} \\ -\ \ 8\ .3\ 6\ 2 \\ \hline 3\ .6\ 3\ 8 \end{array}$$

(b)
$$\begin{array}{r} \cancel{1}\,\overset{14}{\overset{\cancel{4}}{\cancel{5}}}\,\overset{15}{\overset{\cancel{5}}{\cancel{6}}}\,.\overset{\cancel{13}}{\cancel{3}}\,8\ 1 \\ -\ \ 9\ 9\ .8\ 2\ 0 \\ \hline 5\ 6\ .5\ 6\ 1 \end{array}$$

Practice Problem 5 Subtract. **(a)** 19 − 12.579 **(b)** 283.076 − 96.38 ■

EXAMPLE 6 On Tuesday Don Ling filled the gas tank in his car. The odometer read 56,098.5. He drove for four days. The next time he filled the tank, the odometer read 56,420.2. How many miles had he driven?

$$\begin{array}{r} \scriptstyle 3\ 11\ 9\ 12 \\ 56{,}420.2 \\ -\ 56{,}098.5 \\ \hline 321.7 \end{array}$$

He had driven 321.7 miles.

Practice Problem 6 Abdul had his car oil changed when his car odometer read 82,370.9 miles. When he changed oil again, the odometer read 87,160.1 miles. How many miles did he drive between oil changes? ■

EXAMPLE 7 Find the value of x if $x + 3.9 = 14.6$.

Recall that the letter x is a variable. It represents a number that is added to 3.9 to obtain 14.6. We can find the number x if we subtract $14.6 - 3.9$.

$$\begin{array}{r} \scriptstyle 3\ 16 \\ 14.6 \\ -\ 3.9 \\ \hline 10.7 \end{array}$$

Thus $x = 10.7$.

Check

Is this true? If we replace x by 10.7, do we get a true statement?

$$x + 3.9 = 14.6$$
$$10.7 + 3.9 \stackrel{?}{=} 14.6$$
$$14.6 = 14.6 \ \checkmark$$

Practice Problem 7 Find the value of x if $x + 10.8 = 15.3$. ■

Developing Your Study Skills

MAKING A FRIEND IN THE CLASS

Attempt to make a friend in your class. You may find that you enjoy sitting together and drawing support and encouragement from each other. Exchange phone numbers so you can call each other whenever you get stuck in your study. Set up convenient times to study together on a regular basis, to do homework, and to review for exams.

You must not depend on a friend or fellow student to tutor you, do your work for you, or in any way be responsible for your learning. However, you will learn from each other as you seek to master the course. Studying with a friend and comparing notes, methods, and solutions can be very helpful. And it can make learning mathematics a lot more fun!

3.3 Exercises

Add.

1. $$\begin{array}{r} 44.6 \\ +\ 28.2 \\ \hline \end{array}$$

2. $$\begin{array}{r} 18.6 \\ +\ 23.2 \\ \hline \end{array}$$

3. $$\begin{array}{r} 718.98 \\ +\ 496.57 \\ \hline \end{array}$$

4. $$\begin{array}{r} 813.47 \\ +\ 629.86 \\ \hline \end{array}$$

5. $$\begin{array}{r} 2.107 \\ +\ 4.918 \\ \hline \end{array}$$

6. $$\begin{array}{r} 5.306 \\ +\ 8.822 \\ \hline \end{array}$$

7. $$\begin{array}{r} 79.061 \\ +\ 57.783 \\ \hline \end{array}$$

8. $$\begin{array}{r} 26.905 \\ +\ 87.453 \\ \hline \end{array}$$

9. $$\begin{array}{r} 6.5 \\ 12.6 \\ +\ 304.8 \\ \hline \end{array}$$

10. $$\begin{array}{r} 18.2 \\ 7.6 \\ +\ 199.8 \\ \hline \end{array}$$

11. $$\begin{array}{r} 5.6 \\ 9.23 \\ +\ 8.17 \\ \hline \end{array}$$

12. $$\begin{array}{r} 2.65 \\ 3.2 \\ +\ 7.76 \\ \hline \end{array}$$

13. $$\begin{array}{r} 4.9637 \\ 28.12 \\ +\ \ 3.645 \\ \hline \end{array}$$

14. $$\begin{array}{r} 7.0276 \\ 3.451 \\ +\ 16.98 \\ \hline \end{array}$$

15. $$\begin{array}{r} 12. \\ 3.62 \\ +\ 51.8 \\ \hline \end{array}$$

16. $$\begin{array}{r} 13. \\ 4.52 \\ +\ 63.7 \\ \hline \end{array}$$

17. $753.61 + 28.75 + 162.3 + 100.5 + 67.05$

18. $432.51 + 16.08 + 892.1 + 301.2 + 84.07$

Applications

Calculate the perimeter of each triangle.

19.

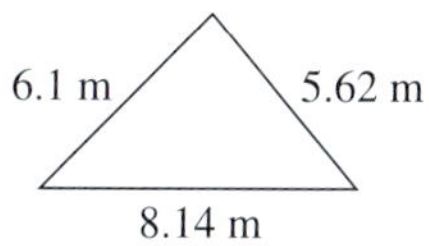

20.

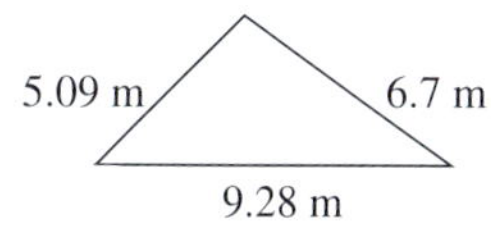

21. Fred and Barney are driving to their high school reunion in New Jersey from their home in Boulder, Colorado. They make purchases of 7.8, 12.1, 9.7, 10.4, 8.8, and 11.3 gallons of fuel. How many gallons of fuel did they buy?

22. Greg, Peter, and Bobby drove from Los Angeles up the Pacific Coast Highway to visit their sister Marcia, who goes to college in Santa Cruz. On the way they purchased 7.7, 8.9, 9.5, and 11.8 gallons of gas. How many total gallons did they purchase?

23. Mick and Keith have arrived in Miami and are going to the beach. They buy sun-block for \$4.99, beverages for \$12.50, sandwiches for \$11.85, towels for \$28.50, bottled water for \$3.29, and two novels for \$16.99. After they got what they needed, what was Mick and Keith's bill for their day at the beach?

24. Brent purchased items at a grocery store just off-campus. They cost \$7.18, \$5.29, \$0.61, \$1.15, and \$1.49. What was the total of Brent's grocery bill?

A portion of a bank checking account deposit slip is shown below. Add the numbers to determine the total deposit in each case. The line drawn between the dollars and the cents column serves as the decimal point.

25.

FOOTHILLS BANK
AUSTIN, TEXAS

DEPOSIT SLIP

	Dollars	Cents
Cash	18	42
Checks	706	15
	21	03
	45	00
	621	37
Total		

26.

SUNSHINE BANK
PHOENIX, ARIZONA

DEPOSIT SLIP

	Dollars	Cents
Cash	52	89
Checks	105	37
	76	04
	25	00
	167	82
Total		

Subtract.

27. $\begin{array}{r} 6.8 \\ -\ 2.9 \\ \hline \end{array}$	**28.** $\begin{array}{r} 3.6 \\ -\ 2.8 \\ \hline \end{array}$	**29.** $\begin{array}{r} 123.51 \\ -\ 96.34 \\ \hline \end{array}$	**30.** $\begin{array}{r} 161.78 \\ -\ 89.29 \\ \hline \end{array}$	**31.** $\begin{array}{r} 76.8 \\ -\ 12.62 \\ \hline \end{array}$	**32.** $\begin{array}{r} 82.5 \\ -\ 43.93 \\ \hline \end{array}$
33. $\begin{array}{r} 586.513 \\ -\ 78.2 \\ \hline \end{array}$	**34.** $\begin{array}{r} 243.967 \\ -\ 84.2 \\ \hline \end{array}$	**35.** $\begin{array}{r} 1.00782 \\ -\ 0.98631 \\ \hline \end{array}$	**36.** $\begin{array}{r} 7.00278 \\ -\ 6.34125 \\ \hline \end{array}$	**37.** $\begin{array}{r} 24.0079 \\ -\ 19.3614 \\ \hline \end{array}$	**38.** $\begin{array}{r} 52.0708 \\ -\ 41.9312 \\ \hline \end{array}$
39. $\begin{array}{r} 8. \\ -\ 1.263 \\ \hline \end{array}$	**40.** $\begin{array}{r} 12. \\ -\ 7.981 \\ \hline \end{array}$	**41.** $\begin{array}{r} 7362.14 \\ -\ 6173.07 \\ \hline \end{array}$	**42.** $\begin{array}{r} 4986.71 \\ -\ 3615.93 \\ \hline \end{array}$	**43.** $\begin{array}{r} 1.5 \\ -\ 0.0365 \\ \hline \end{array}$	**44.** $\begin{array}{r} 2.8 \\ -\ 0.07763 \\ \hline \end{array}$

Applications

Solve each problem.

45. A toilet seat should not cost more than \$29.95. Someone purchased one, without looking at the bill, for \$63.45. How much did that person overpay?

46. Jack and Rita are building a garage. The builder is charging them \$11,597.00. If they pay cash in advance, the builder will reduce the cost by \$1593.25. How much will the garage cost if they pay in advance?

47. During the three months of August, September, and October, baby Kathryn gained 7.675 kilograms. During the three months of November, December, and January, she gained another 9.986 kilograms. How much weight has the baby gained in the last six months?

48. The average length of an Emerald Tree boa constrictor snake is 1.8 meters. A native found one that measured 3.264 meters long. How much longer was this snake than the average Emerald Tree boa constrictor?

49. A child's beginner telescope is priced at \$79.49. The price of a certain professional telescope is \$37,026.65. How much more does the professional telescope cost?

50. Tamika drove on a summer trip. When she began, the odometer read 26,052.3 miles. At the end of the trip, the odometer read 28,715.1 miles. How long was the trip?

51. Malcolm took a taxi from John F. Kennedy Airport in New York to his hotel in the city. His fare was \$47.70 and he tipped the driver \$7.00. How much change should Malcolm get back if he gives the driver a \$100 bill?

52. Michael bought \$76.49 worth of supplies at the hardware store. He gave the clerk \$80.00. How much change should he receive?

53. The outside radius of a pipe is 9.39 centimeters. The inside radius is 7.93 centimeters. What is the thickness of the pipe?

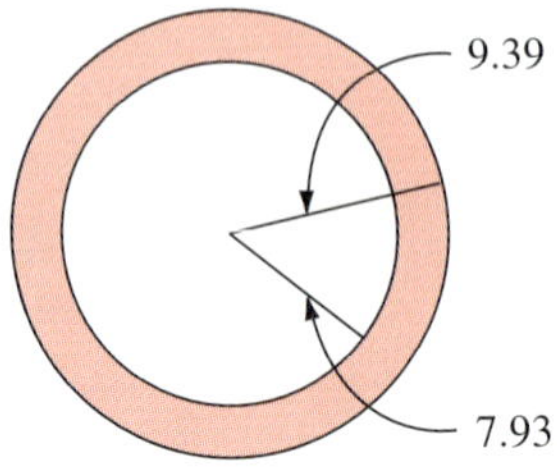

54. An insulated wire measures 12.62 centimeters. The last 0.98 centimeter of the wire is exposed. How long is the part of the wire that is not exposed?

55. The state budget called for $12,563,784.56 in revenue. Unfortunately, the state's actual revenue for the year was only $11,962,375.49. What was the revenue shortage for the state?

★ 56. Everyone is becoming aware of the rapid loss of the Earth's rainforests. Between 1981 and 1990, tropical South America lost a substantial amount of natural resources due to deforestation and development. In 1981, there were 797,100,000 hectares of rainforest. In 1990 there were 729,300,000 hectares. If 1 hectare is approximately equal to 2.47 acres, how much rainforest, in acres, was destroyed?

The federal water safety standard requires that drinking water contain no more than 0.015 milligrams of lead per liter of water.

57. Fred and Donna use water provided by the city for the drinking water in their home. A sample of their tap water contained 0.023 milligrams of lead per liter of water. What is the difference between their sample and the federal safety standard? Is it safe for them to drink the water?

58. Carlos and Maria had the well that supplies their home analyzed for safety. A sample of well water contained 0.0089 milligrams of lead per liter of water. What is the difference between their sample and the federal safety standard? Is it safe for them to drink the water?

The table below shows the income of the United States by industry. Use this table for problems 59 to 62. Write each answer as a decimal and as a whole number. The table values are recorded in billions of dollars.

TYPE OF INDUSTRY	INCOME IN 1980	INCOME IN 1990
Agriculture, forestry, fisheries	61.4	97.1
Mining	43.8	38.1
Construction	126.6	234.4
Communication	48.1	96.4

59. How many more dollars were earned in mining in 1980 than in 1990?

60. How many more dollars were earned in construction in 1990 than in 1980?

61. In 1980, how many more dollars were earned in agriculture, forestry, and fisheries than in communication?

62. In 1990, how many more dollars were earned in communication than in mining?

Find the value of x.

63. $x + 7.1 = 15.5$

64. $x + 4.8 = 23.1$

65. $156.9 + x = 200.6$

66. $210.3 + x = 301.2$

67. $4.162 = x + 2.053$

68. $7.076 = x + 5.602$

Cumulative Review Problems

Multiply.

69. $\begin{array}{r} 2536 \\ \times \quad 8 \\ \hline \end{array}$

70. $\begin{array}{r} 467 \\ \times \ 39 \\ \hline \end{array}$

71. $\dfrac{22}{7} \times \dfrac{49}{50}$

72. $2\dfrac{1}{3} \times 3\dfrac{3}{4}$

Putting Your Skills to Work

PERSONAL CHECKING ACCOUNT

If you have a personal checking account, you will keep a record of your banking on a check register. When the bank sends you a copy of their records (showing the amount in your account), you should compare the bank's total to your register and be sure the two are the same. This is called *balancing* the checkbook.

Regina's checking account balance was \$23.69. She made deposits of \$116.84 on March 3 and \$89.46 on March 15. Then she wrote check 156 to Joe's Market for \$16.80 on March 8 and check 157 to Marshall Field for \$129.95 on March 18. She recorded her transactions as follows:

Check Number	Date	Transaction	Amount of Withdrawal (−)		Amount of Deposit (+)		New Balance	
							23	69
	3/3	Deposit			116	84	140	53
156	3/8	Joe's Market	16	80			123	73
	3/15	Deposit			89	46	213	19
157	3/18	Marshall Field	129	95			83	24

PROBLEMS FOR INDIVIDUAL INVESTIGATION

1. What was Regina's balance on March 18?
2. Regina's balance on March 18 was \$83.24. Use the check register below to record the following transactions Regina made. Determine the new balance after each transaction.
   - **(a)** A monthly service fee of \$3.50 is charged to her account on March 20.
   - **(b)** A deposit of \$301.64 is made on March 22.
   - **(c)** Check 158 is made out to Sears for \$159.95 on March 30.
   - **(d)** A deposit of \$450.36 is made on March 31.
   - **(e)** Check 159 is made out to Gulf Oil for \$46.52 on April 2.
   - **(f)** Check 160 is made out to Western Bell for \$39.64 on April 3.
   - **(g)** Check 161 is made out to IRS for \$561.18 on April 14.
   - **(h)** A bank charge for \$6.95 is billed to Regina for printing new checks on April 18.

	Check Number	Date	Transaction	Amount of Withdrawal (−)		Amount of Deposit (+)		New Balance	
								83	24
(a)									
(b)									
(c)									
(d)									
(e)									
(f)									
(g)									
(h)									

3. On June 30, Regina's balance was \$382.54. On the next Monday, she made an ATM withdrawal for \$120.00 using her bank card. On the next Tuesday, she deposited \$98.96. On the next Wednesday, she used her bank card at an ATM to withdraw \$70.00. Her bank charges her \$0.65 for each ATM withdrawal. What was her final balance?

QUESTIONS FOR COOPERATIVE GROUP INVESTIGATION

Together with some other members of your class, investigate the following.

4. Regina usually uses about 15 checks per month and makes about 5 withdrawals per month from the ATM machine near her apartment. The bank offers two checking plans. The Economy Account costs \$2.00 per month, plus \$0.35 for every check she writes and \$0.65 for each ATM withdrawal. The Advanced Service Account costs \$10.00 per month, but there is no charge for ATM withdrawals and no service charge for checks unless she uses more than 30 checks per month. In that case, the cost is \$0.20 for each additional check. Which type of checking account would be better for Regina? Why?

5. During the month of August, Regina found that she could not balance her checking account. Her records showed she had \$56.78 in the account, but the bank statement listed \$53.53 in the account. Give four possible reasons that could account for this discrepancy. What should Regina do to find out which reason is the cause in her case?

INTERNET CONNECTIONS: Go to ''http://www.prenhall.com/tobey'' to be connected

Site: Balancing Your Checkbook (College Board Online) or a related site

This site gives information about balancing a checkbook. It includes a worksheet that you can use to balance a checkbook. Print a copy of this worksheet and use it to answer the question below.

6. Sara just received a bank statement with an ending balance of \$643.94. The statement does not include a recent deposit of \$275.00, nor does it include two checks she made out for \$28.35 and \$64.89. What should be the balance in her checkbook?

3.4 Multiplication of Decimals

MathPro Video 7 SSM

After studying this section, you will be able to:

1 *Multiply a decimal by a decimal or a whole number.*
2 *Multiply a decimal by a power of 10.*

1 *Multiplying a Decimal by a Decimal or a Whole Number*

We learned previously that the product of two fractions is the product of the numerators over the product of the denominators. For example,

$$\frac{3}{10} \times \frac{7}{100} = \frac{21}{1000}$$

In decimal form this product would be written

$$\underbrace{0.3}_{\text{one decimal place}} \times \underbrace{0.07}_{\text{two decimal places}} = \underbrace{0.021}_{\text{three decimal places}}$$

Multiplication of Decimals

1. Multiply the numbers just as you would multiply whole numbers.
2. Find the sum of the decimal places in the two factors.
3. Place the decimal point in the product so that the product has the same number of decimal places as the sum in step 2. You may need to write zeros to the left of the number found in step 1.

Now use these steps to do the multiplication problem above.

EXAMPLE 1 Multiply. 0.07×0.3

0.07	2 decimal places
× 0.3	1 decimal place
0.021	3 decimal places in product (2 + 1 = 3)

Practice Problem 1 Multiply. 0.09×0.6 ■

When performing the calculation, it is usually easier to place the factor with the fewest number of nonzero digits underneath the other factor.

EXAMPLE 2 Multiply.

(a) 0.8×2.6 **(b)** 0.3×1.672 **(c)** 0.38×0.26 **(d)** 12.64×0.572

(a)

2.6	1 decimal place
× 0.8	1 decimal place
2.08	2 decimal places (1 + 1 = 2)

(b)

1.672	3 decimal places
× 0.3	1 decimal place
0.5016	4 decimal places (3 + 1 = 4)

(c)

0.38	2 decimal places
× 0.26	2 decimal places
228	
76	
0.0988	4 decimal places (2 + 2 = 4)

Note that we need to insert a zero before the 988.

(d)

12.64	2 decimal places
× 0.572	3 decimal places
2528	
8848	
6 320	
7.23008	5 decimal places (2 + 3 = 5)

> **CALCULATOR**
>
> **Multiplying Decimals**
>
> You can use your calculator to multiply a decimal by a decimal. To find 0.08 × 1.53 enter:
>
> 0.08 [×] 1.53 [=]
>
> Display:
>
> [0.1224]

Practice Problem 2 Multiply.

(a) 0.9 × 5.3 **(b)** 2.831 × 0.7 **(c)** 0.47 × 0.28 **(d)** 0.436 × 18.39 ■

When multiplying decimal fractions by a whole number, you need to remember that a whole number has no decimal places.

EXAMPLE 3 Multiply. 5.261 × 45

5.261	3 decimal places
× 45	0 decimal places
26 305	
210 44	
236.745	3 decimal places (3 + 0 = 3)

Practice Problem 3 Multiply. 0.4264 × 38 ■

EXAMPLE 4 Uncle Roger's rectangular front lawn measures 50.6 yards wide and 71.4 yards long. What is the area of the lawn in square yards?

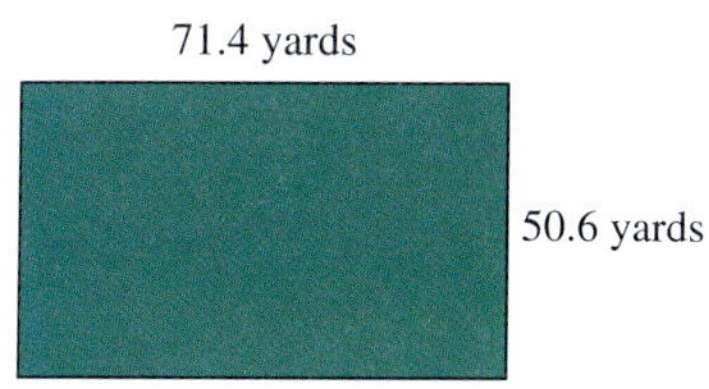

Since the lawn is rectangular in shape, we will use the fact that to find the area of a rectangle we multiply the length by the width.

71.4	(1 decimal place)
× 50.6	(1 decimal place)
42 84	
3570 0	
3612.84	(2 decimal places)

The area of the lawn is 3612.84 square yards.

Practice Problem 4 A rectangular computer chip measures 1.26 millimeters wide and 2.3 millimeters long. What is the area of the chip in square millimeters? ■

2 *Multiplying a Decimal by a Power of 10*

Observe the following pattern.

one zero → Decimal point moved one place to right.

$0.035 \times 10^1 = 0.035 \times 10 = 0.35$

two zeros → Decimal point moved two places to right.

$0.035 \times 10^2 = 0.035 \times 100 = 3.5$

three zeros → Decimal point moved three places to right.

$0.035 \times 10^3 = 0.035 \times 1000 = 35.$

> **Multiplication of a Decimal by a Power of 10**
>
> To multiply a decimal by a power of 10, move the decimal point to the right the same number of places as the number of zeros in the power of 10.

EXAMPLE 5 Multiply. **(a)** 2.671×10 **(b)** 37.85×100

(a) $2.671 \times 10 = 26.71$

one zero — Decimal point moved one place to the right.

(b) $37.85 \times 100 = 3785.$

two zeros — Decimal point moved two places to the right.

Practice Problem 5 Multiply. **(a)** 0.0561×10 **(b)** 1462.37×100 ■

Sometimes it is necessary to add extra zeros before placing the decimal point in the answer.

EXAMPLE 6 Multiply. **(a)** 4.8×1000 **(b)** $0.076 \times 10{,}000$

(a) $4.8 \times 1000 = 4800.$

(three zeros) — Decimal point moved three places to the right. Two extra zeros were needed.

(b) $0.076 \times 10{,}000 = 760.$

(four zeros) — Decimal point moved four places to the right. One extra zero was needed.

Practice Problem 6 Multiply. **(a)** 0.26×1000 **(b)** $5862.89 \times 10{,}000$ ■

If the number that is a power of 10 is in exponent form, move the decimal point to the right the same number of places as the number that is the exponent.

EXAMPLE 7 Multiply. **(a)** 3.68×10^3 **(b)** 0.00027×10^4

exponent of 3 → Decimal point moved three places to right.

(a) $3.68 \times 10^3 = 3680.$

exponent of 4 → Decimal point moved four places to right.

(b) $0.00027 \times 10^4 = 2.7$

Practice Problem 7 Multiply. **(a)** 7.684×10^4 **(b)** 0.00073×10^2 ■

SIDELIGHT Can you devise a quick rule to use when multiplying a decimal fraction by $\frac{1}{10}$, $\frac{1}{100}$, $\frac{1}{1000}$, and so on? How is it like the rules developed in this section? Consider a few examples.

Original Problem	Change Fraction to Decimal	Decimal Multiplication	Observation
$86 \times \frac{1}{10}$	86×0.1	$\begin{array}{r} 86 \\ \times\ 0.1 \\ \hline 8.6 \end{array}$	Decimal point moved one place to left
$86 \times \frac{1}{100}$	86×0.01	$\begin{array}{r} 86 \\ \times\ 0.01 \\ \hline 0.86 \end{array}$	Decimal point moved two places to left
$86 \times \frac{1}{1000}$	86×0.001	$\begin{array}{r} 86 \\ \times\ 0.001 \\ \hline 0.086 \end{array}$	Decimal point moved three places to left

Can you think of a way to describe a rule that you could use in solving this type of problem without going through all the foregoing steps?

You use multiplying by a power of 10 when you change a larger unit to a smaller unit in the metric system of measure.

EXAMPLE 8 Change 2.96 kilometers to meters.

Since we are going from a larger unit to a smaller unit, we multiply. There are 1000 meters in 1 kilometer. Multiply 2.96 by 1000.

$$2.96 \times 1000 = 2960$$

2.96 kilometers is equal to 2960 meters.

Practice Problem 8 Change 156.2 kilometers to meters. ■

3.4 Exercises

Multiply.

1. 0.6×0.2
2. 0.9×0.3
3. 0.12×0.5
4. 0.17×0.4
5. 0.0036×0.8
6. 0.067×0.07
7. 0.079×0.09
8. 0.0034×0.5
9. 0.025×0.081
10. 0.071×0.031
11. 10.97×0.06
12. 18.07×0.05
13. 7986×0.32
14. 5167×0.19
15. 1.892×0.007
16. 2.163×0.008
17. 0.7613×1009
18. 0.6178×5004
19. 9630×0.51
20. 7980×0.46
21. 126×3.5
22. 709×4.6
23. 5030×8.62
24. 8703×5.06
25. 0.07×0.0034
26. 0.026×0.05
27. 0.001×6523.7
28. 0.01×826.75

Applications

29. Melissa earns $5.85 per hour for a 40-hour week. How much does she make in one week?

30. Kenny is making payments on his Ford Escort of $155.40 per month for the next 60 months. How much will he have spent in car payments after he sends in his final payment?

31. Mei Lee works for a company that manufactures electric and electronic equipment and earns $9.55 per hour for a 40-hour week. How much does she earn in one week? (The average wage, for 1987, for U.S. electric/electronic equipment manufacturers was $9.90 per hour for an average of 40.9 hours, for a total of $404.91 per week.)

32. Elva works for a company that manufactures textile products. She earns $7.20 per hour for a 40-hour week. How much does she earn in 1 week? (The average wage, for 1987, for U.S. textile mill industries was $7.18 per hour for an average of 41.9 hours, for a total of $300.84 weekly earnings.)

33. The dimensions of a living room are 20.3 feet and 11.6 feet. What is the area of the living room in square feet?

34. The dimensions of a sheet of paper are 8.2 inches and 10.7 inches. What is the area of the paper in square inches?

35. Dwight is paying off a student loan at Westmont College with payments of $36.90 per month for the next 18 months. How much will he pay off during the next 18 months?

36. Marcia is making car payments to Highfield Center Chevrolet of $230.50 per month for 16 more months. How much will she pay for car payments in the next 16 months?

37. Maria bought 1.4 pounds of bananas at \$0.55 per pound. How much did she pay for the bananas?

38. Angela purchased 2.6 pounds of tomatoes at \$0.65 per pound. How much did she pay for the tomatoes?

39. Steve's car gets approximately 26.4 miles per gallon. His gas tank holds 19.5 gallons. Approximately how many miles can he travel on a full tank of gas?

40. Jim's 4×4 truck gets approximately 18.6 miles per gallon. His gas tank holds 20.5 gallons. Approximately how many miles can he travel on a full tank of gas?

Multiply.

41. 2.86×10

42. 1.98×10

43. 0.701×100

44. 0.236×100

45. 128.65×1000

46. 204.37×1000

47. $5.60982 \times 10{,}000$

48. $1.27986 \times 10{,}000$

49. $280{,}560.2 \times 10^2$

50. 7163.241×10^2

51. 816.32×10^3

52. 763.49×10^4

53. 0.6718×10^3

54. 0.7153×10^4

55. 0.00081376×10^5

56. 0.0007163×10^5

Applications

57. To convert from meters to centimeters, multiply by 100. How many centimeters are in 5.932 meters?

58. A store purchased 100 items at \$19.64 each. How much did the order cost?

59. To convert from kilometers to meters, multiply by 1000. How many meters are in 2.98 kilometers?

60. To convert from kilograms to grams, multiply by 1000. How many grams are in 9.64 kilograms?

61. The city school system pays approximately \$3640.50 per student for schooling. What would the city spend for 10,000 students?

62. An automobile manufacturer makes a profit of \$2984.60 per car. What would be the profit for 10,000 cars?

★ 63. The college is purchasing new carpeting for the learning center. What is the price of a carpet that is 19.6 yards wide and 254.2 yards long if the cost is \$12.50 per square yard?

★ 64. A jewelry store purchased long lengths of gold chain, which will be cut and made into necklaces and bracelets. The store purchased 3220 grams of gold chain at \$3.50 per gram.
   (a) How much did the jewelry store spend?
   (b) If they sell a 28-gram gold necklace for \$17.75 per gram, how much profit will they make on the necklace?

Cumulative Review Problems

Divide. Be sure to include any remainder as part of your answer.

65. $12\overline{)1176}$

66. $14\overline{)1204}$

67. $37\overline{)4629}$

68. $29\overline{)3745}$

3.5 Division of Decimals

MathPro Video 8 SSM

After studying this section, you will be able to:

1 *Divide a decimal by a whole number.*
2 *Divide a decimal by a decimal.*

1 *Dividing a Decimal by a Whole Number*

When you divide a decimal by a whole number, place the decimal point for the quotient directly above the decimal point in the dividend. Then divide as if the numbers were whole numbers.

To divide 26.8 ÷ 4, we place the decimal point of our answer (the quotient) directly *above* the decimal point in the dividend.

$$4\overline{)26.8}$$

The decimal points are directly above each other.

Then we divide as if we were dividing whole numbers.

$$\begin{array}{r} 6.7 \\ 4\overline{)26.8} \\ \underline{24} \\ 2\,8 \\ \underline{2\,8} \\ 0 \end{array}$$

The quotient is 6.7.

The quotient to a problem may have all digits to the right of the decimal point. In some cases you will have to put a zero in the quotient as a "place holder." Thus, if we divide 0.268 ÷ 4,

We have

$$\begin{array}{r} 0.067 \\ 4\overline{)0.268} \\ \underline{24} \\ 28 \\ \underline{28} \\ 0 \end{array}$$

Note that we must have a zero after the decimal point in 0.067.

EXAMPLE 1 Divide. **(a)** $9\overline{)0.3204}$ **(b)** $14\overline{)36.12}$

$$\textbf{(a)}\quad \begin{array}{r} 0.0356 \\ 9\overline{)0.3204} \\ \underline{27} \\ 50 \\ \underline{45} \\ 54 \\ \underline{54} \\ 0 \end{array}$$

Note the zero AFTER the decimal point

$$\textbf{(b)}\quad \begin{array}{r} 2.58 \\ 14\overline{)36.12} \\ \underline{28} \\ 8\,1 \\ \underline{7\,0} \\ 1\,12 \\ \underline{1\,12} \\ 0 \end{array}$$

Practice Problem 1 Divide. **(a)** $7\overline{)1.806}$ **(b)** $16\overline{)0.0928}$ ■

Some division problems do not yield a remainder of zero. In such cases, we may be asked to round off the answer to a specified place. To round off, we carry out the division until our answer contains a digit that is one place to the right of that to which we intend to round. Then we round our answer to the specified place. For example, to round to the nearest thousandth, we carry out the division to the ten-thousandths place. In some division problems, you will need to write in zeros after the dividend so that this division can be carried out.

EXAMPLE 2 Divide and round the quotient to the nearest thousandth.

$$12.67 \div 39$$

We will carry out our division to the ten-thousandths place. Then we round our answer to the nearest thousandth.

```
        0.3248
    39)12.6700 ← Two extra zeros are written here to
       11 7      carry out the division to the required
          97     place.
          78
          190
          156
           340
           312
            28   Note that the remainder is not zero.
```

Now we round 0.3248 to 0.325. The answer is rounded to the nearest thousandth.

Practice Problem 2 Divide and round the quotient to the nearest hundredth. 23.82 ÷ 46 ■

EXAMPLE 3 Maria paid \$5.92 for 16 pounds of tomatoes. How much did she pay per pound?

The cost of one pound of tomatoes equals the total cost, \$5.92, divided by 16 pounds. Thus we will divide.

```
      0.37
   16)5.92
      4 8
      112
      112
```

Maria paid \$0.37 per pound for the tomatoes.

Practice Problem 3 Won Lin will pay off his auto loan for \$3538.75 over 19 months. If the monthly payments are equal, how much will he pay each month? ■

2 Dividing a Decimal by a Decimal

When the divisor is not a whole number, we can convert the division problem to an equivalent problem that has a whole number as a divisor. Think about the reasons why this procedure will work. We will ask you about it after you study Examples 4 and 5.

> **Dividing by a Decimal**
>
> 1. Make the divisor a whole number by moving the decimal point to the right. Mark that position with a caret ($_\wedge$). Count the number of places the decimal point moved.
> 2. Move the decimal point in the dividend to the right the same number of places. Mark that position with a caret.
> 3. Place the decimal point of your answer directly above the caret marking the decimal point of the dividend.
> 4. Divide as with whole numbers.

EXAMPLE 4 **(a)** Divide. $0.08\overline{)1.632}$ **(b)** Divide. $1.352 \div 0.026$

(a) $0.08.\overline{)1.63.2}$ *Move each decimal point two places to the right.*

Place the decimal point of the answer directly above the caret.

$0.08_\wedge\overline{)1.63_\wedge 2}$ *Mark the new position by a caret ($_\wedge$).*

$$\begin{array}{r} 20.4 \\ 0.08_\wedge\overline{)1.63_\wedge 2} \\ \underline{16} \\ 3\ 2 \\ \underline{3\ 2} \\ 0 \end{array}$$

Perform the division.

The answer is 20.4.

(b)
$$\begin{array}{r} 52. \\ 0.026_\wedge\overline{)1.352_\wedge} \\ \underline{1\ 30} \\ 52 \\ \underline{52} \\ 0 \end{array}$$

Move the decimal point three places to the right and mark the new position by a caret ($_\wedge$).

The answer is 52.

Practice Problem 4 Divide. **(a)** $0.09\overline{)0.1008}$ **(b)** $1.702 \div 0.037$ ■

To Think About Why do we move the decimal point so many places to the right in the divisor and the dividend? What rule allows us to do this? How do we know the answer will be valid? We are actually using the property that multiplication of a fraction by 1 leaves the fraction unchanged. This is called the *multiplication identity*. Let us examine Example 4(b) again. We will write $1.352 \div 0.026$ as a fraction.

$\frac{1.352}{0.026} \times 1$ — *Multiplication of a fraction by 1 does not change the value of the fraction.*

$= \frac{1.352}{0.026} \times \frac{1000}{1000}$ — *We know that* $\frac{1000}{1000} = 1$.

$= \frac{1352}{26}$ — *Multiplication by* 1000 *can be done by moving the decimal point three places to the right.*

$= 52$ — *Divide the whole numbers*

Thus in Example 4(b) when we moved the decimal point three places to the right in the divisor and the dividend, we were actually creating an equivalent fraction where numerator and denominator were multiplied by 1000.

EXAMPLE 5 Divide. **(a)** $1.7\overline{)0.0323}$ **(b)** $0.0032\overline{)7.68}$

(a)

```
          0.019
1.7^)0.0^323
         17
        153
        153
          0
```

Move the decimal point in the divisor and dividend one place to the right and mark that position with a caret.

(b)

```
              2400.
0.0032^)7.6800^
         6 4
         1 28
         1 28
           000
```

Note that two extra zeros are needed in the dividend as we move the decimal point four places to the right.

Practice Problem 5 Divide. **(a)** $1.8\overline{)0.0414}$ **(b)** $0.0036\overline{)8.316}$ ■

CALCULATOR

Dividing decimals

You can use your calculator to divide a decimal by a decimal. To find $21.83\overline{)54.53}$ rounded to the nearest hundredth enter:

54.53 [÷] 21.38 [=]

Display:

[2.5505145]

This is an approximation. Some calculators will round to eight digits. The answer rounded to the nearest hundredth is 2.55.

EXAMPLE 6 **(a)** Find $2.9\overline{)431.2}$ rounded to the nearest tenth.

(b) Find $2.17\overline{)0.08}$ rounded to the nearest thousandth.

(a)

```
          148.68
2.9^)431.2^00
     29
     141
     116
      25 2
      23 2
       2 00
       1 74
         260
         232
          28
```

Calculate to the hundredths place and round the answer to the nearest tenth.

The answer rounded to the nearest tenth is 148.7.

(b)

```
         .0368
2.17^)0.08^0000
          6 51
          1 490
          1 302
            1880
            1736
             144
```

Calculate to the ten-thousandths place and then round the answer. *Rounding* 0.0368 *to the nearest thousandth, we obtain* 0.037.

Practice Problem 6 **(a)** Find $3.8\overline{)521.6}$ rounded to the nearest tenth.

(b) Find $8.05\overline{)0.17}$ rounded to the nearest thousandth. ■

EXAMPLE 7 John drove his 1987 Cavalier 420.5 miles to Chicago. He used 14.5 gallons of gas on the trip. How many miles per gallon did his car get on the trip?

To find miles per gallon we need to divide the number of miles, 420.5, by the number of gallons, 14.5.

```
          29.
14.5^)420.5^
      290
      1305
      1305
```

John's car achieved 29 miles per gallon on the trip to Chicago.

Practice Problem 7 Sarah rented a large truck to move to Boston. She drove 454.4 miles yesterday. She used 28.5 gallons of gas on the trip. How many miles per gallon did the rental truck get? ■

EXAMPLE 8 Find the value of n if $0.8 \times n = 2.68$.

Here 0.8 is multiplied by some number n to obtain 2.68. What is this number n? If we divide 2.68 by 0.8, we will find the value of n.

$$\begin{array}{r} 3.35 \\ 0.8_{\wedge})\overline{2.6_{\wedge}80} \\ \underline{2\,4} \\ 2\,8 \\ \underline{2\,4} \\ 40 \\ \underline{40} \end{array}$$

Thus the value of n is 3.35.

Check Is this true? Are we sure the value of $n = 3.35$?

We substitute the value of $n = 3.35$ into the equation to see if it makes the statement true.

$$\begin{aligned} 0.8 \times n &= 2.68 \\ 0.8 \times 3.35 &\stackrel{?}{=} 2.68 \\ 2.68 &= 2.68 \checkmark \quad \text{Yes. It is true.} \end{aligned}$$

Practice Problem 8 Find the value of n if $0.12 \times n = 0.696$. ■

EXAMPLE 9 The level of sulfur dioxide emissions in the air has slowly been decreasing over the last 15 years, as can be seen in the accompanying bar graph. Find the average amount of sulfur dioxide emissions in the air over these four specific years.

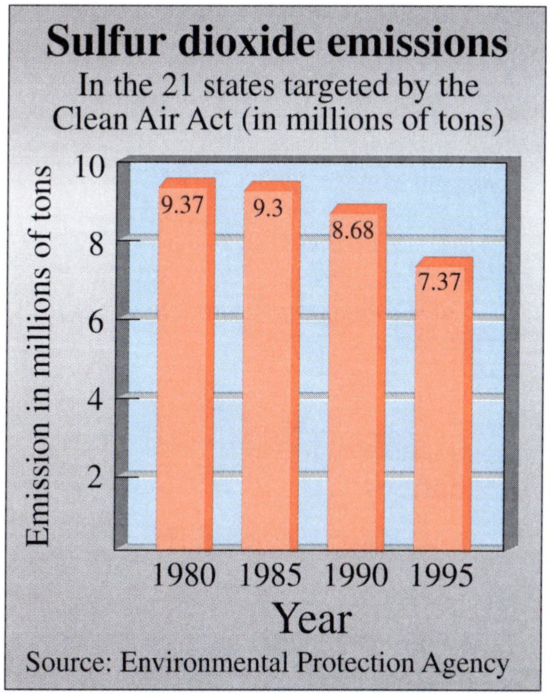

First we take the sum of the four years.

$$\begin{array}{r} 9.37 \\ 9.3 \\ 8.68 \\ +\ 7.37 \\ \hline 34.72 \end{array}$$

Then we divide by four to obtain the average.

$$\begin{array}{r} 8.68 \\ 4)\overline{34.72} \\ \underline{32} \\ 2\,7 \\ \underline{2\,4} \\ 32 \\ \underline{32} \end{array}$$

Thus the yearly average is 8.68 in millions of tons of sulfur dioxide in these 21 states.

Practice Problem 9 Use the accompanying bar graph to find the average level of sulfur dioxide for the three years: 1985, 1990, and 1995. By how much does the 3-year average differ from the 4-year average? ■

3.5 Exercises

Divide until there is a remainder of zero.

1. $6\overline{)12.6}$ **2.** $8\overline{)17.28}$ **3.** $4\overline{)0.1476}$ **4.** $5\overline{)0.0129}$ **5.** $7\overline{)128.17}$

6. $9\overline{)238.86}$ **7.** $64\overline{)3.616}$ **8.** $56\overline{)1.624}$ **9.** $21\overline{)0.0609}$ **10.** $23\overline{)0.0805}$

11. $0.8\overline{)9.76}$ **12.** $0.6\overline{)8.16}$ **13.** $0.5\overline{)32.15}$ **14.** $0.4\overline{)47.28}$ **15.** $0.09\overline{)0.7209}$

16. $0.8113 \div 0.07$ **17.** $75.6 \div 3.6$ **18.** $68.4 \div 3.8$ **19.** $82.8 \div 0.36$ **20.** $40.30 \div 0.31$

Divide and round your answer to the nearest tenth.

21. $7\overline{)36.92}$ **22.** $9\overline{)47.31}$ **23.** $1.8\overline{)4.16}$

24. $1.9\overline{)2.36}$ **25.** $0.95\overline{)32.067}$ **26.** $0.85\overline{)41.901}$

Divide and round your answer to the nearest hundredth.

27. $4\overline{)263.82}$ **28.** $5\overline{)471.03}$ **29.** $1.7\overline{)20.8}$

30. $1.6\overline{)36.5}$ **31.** $0.27\overline{)6.729}$ **32.** $0.41\overline{)8.378}$

Divide and round your answer to the nearest thousandth.

33. $8\overline{)248.162}$ **34.** $6\overline{)409.387}$ **35.** $0.69\overline{)8.45}$ **36.** $0.87\overline{)79.40}$

Divide and round your answer to the nearest whole number.

37. $0.075\overline{)3.729}$ **38.** $0.065\overline{)4.398}$ **39.** $0.55\overline{)7.00}$ **40.** $0.39\overline{)5.00}$

Applications

41. The college cafeteria staff is portioning out butterscotch pudding for the next meal. The cook has 235.6 ounces of pudding and wants to end up with 38 equal portions. How many ounces will each portion have if the cook uses up the entire batch of butterscotch pudding?

42. The Miller family wants to use the latest technology to access the Internet from their home television system. The equipment needed to upgrade their existing equipment will cost $992.76. If the Millers make 12 equal monthly payments, how much will they pay per month?

43. Steve knew that he would sacrifice some gas mileage when he sold his car and bought a 4×4 Ford Explorer. His new truck can go 312 miles on 15.6 gallons of gas. How many miles per gallon does his Explorer achieve?

44. Wally owns a Plymouth Breeze that travels 360 miles on 13.2 gallons of gas. How many miles per gallon does it achieve? (Round your answer to the nearest tenth).

45. The local concert promoter paid a total of \$3388.50 for 9 bands to play in the outdoor festival. Each band received an equal amount. How much did he pay per band?

46. Peter paid \$21.78 for 11 pounds of papayas. How much did he pay per pound? (Round your answer to the nearest cent.)

47. Demitri had a contractor build an outdoor deck for his back porch. He now has \$1131.75 to pay off, and he agreed to pay \$125.75 per month. How many more payments on the outdoor deck must he make?

48. Juanita has to pay off \$4533.13 in car payments that amount to \$146.23 per month. How many more car payments must she make?

49. Pascal works as a supervisor in a billiard supply business. He was inspecting a box of blue cue chalk, which players rub on the end of their playing sticks (cues) to keep from slipping when they hit the ball. Each small square of chalk weighs 7.9 grams. The contents of the entire box of chalk weigh 300.2 grams. How many blocks of chalk are in the box? If the box is labeled CONTENTS: 40 CUE CHALKS, how great an error was made packing the box?

50. Yoshi is working as an inspector for a company that makes snowboards. A Mach 1 snowboard weighs 3.8 kilograms. How many of these snowboards are contained in a box in which the contents weigh 87.40 kilograms? If the box is labeled CONTENTS: 24 SNOWBOARDS, how great an error was made in packing the box?

In each equation find the value of n.

51. $0.7 \times n = 0.0861$

52. $0.6 \times n = 3.948$

53. $1.6 \times n = 110.4$

54. $1.4 \times n = 821.8$

55. $n \times 0.063 = 2.835$

56. $n \times 0.098 = 4.312$

57. $n \times 0.008 = 6.48$

58. $n \times 0.009 = 46.8$

To Think About

Multiply the numerator and denominator of each fraction below by 10,000. Then divide the numerator by the denominator. Is the result the same as if we divided the original numerator by the original denominator? Why?

59. $\dfrac{3.8702}{0.0523}$

60. $\dfrac{2.9356}{0.0716}$

Cumulative Review Problems

61. Add. $\dfrac{3}{8} + 1\dfrac{2}{5}$

62. Subtract. $2\dfrac{13}{16} - 1\dfrac{7}{8}$

63. Multiply. $3\dfrac{1}{2} \times 2\dfrac{1}{6}$

64. Divide. $4\dfrac{1}{3} \div 2\dfrac{3}{5}$

3.6 Converting Fractions to Decimals and Order of Operations

MathPro Video 8 SSM

After studying this section, you will be able to:

1. *Convert any fraction to decimal form.*
2. *Use the correct order of operations when several decimals are combined with various operations.*

1 *Converting a Fraction to a Decimal*

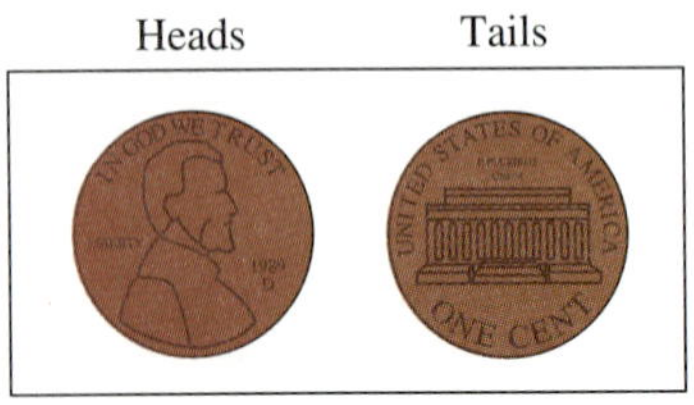

For a coin, heads and tails are two sides of the same object, two ways of identifying a single quantity. For numbers, fractions and decimals are two equivalent forms for the same quantity.

Fraction side $2\frac{1}{2}$ two and one-half	2.5 Decimal side two and five-tenths

Same quantity, different appearance

Every decimal can be expressed as an equivalent fraction. For example,

Decimal form $\Longrightarrow$ fraction form

$$0.75 = \frac{75}{100} \quad \text{or} \quad \frac{3}{4}$$

$$0.5 = \frac{5}{10} \quad \text{or} \quad \frac{1}{2}$$

$$2.5 = 2\frac{5}{10} = 2\frac{1}{2} \quad \text{or} \quad \frac{5}{2}$$

And every fraction can be expressed as an equivalent decimal, as we will learn in this section. For example,

Fraction form $\Longrightarrow$ decimal form

$$\frac{1}{5} = 0.20 \quad \text{or} \quad 0.2$$

$$\frac{3}{8} = 0.375$$

$$\frac{5}{11} = 0.4545\ldots \quad \text{(the ``45'' keeps repeating)}$$

Some of these decimal equivalents are so common that people find it helpful to memorize them. You would be wise to memorize the following equivalents.

$$\frac{1}{2} = 0.5 \qquad \frac{1}{4} = 0.25 \qquad \frac{1}{5} = 0.2 \qquad \frac{1}{10} = 0.1$$

We previously studied how to transfer some fractions with a denominator of 10, 100, 1000, and so on, to decimal form. For example, $\frac{3}{10} = 0.3$ and $\frac{7}{100} = 0.07$. We need to develop a procedure to write other fractions, such as $\frac{3}{8}$ and $\frac{5}{16}$, in decimal form.

Converting a Fraction to an Equivalent Decimal

Divide the denominator into the numerator until

a. the remainder becomes zero, or

b. the remainder repeats itself, or

c. the desired number of decimal places is achieved.

EXAMPLE 1 Write as an equivalent decimal. **(a)** $\frac{3}{8}$ **(b)** $\frac{31}{40}$

Divide the denominator into the numerator until the remainder becomes zero.

(a)
$$\begin{array}{r} 0.375 \\ 8\overline{)3.000} \\ \underline{2\,4} \\ 60 \\ \underline{56} \\ 40 \\ \underline{40} \\ 0 \end{array}$$

Therefore, $\frac{3}{8} = 0.375$

(b)
$$\begin{array}{r} 0.775 \\ 40\overline{)31.000} \\ \underline{280} \\ 300 \\ \underline{280} \\ 200 \\ \underline{200} \\ 0 \end{array}$$

Therefore, $\frac{31}{40} = 0.775$

Practice Problem 1 Write as an equivalent decimal.

(a) $\frac{5}{16}$ **(b)** $\frac{11}{80}$ ■

Decimals such as 0.375 and 0.775 are called **terminating decimals.** When converting $\frac{3}{8}$ to 0.375 or $\frac{31}{40}$ to 0.775, the division operation eventually yields a remainder of zero. Other fractions yield a repeating pattern. For example, $\frac{1}{3} = 0.3333\ldots$ and $\frac{2}{3} = 0.6666\ldots$ have a pattern of repeating digits. Decimals that have a digit or a group of digits that repeats are called **repeating decimals.** We often indicate the repeating pattern with a bar over the repeating group of digits:

$$0.3333\ldots = 0.\overline{3} \qquad 0.74\ 74\ 74\ldots = 0.\overline{74}$$

$$0.218\ 218\ 218\ldots = 0.\overline{218} \qquad 0.8942\ 8942\ldots = 0.\overline{8942}$$

If when converting fractions to decimal form the remainder repeats itself, we know that we have a repeating decimal.

EXAMPLE 2 Write as an equivalent decimal.

(a) $\dfrac{5}{11}$ (b) $\dfrac{13}{22}$ (c) $\dfrac{5}{37}$

(a)

```
        0.4545
   11)5.0000
       4 4
         60   <- remainder repeats
         55
          50
          44
           60  <- remainder repeats
           55
            5
```

Thus $\dfrac{5}{11} = 0.4545\ldots = 0.\overline{45}$

(b)

```
          0.59090
   22)13.00000
        11 0
          2 00   <- repeating remainders
          1 98
            2 00 <- repeating remainders
            1 98
               20
```

Thus $\dfrac{13}{22} = 0.5909090\ldots = 0.5\overline{90}$

Notice that the bar is over the digits 9 and 0 but *not* over the digit 5.

(c)

```
         0.1351
   37)5.0000
       37
       13 0   <- repeating remainders
       11 1
        1 90
        1 85
           50
           37
           13  <- repeating remainders
```

Thus, $\dfrac{5}{37} = 0.135135\ldots = 0.\overline{135}$

CALCULATOR

Fraction to Decimal

You can use a calculator to change $\frac{5}{8}$ to a decimal.

Enter

5 [÷] 8 [=]

The display should read

[0.625]

Try

(a) $\dfrac{17}{25}$ (b) $\dfrac{2}{9}$

(c) $\dfrac{13}{10}$ (d) $\dfrac{15}{19}$

Note: 0.7894737 is an approximation for $\frac{15}{19}$. Some calculators only round to 8 places.

Practice Problem 2 Write as an equivalent decimal.

(a) $\dfrac{7}{11}$ (b) $\dfrac{8}{15}$ (c) $\dfrac{13}{44}$ ■

EXAMPLE 3 Write as an equivalent decimal. (a) $3\dfrac{7}{15}$ (b) $\dfrac{20}{11}$

(a) $3\dfrac{7}{15}$ means $3 + \dfrac{7}{15}$

```
        0.466
   15)7.000
       60
       100
        90
        100
         90
         10
```

Thus $\dfrac{7}{15} = 0.4\overline{6}$ and $3\dfrac{7}{15} = 3.4\overline{6}$

(b)

```
        1.818
   11)20.000
       11
        90
        88
         20
         11
          90
          88
           2
```

Thus $\dfrac{20}{11} = 1.818181\ldots = 1.\overline{81}$

Practice Problem 3 Write as an equivalent decimal.

(a) $2\dfrac{11}{18}$ (b) $\dfrac{28}{27}$ ■

In some cases, the pattern of repeating is quite long. For example,

$$\frac{1}{7} = 0.142857142857\ldots = 0.\overline{142857}$$

Such problems are often rounded to a certain value.

EXAMPLE 4 Express $\frac{5}{7}$ as a decimal rounded to the nearest thousandth.

$$\begin{array}{r} 0.7142 \\ 7\overline{)5.0000} \\ \underline{49} \\ 10 \\ \underline{7} \\ 30 \\ \underline{28} \\ 20 \\ \underline{14} \\ 6 \end{array}$$

Rounding to the nearest thousandth, we round 0.7142 to 0.714. (In repeating form, $\frac{5}{7} = 0.714285714285\ldots = 0.\overline{714285}$.)

Practice Problem 4 Express $\frac{19}{24}$ as a decimal rounded to the nearest thousandth. ■

2 *Order of Operations with Decimals*

The rules for order of operations that we discussed in Section 1.6 apply to operations with decimals.

Order of Operations

Do first	**1.** Perform operations inside parentheses.
	2. Simplify any expressions with exponents.
	3. Multiply or divide from left to right.
Do last	**4.** Add or subtract from left to right.

EXAMPLE 5 Evaluate. $0.5 + 0.7 \times 0.21 - 0.63$

$= 0.5 + 0.147 - 0.63$ We multiplied. $\begin{array}{r} 0.21 \\ \times\ \ 0.7 \\ \hline 0.147 \end{array}$

$= 0.647 - 0.63$ We added. $\begin{array}{r} 0.5 \\ +\ 0.147 \\ \hline 0.647 \end{array}$

$= 0.017$ We subtracted. $\begin{array}{r} 0.647 \\ -\ 0.630 \\ \hline 0.017 \end{array}$

Practice Problem 5 Evaluate. $0.9 - 0.3 + 0.8 \times 0.3$ ■

Sometimes exponents are used with decimals. In such cases, we merely evaluate using repeated multiplication.

$$(0.2)^2 = 0.2 \times 0.2 = 0.04$$
$$(0.2)^3 = 0.2 \times 0.2 \times 0.2 = 0.008$$
$$(0.2)^4 = 0.2 \times 0.2 \times 0.2 \times 0.2 = 0.0016$$

EXAMPLE 6 Evaluate. $(0.3)^3 + 0.6 \times 0.2 + 0.013$

First we need to evaluate $(0.3)^3 = 0.3 \times 0.3 \times 0.3 = 0.027$. Thus

$(0.3)^3 + 0.6 \times 0.2 + 0.013$
$= 0.027 + 0.6 \times 0.2 + 0.013$
$= 0.027 + 0.12 + 0.013 \longleftarrow$ When addends have a different number of decimal places, writing the problem in column form makes adding easier.

$$\begin{array}{r} 0.027 \\ 0.120 \\ +\ 0.013 \\ \hline 0.160 \end{array}$$

$= 0.16$

Practice Problem 6 Evaluate. $0.3 \times 0.5 + (0.4)^3 - 0.036$ ■

In the next example all four steps of the rules for order of operations will be used.

EXAMPLE 7 Evaluate. $(8 - 0.12) \div 2^3 + 5.68 \times 0.1$

$= 7.88 \div 2^3 + 5.68 \times 0.1$ *First do subtraction inside the parentheses.*

$= 7.88 \div 8 + 5.68 \times 0.1$ *Simplify the expression with exponents.*

$= 0.985 + 0.568$ *From left to right do division and multiplication.*

$= 1.553$ *Add the final two numbers.*

Practice Problem 7 Evaluate. $6.56 \div (2 - 0.36) + (8.5 - 8.3)^2$ ■

Developing Your Study Skills

KEEP TRYING

We live in a highly technical world, and you cannot afford to give up on the study of mathematics. Dropping mathematics may prevent you from entering certain career fields that you may find interesting. You may not have to take math courses as high-level as calculus, but such courses as intermediate algebra, finite math, college algebra, and trigonometry may be necessary. Learning mathematics can open new doors for you.

Learning mathematics is a process that takes time and effort. You will find that regular study and daily practice are necessary to strengthen your skills and to help you grow academically. This process will lead you toward success in mathematics. Then, as you become more successful, your confidence in your ability to do mathematics will grow.

3.6 Exercises

Verbal and Writing Skills

1. 0.75 and $\frac{3}{4}$ are different forms for the __________ __________.
2. To convert a fraction to an equivalent decimal, divide the __________ into the numerator.
3. Why is $0.\overline{8942}$ called a repeating decimal?
4. The order of operations for decimals is the same as the order of operations for whole numbers. Write the steps for the order of operations.

Write as an equivalent decimal. If a repeating decimal is obtained, use notation such as $0.\overline{7}$, $0.\overline{16}$, *or* $0.\overline{245}$.

5. $\frac{1}{4}$ **6.** $\frac{3}{4}$ **7.** $\frac{7}{8}$ **8.** $\frac{5}{8}$ **9.** $\frac{7}{16}$ **10.** $\frac{3}{16}$ **11.** $\frac{7}{20}$ **12.** $\frac{3}{40}$

13. $\frac{31}{50}$ **14.** $\frac{23}{25}$ **15.** $\frac{9}{4}$ **16.** $\frac{14}{5}$ **17.** $2\frac{1}{8}$ **18.** $3\frac{13}{16}$ **19.** $1\frac{7}{16}$ **20.** $1\frac{1}{40}$

21. $\frac{2}{3}$ **22.** $\frac{1}{3}$ **23.** $\frac{5}{9}$ **24.** $\frac{7}{9}$ **25.** $\frac{5}{6}$ **26.** $\frac{13}{18}$ **27.** $\frac{10}{11}$ **28.** $\frac{4}{15}$

29. $\frac{14}{15}$ **30.** $\frac{3}{18}$ **31.** $\frac{5}{33}$ **32.** $\frac{7}{22}$ **33.** $\frac{28}{27}$ **34.** $\frac{41}{36}$ **35.** $2\frac{5}{18}$ **36.** $6\frac{1}{6}$

Write as an equivalent decimal or a decimal approximation. Round your answer to the nearest thousandth if needed.

37. $\frac{4}{7}$ **38.** $\frac{13}{27}$ **39.** $\frac{19}{21}$ **40.** $\frac{20}{21}$ **41.** $\frac{7}{48}$ **42.** $\frac{5}{48}$ **43.** $\frac{35}{27}$ **44.** $\frac{37}{23}$

45. $\frac{20}{81}$ **46.** $\frac{7}{82}$ **47.** $\frac{17}{18}$ **48.** $\frac{5}{13}$ **49.** $\frac{22}{7}$ **50.** $\frac{17}{14}$ **51.** $\frac{9}{19}$ **52.** $\frac{11}{17}$

Applications

53. A computer disc is $\frac{5}{24}$ inch thick. Write the thickness as a decimal.

54. In the winter in Maine a car engine will run safely if the radiator solution is $\frac{5}{8}$ antifreeze. Write this value as a decimal.

55. An ice climber had the local mountaineering shop install boot heaters to the back of his hiking boots. The installer drilled a hole $\frac{3}{8}$ inch in diameter. The hole for the boot heater should have been 0.5 inch in diameter. Is the hole too large or too small? By how much?

56. A master carpenter is re-creating a room for the set of a movie being filmed. He is using a burled maple veneer $\frac{7}{16}$ inch thick. The designer specified maple veneer 0.45 inch thick. Is the veneer he is using too thick or too thin? By how much?

57. A machinist in a factory is using sheet metal $\frac{17}{32}$ inch thick. The plans call for sheet metal 0.53 inch thick. Is the metal he is using too thick or too thin? By how much?

58. An auto body shop technician drilled a hole $\frac{7}{16}$ inch in diameter. The hole should have been 0.44 inch in diameter. Is the hole too large or too small? By how much?

Evaluate.

59. $(0.2)^3 + 5.9 \times 1.3 - 2.6$

60. $(0.3)^2 + 0.6 \times (0.2 + 5.8)$

61. $12.2 \times 9.4 - 2.68 + 1.6 \div 0.8$

62. $2.3 \times 12.6 - 1.98 + 12.6 \div 1.4$

63. $12 \div 0.03 - 15 \times (0.6 + 0.7)^2$

64. $61.95 \div 1.05 + 3.6 \times (2.1 + 0.31)$

65. $116.32 + (0.12)^2 + 18.06 \times 2.2$

66. $105.08 + (0.21)^2 - 0.05 \times 123.4$

67. $(1.1)^3 + 2.6 \div 0.13 + 0.083$

68. $(1.1)^3 + 8.6 \div 2.15 - 0.086$

69. $(56.3 - 56.1)^2 \div 0.016$

70. $(23.8 - 23.5)^2 \div 0.36$

71. $(0.5)^3 + (3 - 2.6) \times 0.5$

72. $(0.6)^3 + (7 - 6.3) \times 0.07$

73. $(0.76 + 4.24) \div 0.25 + 8.6$

74. $(2.4)^2 + 3.6 \div (1.2 - 0.7)$

Evaluate.

★ 75. $(1.6)^3 + (2.4)^2 + 18.666 \div 3.05 + 4.86$

★ 76. $5.9 \times 3.6 \times 2.4 - 0.1 \times 0.2 \times 0.3 \times 0.4$

Write each fraction as a decimal. Round your answer to six decimal places.

★ 77. $\dfrac{5236}{8921}$

★ 78. $\dfrac{17{,}359}{19{,}826}$

To Think About

★ 79. Subtract. $0.\overline{16} - 0.00\overline{16}$
- **(a)** What do you obtain?
- **(b)** Now subtract $0.\overline{16} - 0.01\overline{6}$. What do you obtain?
- **(c)** What is different about these results?

★ 80. Subtract. $1.\overline{89} - 0.01\overline{89}$
- **(a)** What do you obtain?
- **(b)** Now subtract $1.\overline{89} - 0.18\overline{9}$. What do you obtain?
- **(c)** What is different about these results?

Cumulative Review Problems

81. What is the area of a rectangle that measures 12 feet by 26 feet?

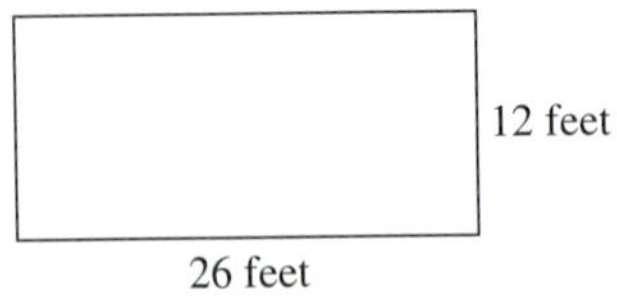

82. In the 1996 presidential election, 1,070,990 voters in Virginia voted for President Clinton and 1,071,859 voters in Wisconsin voted for President Clinton. What was the difference in the number of votes for President Clinton between these two states?

83. If a person makes deposits of \$56, \$81, \$42, and \$198, what is the total amount deposited in the account?

84. If Central Texas Catering equally divides \$3320 in tips from this weekend among its 40 employees, how much will each person receive?

3.7 Applied Problems Using Decimals

After studying this section, you will be able to:

1 *Solve applied problems using operations with decimals.*

MathPro Video 9 SSM

1 *Solving Applied Problems Using Operations with Decimals*

We use the basic plan of solving applied problems that we discussed in Section 1.8 and Section 2.9. Let us review how we analyze applied problem situations.

1. *Understand the problem.*
2. *Solve and state the answer.*
3. *Check.*

EXAMPLE 1 Brendon bought a camera for \$89.59, and the sales tax was \$4.48. He also bought a camera case for \$16.99, and the sales tax was \$0.85. He brought \$120 for the purchase. How much money did he have left over?

1. *Understand the problem.*
   Fill in the Mathematics Blueprint for Problem Solving.

MATHEMATICS BLUEPRINT FOR PROBLEM SOLVING

Gather the Facts	What Am I Asked to Do?	How Do I Proceed?	Key Points to Remember
Camera cost: \$89.59 Tax on camera: \$4.48 Camera case cost: \$16.99 Tax on case: \$0.85 Brendon has \$120 to spend.	Find out how much money Brendon has left after making these purchases.	Add the cost of each item and the tax on each item to obtain the total purchase price. Subtract that total from \$120.	Be sure to line up the decimal points when adding and subtracting the decimal values.

2. *Solve and state the answer.*
   Add the four amounts.

$$\begin{array}{r} \$89.59 \\ 4.48 \\ 16.99 \\ +\quad 0.85 \\ \hline \$111.91 \end{array}$$

Subtract all expenses from \$120.

$$\begin{array}{r} \$120.00 \\ -\quad 111.91 \\ \hline \$\quad 8.09 \end{array}$$

Brendon has \$8.09 left after purchasing the camera and the case and paying the tax on each.

3. *Check.*

Estimate the answer to see if it is reasonable. First, we take each of the four amounts and round them to one significant digit. Then we add the four numbers together.

$$\$90 + \$4 + \$20 + \$1 = \$115$$

Next, we subtract our estimated sum from \$120. Since the calculation is so easy to do, we do not estimate the subtraction. Rather we do the calculation directly.

$$\begin{array}{r} \$120 \\ -\ 115 \\ \hline \$5 \end{array}$$

Thus we estimate Brendon will have about \$5 left after the purchase. Our answer of \$8.09 is fairly close to the estimate, so our answer is reasonable. ✓

Practice Problem 1 Renée purchased a coat for \$198.95, and the sales tax was \$9.95. She also purchased boots for \$59.95, and the sales tax was \$3.00. She brought \$280 to make the purchase. How much money did she have left over? ■

In the United States for almost all jobs where you are paid an hourly wage, if you work more than 40 hours in one week, you should be paid overtime, at a rate of 1.5 times the normal hourly rate, for the extra hours worked in that week. The next problem deals with overtime wages.

EXAMPLE 2 A laborer is paid \$7.38 per hour for a 40-hour week and 1.5 times that wage for any hours worked beyond the standard 40. If he works 47 hours in a week, what will he earn?

1. *Understand the problem.*

MATHEMATICS BLUEPRINT FOR PROBLEM SOLVING

Gather the Facts	What Am I Asked to Do?	How Do I Proceed?	Key Points to Remember
He works 47 hours. He gets paid \$7.38 per hour for 40 hours. He gets paid 1.5 × \$7.38 per hour for 7 hours.	Find the earnings of the laborer if he works 47 hours in one week.	Add the earnings of 40 hours at \$7.38 per hour to the earnings of 7 hours at overtime pay.	Multiply 1.5 × \$7.38 to find the pay he earns for overtime.

2. *Solve and state the answer.*

We want to compute his regular pay and his overtime pay and add the results.

Regular pay + Overtime pay = Total pay

Regular pay Calculate his pay for 40 hours of work.

$$\begin{array}{r} 7.38 \\ \times\ \ 40 \\ \hline 295.20 \end{array}$$

He earns \$295.20 at \$7.38 per hour.

Overtime pay Calculate his overtime pay rate. This is 7.38 × 1.5.

$$\begin{array}{r} 7.38 \\ \times\ \ 1.5 \\ \hline 3\ 690 \\ 7\ 38\ \ \\ \hline 11.070 \end{array}$$

He earns $11.07 per hour in overtime.

Calculate how much he earned doing 7 hours of overtime work.

$$\begin{array}{r} 11.07 \\ \times\ \ \ \ \ 7 \\ \hline 77.49 \end{array}$$

For 7 hours overtime he earns $77.49.

Total pay Add the two amounts.

\$295.20	Regular 40-hour week earnings
+ 77.49	Overtime earnings
\$372.69	Total earnings

The total earnings of the laborer for a 47-hour workweek were $372.69.

3. *Check.*

Estimate his regular pay.

$$40 \times \$7 = \$280$$

Estimate his overtime rate of pay, and then his overtime pay.

$$1.5 \times \$7 = \$10.50$$

$$7 \times \$11 = \$77$$

Then add.

$$\begin{array}{r} \$280 \\ +\ \ \ 77 \\ \hline \$357 \end{array}$$

Our estimate of $357 is close to our answer of $372.69. Our answer is reasonable. ✓

Practice Problem 2 Melinda works for the phone company as a line repair technician. She earns $9.36 per hour. She worked 51 hours last week. If she gets time and a half for all hours worked above 40 hours per week, how much did she earn last week? ■

EXAMPLE 3 A chemist is testing 36.85 liters of cleaning fluid. She wishes to pour it into several smaller containers that each hold 0.67 liters of fluid. **(a)** How many containers will she need? **(b)** If each liter of this fluid costs \$3.50, how much does the cleaning fluid in one container cost? (Round your answer to the nearest cent.)

MATHEMATICS BLUEPRINT FOR PROBLEM SOLVING

Gather the Facts	What Am I Asked to Do?	How Do I Proceed?	Key Points to Remember
The total amount of cleaning fluid is 36.85 liters. Each small container holds 0.67 liters. Each liter of fluid costs \$3.50.	(a) Find out how many containers the chemist needs. (b) Find the cost of cleaning fluid in each small container.	(a) Divide the total, 36.85, by the amount in each small container, 0.67, to find the number of containers. (b) Multiply the cost of one liter, \$3.50, by the amount of liters in one container, 0.67.	If you are not clear as to what to do at any stage of the problem, then do a similar, simpler problem.

(a) How many containers will the chemist need?

She has 36.85 liters of cleaning fluid and she wants to put it into several equal-sized containers each holding 0.67 liter. Suppose we are not sure what to do. Let's do a similar, simpler problem. If we had 40 liters of cleaning fluid and we wanted to put it into little containers each holding 0.5 liter, what would we do? Since two little containers would only hold 1 liter, we would need 80 containers. We know that $40 \div 0.5 = 80$. So we see that, in general, we divide the total number of liters by the amount in the small container. Thus $36.85 \div 0.67$ will give us the number of containers in this case.

$$\begin{array}{r} 55. \\ 0.67_\wedge\overline{)36.85_\wedge} \\ \underline{33\ 5} \\ 3\ 35 \\ \underline{3\ 35} \end{array}$$

The chemist will need 55 containers to hold this amount of cleaning fluid.

(b) How much does the cleaning fluid in each container cost? Each container will hold only 0.67 liter. If one liter costs \$3.50, then to find the cost of one container we multiply $0.67 \times \$3.50$.

$$\begin{array}{r} 3.50 \\ \times\ 0.67 \\ \hline 2450 \\ 2100 \\ \hline 2.3450 \end{array}$$

We round our answer to the nearest cent. Thus each container would cost \$2.35.

Check.

(a) Is it really true that 55 containers each holding 0.67 liter will hold a total of 36.85 liters? To check we multiply.

$$\begin{array}{r} 55 \\ \times\ 0.67 \\ \hline 385 \\ 330 \\ \hline 36.85 \end{array} \checkmark$$

(b) One liter of cleaning fluid costs \$3.50. We would expect the cost of 0.67 liter to be less than \$3.50. \$2.35 is less than \$3.50. ✓

We use estimation to check more closely.

$$\begin{array}{rcr} \$3.50 & \longrightarrow & \$4.00 \\ \times\ \ 0.67 & \longrightarrow & \times\ \ \ 0.7 \\ \hline & & \$2.800 \end{array}$$

\$2.80 is fairly close to \$2.35. Our answer is reasonable. ✓

Practice Problem 3 A butcher divides 17.4 pounds of prime steak into small equal-sized packages. Each package contains 1.45 pounds of prime steak. **(a)** How many packages of steak will he have? **(b)** Prime steak sells for \$4.60 per pound. How much will each package of prime steak cost? ■

Developing Your Study Skills

APPLICATIONS OR WORD PROBLEMS

Applications or word problems are the very life of mathematics! They are the reason for doing mathematics, because they teach you how to put into use the mathematical skills you have developed. Learning mathematics without ever doing word problems is similar to learning all the skills of a sport without ever playing a game or learning all the notes on an instrument without ever playing a song.

The key to success is practice. Make yourself do as many problems as you can. You may not be able to do them all correctly at first, but keep trying. Do not give up whenever you reach a difficult one. If you cannot solve it, just try another one. Then come back and try it again later. Ask for help from your teacher or the tutoring lab. Ask other classmates how they solved the problem.

A misconception among students when they begin studying word problems is that each problem is different. At first the problems may seem this way, but as you practice more and more, you will begin to see the similarities, the different "types." You will see patterns in solving problems, which will enable you to solve problems of a given type more easily.

3.7 Exercises

Applications

1. A single-engine plane has a passenger weight limit enforced by the Federal Aviation Association. A pilot asked his passengers to step on the scale to check their respective weights. The four passengers weighed 167.2 pounds, 136.7 pounds, 99.3 pounds, and 218 pounds. What was the total weight of the passengers?

2. Angela buys costume jewelry and watches at a wholesale price, and resells them in her retail shop. During the last four weeks, she spent $412.65, $78.90, $331.47, and $146.03. How much did she spend on merchandise?

3. The local baseball team took in $11,022.45 in ticket sales yesterday. Today the team took in $19,891.33. How much more money did the box office receive today than yesterday?

4. Renee LeBlanc owns a mail-order company. Last year the company made a profit of $233,598.72. This year the company made a profit of $314,255.33. How much more profit did the company make this year?

5. A parking lot in front of the Orlando Outback Steak House measures 103.4 meters long by 76.3 meters wide. What is the area in square meters of this parking lot?

6. A mouse pad measures 22.4 centimeters by 26.3 centimeters. What is the area of the mouse pad in square centimeters?

7. Cynthia is making holiday cookies, which she will give as gifts to her friends and co-workers. She has made 34.5 pounds of cookies, which need to be divided up into equal packages containing 0.75 pound of cookies. How many packages can she make?

8. Hans is making gourmet chocolate in Switzerland. He has 11.52 liters of liquid white chocolate that will be poured into molds that hold 0.12 liter. How many individual molds can Hans make with his 11.52 liters of liquid white chocolate?

9. Donna went grocery shopping with her three children. She spent $71.32 on groceries. On the way out she realized that she had forgotten to buy milk. She went back in and paid $2.16 for milk. Just as she got back to the car, she realized that there was only one egg in the house, so she went back in and found out that eggs cost $1.19. Donna started out with $75.00 in her wallet. Did she have enough to buy the eggs? How much was left over?

10. David is planning to propose to his girlfriend today. He bought perfume for her for $68.49 and paid $3.94 in tax. Next, he went to the florist and bought a huge bouquet of assorted Holland flowers for $72.39 and paid $4.16 in tax. If he started out with $150.00 in his wallet, how much money did he have left over?

11. One year in Mount Waialeale, Hawaii, considered the "rainiest place in the world," the yearly rainfall totaled 11.68 meters. The next year, the yearly rainfall on this mountain totaled 10.42 meters. The third year it was 12.67 meters. On average, how much rain fell on Mount Waialeale, Hawaii, per year.

12. Emma and Jennie took a trip in their Chrysler Concorde from Saskatoon, Saskatchewan, to calgary, Alberta, in Canada to check out the glacier lakes. When they left, their odometer read 54,089. When they returned home, the odometer read 55,401. They used 65.6 gallons of gas. How many miles per gallon did they get on the trip? (Round your answer to the nearest tenth.)

13. The York Steak House uses 3.5 boxes of customized paper towels per day for its customers. If the master box holds 42 regular boxes, how many days will the master box last?

14. Hanu's outdoor gas grill uses 2.4 pounds of propane per hour. If the tank holds 20.64 pounds of propane, how many hours can the barbecue run on a full tank?

15. The local Police Athletic League raised enough money to renovate the local youth hall and turn it into a coffeehouse/activity center so that there is a safe place to hang out. The room that holds the ping-pong table needs 43.9 square yards of new carpeting. The entryway needs 11.3 square yards and the stage/seating area needs 63.4 yards. The carpeting will cost $10.65 per square yard. What will be the total bill for carpeting these three areas of the coffeehouse?

16. The Williams family is installing insulation in their house. They are installing 98.6 square yards of insulation in the attic, 24.3 square yards of insulation in the basement, and 36.1 square yards of insulation in the family room. The insulation costs $6.40 per square yard. How much will all the insulation cost?

17. An electrician is paid $14.30 per hour for a 40-hour week. She is paid time and a half for overtime (1.5 times the hourly wage) for every hour more than 40 hours worked in the same week. If she works 48 hours in one week, what will she earn for that week?

18. Shirley works at the local pizza franchise and is paid $6.20 per hour for a 40-hour week. She is paid time and a half for overtime, (1.5 times the hourly wage), for every hour more than 40 hours worked in the same week. If she works 52 hours in one week, what will she earn for that week?

19. Barbara had $420.13 in her savings account last month. Since then she has made deposits of $116.32 and $318.57. The bank credited her with interest of $1.86 for the month. She wrote three checks, for $16.50, $36.89, and $376.94. What is her new balance? (Assume that no bank fees have been charged.)

20. Harvey had $113.08 in his checking account last month. He made two deposits: $612.83 and $86.54. The bank deducted a fee this month of $4.60. He made out three checks: $100.00, $216.34, and $398.92. What will his new balance be?

21. Charlie borrowed $11,500 to purchase a new car. His loan requires him to pay $288.65 each month over the next 60 months (5 years). How much will he pay over the 5 years? How much more will he pay back than the amount of the loan?

22. Hector and Junita borrowed $80,000 to buy their new home. They make monthly payments to the bank of $450.25 to repay the loan. They will be making these monthly payments for the next 30 years. How much money will they pay to the bank in the next 30 years? How much more will they pay back than they borrowed?

23. A famous athletic shoe store at the Heartland Shopping Mall buys basketball shoes at a cost of $7920 for 12 dozen pairs. They sell the sneakers for $68 per pair. How much profit will they make if they sell all 144 pairs?

24. Concert T-shirts are usually sold for $24.00 at the show. If the tour promoter buys 5000 shirts for $56,250 and sells every shirt, how much profit will the promoter realize?

25. The EPA standard for safe drinking water is a maximum of 1.3 milligrams of copper per liter of water. A study was conducted on a sample of 7 liters of water drawn from Jeff Slater's house. The analysis revealed 8.06 milligrams of copper in the sample. Is the water safe or not? By how much?

26. The EPA standard for safe drinking water is a maximum of 0.015 milligram of lead per liter of water. A study was conducted on 6 liters of water from West Towers Dormitory. The analysis revealed 0.0795 milligram of lead in the sample. Is the water safe or not? By how much?

27. A jet fuel tank containing 17316.8 gallons is being emptied at the rate of 126.4 gallons per minute. How many minutes will it take to empty the tank?

28. If General Motors sells 6340 new Chevrolet Camaros next month and makes a profit of $2318.20 on each car, what profit will GM make on Camaros next month?

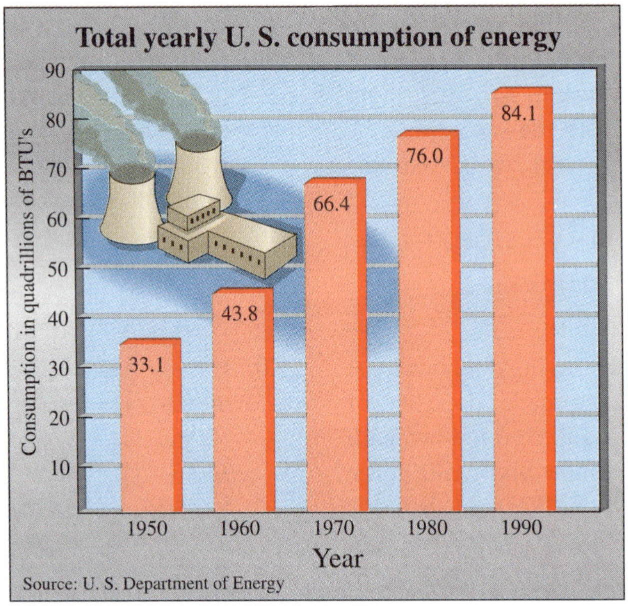

Use the bar graph above to answer questions 29–32.

29. How many more Btus were consumed in the United States during the year 1990 than in the year 1970?

30. What was the greatest increase in consumption of energy in a 10-year period? When did it occur?

31. What was the average consumption of energy per year in the United States for the 20-year period including the years 1950, 1960, and 1970? Write your answer in quadrillion Btus and then write your answer in Btus. (Remember a quadrillion is 1000 trillions.)

32. What was the average consumption of energy per year in the United States for the 20-year period including the years 1970, 1980, and 1990? Write your answer in quadrillion Btus and then write your answer in Btus. (Remember a quadrillion is 1000 trillions.)

Cumulative Review Problems

Add.

33. $\frac{1}{5} + \frac{3}{7} + \frac{1}{2}$

34. $\frac{10}{17} + \frac{1}{34} - \frac{5}{17}$

Multiply.

35. $\frac{5}{12} \times \frac{36}{27}$

36. $2\frac{2}{3} \times 8\frac{1}{4}$

Chapter Organizer

Topic	Procedure	Examples
Word names for decimals, p. 178.	Hundreds: 3 Tens: 4 Ones: 1 Decimal Point: . Tenths: 6 Hundredths: 7 Thousandths: 8 Ten Thousandths: 3	The word name for 341.6783 is three hundred forty-one and six thousand seven hundred eighty-three ten thousandths.
Writing a decimal as a fraction, p. 180.	**1.** Read the decimal in words. **2.** Write it in fraction form. **3.** Reduce if possible.	Write 0.36 as a fraction. **1.** 0.36 is read thirty-six hundredths **2.** Write the fractional form. $\frac{36}{100}$ **3.** Reduce. $\frac{36}{100} = \frac{9}{25}$
Determining which of two decimals is larger, p. 184.	**1.** Start at the left and compare corresponding digits. Write in extra zeros if needed. **2.** When two digits are different, the larger number is the one with the larger digit.	Which is larger? 0.138 or 0.13 0.138 ? 0.130 $8 > 0$ So $0.138 > 0.130$.
Rounding decimals, p. 186.	**1.** Locate the place (units, tenths, hundredths, etc.) for which the round-off is required. **2.** If the first digit to the right of this place is less than 5, drop it and all the digits to the right of it. **3.** If the first digit to the right of this place is 5 or greater, increase the digit by one. Drop all digits to the right.	Round to the nearest hundredth. 0.8652 0.87 Round to the nearest thousandth. 0.21648 0.216
Adding and subtracting decimals, pp. 190.	**1.** Write the numbers vertically and line up the decimal points. Extra zeros may be written to the right of the decimal points if needed. **2.** Add or subtract all the digits with the same place value, starting with the right column, moving to the left. Use carrying or borrowing as needed. **3.** Place the decimal point of the sum in line with the decimal points of all the numbers added.	Add. 36.3 + 8.007 + 5.26 $\begin{array}{r} \overset{1}{3}6.300 \\ 8.007 \\ +\ 5.260 \\ \hline 49.567 \end{array}$ Subtract. 82.5 − 36.843 $\begin{array}{r} \overset{7}{\not 8}\ \overset{11}{\not 2}.\overset{14}{\not 5}\ \overset{9}{\not 0}\ \overset{10}{\not 0} \\ -\ 3\ 6.8\ 4\ 3 \\ \hline 4\ 5.6\ 5\ 7 \end{array}$
Multiplying decimals, p. 200.	**1.** Multiply the numbers just as you would multiply whole numbers. **2.** Find the sum of the decimal places in the two factors. **3.** Place the decimal point in the product so that the product has the same number of decimal places as the sum in step 2. You may need to insert zeros to the left of the number found in step 1.	Multiply. $\begin{array}{r} 0.2 \\ \times\ 0.6 \\ \hline 0.12 \end{array}$ $\begin{array}{r} 0.3174 \\ \times\ \ 0.8 \\ \hline 0.25392 \end{array}$ $\begin{array}{r} 0.0064 \\ \times\ \ 0.21 \\ \hline 64 \\ 128 \\ \hline 0.001344 \end{array}$ $\begin{array}{r} 1364 \\ \times\ \ 0.7 \\ \hline 954.8 \end{array}$
Multiplying a decimal by a power of 10, p. 202.	Move the decimal point to the right the same number of places as there are zeros in the power of 10. (Sometimes it is necessary to write extra zeros before placing the decimal point in the answer.)	Multiply. $5.623 \times 10 = 56.23$ $0.597 \times 10^4 = 5970$ $0.0082 \times 1000 = 8.2$ $0.075 \times 10^6 = 75{,}000$ $28.93 \times 10^2 = 2893$

Chapter Organizer

Topic	Procedure	Examples
Dividing by a decimal, p. 208.	**1.** Make the divisor a whole number by moving the decimal point to the right. Mark that position with a caret ($_\wedge$). **2.** Move the decimal point in the dividend to the right the same number of places. Mark that position with a caret. **3.** Place the decimal point of your answer directly above the caret in the dividend. **4.** Divide as with whole numbers.	Divide. **(a)** $0.06\overline{)0.162}$ **(b)** $0.003\overline{)85.8}$ **(a)** $0.06_\wedge\overline{)0.16_\wedge2}$ = 2.7 12 42 42 0 **(b)** $0.003_\wedge\overline{)85.800_\wedge}$ = 28 600. 6 25 24 18 18 0
Converting a fraction to a decimal, p. 214.	Divide the denominator into the numerator until **1.** the remainder is zero, or **2.** the decimal repeats itself, or **3.** the desired number of decimal places is achieved.	Find the decimal equivalent for **(a)** $\frac{13}{22}$ **(b)** $\frac{5}{7}$, rounded to the nearest ten thousandth **(a)** $22\overline{)13.0000}$ = 0.5909 110 200 198 200 198 2 $\frac{13}{22} = 0.5\overline{90}$ or 0.5909090 . . . **(b)** $7\overline{)5.00000}$ = 0.71428 49 10 7 30 28 20 14 60 56 4 0.71428 rounded to the nearest ten thousandth is 0.7143.
Order of operations with decimal numbers, p. 217.	Same as order of operations of whole numbers. **1.** Perform operation inside parentheses. **2.** Simplify any expressions with exponents. **3.** Multiply or divide from left to right. **4.** Add or subtract from left to right.	Evaluate. $(0.4)^3 + 1.26 \div 0.12 - 0.12 \times (1.3 - 1.1)$ $(0.4)^3 + 1.26 \div 0.12 - 0.12 \times 0.2$ $0.064 + 1.26 \div 0.12 - 0.12 \times 0.2$ $0.064 + 10.5 - 0.024$ $10.564 - 0.024$ 10.54

Using the Mathematical Blueprint for Problem Solving, p. 221.

In solving an applied problem with decimals, students may find it helpful to complete the following steps. You will not use all the steps all of the time. Choose the steps that best fit the conditions of the problem.

1. *Understand the problem.*
   - **(a)** Read the problem carefully.
   - **(b)** Draw a picture if this helps you to visualize the situation. Think about what facts you are given and what you are asked to find.
   - **(c)** It may help to write a similar, simpler problem to get started and to determine what operation to use.
   - **(d)** Use the Mathematics Blueprint for Problem Solving to organize your work. Follow these four parts.
     - **1.** Gather the facts. (Write down specific values given in the problem.)
     - **2.** What am I asked to do? (Identify what you must obtain for an answer.)
     - **3.** How do I proceed? (What calculations need to be done?)
     - **4.** Key points to remember. (Record any facts, warnings, formulas, or concepts you think will be important as you solve the problem.)

2. *Solve and state the answer.*
   - **(a)** Perform the necessary calculations.
   - **(b)** State the answer, including the unit of measure.

3. *Check.*
   - **(a)** Estimate the answer to the problem. Compare this estimate to the calculated value. Is your answer reasonable?
   - **(b)** Repeat your calculations.
   - **(c)** Work backward from your answer. Do you arrive at the original conditions of the problem?

Example

Fred has a rectangular living room that is 3.5 yards wide and 6.8 yards long. He has a hallway that is 1.8 yards wide and 3.5 yards long. He wants to carpet each area using carpeting that costs $12.50 per square yard. What will the carpeting cost him?

1. *Understand the problem.*
It is helpful to draw a sketch.

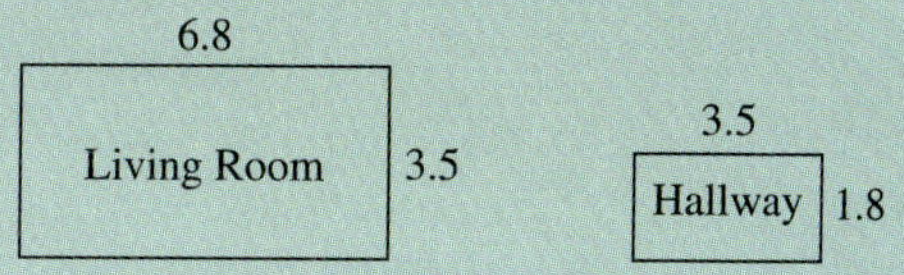

MATHEMATICS BLUEPRINT FOR PROBLEM SOLVING

Gather the Facts	What Am I Asked to Do?	How Do I Proceed?	Key Points to Remember
Living room: 6.8 yards by 3.5 yards Hallway: 3.5 yards by 1.8 yards Cost of carpet: $12.50 per square yard	Find out what the carpeting will cost Fred.	Find the area of each room. Add the two areas. Multiply the total area by $12.50.	Multiply the length by the width to get the area of the room. Remember, area is measured in square yards.

To find the area of each room, we multiply the dimensions for each room.

Living room $6.8 \times 3.5 = 23.80$ square yards   *Hallway* $3.5 \times 1.8 = 6.30$ square yards

Add the two areas.

$$\begin{array}{r} 23.80 \\ +\ 6.30 \\ \hline 30.10 \end{array} \text{ square yards}$$

Multiply the total area by the cost per square yard.

$$30.1 \times 12.50 = \$376.25$$

Estimate to check. You may be able to do some of this mentally.

$7 \times 4 = 28$ square yards $4 \times 2 = 8$ square yards

$$\begin{array}{r} 28 \\ +\ 8 \\ \hline 36 \end{array}$$

$36 \times 10 = \$360$ $360 is close to $376.25. ✓

Chapter 3 Review Problems

3.1 *Write a word name for each decimal.*

1. 13.672

2. 0.00084

Write each fraction as a decimal.

3. $\frac{7}{10}$

4. $\frac{81}{100}$

5. $1\frac{523}{1000}$

6. $\frac{79}{10{,}000}$

Write each decimal as a fraction or a mixed number.

7. 0.17

8. 0.365

9. 26.88

10. 1.00025

3.2 *Fill in the blank with <, =, or >.*

11. $\frac{13}{100}$ ________ 0.103

12. 0.716 ________ 0.706

Arrange the set of numbers from smallest to largest.

13. 0.981, 0.918, 0.98, 0.901

14. 5.62, 5.2, 5.6, 5.26, 5.59

15. Round to the nearest tenth. 0.613

16. Round to the nearest hundredth. 19.2076

17. Round to the nearest thousandth. 1.09952

18. Round to the nearest dollar. $156.48

3.3 *Add.*

19.
$$\begin{array}{r} 1.8 \\ 2.603 \\ 0.52 \\ +\ 1.716 \\ \hline \end{array}$$

20.
$$\begin{array}{r} 9.6 \\ 11.5 \\ 21.8 \\ +\ 34.7 \\ \hline \end{array}$$

21. Fred needs 2.9 gallons of paint for the living room, 1.6 gallons for the kitchen, and 4 gallons for the bedrooms.

(a) Exactly how many gallons and decimal parts of a gallon does he need?

(b) How many whole gallons should he buy if he allows for a little extra?

Subtract.

22.
$$\begin{array}{r} 5.19 \\ -\ 1.296 \\ \hline \end{array}$$

23.
$$\begin{array}{r} 199.703 \\ -\ 108.964 \\ \hline \end{array}$$

24. Serina drove to Cincinnati. The car odometer read 86,804.7 miles when she started and 87,041.6 miles when she got there. How long was the trip?

3.4 *Multiply.*

25.
$$\begin{array}{r} 0.76 \\ \times\ 0.03 \\ \hline \end{array}$$

26.
$$\begin{array}{r} 1.2874 \\ \times\ \quad 0.7 \\ \hline \end{array}$$

27.
$$\begin{array}{r} 0.026 \\ \times\ 0.014 \\ \hline \end{array}$$

28.
$$\begin{array}{r} 126.83 \\ \times\ \quad 7 \\ \hline \end{array}$$

29.
$$\begin{array}{r} 54 \\ \times\ 1.6 \\ \hline \end{array}$$

30.
$$\begin{array}{r} 7053 \\ \times\ 0.34 \\ \hline \end{array}$$

31. 7.86×10

32. 156.371×100

33. 0.000613×10^3

34. 1.2354×10^5

35. Hillary purchased 3.6 pounds of bananas at $0.25 per pound. How much did the purchase cost?

3.5 *Divide until there is a remainder of zero.*

36. $0.4\overline{)9630}$

37. $0.07\overline{)0.0001806}$

38. $5.2\overline{)191.36}$

39. $8\overline{)1863.2}$

40. Divide and round your answer to the nearest tenth.

$$1.3\overline{)746.75}$$

41. Divide and round your answer to the nearest thousandth.

$$0.06\overline{)0.003539}$$

42. My stereo cost \$283.95 and it lasted for five years. I had it repaired once during the five years for \$49.82. If I spread the total cost over five years, how much did it cost me per year for the stereo? (Round to the nearest cent.)

3.6 *Write each fraction as an equivalent decimal.*

43. $\frac{5}{18}$

44. $\frac{7}{40}$

45. $1\frac{5}{6}$

46. $\frac{15}{16}$

Write each fraction as a decimal rounded to the nearest thousandth.

47. $\frac{11}{14}$

48. $2\frac{5}{17}$

Evaluate by doing the operations in proper order.

49. $1.6 \times 2.3 + 0.4 - 0.6 \times 0.8$

50. $2.5 \div 0.5 \times 3.2 + 1.8 - 3.7$

51. $0.03 + (1.2)^2 - 5.3 \times 0.06$

52. $(1.02)^3 + 5.76 \div 1.2 \times 0.05$

53. $6.63 + 8.24 \div (5.76 - 5.68) - 22.5$

54. $(6 - 5.25)^2 \div 0.25 + 3.75$

Mixed Practice

Calculate each of the following.

55. 0.38×0.56

56. $1.38 - 0.83$

57. $0.204 \div 0.34$

58. $8.03 + 0.562 + 4.53$

59. $2398.26 - 1959.07$

60. 67.036×0.006

61. $0.061 + 0.0023 + 0.777$

62. $110.72 \div 1.6$

63. $8 \div 0.4 + 0.1 \times (0.2)^2$

64. $(3.8 - 2.8)^3 \div (0.5 + 0.3)$

Applications

3.7 *Solve each problem.*

65. An office manager paid $18.50 for a desk pad set. She paid $29.95 for a lamp. She also paid $2.42 in sales tax for the two items. She took $60 from petty cash to make the purchase at a local supply store. How much will she have left after the purchase?

66. Terri had a balance of $62.36 in her checking account. She made deposits of $108.19 and $73.56. She made out checks for $102.14, $37.56, and $19.95. What will her new balance be?

67. Jill earns $6.88 per hour in her new job. She earns 1.5 times that amount for every hour over 40 hours she works per week. She worked 49 hours last week. How much did she earn?

68. Dan drove to the mountains. His odometer read 26,005.8 miles at the start, and 26,325.8 miles at the end of the trip. He used 12.9 gallons of gas on the trip. How many miles per gallon did his car get? (Round your answer to the nearest tenth.)

69. Robert is considering buying a car and making installment payments of $189.60 for 48 months. The cash price of the car is $6930.50. How much extra does he pay if he uses the installment plan instead of buying the car with one payment?

70. Mr. Zeno has a choice of working as an assistant manager at ABC Company at $315.00 per week or receiving an hourly salary of $8.26 per hour at a different company. He learned from several previous assistant managers that they usually worked 38 hours per week. At which company will he probably earn more money?

71. The EPA standard for safe drinking water is a maximum of 0.002 milligram of mercury in one liter of water. The town wells at Winchester were tested. The test was done on 12 liters of water. The entire 12-liter sample contained 0.03 milligram of mercury. Is the water safe or not? By how much?

72. A chemist wishes to take a sample of 0.322 liter of acid and place it in small test tubes each containing 0.046 liter. How many test tubes will be needed?

Developing Your Study Skills

EXAM TIME: HOW TO REVIEW

Reviewing adequately for an exam enables you to bring together the concepts you have learned over several sections. For your review, you will need to:

1. Reread your textbook. Make a list of any terms, rules, or formulas you need to know for the exam. Be sure you understand them all.
2. Reread your notes. Go over returned homework and quizzes. Redo the problems you missed.
3. Practice some of each type of problem covered in the chapter(s) you are to be tested on.
4. Use the end-of-chapter materials provided in your textbook. Read carefully through the Chapter Organizer. Do the Review Problems. Take the Chapter Test. When you are finished, check your answers. Redo any problems you missed.
5. Get help if any concepts give you difficulty.

Chapter 3 Test

1. Write a word name for the decimal. 0.157

2. Write as a decimal. $\frac{3,977}{10,000}$

Write in fractional notation. Reduce whenever possible.

3. 7.15

4. 0.261

5. Arrange from smallest to largest.
2.19, 2.91, 2.9, 2.907

6. Round to the nearest hundredth.
78.6562

7. Round to the nearest ten thousandth.
0.0341752

Add.

8. $\begin{array}{r} 42.718 \\ 0.191 \\ +\ 1.08 \\ \hline \end{array}$

9. $17 + 2.1 + 16.8 + 0.04 + 1.59$

Subtract.

10. $\begin{array}{r} 1.0075 \\ -\ 0.9096 \\ \hline \end{array}$

11. $72.3 - 1.145$

Multiply.

12. $\begin{array}{r} 8.31 \\ \times\ 0.07 \\ \hline \end{array}$

13. 2.189×100

1. __________
2. __________
3. __________
4. __________
5. __________
6. __________
7. __________
8. __________
9. __________
10. __________
11. __________
12. __________
13. __________

14.

15.

16.

17.

18.

19.

20.

21.

22.

Divide.

14. $0.004\overline{)0.1028}$

15. $0.69\overline{)32.43}$

Write as a decimal.

16. $\frac{11}{9}$

17. $\frac{9}{16}$

Perform the operations in the proper order to evaluate.

18. $(0.3)^3 + 1.02 \div 0.5 - 0.58$

19. $19.36 \div (0.24 + 0.26) \times (0.4)^2$

20. A beef roast weighing 7.8 pounds costs $4.25 per pound. How much will the roast cost?

21. Frank traveled from the city to the shore. His odometer read 42,620.5 miles at the start and 42,780.5 at the end of the trip. He used 8.5 gallons of gas. How many miles per gallon did his car achieve? Round to the nearest tenth.

22. The rainfall for March in Central City was 8.01 centimeters, for April, 5.03 centimeters, and for May, 8.53 centimeters. The normal rainfall for these three months is 25 centimeters. How much less rain fell during these 3 months than usual; that is, how does this year's figure compare with the figure for normal rainfall?

Cumulative Test for Chapters 1–3

Approximately one-half of this test is based on Chapter 3 material. The remainder is based on material covered in Chapters 1 and 2.

1. Write in words. 38,056,954

2. Add.
$$\begin{array}{r} 156{,}028 \\ 301{,}579 \\ +\ 21{,}980 \\ \hline \end{array}$$

3. Subtract.
$$\begin{array}{r} 1{,}091{,}000 \\ -\ 1{,}036{,}520 \\ \hline \end{array}$$

4. Multiply.
$$\begin{array}{r} 589 \\ \times\ 67 \\ \hline \end{array}$$

5. Divide. $17\overline{)4386}$

6. Evaluate. $20 \div 4 + 2^5 - 7 \times 3$

7. Reduce. $\frac{33}{88}$

8. Add. $4\frac{1}{3} + 3\frac{1}{6}$

9. Subtract. $\frac{23}{35} - \frac{2}{5}$

10. Multiply. $\frac{7}{10} \times \frac{5}{3}$

11. Divide. $52 \div 3\frac{1}{4}$

12. Divide. $1\frac{3}{8} \div \frac{5}{12}$

13. Estimate. $58{,}216 \times 438{,}207$

14. Write as a decimal. $\frac{571}{1000}$

15. Arrange from smallest to largest.
2.1, 20.1, 2.01, 2.12, 2.11

16. Round to the nearest thousandth.
26.07984

1. ______

2. ______

3. ______

4. ______

5. ______

6. ______

7. ______

8. ______

9. ______

10. ______

11. ______

12. ______

13. ______

14. ______

15. ______

16. ______

17.

18.

19.

20.

21.

22.

23.

24.

25.

17.
$$\begin{array}{r} 1.9 \\ 2.36 \\ 15.2 \\ +\ 0.08 \\ \hline \end{array}$$

18.
$$\begin{array}{r} 28.007 \\ -\ 19.368 \\ \hline \end{array}$$

19.
$$\begin{array}{r} 56.8 \\ \times\ 0.02 \\ \hline \end{array}$$

20. 365.123×100

21. $0.06\overline{)0.06348}$

22. Write as a decimal. $\frac{13}{16}$

23. Perform the operations in the proper order.

$$1.44 \div 0.12 + (0.3)^3 + 1.57$$

24. A car gets 28.5 miles per gallon. Its gas tank holds 16.0 gallons of gas. Approximately how many miles can the car travel on one tank of gas?

25. Sue's savings account balance is \$199.36. This month she earned interest of \$1.03. She deposited \$166.35 and \$93.50. She withdrew money three times, in the amounts for \$90.00, \$37.49, and \$137.18. What will her balance be at the start of the next month?

CHAPTER 4

Ratio and Proportion

Have you ever lived in or visited a city that was so crowded that you could hardly move? Other cities and towns have many people, but somehow in them you never feel crowded.

Mathematics helps us to be more precise about this "crowded" feeling in a city. If you were to compare two cities, could you make some mathematical calculations to determine which of the two was more densely populated? Turn to the Putting Your Skills to Work on page 270.

Pretest Chapter 4

If you are familiar with the topics in this chapter, take this test now. Check your answers with those in the back of the book. If an answer was wrong or you couldn't do a problem, study the appropriate section of the chapter.

If you are not familiar with the topics in this chapter, don't take this test now. Instead, study the examples, work the practice problems, and then take the test.

This test will help you identify which concepts you have mastered and which you need to study further.

Section 4.1

Write each ratio *in simplest form.*

1. 13 to 18 **2.** 44 to 220

3. \$72 to \$16 **4.** 121 kilograms to 132 kilograms

5. Sam's take-home pay is \$240 per week. \$70 per week is withheld for federal taxes and \$22 per week is withheld for state taxes.

(a) Find the ratio of federal withholding to take-home pay.

(b) Find the ratio of state withholding to take-home pay.

Write each rate *in simplest form.*

6. \$164 for 48 cabinets

7. 620 gallons of water for each 840 square feet of lawn

Write as a unit rate. *Round to the nearest tenth if necessary.*

8. 122 miles are traveled in 4 hours. What is the rate in miles per hour?

9. 16 radios are purchased for \$592. What is the cost per radio?

Section 4.2

Write a proportion for each of the following.

10. 13 is to 40 as 39 is to 120

11. 42 is to 78 as 21 is to 39

Determine if each proportion is true or false.

12. $\frac{14}{31} = \frac{42}{93}$ **13.** $\frac{17}{33} = \frac{19}{45}$

1. ______
2. ______
3. ______
4. ______
5. (a) ______
   (b) ______
6. ______
7. ______
8. ______
9. ______
10. ______
11. ______
12. ______
13. ______

Section 4.3

Solve for n in each equation.

14. $9 \times n = 153$ **15.** $234 = 13 \times n$

Solve for n in each proportion.

16. $\frac{36}{16} = \frac{9}{n}$ **17.** $\frac{3}{144} = \frac{n}{336}$

18. $\frac{n}{900} = \frac{15}{22.5}$

Section 4.4

Solve each problem using a proportion.

19. A recipe for six portions calls for 1.5 cups of flour. How many cups of flour are needed to make 14 portions?

20. Maria's car can travel 81 miles on 2 gallons of gas. How many can she travel on 9 gallons of gas?

21. Two cities are 5 inches apart on a map, but the actual distance between them is 365 miles. What is the actual distance between two other cities that are 2 inches apart on the map?

22. A shipment of 121 light bulbs had 6 defective bulbs. How many defective bulbs would we expect, at the same rate, in a shipment of 1089 light bulbs?

14. ______________

15. ______________

16. ______________

17. ______________

18. ______________

19. ______________

20. ______________

21. ______________

22. ______________

4.1 Ratios and Rates

MathPro Video 9 SSM

After studying this section, you will be able to:

1 *Use a ratio to compare two quantities with the same unit.*
2 *Use a rate to compare two quantities with different units.*

1 *Using a Ratio to Compare Two Quantities with the Same Unit*

Assume that you earn 13 dollars an hour and your friend earns 10 dollars per hour. The *ratio* 13:10 compares what you and your friend make. This ratio means that for every 13 dollars you earn your friend earns 10. The *rate* you are paid is 13 dollars per hour, which compares 13 dollars to 1 hour. In this section we see how to use both rates and ratios to solve many everyday problems.

Suppose that we want to compare an object weighing 20 pounds to an object weighing 23 pounds. The ratio of their weight would be 20 to 23. A **ratio** is the comparison of two quantities that have the *same units.*

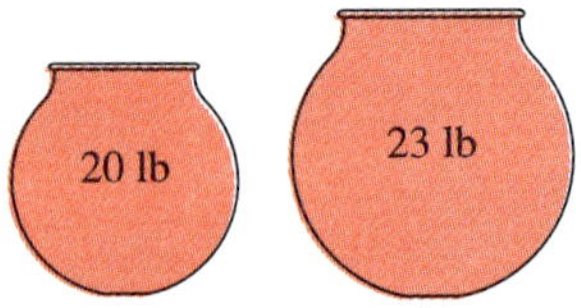

(a) (b)

We can express the ratio three ways.

We can write "the ratio of 20 to 23."
We can write "20:23" using a colon.
We can write "$\frac{20}{23}$" using a fraction.

All are valid ways to *compare* 20 to 23. Each is read as "20 to 23."

> We always want to write a ratio in simplest form. A ratio is in *simplest form* when the two numbers do not have a common factor.

EXAMPLE 1 Write each ratio in simplest form. Express your answer as a fraction.

(a) The ratio of 15 hours to 20 hours
(b) The ratio of 36 hours to 30 hours
(c) 125:150

(a) $\frac{15}{20} = \frac{3}{4}$ **(b)** $\frac{36}{30} = \frac{6}{5}$ **(c)** $\frac{125}{150} = \frac{5}{6}$

Notice that in each case the two numbers *do* have a common factor. When we form the fraction—that is, the ratio—we take the extra step of *reducing* the fraction.

Practice Problem 1 Write each ratio in simplest form. Express your answer as a fraction.

(a) The ratio of 36 feet to 40 feet
(b) The ratio of 18 feet to 15 feet
(c) 220:270 ■

EXAMPLE 2 Martin earns \$350 weekly. However, he takes home only \$250 per week in his paycheck.

\$350.00	gross pay (what Martin earns)	
45.00	withheld for federal tax	(what is taken out of Martin's earnings)
20.00	withheld for state tax	
35.00	withheld for retirement	
\$250.00	take-home pay (what Martin has left)	

(a) What is the ratio of the amount withheld for federal tax to gross pay?

(b) What is the ratio of the amount withheld for state tax to the amount withheld for federal tax?

(a) The ratio of the amount withheld for federal tax compared to gross pay is

$$\frac{45}{350} = \frac{9}{70}$$

(b) The ratio of the amount withheld for state tax compared to the amount withheld for federal tax is

$$\frac{20}{45} = \frac{4}{9}$$

Practice Problem 2 Professor Fowler has 90 students in his course on introductory psychology. The class has 18 freshman students, 25 sophomore students, 36 junior students, and 11 senior students.

(a) What is the ratio of freshman students to junior students?

(b) What is the ratio of sophomore students to the entire number of students in the class? ■

To Think About Perhaps you have heard statements like "a certain jet plane travels at Mach 2.2." What does that mean? A Mach number is a ratio that compares the velocity (speed) of an object to the velocity of sound. Sound travels at about 330 meters per second. This ratio is written in decimal form.

What is the Mach number of a jet traveling at 690 meters per second?

$$\text{Mach number of jet} = \frac{690 \text{ meters per second}}{330 \text{ meters per second}}$$

$$= 2.09090909\ldots \text{ or } 2.\overline{09}$$

Rounded to the nearest tenth, the Mach number of the jet is 2.1.

Problems 71 and 72 in Exercises 4.1 deal with Mach numbers.

2 *Using a Rate to Compare Two Quantities with Different Units*

A **rate** is a comparison of two quantities with *different units*. Usually, to avoid misunderstanding, we express a rate as a fraction with the units included.

EXAMPLE 3 John's car traveled 371 miles on 14 gallons of gasoline. What is the rate of miles to gallons?

The rate is

$$\frac{371 \text{ miles}}{14 \text{ gallons}} = \frac{53 \text{ miles}}{2 \text{ gallons}}$$

Practice Problem 3 A farmer is charged a $44 storage fee for every 900 tons of grain he stores. What is the rate of the storage fee in dollars to tons of grain? ■

Often we want to know the rate for a single unit, which is the unit rate. A **unit rate** is a rate in which the denominator is the number 1.

EXAMPLE 4 A car traveled 301 miles in 7 hours. Find the unit rate.

$\frac{301}{7}$ can be simplified. We divide $301 \div 7 = 43$.

Thus

$$\frac{301 \text{ miles}}{7 \text{ hours}} = \frac{43 \text{ miles}}{1 \text{ hour}}$$

The denominator is 1. We write our answer as 43 miles/hour. The fraction line is read as the word "per," so our answer here is read "43 miles per hour." "Per" means "for every," so a rate of 43 miles per hour means 43 miles traveled for every hour traveled.

Practice Problem 4 A car traveled 212 miles in 4 hours. Find the unit rate. ■

EXAMPLE 5 A grocer purchased 200 pounds of apples for \$68. He sold the 200 pounds of apples for \$86. How much profit did he make per pound of apples?

$$\begin{array}{rl} \$86 & \text{selling price} \\ -\ 68 & \text{cost} \\ \hline \$18 & \text{profit} \end{array}$$

The rate that compares profit to pounds of apples sold is $\dfrac{18 \text{ dollars}}{200 \text{ pounds}}$. We will divide $18 \div 200$.

$$\begin{array}{r} 0.09 \\ 200\overline{)18.00} \\ \underline{18\ 00} \\ 0 \end{array}$$

The unit rate of profit is \$0.09 per pound.

Practice Problem 5 A retailer purchased 120 nickel-cadmium batteries for flashlights for \$129.60. She sold them for \$170.40. What was her profit per battery? ■

EXAMPLE 6 Hamburger at a local butcher is packaged in large and extra-large packages. A large package costs \$7.86 for 6 pounds and an extra-large package is \$10.08 for 8 pounds.

(a) What is the unit rate in dollars per pound for each size package?

(b) How much per pound does a consumer save by buying the extra-large package?

(a) $\dfrac{7.86 \text{ dollars}}{6 \text{ pounds}} = \$1.31/\text{pound}$ for the large package

$\dfrac{10.08 \text{ dollars}}{8 \text{ pounds}} = \$1.26/\text{pound}$ for the extra-large package

(b) $\begin{array}{r} \$1.31 \\ -\ 1.26 \\ \hline \$0.05 \end{array}$ A person saves \$0.05/pound by buying the extra-large package.

Practice Problem 6 A 12-ounce package of Fred's favorite cereal costs \$2.04. A 20-ounce package of the same cereal costs \$2.80.

(a) What is the cost per ounce of each size of cereal?

(b) How much per ounce would Fred save by buying the larger size? ■

4.1 Exercises

Verbal and Writing Skills

1. A ____________ is a comparison of two quantities that have the same unit.
2. A rate is a comparison of two quantities that have ____________ units.
3. The ratio 5:8 is read ____________.
4. Marion compares the number of loaves of bread she bakes to the number of pounds of flour she needs to make the bread. Is this a ratio or a rate? Why?

Write each ratio in simplest form. Express your answer as a fraction.

5. 18:24
6. 15:12
7. 21:18
8. 42:51
9. 55:121
10. 78:104
11. 150:225
12. 360:480
13. 60 to 64
14. 33 to 57
15. 28 to 42
16. 21 to 98
17. 32 to 20
18. 90 to 54
19. 8 ounces to 12 ounces
20. 24 minutes to 52 minutes
21. 16 feet to 24 feet
22. 20 miles to 45 miles
23. 50 years to 85 years
24. 39 kilograms to 26 kilograms
25. \$86 to \$120
26. \$54 to \$63
27. 153 inches to 17 inches
28. 143 feet to 13 feet
29. 91 tons to 133 tons
30. 99 pounds to 143 pounds

Use the following table to answer problems 31–34.

ROBIN'S WEEKLY PAYCHECK

Total (Gross) Pay	Federal Withholding	State Withholding	Retirement	Insurance	Take-Home Pay
\$285	\$35	\$20	\$28	\$16	\$165

31. What is the ratio of take-home pay to total (gross) pay?
32. What is the ratio of retirement to insurance?
33. What is the ratio of federal withholding to take-home pay?
34. What is the ratio of retirement to total (gross) pay?

An automobile insurance company prepared the following analysis for its clients. Use this table for problems 35–38.

Analysis of number of years that four-door sedans are driven

Sedans that lasted 2 years or less	Sedans that lasted 4 years or less but more than 2 years	Sedans that lasted 6 years or less but more than 4 years	Sedans that lasted more than 6 years	Total number of sedans
205	255	450	315	1225

35. What is the ratio of sedans that lasted two years or less to the total number of sedans?

36. What is the ratio of sedans that lasted more than six years to the total number of sedans?

37. What is the ratio of the number of sedans that lasted six years or less but more than four years to the number of sedans that lasted two years or less?

38. What is the ratio of the number of sedans that lasted more than six years to the number of sedans that lasted four years or less but more than two years?

39. A builder constructs a new home for \$128,000. His labor costs were \$72,000. What is the ratio of labor cost to total cost?

40. A company produces a new-model television set for a total cost of \$405.00. The research and development cost for each set is \$90. What is the ratio of research and development cost to the total cost?

Write as a rate in simplest form.

41. \$60 for 18 plants

42. \$135 for 20 cabinets

43. \$170 for 12 bushes

44. 98 pounds for 22 people

45. 310 gallons of water for every 625 square feet of lawn

46. 460 gallons of water for every 825 square feet of lawn

47. 6150 revolutions for every 15 miles

48. 9540 revolutions for every 18 miles

49. Snowplow 18 miles of road every 8 hours

50. 48 pies for 340 people

Write as a unit rate. Round to the nearest tenth when necessary.

51. Earn \$520 in 40 hours

52. Earn \$304 in 38 hours

53. Travel 192 miles on 12 gallons of gas

54. Travel 322 miles on 14 gallons of gas

55. Pump 2480 gallons in 16 hours

56. Pump 2880 gallons in 24 hours

57. 2760 words on 12 pages

58. 1935 words on 9 pages

59. Travel 3619 kilometers in 7 hours

60. Travel 2992 kilometers in 11 hours

Applications

61. \$3870 was spent for 129 shares of Mattel stock. Find the cost per share.

62. \$6150 was spent for 150 shares of Polaroid stock. Find the cost per share.

63. A jeweler purchased 245 watches for \$6370. She sold them for \$9800. How much profit did she make per watch?

64. A retailer purchased 24 compact disc players for \$3380. She sold them for \$4030. How much profit did she make per compact disc player? Round your answers to the nearest dollar.

65. A 16-ounce box of dry pasta costs \$1.28. A 24-ounce box of the same pasta costs \$1.68.
(a) What is the cost of each box of pasta per ounce?
(b) How much does the educated consumer save by buying the larger box?
(c) How much does the consumer save by buying 2 large boxes instead of 3 small boxes?

66. A 16-ounce can of beef stew costs \$2.88. A 26-ounce can of the same beef stew costs \$4.16.
(a) What is the cost of each can of stew per ounce?
(b) How much does the consumer save per ounce by buying the larger can?

67. Dr. Robert Tobey completed a count of two herds of moose in the central regions of Alaska. He recorded 3978 moose on the North Slope and 5520 moose on the South Slope. There are 306 acres on the North Slope and 460 acres on the South Slope.

(a) How many moose per acre were found on the North Slope?

(b) How many moose per acre were found on the South Slope?

(c) In which region are the moose more closely crowded together?

68. In Melbourne, Australia, 27,900 people live in the suburb of St. Kilda and 38,700 live in the suburb of Caulfield. The area of St. Kilda is 6500 acres. The area of Caulfield is 9200 acres. Round your answers to the nearest tenth.

(a) How many people per acre live in St. Kilda?

(b) How many people per acre live in Caulfield?

(c) Which suburb is more crowded?

★ **69.** Mr. Jenson bought 525 shares of Zenith stock for \$12,876.50. How much did he pay per share? Round your answer to the nearest cent.

★ **70.** A Kentucky horse breeder bought 59 horses for \$4,350,500. How much did she pay per horse? Round to the nearest cent.

To Think About

For Problems 71 and 72, recall that the speed of sound is about 330 meters per second. Round your answers to the nearest tenth.

71. A jet plane was originally designed to fly at 750 meters per second. It was modified to fly at 810 meters per second. By how much was its Mach number increased?

72. A rocket was first flown at 1960 meters per second. It proved unstable and unreliable at that speed. It is now flown at a maximum of 1920 meters per second. By how much was its Mach number decreased?

Cumulative Review Problems

Calculate.

73. $2\frac{1}{4} + \frac{3}{8}$

74. $\frac{5}{7} \div \frac{3}{21}$

75. $\frac{8}{23} \times \frac{5}{16}$

76. $3\frac{1}{16} - 2\frac{1}{24}$

77. A room 12 yards × 5.2 yards had a carpet installed. The bill was \$764.40. What was the cost of the installed carpet per yard?

78. An electronics superstore bought 1050 computer games for \$23 each. How much did the store pay in all for these games? The store sold the games for \$39 each. How much profit did the store make?

4.2 The Concept of Proportions

MathPro Video 9 SSM

After studying this section, you will be able to:

1. *Write a proportion.*
2. *Determine if a statement is a proportion.*

1 *Writing a Proportion*

A **proportion** states that two ratios or two rates are equal. For example, $\frac{5}{8} = \frac{15}{24}$ is a proportion and $\frac{7 \text{ feet}}{8 \text{ dollars}} = \frac{35 \text{ feet}}{40 \text{ dollars}}$ is also a proportion. A proportion can be read two ways. The proportion $\frac{5}{8} = \frac{15}{24}$ can be read "five eighths equals fifteen twenty-fourths," or it can be read "five *is to* eight *as* fifteen *is to* twenty-four."

EXAMPLE 1 Write the proportion 5 is to 7 as 15 is to 21.

$$\frac{5}{7} = \frac{15}{21}$$

Practice Problem 1 Write the proportion 6 is to 8 as 9 is to 12. ■

EXAMPLE 2 Write a proportion to express the following: If 4 rolls of wallpaper measure 300 feet, then 8 rolls of wallpaper will measure 600 feet.

When you write a proportion, order is important. Be sure that the units for each ratio are in the same position in the fraction.

$$\frac{4 \text{ rolls}}{300 \text{ feet}} = \frac{8 \text{ rolls}}{600 \text{ feet}}$$

Practice Problem 2 Write a proportion to express the following: If it takes 2 hours to drive 72 miles, then it will take 3 hours to drive 108 miles. ■

2 *Determining If a Statement Is a Proportion*

By definition, a proportion states that two ratios are equal. $\frac{2}{7} = \frac{4}{14}$ is a proportion because $\frac{2}{7}$ and $\frac{4}{14}$ are equivalent fractions. You might say that $\frac{2}{7} = \frac{4}{14}$ is a *true* proportion. It is easy enough to see that $\frac{2}{7} = \frac{4}{14}$ is true. However, is $\frac{4}{14} = \frac{6}{21}$ true? Is $\frac{4}{14} = \frac{6}{21}$ a proportion? To determine whether or not a statement is a proportion, we use the equality test for fractions.

Equality Test for Fractions

For any two fractions where $b \neq 0$, $d \neq 0$

$$\text{If } \frac{a}{b} = \frac{c}{d}, \text{ then } a \times d = b \times c$$

Thus, to see if $\frac{4}{14} = \frac{6}{21}$, we can multiply

$$\frac{4}{14} = \frac{6}{21}$$

$14 \times 6 = 84$
$4 \times 21 = 84$

Products are equal.

$\frac{4}{14} = \frac{6}{21}$ is true. $\frac{4}{14} = \frac{6}{21}$ is a proportion.

EXAMPLE 3 Determine which of the following is a proportion.

(a) $\frac{14}{18} \stackrel{?}{=} \frac{35}{45}$ **(b)** $\frac{16}{21} \stackrel{?}{=} \frac{174}{231}$

(a) $\frac{14}{18} \stackrel{?}{=} \frac{35}{45}$

$18 \times 35 = 630$

products equal

$14 \times 45 = 630$

Thus $\frac{14}{18} = \frac{35}{45}$. This is a proportion.

(b) $\frac{16}{21} \stackrel{?}{=} \frac{174}{231}$

$21 \times 174 = 3654$

products not equal

$16 \times 231 = 3696$

Thus $\frac{16}{21} \neq \frac{174}{231}$. This is *not* a proportion.

Practice Problem 3 Determine which of the following is a proportion.

(a) $\frac{10}{18} \stackrel{?}{=} \frac{25}{45}$ **(b)** $\frac{42}{100} \stackrel{?}{=} \frac{22}{55}$ ■

Proportions may involve fractions or decimals.

EXAMPLE 4 Determine which of the following is a proportion.

(a) $\frac{5.5}{7} \stackrel{?}{=} \frac{33}{42}$ **(b)** $\frac{5}{8\frac{3}{4}} \stackrel{?}{=} \frac{40}{72}$

(a) $\frac{5.5}{7} \stackrel{?}{=} \frac{33}{42}$

$7 \times 33 = 231$

products equal

$5.5 \times 42 = 231$

Thus $\frac{5.5}{7} = \frac{33}{42}$. This is a proportion.

(b) $\dfrac{5}{8\frac{3}{4}} \stackrel{?}{=} \dfrac{40}{72}$. To multiply $8\frac{3}{4} \times 40$ we use

$$\frac{35}{\not{4}_{1}} \times \overset{10}{\not{40}} = 35 \times 10 = 350$$

Thus

$$\frac{5}{8\frac{3}{4}} \times \frac{40}{72} \qquad 8\frac{3}{4} \times 40 = 350 \;\; \text{products not equal} \;\; 5 \times 72 = 360$$

Thus $\dfrac{5}{8\frac{3}{4}} \neq \dfrac{40}{72}$. This is *not* a proportion.

Practice Problem 4 Determine which of the following is a proportion.

(a) $\dfrac{2.4}{3} \stackrel{?}{=} \dfrac{12}{15}$ **(b)** $\dfrac{2\frac{1}{3}}{6} \stackrel{?}{=} \dfrac{14}{38}$ ■

EXAMPLE 5 **(a)** Is the rate $\dfrac{\$86}{13 \text{ tons}}$ equal to the rate $\dfrac{\$79}{12 \text{ tons}}$?

(b) Is the rate $\dfrac{3 \text{ American dollars}}{2 \text{ British pounds}}$ equal to the rate $\dfrac{27 \text{ American dollars}}{18 \text{ British pounds}}$?

(a) We want to know whether $\dfrac{86}{13} \stackrel{?}{=} \dfrac{79}{12}$.

$$\frac{86}{13} \times \frac{79}{12} \qquad 13 \times 79 = 1027 \;\; \text{products not equal} \;\; 86 \times 12 = 1032$$

Thus the two rates are not equal. This is not a proportion.

(b) We want to know whether $\dfrac{3}{2} \stackrel{?}{=} \dfrac{27}{18}$.

$$\frac{3}{2} \times \frac{27}{18} \qquad 2 \times 27 = 54 \;\; \text{products equal} \;\; 3 \times 18 = 54$$

Thus the two rates are equal. This is a proportion.

Practice Problem 5

(a) Is the rate $\dfrac{1260 \text{ words}}{7 \text{ pages}}$ equal to the rate $\dfrac{3530 \text{ words}}{20 \text{ pages}}$?

(b) Is the rate $\dfrac{2 \text{ American dollars}}{11 \text{ French francs}}$ equal to the rate $\dfrac{16 \text{ American dollars}}{88 \text{ French francs}}$? ■

4.2 Exercises

Verbal and Writing Skills

1. A proportion states that two ratios or rates are ____________.

2. Explain in your own words how we use the equality test for fractions to determine if a statement is a proportion. Give an example.

Write a proportion for each.

3. 48 is to 32 as 3 is to 2.

4. 40 is to 36 as 10 is to 9.

5. 8 is to 3 as 32 is to 12.

6. 32 is to 48 as 18 is to 27.

7. 20 is to 36 as 5 is to 9.

8. 42 is to 48 as 14 is to 16.

9. 27 is to 15 as 9 is to 5.

10. 42 is to 28 as 6 is to 4.

11. 22 is to 30 as 11 is to 15.

12. 75 is to 100 as 9 is to 12.

13. 45 is to 135 as 9 is to 27.

14. 44 is to 121 as 8 is to 22.

15. 5.5 is to 10 as 11 is to 20.

16. 6.5 is to 14 as 13 is to 28.

Applications

Write a proportion for each.

17. If 12 pounds of flour cost \$4, then 33 pounds will cost \$11.

18. If we travel 110 miles on Interstate I-90 in 2 hours, then we will travel 165 miles in 3 hours.

19. If Ty hit 10 home runs in 45 baseball games, then he should hit 36 home runs in 162 games.

20. If Amelia drove 450 miles in her Toyota Camry on 25 gallons of gas, then she should drive 1080 miles on 60 gallons of gas.

21. If 20 pounds of pistachio nuts cost \$75, then 30 pounds will cost \$112.50.

22. If 3 credit hours at El Paso Community College costs \$525, then 7 credit hours should cost \$1225.

23. Last year the Little League Baseball finals held 2200 people on 100 benches. This year we will have 2750 on 125 benches.

24. There are 3 teaching assistants for every 40 children in the elementary school. If we have 280 students, then we will have 21 teaching assistants.

25. If 16 pounds of fertilizer covers 1520 square feet of lawn, then 19 pounds of fertilizer should cover 1805 square feet of lawn.

26. When New City had 4800 people, it had 3 restaurants. Now New City has 11,200 people, so it should have 7 restaurants.

Determine which of the following is a proportion.

27. $\frac{21}{35} \stackrel{?}{=} \frac{15}{25}$

28. $\frac{12}{42} \stackrel{?}{=} \frac{10}{35}$

29. $\frac{18}{12} \stackrel{?}{=} \frac{42}{28}$

30. $\frac{14}{10} \stackrel{?}{=} \frac{56}{40}$

31. $\frac{48}{56} \stackrel{?}{=} \frac{40}{45}$

32. $\frac{27}{45} \stackrel{?}{=} \frac{22}{40}$

33. $\frac{9}{12} \stackrel{?}{=} \frac{15}{20}$

34. $\frac{8}{9} \stackrel{?}{=} \frac{48}{54}$

35. $\frac{99}{100} \stackrel{?}{=} \frac{49}{50}$

36. $\frac{17}{75} \stackrel{?}{=} \frac{22}{100}$

37. $\frac{315}{2100} \stackrel{?}{=} \frac{15}{100}$

38. $\frac{102}{120} \stackrel{?}{=} \frac{85}{100}$

39. $\frac{6}{14} \stackrel{?}{=} \frac{4.5}{10.5}$

40. $\frac{7}{9} \stackrel{?}{=} \frac{10.5}{13.5}$

41. $\frac{11}{12} \stackrel{?}{=} \frac{9.5}{10}$

42. $\frac{3}{17} \stackrel{?}{=} \frac{4.5}{24.5}$

43. $\frac{2}{4\frac{1}{3}} \stackrel{?}{=} \frac{6}{13}$

44. $\frac{2\frac{1}{3}}{3} \stackrel{?}{=} \frac{7}{15}$

45. $\frac{8}{19} \stackrel{?}{=} \frac{2}{4\frac{3}{4}}$

46. $\frac{9}{22} \stackrel{?}{=} \frac{3}{7\frac{1}{3}}$

47. $\frac{\text{55 feet}}{\text{4 rolls}} \stackrel{?}{=} \frac{\text{220 feet}}{\text{16 rolls}}$

48. $\frac{\text{68 feet}}{\text{3 rolls}} \stackrel{?}{=} \frac{\text{204 feet}}{\text{9 rolls}}$

49. $\frac{\text{286 gallons}}{\text{12 acres}} \stackrel{?}{=} \frac{\text{429 gallons}}{\text{18 acres}}$

50. $\frac{\text{166 gallons}}{\text{14 acres}} \stackrel{?}{=} \frac{\text{249 gallons}}{\text{21 acres}}$

51. $\frac{\text{48 points}}{\text{56 games}} \stackrel{?}{=} \frac{\text{40 points}}{\text{45 games}}$

52. $\frac{\text{27 points}}{\text{45 games}} \stackrel{?}{=} \frac{\text{22 points}}{\text{40 games}}$

Applications

53. At the Amy Grant concert on Friday there were 9600 female fans and 8200 male fans. The concert on Saturday had 12,480 female fans and 10,660 male fans. Is the ratio of female fans to male fans the same for each night of the concert?

54. In the beginner scuba diving course, there are 96 male students and 54 female students. In the intermediate scuba diving course, there are 144 male students and 81 female students. Is the ratio of male students to female students the same in each scuba class?

55. A machine folds 730 boxes in 6 hours. Another machine folds 1090 boxes in 9 hours. Do they fold boxes at the same rate?

56. A car traveled 550 miles in 15 hours. A bus traveled 230 miles in 6 hours. Did they travel at the same rate?

To Think About

57. Determine if $\frac{63}{161} = \frac{171}{437}$.

(a) By reducing each side to lowest terms.

(b) By using the equality test for fractions.

(c) Which method was faster? Why?

58. Determine if $\frac{169}{221} = \frac{247}{323}$.

(a) By reducing each side to lowest terms.

(b) By using the equality test for fractions.

(c) Which method was faster? Why?

Cumulative Review Problems

Perform each calculation.

59. $9.6 + 7.8 + 2.56 + 3.004 + 0.1765$

60. 2.83×5.002

61. $\begin{array}{r} 29{,}366.215 \\ -\ 28{,}963.807 \\ \hline \end{array}$

62. $7.03\overline{)181.374}$

4.3 Solving Proportions

MathPro Video 10 SSM

After studying this section, you will be able to:

1 *Understand the idea of a variable and be able to solve for n in an equation of the form $a \times n = b$.*

2 *Find the missing number in a proportion.*

1 Solving for a Variable in an Equation of the Form $a \times n = b$

Consider this expression: ''3 times a number yields 15. What is the number?'' We could write this as

$$3 \times \boxed{?} = 15$$

and guess that the number $\boxed{?} = 5$. There is a better way of solving this problem, a way that eliminates the guesswork. We will begin by using a **variable.** That is, we will use a letter to represent a number we do not yet know. We briefly used variables in Chapters 1–3. Now we use them more extensively.

Let the letter n represent the unknown number. We write

$$3 \times n = 15$$

This is called an **equation.** An equation has an equals sign. We want to find the number n in this equation without guessing. We will not change the value of n in the equation if we divide both sides of the equation by 3. Thus if

$$3 \times n = 15$$

we can say

$$\frac{3 \times n}{3} = \frac{15}{3}$$

which is

$$\frac{3}{3} \times n = 5$$

or

$$1 \times n = 5$$

Since $1 \times$ any number is the same number, we know that $n = 5$. Any equation of the form $a \times n = b$ can be solved in this way. We divide both sides of an equation of the form $a \times n = b$ by the number that is multiplied by n. (This method will not work for $3 + n = 15$, since here the 3 is added to n and not multiplied by n.)

EXAMPLE 1 Solve each equation to find the value of n.

(a) $16 \times n = 80$ **(b)** $24 \times n = 240$

(a) $16 \times n = 80$

$$\frac{16 \times n}{16} = \frac{80}{16}$$ Divide each side by 16.

$n = 5$ Because $16 \div 16 = 1$ and $80 \div 16 = 5$.

(b) $24 \times n = 240$

$$\frac{24 \times n}{24} = \frac{240}{24}$$ Divide each side by 24.

$n = 10$ Because $24 \div 24 = 1$ and $240 \div 24 = 10$.

Practice Problem 1 Solve each equation to find the value of n.

(a) $5 \times n = 45$ **(b)** $7 \times n = 84$ ■

The same procedure is followed if the variable n is on the right side of the equation.

EXAMPLE 2 Solve each equation to find the value of n.

(a) $66 = 11 \times n$ **(b)** $143 = 13 \times n$

(a) $66 = 11 \times n$

$$\frac{66}{11} = \frac{11 \times n}{11}$$ Divide each side by 11.

$$6 = n$$

(b) $143 = 13 \times n$

$$\frac{143}{13} = \frac{13 \times n}{13}$$ Divide each side by 13.

$$11 = n$$

Practice Problem 2 Solve each equation to find the value of n.

(a) $108 = 9 \times n$ **(b)** $210 = 14 \times n$ ■

The numbers in the equations are not always whole numbers, and the answer to an equation is not always a whole number.

EXAMPLE 3 Solve each equation to find the value of n.

(a) $16 \times n = 56$ **(b)** $18.2 = 2.6 \times n$

(a) $16 \times n = 56$

$$\frac{16 \times n}{16} = \frac{56}{16}$$ Divide each side by 16.

$$n = 3.5$$

```
    3.5
16)56.0
   48
    8 0
    8 0
      0
```

(b) $18.2 = 2.6 \times n$

$$\frac{18.2}{2.6} = \frac{2.6 \times n}{2.6}$$ Divide each side by 2.6.

$$7 = n$$

```
         7.
2.6^)18.2^
     18 2
        0
```

Practice Problem 3 Solve each equation to find the value of n.

(a) $15 \times n = 63$ **(b)** $39.2 = 5.6 \times n$ ■

2 *Finding the Missing Number in a Proportion*

Sometimes one of the pieces of a proportion is unknown. We can use an equation such as $a \times n = b$ and solve for n to find the unknown quantity. Suppose that we want to know the value of n in the proportion

$$\frac{5}{12} = \frac{n}{144}$$

Since this is a proportion, we know that $5 \times 144 = 12 \times n$. Simplifying, we have

$$720 = 12 \times n$$

Next we divide both sides by 12.

$$\frac{720}{12} = \frac{12 \times n}{12}$$

$$60 = n$$

We check to see if this is correct. Do we have a true proportion?

$$\frac{5}{12} \stackrel{?}{=} \frac{60}{144}$$

$$\frac{5}{12} \times \frac{60}{144} \qquad \begin{array}{l} 12 \times 60 = 720 \\ 5 \times 144 = 720 \end{array} \quad \text{products equal}$$

Thus $\frac{5}{12} = \frac{60}{144}$ is true. We have checked our answer.

To Solve for a Missing Number in a Proportion

1. Find the two diagonal products (also called "cross-multiplying" or "forming a cross product").
2. Divide each side of the equation by the number multiplied by n.
3. Simplify the result.
4. Check your answer.

EXAMPLE 4 Find the value of n in $\frac{25}{4} = \frac{n}{12}$.

$25 \times 12 = 4 \times n$ Find the two diagonal products.

$300 = 4 \times n$

$\frac{300}{4} = \frac{4 \times n}{4}$ *Divide each side by* 4.

$75 = n$

Check whether the proportion is true.

$$\frac{25}{4} \stackrel{?}{=} \frac{75}{12}$$

$$25 \times 12 \stackrel{?}{=} 4 \times 75$$

$$300 = 300 \checkmark$$

The proportion is true. The answer $n = 75$ is correct.

Practice Problem 4 Find the value of n in $\frac{24}{n} = \frac{3}{7}$. ■

Some answers will be exact values. In other cases we will obtain answers that are rounded to a certain decimal place. If a value is rounded we sometimes use the $\approx$ symbol which indicates "is approximately equal to."

The answer to the next problem is not a whole number.

EXAMPLE 5 Find the value of n in $\frac{125}{2} = \frac{150}{n}$.

$$125 \times n = 2 \times 150 \qquad \text{Find the two diagonal products.}$$

$$125 \times n = 300$$

$$\frac{125 \times n}{125} = \frac{300}{125} \qquad \text{Divide each side by 125.}$$

$$n = 2.4$$

$$\begin{array}{r} 2.4 \\ 125\overline{)300.0} \\ \underline{250} \\ 50\ 0 \\ \underline{50\ 0} \\ 0 \end{array}$$

Check. $\frac{125}{2} \stackrel{?}{=} \frac{150}{2.4}$

$$125 \times 2.4 \stackrel{?}{=} 2 \times 150$$

$$300 = 300 \quad \checkmark$$

Practice Problem 5 Find the value of n in $\frac{176}{4} = \frac{286}{n}$. ■

In applied situations it is helpful to write the units of measure in the proportion. Remember, order is important. The same units should be in the same position in each fraction.

EXAMPLE 6 If 5 grams of a nonicing additive are placed in 8 liters of diesel fuel, how many grams n should be added to 12 liters of diesel fuel?

We need to find the value of n in $\frac{n \text{ grams}}{12 \text{ liters}} = \frac{5 \text{ grams}}{8 \text{ liters}}$.

$$8 \times n = 12 \times 5$$

$$8 \times n = 60$$

$$\frac{8 \times n}{8} = \frac{60}{8}$$

$$n = 7.5$$

The answer is 7.5 grams. 7.5 grams of the additive should be added to 12 liters of the diesel fuel.

Check. $\frac{7.5 \text{ grams}}{12 \text{ liters}} = \frac{5 \text{ grams}}{8 \text{ liters}}$

$$7.5 \times 8 \stackrel{?}{=} 12 \times 5$$

$$60 = 60 \quad \checkmark$$

Practice Problem 6 Find the value of n in $\frac{80 \text{ dollars}}{5 \text{ tons}} = \frac{n \text{ dollars}}{6 \text{ tons}}$. ■

Some proportions contain decimals.

EXAMPLE 7 Find the value of n in $\dfrac{141.75 \text{ miles}}{4.5 \text{ hours}} = \dfrac{63 \text{ miles}}{n \text{ hours}}$.

$$141.75 \times n = 63 \times 4.5$$

$$141.75 \times n = 283.5$$

$$\frac{141.75 \times n}{141.75} = \frac{283.5}{141.75}$$

$$n = 2$$

The answer is $n = 2$. The check is up to you.

Practice Problem 7 Find the value of n in $\dfrac{264.25 \text{ meters}}{3.5 \text{ seconds}} = \dfrac{n \text{ meters}}{2 \text{ seconds}}$. ■

To Think About Suppose that the proportion contains fractions or mixed numbers. Could you still follow all the steps? For example, find n when

$$\frac{n}{3\frac{1}{4}} = \frac{5\frac{1}{6}}{2\frac{1}{3}} \qquad \text{We would have} \qquad 2\frac{1}{3} \times n = 5\frac{1}{6} \times 3\frac{1}{4}$$

This can be written as

$$\frac{7}{3} \times n = \frac{31}{6} \times \frac{13}{4}$$

$$\boxed{\frac{7}{3} \times n = \frac{403}{24}} \qquad \text{equation (1)}$$

Now we divide each side of equation (1) by $\frac{7}{3}$. Why?

$$\frac{\frac{7}{3} \times n}{\frac{7}{3}} = \frac{\frac{403}{24}}{\frac{7}{3}}$$

Be careful here. The right-hand side means $\frac{403}{24} \div \frac{7}{3}$, which we evaluate by *inverting* the second fraction and multiplying.

$$\frac{403}{\underset{8}{\cancel{24}}} \times \frac{\overset{1}{\cancel{3}}}{7} = \frac{403}{56}$$

Thus $n = \frac{403}{56}$ or $7\frac{11}{56}$. Think about all the steps to solving this problem. Can you follow them? There is another way to do the problem. We could multiply each side of equation (1) by $\frac{3}{7}$. Try it. Why does this work? Now try problems 65–68 in Exercises 4.3.

4.3 Exercises

Solve each equation to find the value of n.

1. $5 \times n = 40$ **2.** $7 \times n = 56$ **3.** $76 = 2 \times n$ **4.** $84 = 3 \times n$ **5.** $6 \times n = 96$

6. $8 \times n = 136$ **7.** $117 = 9 \times n$ **8.** $119 = 7 \times n$ **9.** $7 \times n = 182$ **10.** $6 \times n = 144$

11. $3 \times n = 16.8$ **12.** $2 \times n = 19.6$ **13.** $2.6 \times n = 13$ **14.** $3.2 \times n = 48$ **15.** $40.6 = 5.8 \times n$

16. $50.4 = 6.3 \times n$ **17.** $260 \times n = 1430$ **18.** $180 \times n = 936$

Find the value of n in each proportion. Check your answer.

19. $\frac{n}{20} = \frac{3}{4}$ **20.** $\frac{n}{28} = \frac{3}{7}$ **21.** $\frac{6}{n} = \frac{3}{8}$ **22.** $\frac{4}{n} = \frac{2}{7}$ **23.** $\frac{12}{40} = \frac{n}{25}$ **24.** $\frac{13}{30} = \frac{n}{15}$ **25.** $\frac{25}{100} = \frac{8}{n}$

26. $\frac{18}{150} = \frac{6}{n}$ **27.** $\frac{n}{32} = \frac{3}{4}$ **28.** $\frac{n}{8} = \frac{225}{3}$ **29.** $\frac{n}{9} = \frac{49}{63}$ **30.** $\frac{n}{42} = \frac{5}{6}$ **31.** $\frac{18}{n} = \frac{3}{11}$ **32.** $\frac{16}{n} = \frac{2}{7}$

Find the value of n in each proportion. Round your answer to the nearest tenth when necessary.

33. $\frac{n}{5} = \frac{7}{4}$ **34.** $\frac{n}{9} = \frac{5}{4}$ **35.** $\frac{12}{n} = \frac{3}{5}$ **36.** $\frac{21}{n} = \frac{2}{3}$ **37.** $\frac{35}{n} = \frac{7}{5}$

38. $\frac{35}{n} = \frac{4}{3}$ **39.** $\frac{9}{26} = \frac{n}{52}$ **40.** $\frac{12}{16} = \frac{n}{40}$ **41.** $\frac{12}{8} = \frac{21}{n}$ **42.** $\frac{15}{12} = \frac{10}{n}$

43. $\frac{n}{18} = \frac{3.5}{1}$ **44.** $\frac{n}{36} = \frac{4.5}{1}$ **45.** $\frac{2.5}{n} = \frac{0.5}{10}$ **46.** $\frac{1.5}{n} = \frac{0.3}{8}$ **47.** $\frac{3}{4} = \frac{n}{3.8}$

48. $\frac{7}{8} = \frac{n}{4.2}$ **49.** $\frac{12.5}{16} = \frac{n}{12}$ **50.** $\frac{13.8}{15} = \frac{n}{6}$ **51.** $\frac{16}{100} = \frac{5}{n}$ **52.** $\frac{22\frac{2}{9}}{100} = \frac{2}{n}$

Applications

Find the value of n. Round to the nearest hundredth when necessary.

53. $\frac{n \text{ grams}}{10 \text{ liters}} = \frac{7 \text{ grams}}{25 \text{ liters}}$

54. $\frac{n \text{ grams}}{14 \text{ liters}} = \frac{6 \text{ grams}}{22 \text{ liters}}$

55. $\frac{76 \text{ dollars}}{5 \text{ tons}} = \frac{n \text{ dollars}}{8 \text{ tons}}$

56. $\frac{84 \text{ dollars}}{10 \text{ tons}} = \frac{n \text{ dollars}}{7 \text{ tons}}$

57. $\frac{50 \text{ gallons}}{12 \text{ acres}} = \frac{36 \text{ gallons}}{n \text{ acres}}$

58. $\frac{80 \text{ gallons}}{26 \text{ acres}} = \frac{42 \text{ gallons}}{n \text{ acres}}$

59. $\frac{10 \text{ miles}}{16.1 \text{ kilometers}} = \frac{n \text{ miles}}{7 \text{ kilometers}}$

60. $\frac{3 \text{ kilometers}}{1.86 \text{ miles}} = \frac{n \text{ kilometers}}{4 \text{ miles}}$

61. $\frac{5 \text{ pounds}}{2.27 \text{ kilograms}} = \frac{3 \text{ pounds}}{n \text{ kilograms}}$

62. $\frac{6 \text{ ounces}}{170.1 \text{ grams}} = \frac{5 \text{ ounces}}{n \text{ grams}}$

63. $\frac{6.3 \text{ acres}}{2 \text{ people}} = \frac{n \text{ acres}}{5 \text{ people}}$

64. $\frac{136 \text{ miles}}{9 \text{ gallons}} = \frac{n \text{ miles}}{10.5 \text{ gallons}}$

To Think About

Study the "To Think About" example in the text. Then solve for n in each of the following problems. Express n as a mixed number.

65. $\frac{n}{2\frac{1}{3}} = \frac{4\frac{5}{6}}{3\frac{1}{9}}$

66. $\frac{n}{7\frac{1}{4}} = \frac{2\frac{1}{5}}{4\frac{1}{8}}$

67. $\frac{8\frac{1}{6}}{n} = \frac{5\frac{1}{2}}{7\frac{1}{3}}$

68. $\frac{9\frac{3}{4}}{n} = \frac{8\frac{1}{2}}{4\frac{1}{3}}$

Cumulative Review Problems

Evaluate by doing each operation in the proper order.

69. $4^3 + 20 \div 5 + 6 \times 3 - 5 \times 2$

70. $(1.6)^2 - 0.12 \times 3.5 + 36.8 \div 2.5$

71. Write a word name for the decimal 0.563

72. Write in decimal notation thirty-four ten thousandths

4.4 Applications of Proportions

After studying this section, you will be able to:

1 *Solve applied problems using proportions.*

MathPro Video 10 SSM

1 *Solving Applied Problems Using Proportions*

Let us examine a variety of applied problems that can be solved by proportions.

EXAMPLE 1 A recent citywide survey discovered that 37 pairs of eyeglasses in a sample of 120 pairs of eyeglasses were defective. If this rate remains the same each year, how many of the 36,000 pairs of eyeglasses made in the city each year are defective?

We will use the letter n to represent the number of defective eyeglasses in the city.

$$\underbrace{\text{We compare the sample}} \quad \text{to} \quad \underbrace{\text{the total number in the city}}$$

$$\frac{37 \text{ defective pairs}}{120 \text{ total pairs of eyeglasses}} = \frac{n \text{ defective pairs}}{36{,}000 \text{ total pairs of eyeglasses}}$$

$$37 \times 36{,}000 = 120 \times n \qquad \textit{Form the cross product.}$$

$$1{,}332{,}000 = 120 \times n \qquad \textit{Simplify.}$$

$$\frac{1{,}332{,}000}{120} = \frac{120 \times n}{120} \qquad \textit{Divide each side by } 120.$$

$$11{,}100 = n$$

Thus, if the rate of defective eyeglasses holds steady, there are about 11,100 defective pairs of eyeglasses made in the city each year.

Practice Problem 1 An automobile assembly line produced 243 engines yesterday, of which 27 were defective. If the same rate is true each day, how many of the 4131 engines produced this month would be defective? ■

EXAMPLE 2 A survey conducted on a college campus showed that approximately 3 out of every 11 students were left-handed. If there are 2310 students on campus, how many are left-handed?

Let n represent the number of left-handed students.

$$\underbrace{\text{We compare the measured rate}} \quad \text{to} \quad \underbrace{\text{the campus-wide situation}}$$

$$\frac{3 \text{ left-handed students in survey}}{11 \text{ students in survey}} = \frac{n \text{ left-handed students on campus}}{2310 \text{ total students on campus}}$$

We assume that the measured rate may be applied to the entire campus.

$$3 \times 2310 = 11 \times n \quad \textit{Form the cross product.}$$

$$6930 = 11 \times n \quad \textit{Simplify.}$$

$$\frac{6930}{11} = \frac{11 \times n}{11} \quad \textit{Divide each side by } 11.$$

$$630 = n$$

Thus there are approximately 630 left-handed students on campus.

Practice Problem 2 A survey conducted on campus showed that 5 out of every 19 students had quit smoking during the last five years. Approximately how many of the 7410 students now on campus quit smoking in the last five years? ■

Looking back at Example 2, perhaps it occurred to you that the fractions in the proportion could be set up in an alternative way. You can set up this problem in several different ways as long as the units are in correctly corresponding positions. It would be correct to set up the problem in the form

$$\frac{\text{left-handed students in survey}}{\text{left-handed students on campus}} = \frac{\text{total students in survey}}{\text{total students on campus}}$$

or

$$\frac{\text{total students on campus}}{\text{total students in survey}} = \frac{\text{left-handed students on campus}}{\text{left-handed students in survey}}$$

But we **cannot** set up the problem this way.

$$\frac{\text{left-handed students}}{\text{total students}} \neq \frac{\text{total students}}{\text{left-handed students}}$$

This is *not* correct. Do you see why? See Exercises 4.4, problems 35–38.

EXAMPLE 3 Ted's car can go 245 miles on 7 gallons of gas. Ted wants to take a trip of 455 miles. Approximately how many gallons of gas will this take?

Let n = the unknown number of gallons.

$$\frac{245 \text{ miles}}{7 \text{ gallons}} = \frac{455 \text{ miles}}{n \text{ gallons}}$$

$$245 \times n = 7 \times 455 \quad \textit{Form the cross product.}$$

$$245 \times n = 3185 \quad \textit{Simplify.}$$

$$\frac{245 \times n}{245} = \frac{3185}{245}$$

$$n = 13$$

Ted will need approximately 13 gallons of gas for the trip.

Practice Problem 3 Cindy's car travels 234 miles on 9 gallons of gas. How many gallons of gas will Cindy need to take a 312-mile trip? ■

EXAMPLE 4 In a certain gear, Alice's 10-speed bicycle has a gear ratio of 3 revolutions of the pedal for every 2 revolutions of the bicycle wheel. If her bicycle wheel is turning at 65 revolutions per minute, how many times must she pedal per minute?

Let n = the number of revolutions of the pedal.

$$\frac{3 \text{ revolutions of the pedal}}{2 \text{ revolutions of the wheel}} = \frac{n \text{ revolutions of the pedal}}{65 \text{ revolutions of the wheel}}$$

$3 \times 65 = 2 \times n$ *Cross-multiply.*

$195 = 2 \times n$ *Simplify.*

$\frac{195}{2} = \frac{2 \times n}{2}$ *Divide both sides by 2.*

$97.5 = n$

Alice will pedal at the rate of 97.5 revolutions in 1 minute.

Practice Problem 4 Alicia must pedal at 80 revolutions per minute to ride her bicycle at 16 miles per hour. If she pedals at 90 revolutions per minute, how fast will she be riding? ■

EXAMPLE 5 Tim operates a bicycle rental center during the summer months on the island of Martha's Vineyard. He discovered that when the ferryboats brought 8500 passengers a day to the island his center rented 340 bicycles a day. Next summer the ferryboats plan to bring 10,300 passengers a day to the island. How many bicycles a day should Tim plan to rent?

Two important cautions are necessary before we solve the proportion. We need to be sure that the bicycle rentals are directly related to the number of people on the ferryboat. (Presumably, people who fly to the island or who take small pleasure boats to the island also rent bicycles.) Next we need to be sure that the people who represent the increase in passengers per day would be as likely to rent bicycles as the present number of passengers do. For example, if the new visitors to the island are all senior citizens, they are not as likely to rent bicycles as younger people would be. If we assume those two conditions are satisfied, then we can solve the problem as follows.

$$\frac{8500 \text{ passengers per day now}}{340 \text{ bike rentals per day now}} = \frac{10{,}300 \text{ passengers per day later}}{n \text{ bike rentals per day later}}$$

$$8500 \times n = 340 \times 10{,}300$$

$$8500 \times n = 3{,}502{,}000$$

$$\frac{8500 \times n}{8500} = \frac{3{,}502{,}000}{8500}$$

$$n = 412$$

If the two conditions are satisfied, we would predict 412 bicycle rentals.

Practice Problem 5 For every 4050 people who walk into Tom's Souvenir Shop, 729 make a purchase. Assuming the same conditions, if 5500 people walk into Tom's Souvenir Shop, how many people may be expected to make a purchase? ■

Biologists and others who observe or protect wildlife sometimes use the capture-mark-recapture method to determine how many animals are in a certain region. In this approach some animals are caught and tagged in a way that does not harm them. They are then released into the wild, where they mix with their kind. It is assumed (usually correctly) that the tagged animals will mix throughout the entire population in that region, so that when they are recaptured in a future sample, the biologists can use them to make reasonable estimates about the total population. We will employ the capture-mark-recapture method in the next example.

EXAMPLE 6 A biologist catches 42 fish in a lake and tags them. She then quickly returns them to the lake. In a few days she catches a new sample of 50 fish. Of those 50 fish, 7 have her tag. Approximately how many fish are in the lake?

$$\frac{42 \text{ fish tagged in 1st sample}}{n \text{ fish in lake}} = \frac{7 \text{ fish tagged in 2nd sample}}{50 \text{ fish caught in 2nd sample}}$$

$$42 \times 50 = 7 \times n$$

$$2100 = 7 \times n$$

$$\frac{2100}{7} = \frac{7 \times n}{7}$$

$$300 = n$$

Assuming that no tagged fish died and that the tagged fish mixed throughout the population of fish in the lake, we estimate the number of fish in the lake as 300.

Practice Problem 6 A park ranger in Alaska captures and tags 50 bears. He then releases them to range through the forest. Sometime later he captures 50 bears. Of the 50, 4 have tags from the previous capture. Estimate the number of bears in the forest. ■

Developing Your Study Skills

GETTING HELP

Getting the right kind of help at the right time can be a key ingredient in being successful in mathematics. When you have gone to class on a regular basis, taken careful notes, methodically read your textbook, and diligently done your homework—in other words, when you have made every effort possible to learn the mathematics—you may still find that you are having difficulty. If this is the case, then you need to seek help. Make an appointment with your instructor to find out what help is available to you. The instructor, tutoring services, a mathematics lab, videotapes, and computer software may be among resources you can draw on.

4.4 Exercises

Applications

1. The policy at the Collonaide Hotel is to have 19 desserts for every 16 people if a buffet is being served. If the Saturday buffet has 320 people, how many desserts must be available?

2. An automobile dealership has found that for every 140 cars sold, 23 will be brought back to the dealer for major repairs. If the dealership sells 980 cars this year, approximately how many cars will be brought back for major repairs?

3. A coffee bar is constantly making "shots" of espresso coffee. For every 90 shots of coffee that are purchased, 5 are thrown away because they get too cold while the milk for cappuccino is being steamed. If the coffee bar sells 162 shots of coffee per day, how many shots will be poured down the sink?

4. Cynthia's car uses 0.5 quarts of oil for every 850 miles she drives. She drove 12,325 miles last year. At that rate, how many quarts of oil did she use?

5. A cosmetics company uses the ratio of 35 drops of clear polish to 4 drops of pigment when it makes red nail polish. How many drops of pigment should it take if the company has a vat containing 8015 drops of clear polish?

6. When Donna lived in Paris, France, the exchange rate was 8 French francs for every U.S. dollar. If she exchanged 355 U.S. dollars, how many French francs should she receive?

7. Jason went to Canada when the exchange rate was $7.50 U.S. for every $10.50 Canadian. When he exchanges $240.00 U.S., how many Canadian dollars should he receive?

8. When Douglas flew to London, the exchange rate was 7 British pounds for every 15 U.S. dollars. Douglas exchanged 255 U.S. dollars. How many British pounds did he receive?

9. At the reclining stationary bike at a health club, Angela pedals at 80 revolutions per minute (rpm) to travel at an indicated speed of 18 miles per hour. If she were to pedal at 75 rpm, how fast would be her indicated speed? Round your answer to the nearest tenth.

10. Rod uses a computerized rowing machine at the gymnasium where he works out. He completes 17 strokes per minute and the machine indicates a speed of 5 miles per hour. If he were to increase his stroke to 34 strokes per minute, how fast would the machine indicate is his rowing speed?

In problems 11 and 12 the shadow of one object is in the same ratio to the height of that object as a second shadow is to the height of a second object at the same time of day.

11. A pro football offensive tackle who stands 6.5 feet tall casts a 5-foot shadow. At the same time, the football stadium, which he is standing next to, casts a shadow of 152 feet. How tall is the stadium? Round your answer to the nearest tenth.

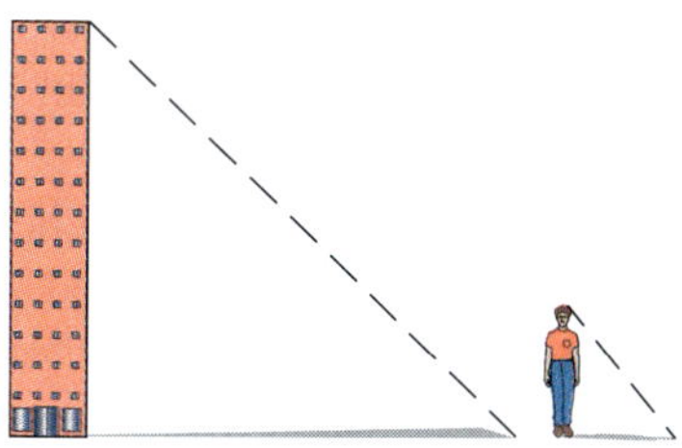

12. A souvenir life-sized cardboard cut-out of Queen Elizabeth (for the tourists) sits outside of a London office building. It is 5 feet tall, and casts a shadow that is 3 feet long. The office building casts a shadow that is 165 feet long. How tall is the building?

13. Jacqueline can write 4 pages of a test book (blue book) in 12 minutes. How long would it take her to write an essay that takes up 11 pages of a blue book?

14. Marcia can type the final copy of a 4-page psychology paper in 27 minutes. How long will it take her to type a 15-page paper? Round your answer to the nearest minute.

15. On Dr. Jennings' map of Antarctica, the scale states that 4 inches represent 250 miles of actual distance. Two Antarctic mountains are 5.7 inches apart on the map. What is the approximate distance in miles between the two mountains? Round your answer to the nearest mile.

16. On a tour guide map of Madagascar, the scale states that 3 inches represent 125 miles. Two beaches are 5.2 inches apart on the map. What is the approximate distance in miles between the two beaches? Round your answer to the nearest mile.

17. A 100-watt stereo system needs copper speaker wire that is 30 millimeters thick to handle the output of sound clearly. How thick would the speaker wire need to be if you had a 140-watt stereo and you wanted the same ratio of watts to millimeters?

18. An electrical engineer specializing in rock concerts uses a 21-foot wire with 0.6 ohm of resistance. (The electrical resistance in a wire is measured in ohms.) How much resistance would be in the same type of wire 230 feet long? Round your answer to the nearest tenth.

19. Jenny's blueberry pancake recipe uses 5 cups of blueberries for every 12 people. How many cups of blueberries does she need for the same recipe prepared for 28 people?

20. In his lasagna recipe, Giovanni uses 2 cups of marinara sauce for every seven people. How much sauce will he need to make lasagna for 62 people?

21. A Mercedes-Benz salesperson sold 7 cars after having spoken to 24 customers. To maintain the same ratio of sales, how many Mercedes-Benz vehicles will she have to sell if she meets with 50 potential customers? Round to the nearest whole number.

22. The Lincoln Lancers basketball team has won nine out of every eleven games played over the last few years. The team has played 143 times. How many times have they won?

23. A tennis player makes 14 unforced errors during 110 played points. At that rate, how many errors would he make in several matches where he and his opponent played 275 points?

24. A baseball pitcher gave up 52 earned runs in 260 innings of pitching. At that rate, how many runs would he give up in a 9-inning game? (This decimal is called the pitcher's *earned run average*.)

25. The Chicago-O'Hare Taxi Drivers Association has discovered that for every 7 plane flights arriving at O'Hare Airport, 79 people need a taxi ride from the airport. If tomorrow 434 plane flights are scheduled to arrive at O'Hare, how many people would you estimate will need a taxi ride from the airport?

26. The local super drugstore has determined that for every 30 customers who come into the store and buy shampoo, 24 of them will also buy conditioner/cream rinse. If the average number of customers who buy shampoo is 830 per week, how many people per week will buy shampoo and conditioner? Round to the nearest whole number.

27. In Kenya, a worker at the game preserve captures 26 giraffes, tags them, and then releases them back into the preserve. The next month, he captures 18 giraffes and finds that 6 of them have already been tagged. Estimate the number of giraffes on the preserve.

28. A ornithologist is studying hawks in the Adirondack Mountains. She catches 24 hawks over a period of one month, tags them, and frees them back into the wild. The next month, she catches 20 hawks and finds that 12 are already tagged. Estimate the number of hawks in this part of the mountains.

29. Bill and Shirley Grant are raising tomatoes to sell at the local co-op. The farm has a yield of 425 pounds of tomatoes for every 3 acres. The farm has 14 acres of good tomatoes. The crop this year should bring \$1.80 per pound for each pound of tomatoes. How much will the Grants get from the sale of the tomato crop?

30. A paint manufacturer suggests 2 gallons of flat latex paint for every 750 square feet of wall. A painter is going to paint 7875 square feet of wall in a Dallas office building with paint that costs \$8.50 per gallon. How much will the painter spend for the paint?

31. An 8-foot oak table weighs 86 pounds. The same type of table also comes in a 12-foot model. A delivery truck loads seven of the 12-foot models into a truck. How much does the cargo weigh?

32. A 40-foot roll of fence wire weighs 37 pounds. The same type of fence wire also comes in 200-foot rolls. Frank put six of the 200-foot rolls in his pickup truck. How much did his cargo weigh?

33. It has been found that 22 out of every 100 people in the United States mailed a survey will respond. If 15,500 are mailed a survey, estimate how many will respond.

34. A study shows that with a certain breed of domesticated birds, 46 out of 80 will be able to mimick their masters by "talking." If there are 2200 such birds as pets in the United States, estimate how many of them will be able to talk.

To Think About

The speed 60 miles per hour is equivalent to the speed 88 feet per second. A car is traveling at 49 miles per hour. Several students attempt to solve the problem of how many feet per second the car is traveling.

35. Student A uses the proportion $\frac{88}{n} = \frac{60}{49}$.

Is student A right? Why?

36. Student B uses the proportion $\frac{60}{88} = \frac{49}{n}$.

Is student B right? Why?

37. Student C uses the proportion $\frac{n}{88} = \frac{60}{49}$.

Is student C right? Why?

38. Student D uses the proportion $\frac{60}{88} = \frac{n}{49}$.

Is student D right? Why?

Cumulative Review Problems

39. Round to the nearest hundred. 56,179

40. Round to the nearest ten thousand. 196,379,910

41. Round to the nearest tenth. 56.148

42. Round to the nearest hundredth. 1.96341

43. Round to the nearest ten thousandth. 0.07615382

44. Round to the nearest thousandth. 598.32149

Putting Your Skills to Work

HOW CROWDED IS THIS CITY? A MATHEMATICAL APPROACH

People can have different impressions as to how crowded a city is. Hardly anyone can travel to New York City and not have the impression that it is extremely crowded. But what about other cities? How do you measure how ''crowded'' a city really is?

Mathematically, a good way to approach the problem is to calculate the number of people per square mile. The following chart lists the population of area in square miles of several major cities in the United States. Source: Bureau of the Census

City	1990 population	Area in square miles
Anaheim, California	266,406	41
Anchorage, Alaska	226,338	1732
Austin, Texas	465,648	116
Boston, Massachusetts	574,283	46
Chicago, Illinois	2,783,726	228
Cincinnati, Ohio	364,114	78
Detroit, Michigan	1,027,974	136
New Orleans, Louisiana	496,938	199
New York City, New York	7,322,564	301
Phoenix, Arizona	983,403	324

PROBLEMS FOR INDIVIDUAL INVESTIGATION

Round all answers to the nearest whole number.

1. Find the number of people per square mile for Cincinnati and Austin. According to your calculations which city is more densely populated?

2. Find the number of people per square mile for Chicago and Boston. According to your calculations which city is more densely populated?

PROBLEMS FOR GROUP INVESTIGATION AND ANALYSIS

Together with some members of your class investigate the following.

3. The 1980 census figures for Anaheim are 219,494 and for Detroit 1,203,368. Assuming that the ratio of change from 1980 to 1990 will be the same for 1990 to 2000, what will be the estimated number of people per square mile for Anaheim and for Detroit in the year 2000?

4. The 1980 census figures for New Orleans are 557,927 and for Phoenix 739,704. Assuming that the ratio of change from 1980 to 1990 will continue to be the same for the period 1990 to 2000, what will be the estimated number of people per square mile in each of these cities in the year 2000?

INTERNET CONNECTIONS: Go to ''http://www.prenhall.com/tobey'' to be connected

Site: U. S. Census Bureau County Population Census Counts or a related site

5. Find the number of people in your county in 1980 and in 1990. Then find the area of your county.

6. Calculate the number of people per square mile for your county in 1980 and 1990. Assuming that the ratio of change from 1990 to 2000 will be the same as it was for 1980 to 1990, estimate the number of people per square mile in your county in 2000.

Chapter Organizer

Topic	Procedure	Examples
Forming a ratio, p. 242.	A *ratio* is the comparison of two quantities that have the same units. A ratio is usually expressed as a fraction. The fractions should be in reduced form.	**1.** Find the ratio of 20 books to 35 books. $\frac{20}{35} = \frac{4}{7}$ **2.** Find the ratio in simplest form of 88:99. $\frac{88}{99} = \frac{8}{9}$ **3.** Bob earns \$250 each week, but \$15 is withheld for medical insurance. Find the ratio of medical insurance to total pay. $\frac{\$15}{\$250} = \frac{3}{50}$
Forming a rate, p. 244.	A *rate* is a comparison of two quantities that have different units. A rate is usually expressed as a fraction in reduced form.	A college has 2520 students with 154 faculty. What is the rate of students to faculty? $\frac{2520 \text{ students}}{154 \text{ faculty}} = \frac{180 \text{ students}}{11 \text{ faculty}}$
Forming a unit rate, p. 244.	A *unit rate* is a rate with a denominator of 1. Divide the denominator into the numerator to obtain the unit rate.	A car traveled 416 miles in 8 hours. Find the unit rate. $\frac{416 \text{ miles}}{8 \text{ hours}} = 52 \text{ miles/hour}$ Bob spread 50 pounds of fertilizer over 1870 square feet of land. Find the unit rate of square feet per pound. $\frac{1870 \text{ square feet}}{50 \text{ pounds}} = 37.4 \text{ square feet/pound}$
Writing proportions, p. 250.	A *proportion* is a statement that two rates or two ratios are equal. A proportion statement *a* is to *b* as *c* is to *d* can be written $\frac{a}{b} = \frac{c}{d}$	Write a proportion for 17 is to 34 as 13 is to 26. $\frac{17}{34} = \frac{13}{26}$
Determining if proportions are true or false, p. 250.	For any two fractions where $b \neq 0$, $d \neq 0$. $\frac{a}{b} = \frac{c}{d}$ if and only if $a \times d = b \times c$	**1.** Is the proportion $\frac{7}{56} = \frac{3}{24}$ true or false? $7 \times 24 \stackrel{?}{=} 56 \times 3$ $168 = 168$ ✓ The proportion is *true.* **2.** Is the proportion $\frac{64 \text{ gallons}}{5 \text{ acres}} = \frac{89 \text{ gallons}}{7 \text{ acres}}$ true or false? $64 \times 7 \stackrel{?}{=} 5 \times 89$ $448 \neq 445$ The proportion is *false.*

Chapter Organizer

Topic	Procedure	Examples
Solving a proportion, p. 257.	To solve a proportion where the value n is not known: **1.** Cross-multiply. **2.** Divide both sides of the equation by the number multiplied by n.	**1.** Solve for n. $\frac{17}{n} = \frac{51}{9}$ $17 \times 9 = 51 \times n$ Cross-multiply. $153 = 51 \times n$ Simplify. $\frac{153}{51} = \frac{51 \times n}{51}$ Divide by 51. $3 = n$
Solving applied problems, p. 263.	**1.** Write a proportion with n representing the unknown value. **2.** Solve the proportion.	Bob purchased eight notebooks for \$19. How much would 14 notebooks cost? $\frac{8 \text{ notebooks}}{\$19} = \frac{14 \text{ notebooks}}{n}$ $8 \times n = 19 \times 14$ $8 \times n = 266$ $\frac{8 \times n}{8} = \frac{266}{8}$ $n = 33.25$ The 14 notebooks would cost \$33.25.

Chapter 4 Review Problems

4.1 *Write each ratio in simplest form. Express your answer as a fraction.*

1. 88:40 **2.** 65:39 **3.** 28:35 **4.** 250:475 **5.** 50 to 124 **6.** 27 to 81

7. 156 to 441 **8.** 280 to 651 **9.** 26 tons to 65 tons

10. 34 tons to 170 tons **11.** 150 kilograms to 200 kilograms **12.** 115 grams to 130 grams

Bob earns \$215 per week and has \$35 per week withheld for federal taxes and \$20 per week withheld for state taxes.

13. Write the ratio of federal taxes withheld to earned income.

14. Write the ratio of state taxes withheld to earned income.

Write as a rate in simplest form.

15. 10 gallons of water for every 18 people

16. 44 revolutions every 121 minutes

17. 188 vibrations every 16 seconds

18. 12 cups of flour for every 38 people

Write as a unit rate. Round to the nearest tenth when necessary.

19. \$2125 was paid for 125 shares of stock. Find the cost per share.

20. \$2244 was paid for 132 chairs. Find the cost per chair.

21. \$742.50 was spent for 55 square yards of carpet. Find the cost per square yard.

22. The Baseball Boosters Club spent \$768.80 for 62 tickets to the ball game. Find the cost per ticket.

23. A 4-ounce jar of instant coffee costs \$2.96. A 9-ounce jar of the same brand of instant coffee costs \$5.22.

(a) What is the cost per ounce of the 4-ounce jar?

(b) What is the cost per ounce of the 9-ounce jar?

(c) How much per ounce do you save by buying the larger jar?

24. A 12.5-ounce can of white tuna costs \$2.75. A 7.0-ounce can of the same brand of white tuna costs \$1.75.

(a) What is the cost per ounce of the large can?

(b) What is the cost per ounce of the small can?

(c) How much per ounce do you save by buying the larger can?

4.2 *Write as a proportion.*

25. 12 is to 48 as 7 is to 28

26. 10 is to 21 as 30 is to 63

27. 7.5 is to 45 as 22.5 is to 135

28. 8.6 is to 43 as 17.2 is to 86

29. 136 is to 17 as 408 is to 51

30. 117 is to 61 as 351 is to 183

31. If 15 pounds cost \$4.50, then 27 pounds will cost \$8.10.

32. If 3 buses can transport 138 passengers, then 5 buses can transport 230 passengers.

Determine if each proportion is true or false.

33. $\frac{16}{48} = \frac{2}{12}$ **34.** $\frac{20}{25} = \frac{8}{10}$ **35.** $\frac{24}{20} = \frac{18}{15}$ **36.** $\frac{84}{48} = \frac{14}{8}$ **37.** $\frac{37}{33} = \frac{22}{19}$ **38.** $\frac{15}{18} = \frac{18}{22}$

39. $\frac{84 \text{ miles}}{7 \text{ gallons}} = \frac{108 \text{ miles}}{9 \text{ gallons}}$ **40.** $\frac{156 \text{ revolutions}}{6 \text{ minutes}} = \frac{181 \text{ revolutions}}{7 \text{ minutes}}$

41. $\frac{1.6 \text{ pounds}}{32 \text{ feet}} = \frac{4.8 \text{ pounds}}{96 \text{ feet}}$ **42.** $\frac{3.9 \text{ pounds}}{45 \text{ feet}} = \frac{7.9 \text{ pounds}}{90 \text{ feet}}$

4.3 *Solve for n in each equation.*

43. $7 \times n = 161$ **44.** $8 \times n = 256$ **45.** $558 = 18 \times n$ **46.** $663 = 39 \times n$

Solve each proportion. Round to the nearest tenth when necessary.

47. $\frac{3}{11} = \frac{9}{n}$ **48.** $\frac{2}{7} = \frac{12}{n}$ **49.** $\frac{n}{28} = \frac{6}{24}$ **50.** $\frac{n}{32} = \frac{15}{20}$ **51.** $\frac{3}{7} = \frac{n}{9}$ **52.** $\frac{4}{9} = \frac{n}{8}$

53. $\frac{54}{72} = \frac{n}{4}$ **54.** $\frac{45}{135} = \frac{n}{3}$ **55.** $\frac{6}{n} = \frac{2}{29}$ **56.** $\frac{8}{n} = \frac{2}{81}$ **57.** $\frac{25}{7} = \frac{60}{n}$ **58.** $\frac{60}{9} = \frac{31}{n}$

59. $\frac{35 \text{ miles}}{28 \text{ gallons}} = \frac{15 \text{ miles}}{n \text{ gallons}}$ **60.** $\frac{8 \text{ defective parts}}{100 \text{ perfect parts}} = \frac{44 \text{ defective parts}}{n \text{ perfect parts}}$

61. $\frac{7 \text{ tons}}{5.5 \text{ horsepower}} = \frac{16 \text{ tons}}{n \text{ horsepower}}$ **62.** $\frac{27 \text{ feet}}{4 \text{ quarts}} = \frac{30.5 \text{ feet}}{n \text{ quarts}}$

4.4 *Solve each problem by using a proportion. Round your answer to the nearest hundredth when necessary.*

63. The school volunteers used 3 gallons of paint to paint two rooms. How many gallons would they need to paint 10 rooms of the same size?

64. Fred paid \$77 for three chairs. How much would he pay for nine chairs?

65. Chaneyville Hospital has 84 patients for seven registered nurses. To keep this ratio, how many registered nurses would be needed for 108 patients?

66. South Park College has 1680 students and 100 faculty. To keep this ratio, how many students should the college have if it increases the faculty to 130?

67. When Marguerite traveled, the rate of French francs to American dollars was 24 francs to 5 dollars. How many francs did Marguerite receive for 420 dollars?

68. A catering service recommends providing 7.5 pounds of cold cuts for every 18 people at the reception. The men's glee club is having a reception for 120 people. How many pounds of cold cuts should they order?

69. Two cities located 225 miles apart appear 3 inches apart on a map. If two other cities appear 8 inches apart on the map, how many miles apart are the cities?

70. If Melissa pedals her bicycle at 84 revolutions per minute, she travels at 14 miles per hour. How fast does she go if she pedals at 96 revolutions per minute?

71. Michael conducted a science experiment and found that sound travels in air at 34,720 feet in 31 seconds. How many feet would sound travel in 60 seconds?

72. After college, Roberto began a chain of Pizza Villas in his home state. He eventually became a millionaire. He told his friends that for every \$3 he invested he earned \$985 in 10 years. How much did he invest to earn \$1 million?

73. In the setting sun, a 6-foot man casts a shadow 16 feet long. At the same time a building casts a shadow of 320 feet. How tall is the building?

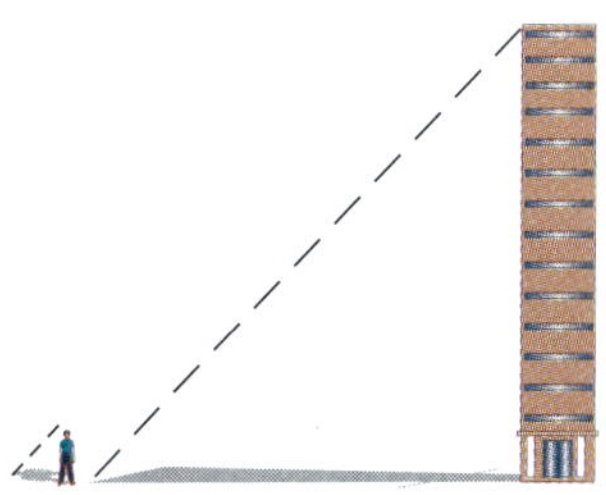

74. During the first 680 miles of a trip, Cindy and Melinda used 26 gallons of gas. They need to travel 200 more miles. Assuming that the car has the same rate of gas consumption,

(a) How many more gallons of gas will they need?

(b) If gas costs \$1.15 per gallon, what will fuel cost them for the last 200 miles?

Putting Your Skills to Work

BUDGET PROBLEMS

Whether you are a full-time college student or the president of a small company, the issue of balancing the budget often comes up. Often, when faced with price increases, people are required to make decisions about future spending practices based on a careful analysis of how the money has been spent in the past. After John spent his first year at Wheaton College he analyzed his first-year expenses and created a spending plan for the second year.

Budget Categories	Actual Budget of First-Year College Expenses	Proposed Budget of Second-Year College Expenses
Tuition	12,350	13,500
Room	2,400	2,700
Board	2,700	3,000
Telephone	250	?
Clothing	650	?
Transportation	500	?
Entertainment	800	?
Churches and Charities	350	380
Total Expenses	20,000	22,000

PROBLEMS FOR INDIVIDUAL INVESTIGATION

Use this budget to answer the following questions.

1. Is the ratio of Tuition costs to Total Expenses higher for the second year than for the first year?
2. After he made up the proposed budget for the second year, John realized that his Total Expenses could only increase in the same ratio as the ratio of increase of the Tuition. On that basis, what should be his proposed budget for Total Expenses for the second year? (Round your answer to the nearest whole dollar.)

PROBLEMS FOR GROUP INVESTIGATION AND COOPERATIVE GROUP ACTIVITY

3. John allowed that his Telephone expenses would double for the second year, but his Clothing and Transportation expenses would be the same for the second year. Using your answer for question **2**, determine what his new entertainment budget will be for the second year.
4. Of the four categories Tuition, Room, Board, and Churches and Charities, which category had the greatest ratio of increase from the first year to the second year? Which had the least ratio of increase?

INTERNET CONNECTIONS: Go to ''http://www.prenhall.com/tobey'' to be connected

Site: The Government vs. a Personal Budget or a related site

This Australian site compares a personal budget to the budget of a government. Read the information about the waitress, Maxine, and answer the following questions.

5. What is the ratio of Maxine's entertainment costs to clothing costs? What is the ratio of Maxine's rent to her income?
6. What is the ratio of Maxine's total expenses to her income? How much money can Maxine save each week? How long will it take her to save $3000?

Write as a ratio *in simplest form.*

1. 18:52

2. 70 to 185

Write as a rate *in simplest form. Express your answer as a fraction.*

3. 784 miles per 24 gallons

4. 2100 square feet per 45 pounds

Write as a unit rate. *Round to the nearest hundredth when necessary.*

5. 19 tons in five days

6. $57.96 for 7 hours

7. 5400 feet per 22 telephone poles

8. $9373 for 110 shares of stock

Write as a proportion.

9. 17 is to 29 as 51 is to 87.

10. 12 is to 19 as 18 is to 28.5.

11. 490 miles is to 21 gallons as 280 miles is to 12 gallons.

12. 5 tablespoons of flour is to 18 people as 15 tablespoons of flour is to 54 people.

Determine if each proportion is true or false.

13. $\frac{50}{24} = \frac{34}{16}$

14. $\frac{18.4}{20} = \frac{46}{50}$

15. $\frac{32 \text{ smokers}}{46 \text{ nonsmokers}} = \frac{160 \text{ smokers}}{230 \text{ nonsmokers}}$

16. $\frac{\$0.74}{16 \text{ ounces}} = \frac{\$1.84}{40 \text{ ounces}}$

1. ________
2. ________
3. ________
4. ________
5. ________
6. ________
7. ________
8. ________
9. ________
10. ________
11. ________
12. ________
13. ________
14. ________
15. ________
16. ________

17. ______

18. ______

19. ______

20. ______

21. ______

22. ______

23. ______

24. ______

25. ______

26. ______

27. ______

28. ______

29. ______

30. ______

Solve each proportion. Round to the nearest tenth when necessary.

17. $\frac{n}{20} = \frac{4}{5}$

18. $\frac{9}{2} = \frac{63}{n}$

19. $\frac{33}{n} = \frac{11}{4}$

20. $\frac{4.2}{11} = \frac{n}{77}$

21. $\frac{45 \text{ women}}{15 \text{ men}} = \frac{n \text{ women}}{40 \text{ men}}$

22. $\frac{3.5 \text{ ounces}}{4.2 \text{ grams}} = \frac{7 \text{ ounces}}{n \text{ grams}}$

23. $\frac{n \text{ inches of snow}}{14 \text{ inches of rain}} = \frac{12 \text{ inches of snow}}{1.4 \text{ inches of rain}}$

24. $\frac{28 \text{ pounds of bananas}}{\$n} = \frac{3 \text{ pounds of bananas}}{\$0.55}$

Solve each problem by using a proportion. Round your answer to the nearest hundredth when necessary.

25. Bob's recipe for pancakes calls for 3 eggs and will serve 11 people. If he wants to feed 22 people, how many eggs will he need?

26. A steel cable 42 feet long weighs 170 pounds. How much will 20 feet of this cable weigh?

27. If 9 inches on a map represents 57 miles what distance does 3 inches represent?

28. John and Nancy found it would cost \$240 per year to fertilize their front lawn of 4000 square feet. How much would it cost to fertilize 6000 square feet?

29. If Jenny's car uses 1.5 quarts of oil every 3000 miles, how many quarts will it use in 8000 miles?

30. Stephen traveled 570 kilometers in 9 hours. At this rate, how far could he go in 11 hours?

Cumulative Test for Chapters 1–4

Approximately one-half of this test is based on Chapter 4 material. The remainder is based on material covered in Chapters 1–3.

1. Write in words. 26,597,089

2. Add. $\begin{array}{r} 86 \\ 124 \\ 38 \\ 107 \\ +\ 56 \\ \hline \end{array}$

3. Multiply. $\begin{array}{r} 208 \\ \times\ 67 \\ \hline \end{array}$

4. Divide. $23\overline{)1564}$

5. Add. $\frac{1}{5} + \frac{1}{8} + \frac{3}{4}$

6. Subtract. $2\frac{1}{5} - 1\frac{3}{7}$

7. Multiply. $4\frac{1}{2} \times 3\frac{1}{4}$

8. Divide. $\dfrac{2\frac{1}{8}}{\frac{3}{4}}$

9. Round to the nearest hundredth. 163.578314

10. Subtract. $\begin{array}{r} 12.1 \\ -\ 3.8416 \\ \hline \end{array}$

11. Multiply. $\begin{array}{r} 0.8163 \\ \times\ 0.22 \\ \hline \end{array}$

12. Divide. $0.06\overline{)0.84}$

13. Write a *ratio* in simplest form to compare 81 miles to 27 miles.

14. Write as a *rate* in simplest form. $1.68 for 12 bananas

15. Write as a *unit rate* in simplest form. 12 yen for 3 pesos

1. ______
2. ______
3. ______
4. ______
5. ______
6. ______
7. ______
8. ______
9. ______
10. ______
11. ______
12. ______
13. ______
14. ______
15. ______

16. ____

17. ____

18. ____

19. ____

20. ____

21. ____

22. ____

23. ____

24. ____

25. ____

26. ____

Determine if each proportion is true or false.

16. $\frac{12}{17} = \frac{30}{42.5}$

17. $\frac{5}{7} = \frac{15}{21}$

Solve each proportion. Round to the nearest tenth when necessary.

18. $\frac{9}{2.1} = \frac{n}{0.7}$

19. $\frac{50}{20} = \frac{5}{n}$

20. $\frac{n}{56} = \frac{16}{7}$

21. $\frac{7}{n} = \frac{28}{36}$

22. $\frac{n}{11} = \frac{5}{16}$

23. $\frac{3\frac{1}{3}}{7} = \frac{10}{n}$

Solve each problem by using a proportion. Round your answers to the nearest hundredth when necessary.

24. Two cities that are located 300 miles apart appear 4 inches apart on a map. If two other cities are 625 miles apart, how far apart will they appear on the same map?

25. Last week Bob earned \$84. He had \$7 withheld for federal income tax. He earned \$9000 last year. Assuming the same rate, how much did he have withheld for federal income tax last year?

26. Mom's lasagna recipe feeds 14 people and calls for 3.5 pounds of sausage. If she wants to feed 20 people, how much sausage does she need?

CHAPTER 5

Percent

Do you have any idea how many soft drinks are consumed per person in the United States? Is it more than the amount of coffee consumed per person or less? What are the future trends? Some of the answers may surprise you. Turn to the Putting Your Skills to Work on page 324.

Pretest Chapter 5

1. ______	19. ______
2. ______	20. ______
3. ______	21. ______
4. ______	22. ______
5. ______	23. ______
6. ______	24. ______
7. ______	25. ______
8. ______	26. ______
9. ______	27. ______
10. ______	28. ______
11. ______	29. ______
12. ______	30. ______
13. ______	31. ______
14. ______	32. ______
15. ______	33. ______
16. ______	34. ______
17. ______	35. ______
18. ______	36. ______

If you are familiar with the topics in this chapter, take this test now. Check your answers with those in the back of the book. If an answer was wrong or you couldn't do a problem, study the appropriate section of the chapter.

If you are not familiar with the topics in this chapter, don't take this test now. Instead, study the examples, work the practice problems, and then take the test.

This test will help you identify which concepts you have mastered and which you need to study further.

Section 5.1

Write as a percent.

1. 0.13 **2.** 0.21 **3.** 0.185 **4.** 0.372

5. 1.34 **6.** 8.94 **7.** 0.002 **8.** 0.004

9. $\frac{17}{100}$ **10.** $\frac{27}{100}$ **11.** $\frac{13.4}{100}$ **12.** $\frac{19.8}{100}$

13. $\frac{6\frac{1}{2}}{100}$ **14.** $\frac{1\frac{3}{8}}{100}$

Section 5.2

Change each fraction to a percent. Round to the nearest hundredth of a percent when necessary.

15. $\frac{6}{10}$ **16.** $\frac{1}{40}$ **17.** $\frac{23}{20}$ **18.** $\frac{17}{16}$

19. $\frac{5}{7}$ **20.** $\frac{2}{7}$ **21.** $\frac{22}{23}$ **22.** $\frac{13}{19}$

23. $4\frac{2}{5}$ **24.** $2\frac{3}{4}$ **25.** $\frac{1}{300}$ **26.** $\frac{1}{400}$

Write each percent as a fraction in simplified form.

27. 22% **28.** 38% **29.** 53% **30.** 41%

31. 150% **32.** 160% **33.** $6\frac{1}{3}\%$ **34.** $4\frac{2}{3}\%$

35. $51\frac{1}{4}\%$ **36.** $43\frac{3}{4}\%$

Section 5.3

Solve each percent problem. Round to the nearest hundredth when necessary.

37. Find 24% of 230.

38. What is 52% of 43?

39. 68 is what percent of 72?

40. What percent of 22 is 18?

41. 8% of what number is 240?

42. 296 is 16% of what number?

Section 5.4

Answer each question. Round to the nearest hundredth when necessary.

43. The home team won 24 of 38 basketball games. What percent of the basketball games did the home team win?

44. A salesperson earns a commission rate of 24%. How much commission would be paid if the salesperson sold $22,500 worth of goods?

45. Robert paid sales tax of $0.72 on his dinner. The sales tax rate is 5%. What was the cost of his dinner without the tax?

46. A television set that sold originally for $480 was sold for $336. What was the percent decrease?

37. ____________

38. ____________

39. ____________

40. ____________

41. ____________

42. ____________

43. ____________

44. ____________

45. ____________

46. ____________

5.1 Understanding Percent

MathPro Video 11 SSM

After studying this section, you will be able to:

1 *Express a fraction whose denominator is 100 as a percent.*
2 *Write a percent as a decimal.*
3 *Write a decimal as a percent.*

1 *Expressing a Fraction Whose Denominator Is 100 as a Percent*

"My raise came through. I got a 6% increase!"

"The leading economic indicators show inflation rising at a 4.3% rate."

"Babe Ruth and Hank Aaron each hit quite a few home runs. But I wonder who has the higher percentage of home runs per at bat?"

We use percents often in our everyday lives. In business, in sports, in shopping, and in many areas of life, percentages play an important role. In this section we introduce the idea of percent, which means "*per centum*" or "per hundred." We then show how to use percentages.

In previous chapters, when we described parts of a whole, we used fractions or decimals. Using a percent is another way to describe a part of a whole. Percents can be described as ratios whose denominators are 100. The word *percent* means per 100. This sketch has 100 rectangles.

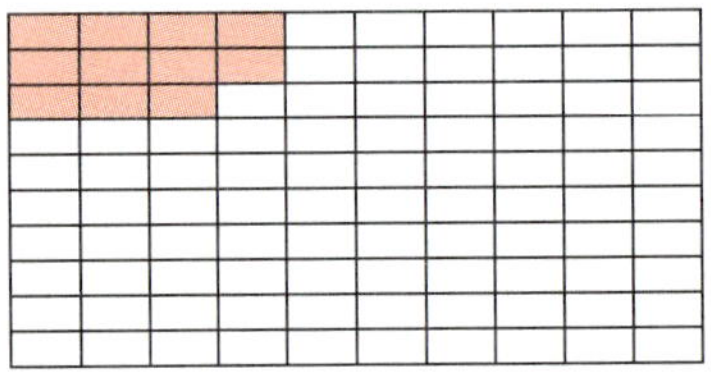

Of 100 rectangles, 11 are shaded. We can say that 11 percent of the whole is shaded. We use the symbol % for percent. It means "parts per 100." When we write 11 percent as 11%, we understand that it means 11 parts per one hundred, or, as a fraction, $\frac{11}{100}$.

EXAMPLE 1 Write each as a percent.

(a) 17 out of 100 students were left-handed.

(b) 31 out of 100 library books returned were overdue.

(c) 9 out of 100 male students were taller than 6 feet.

(d) 86 out of 100 college students had at least one parent who attended college.

(a) $\frac{17}{100} = 17\%$ **(b)** $\frac{31}{100} = 31\%$

(c) $\frac{9}{100} = 9\%$ **(d)** $\frac{86}{100} = 86\%$

Practice Problem 1 Write each as a percent.

(a) 51 out of 100 students in the class were women.

(b) 68 out of 100 cars in the parking lot have front-wheel drive.

(c) 7 out of 100 students in the dorm quit smoking.

(d) 26 out of 100 students did not vote in class elections. ■

Some percents are larger than 100. Consider the following situations.

EXAMPLE 2

(a) Write $\frac{386}{100}$ as a percent.

(b) Last year's enrollment in psychology classes was 100 students. This year's enrollment is 136. Write this year's enrollment as a percent of last year's.

(c) Twenty years ago four car tires for a full-sized car cost \$100. Now the average price for four car tires for a full-sized car is \$270. Write the present cost as a percent of the cost 20 years ago.

(a) $\frac{386}{100} = 386\%$

(b) The ratio is $\frac{136 \text{ students this year}}{100 \text{ students last year}}$. $\frac{136}{100} = 136\%$

This year's enrollment is 136% of last year's enrollment.

(c) The ratio is $\frac{\$270 \text{ for four tires now}}{\$100 \text{ for four tires then}}$. $\frac{270}{100} = 270\%$

The present cost of four car tires for a full-sized car is 270% of the cost 20 years ago.

Practice Problem 2

(a) Write $\frac{238}{100}$ as a percent.

(b) Write $\frac{505}{100}$ as a percent.

(c) Last year 100 students tried out for varsity baseball. This year 121 students tried out. Write this year's number as a percent of last year's number. ■

Some percents are smaller than 1%.

$\frac{0.7}{100}$ can be written as 0.7%.

$\frac{0.3}{100}$ can be written as 0.3%.

$\frac{0.04}{100}$ can be written as 0.04%.

EXAMPLE 3 Write each as a percent.

(a) $\frac{0.9}{100}$ **(b)** $\frac{0.2}{100}$ **(c)** $\frac{0.07}{100}$

(a) $\frac{0.9}{100} = 0.9\%$ **(b)** $\frac{0.2}{100} = 0.2\%$ **(c)** $\frac{0.07}{100} = 0.07\%$

Practice Problem 3 Write each as a percent.

(a) $\frac{0.5}{100}$ **(b)** $\frac{0.06}{100}$ **(c)** $\frac{0.003}{100}$ ■

2 *Writing a Percent as a Decimal*

Suppose we have a percent such as 59%. What would be the equivalent in decimal form? Using our definition of percent, $59\% = \frac{59}{100}$. This fraction could be written in decimal form as 0.59. In a similar way, we could write 21% as $\frac{21}{100} = 0.21$. This pattern allows us to quickly change the form of a number from a percent to a fraction whose denominator is 100 to a decimal.

EXAMPLE 4 Write each percent as a decimal.

(a) 38% **(b)** 6% **(c)** 70%

(a) $38\% = \frac{38}{100} = 0.38$ **(b)** $6\% = \frac{6}{100} = 0.06$

(c) $70\% = \frac{70}{100} = 0.70$ or 0.7

The final zero is not really necessary.

Practice Problem 4 Write each percent as a decimal.

(a) 47% **(b)** 2% **(c)** 90% ■

The results of Example 4 suggest the following rule.

Changing a Percent to a Decimal

1. Move the decimal point two places to the left.
2. Drop the % symbol.

EXAMPLE 5 Write each percent as a decimal.

(a) 26.9% **(b)** 7.2% **(c)** 0.13% **(d)** 158%

In each case, we move the decimal point two places to the left and drop the percent symbol.

(a) 26.9% = 0.269

(b) 7.2% = 0.072 *Note we need to add an extra zero to the left of the seven.*

(c) 0.13% = 0.0013 *Here we added two zeros to the left of the* 1.

(d) 158% = 1.58

Practice Problem 5 Write each percent as a decimal.

(a) 80.6% **(b)** 2.5% **(c)** 0.29% **(d)** 231% ■

3 *Writing a Decimal as a Percent*

In Example 4(a) we changed 38% to $\frac{38}{100}$ to 0.38. We can reverse the process. We can start with $0.38 = \frac{38}{100} = 38\%$. Study all the parts of Examples 4 and 5. You will see that the steps are reversible. Thus 0.38 = 38%, 0.06 = 6%, 0.70 = 70%, 0.269 = 26.9%, 0.072 = 7.2%, 0.0013 = 0.13%, and 1.58 = 158%. In each part we see that the decimal point is moved two places to the right. Then the percent symbol is written after the number.

Changing a Decimal to a Percent

1. Move the decimal point two places to the right.
2. Then write the % symbol at the end of the number.

EXAMPLE 6 Write each decimal as a percent.

(a) 0.47 **(b)** 0.08 **(c)** 6.31 **(d)** 0.055 **(e)** 0.001

In each part we move the decimal point two places to the right and write the percent symbol at the end of the number.

(a) 0.47 = 47% **(b)** 0.08 = 8% **(c)** 6.31 = 631%
(d) 0.055 = 5.5% **(e)** 0.001 = 0.1%

Practice Problem 6 Write each decimal as a percent.

(a) 0.78 **(b)** 0.02 **(c)** 5.07 **(d)** 0.029 **(e)** 0.006 ■

CALCULATOR

Percent to Decimal

You can use a calculator to change 52% to a decimal.
Enter

52 [%]

The display should read

[0.52]

Try

(a) 46% **(b)** 137%
(c) 9.3% **(d)** 6%

Note: The calculator divides by 100 when the percent key is pressed. If you do not have a [%] key then you can use the keystrokes [÷] 100 [=].

To Think About What is really happening when we change a decimal to a percent? Suppose that we wanted to change 0.59 to a percent.

$$0.59 = \frac{59}{100} \qquad \textit{Definition of a decimal.}$$

$$= 59 \times \frac{1}{100} \qquad \textit{Definition of multiplying fractions.}$$

$$= 59 \text{ percent} \qquad \textit{Because ``per 100'' means percent.}$$

$$= 59\% \qquad \textit{Writing the symbol for percent.}$$

Can you see why each step is valid? Since we know the reason behind each step, we know we can always move the decimal point two places to the right and write the percent symbol. See Exercises 5.1, problems 77 and 78.

5.1 Exercises

Verbal and Writing Skills

1. In this section we introduced percent, which means "per centum" or "per ______________."

2. The number 1 written as a percent is ______________.

3. To change a percent to a decimal, move the decimal point ______________ places to the ______________. ______________ the % symbol.

4. To change a decimal to a percent, move the decimal point ______________ places to the ______________. ______________ the % symbol at the end of the number.

Write each fraction as a percent.

5. $\frac{45}{100}$ **6.** $\frac{48}{100}$ **7.** $\frac{7}{100}$ **8.** $\frac{9}{100}$ **9.** $\frac{80}{100}$ **10.** $\frac{60}{100}$

11. $\frac{245}{100}$ **12.** $\frac{110}{100}$ **13.** $\frac{5.3}{100}$ **14.** $\frac{8.3}{100}$ **15.** $\frac{0.6}{100}$ **16.** $\frac{0.4}{100}$

Applications

Write a percent to express each of the following.

17. 29 out of 100 power boats had a radar navigation system.

18. 13 out of 100 students in the class were seniors.

19. 78 out of 100 new car owners were satisfied with their purchase.

20. 64 out of 100 college sophomores are confident of their selection of a major.

21. 32 out of 100 people who ate out chose to eat at a fast-food restaurant.

22. 8 out of 100 people overdraw their checking account at least twice per year.

Write each percent as a decimal.

23. 51% **24.** 42% **25.** 7% **26.** 6% **27.** 20% **28.** 40% **29.** 43.6% **30.** 81.5%

31. 0.3% **32.** 0.9% **33.** 0.72% **34.** 0.61% **35.** 126% **36.** 175% **37.** 366% **38.** 398%

Write each decimal as a percent.

39. 0.74 **40.** 0.66 **41.** 0.50 **42.** 0.40 **43.** 0.08 **44.** 0.03 **45.** 0.563 **46.** 0.408

47. 0.002 **48.** 0.009 **49.** 0.0057 **50.** 0.0026 **51.** 1.35 **52.** 1.86 **53.** 2.72 **54.** 3.04

55. Bob paid 0.27 of his income for federal income tax.

56. Sally spends 0.31 of her income for housing.

Mixed Practice

Write as a percent.

57. $\frac{36}{100}$ **58.** 0.25 **59.** $\frac{143}{100}$ **60.** 1.48 **61.** 0.3 **62.** $\frac{40}{100}$ **63.** $\frac{0.5}{100}$ **64.** 0.005

Write as a decimal.

65. 62% **66.** $\frac{12}{100}$ **67.** 128% **68.** $\frac{210}{100}$ **69.** 0.5% **70.** $\frac{0.8}{100}$ **71.** $\frac{80}{100}$ **72.** 70%

Applications

Write as a percent.

73. For every 100 people who voted in the 1996 presidential election, 49 voted for Bill Clinton.

74. For every 100 people who voted in the 1996 presidential election, 41 voted for Robert Dole.

75. In Alaska, 0.03413 of the state is covered by water.

76. In Florida, 0.07689 of the state is covered by water.

To Think About

77. Suppose that we want to change 36% to 0.36 by moving the decimal point two places to the left and dropping the % symbol. Explain the steps to show what is really involved in changing 36% to 0.36. Why does the rule work?

78. Suppose that we want to change 10.65 to 1065%. Give a complete explanation of the steps.

Take the value in each case and write it (a) as a decimal, (b) as a fraction with a denominator of 100, and (c) as a reduced fraction.

79. 55,562% **80.** 60,724%

Cumulative Review Problems

Write as a fraction in simplest form.

81. 0.56 **82.** 0.72

Write as a decimal.

83. $\frac{11}{16}$ **84.** $\frac{13}{16}$

5.2 Changing Between Percents, Decimals, and Fractions

MathPro Video 11 SSM

After studying this section, you will be able to:

1 *Change a percent to a fraction.*
2 *Change a fraction to a percent.*
3 *Change a percent, a decimal, or a fraction to equivalent forms.*

1 *Changing a Percent to a Fraction*

By using the definition of percent, we can write any percent as a fraction whose denominator is 100. Thus when we change a percent to a fraction, we remove the percent symbol and write the number over 100. If possible, we simplify the fraction.

EXAMPLE 1 Write as a fraction in simplest form.

(a) 37% **(b)** 75% **(c)** 2%

(a) $37\% = \frac{37}{100}$ **(b)** $75\% = \frac{75}{100} = \frac{3}{4}$ **(c)** $2\% = \frac{2}{100} = \frac{1}{50}$

Practice Problem 1 Write as a fraction in simplest form.

(a) 71% **(b)** 25% **(c)** 8% ■

In some cases, it may be helpful to write the percent as a decimal before you write it as a fraction in simplest form.

EXAMPLE 2 Write as a fraction in simplest form.

(a) 43.5% **(b)** 5.8% **(c)** 36.75%

(a) $43.5\% = 0.435$ *Change the percent to a decimal.*

$= \frac{435}{1000}$ *Change the decimal to a fraction.*

$= \frac{87}{200}$ *Reduce the fraction.*

(b) $5.8\% = 0.058 = \frac{58}{1000} = \frac{29}{500}$

(c) $36.75\% = 0.3675 = \frac{3675}{10{,}000} = \frac{147}{400}$

Practice Problem 2 Write each percent as a fraction in simplest form.

(a) 8.4% **(b)** 55.25% **(c)** 28.5% ■

If the percent is greater than 100%, the simplified fraction is usually changed to a mixed number.

EXAMPLE 3 Write each percent as a mixed number.

(a) 225% **(b)** 138%

(a) $225\% = 2.25 = 2\frac{25}{100} = 2\frac{1}{4}$ **(b)** $138\% = 1.38 = 1\frac{38}{100} = 1\frac{19}{50}$

Practice Problem 3 Write each percent as a mixed number.

(a) 170% **(b)** 288% ■

Sometimes a percent is not a whole number, such as 9% or 10%. Instead, it contains a fraction, such as $9\frac{1}{12}\%$ or $9\frac{3}{8}\%$. Extra steps will be needed to write such a percent as a simplified fraction.

EXAMPLE 4 Convert $9\frac{3}{8}\%$ to a fraction in simplest form.

$9\frac{3}{8}\% = \dfrac{9\frac{3}{8}}{100}$ *Change the percent to a fraction.*

$= 9\frac{3}{8} \div \frac{100}{1}$ *Write the division horizontally.* $\dfrac{9\frac{3}{8}}{100}$ *means* $9\frac{3}{8}$ *divided by* 100.

$= \frac{75}{8} \div \frac{100}{1}$ *Write* $9\frac{3}{8}$ *as a fraction.*

$= \frac{75}{8} \times \frac{1}{100}$ *Use the definition of division of fractions.*

$= \frac{\cancelto{3}{75}}{8} \times \frac{1}{\cancelto{4}{100}}$ *Simplify.*

$= \frac{3}{32}$ *Reduce the fraction.*

Practice Problem 4 Convert $7\frac{5}{8}\%$ to a fraction in simplest form. ■

EXAMPLE 5 In the fiscal 1994 budget of the United States, approximately $20\frac{3}{8}\%$ of the budget was designated for defense. Write this percent as a fraction.

$$20\frac{3}{8}\% = \frac{20\frac{3}{8}}{100} = 20\frac{3}{8} \div 100 = \frac{163}{8} \times \frac{1}{100} = \frac{163}{800}$$

Thus we could say $\frac{163}{800}$ of the fiscal 1994 budget was designated for defense. That is, for every \$800 in the budget, \$163 was spent for defense.

Practice Problem 5 In the fiscal 1994 budget of the United States, approximately $21\frac{1}{4}\%$ was designated for Health and Human Services. Write this percent as a fraction. ■

Certain percents occur very often, especially in money matters. Here are some common equivalents that you may already know. If not, be sure to memorize them.

$$25\% = \frac{1}{4} \qquad 33\frac{1}{3}\% = \frac{1}{3} \qquad 10\% = \frac{1}{10}$$

$$50\% = \frac{1}{2} \qquad 66\frac{2}{3}\% = \frac{2}{3}$$

$$75\% = \frac{3}{4}$$

2 *Changing a Fraction to a Percent*

A convenient way to change a fraction to a percent is to write the fraction in decimal form first and then convert the decimal to a percent.

EXAMPLE 6 Write $\frac{3}{8}$ as a percent.

We see that $\frac{3}{8} = 0.375$ by dividing out $3 \div 8$.

```
   0.375
8)3.000
  24
   60
   56
    40
    40
     0
```

Thus $\frac{3}{8} = 0.375 = 37.5\%$.

Practice Problem 6 Write $\frac{5}{8}$ as a percent. ■

EXAMPLE 7 Write each fraction as a percent.

(a) $\frac{4}{5}$ **(b)** $\frac{7}{40}$ **(c)** $\frac{39}{50}$

(a) $\frac{4}{5} = 0.8 = 80\%$ **(b)** $\frac{7}{40} = 0.175 = 17.5\%$

(c) $\frac{39}{50} = 0.78 = 78\%$

Practice Problem 7 Write each fraction as a percent.

(a) $\frac{3}{5}$ **(b)** $\frac{21}{25}$ **(c)** $\frac{7}{16}$ ■

Changing some fractions to decimal form results in an infinitely repeating decimal. In such cases, we usually round to the nearest hundredth of a percent.

EXAMPLE 8 Write each fraction as a percent. Round to the nearest hundredth of a percent.

(a) $\frac{1}{6}$ **(b)** $\frac{15}{33}$

(a) $\frac{1}{6} = 0.16666\ldots$ by dividing out $1 \div 6$.

$$\begin{array}{r} 0.166 \\ 6\overline{)1.000} \\ \underline{6} \\ 40 \\ \underline{36} \\ 40 \end{array}$$

The repeating quotient can be written as $0.1\overline{6}$. If we round the decimal to the nearest ten thousandth, we have $\frac{1}{6} = 0.1667$. If we change this to a percent, we have

$$\frac{1}{6} = 16.67\%$$

This is correct to the nearest hundredth of a percent.

(b) By dividing out $15 \div 33$, we see that $\frac{15}{33} = 0.45454545\ldots$ This can be written as $0.\overline{45}$. If we round to the nearest ten-thousandth, we have

$$\frac{15}{33} = 0.4545 = 45.45\%$$

This rounded value is correct to the nearest hundredth of a percent.

Practice Problem 8 Write each fraction as a percent. Round to the nearest hundredth of a percent.

(a) $\frac{7}{9}$ **(b)** $\frac{19}{30}$ ■

In some cases, a percent is written with a fraction.

EXAMPLE 9 Express $\frac{11}{12}$ as a percent containing a fraction.

We will terminate the division after two steps, and write the remainder in fraction form.

$$\begin{array}{r} 0.91 \\ 12\overline{)11.00} \\ \underline{108} \\ 20 \\ \underline{12} \\ 8 \end{array}$$

This division tells us that we can write

$$\frac{11}{12} \quad \text{as} \quad 0.91\frac{8}{12} \quad \text{or} \quad 0.91\frac{2}{3}$$

We now have a decimal with a fraction. When we express this decimal as a percent, we move the decimal point two places to the right. We do not write the decimal in front of the fraction.

$$0.91\frac{2}{3} = 91\frac{2}{3}\%$$

Practice Problem 9 Express $\frac{7}{12}$ as a percent containing a fraction. ■

3 *Changing a Percent, a Decimal, or a Fraction to Equivalent Forms*

We have seen so far that a fraction, a decimal, and a percent are three different forms (notations) for the same number. We can illustrate this in a chart.

EXAMPLE 10 Complete the following table of equivalent notations. Round decimals to the nearest ten thousandth. Round percents to the nearest hundredth of a percent.

Fraction	Decimal	Percent
$\frac{11}{16}$		
	0.265	
		$17\frac{1}{5}\%$

Begin with the first row. The number is written as a fraction. We will change the fraction to a decimal and then to a percent.

The Fraction is changed to a Decimal is changed to a Percent

$$\frac{11}{16} \longrightarrow 16\overline{)11.0000}^{\,0.6875} \longrightarrow 68.75\%$$

In the second row the number is written as a decimal. This can easily be written as a percent.

$$0.265 \longrightarrow 26.5\%$$

To write the number as a fraction, write 0.265 as a fraction and simplify.

$$0.265 \downarrow \frac{265}{1000} \longrightarrow \frac{53}{200}$$

In the third row the number is written as a percent. Proceed from right to left—that is, write the number as a decimal and then as a fraction.

$$17\frac{1}{5}\% \rightarrow \frac{17\frac{1}{5}}{100}$$

$$\downarrow$$

$$\frac{86}{5} \times \frac{1}{100}$$

$$\downarrow$$

$$0.172 \leftarrow \frac{86}{500} \qquad \textit{Divide.}\ 500\overline{)86.000}^{\,0.172}$$

and

$$\frac{43}{250} \leftarrow \frac{86}{500}$$

Thus the completed table is as follows.

Fraction	Decimal	Percent
$\frac{11}{16}$	0.6875	68.75%
$\frac{53}{200}$	0.265	26.5%
$\frac{43}{250}$	0.172	$17\frac{1}{5}\%$

Practice Problem 10 Complete the following table of equivalent notations. Round decimals to the nearest ten thousandth. Round percents to the nearest hundredth of a percent.

Fraction	Decimal	Percent
$\frac{23}{99}$		
	0.516	
		$38\frac{4}{5}\%$

■

ALTERNATIVE METHOD Another way to convert a fraction to a percent is to use a proportion. To change $\frac{7}{8}$ to a percent, write the proportion

$$\frac{7}{8} = \frac{n}{100}$$

$$7 \times 100 = 8 \times n \qquad \textit{Cross-multiply.}$$

$$700 = 8 \times n \qquad \textit{Simplify.}$$

$$\frac{700}{8} = \frac{8 \times n}{8} \qquad \textit{Divide each side by 8.}$$

$$87.5 = n \qquad \textit{Simplify.}$$

Thus $\frac{7}{8} = 87.5\%$. You will use this approach in Exercises 5.2, problems 93 and 94.

CALCULATOR

Fraction to Decimal

You can use a calculator to change $\frac{3}{5}$ to a decimal.

Enter

3 [÷] 5 [=]

The display should read

[0.6]

Try

(a) $\frac{17}{25}$ (b) $\frac{2}{9}$

(c) $\frac{13}{10}$ (d) $\frac{15}{19}$

Note: 0.7894737 is an approximation for $\frac{15}{19}$. Some calculators only round to 8 places.

Developing Your Study Skills

READING THE TEXTBOOK

Homework time each day should begin with the careful reading of the section(s) assigned in your textbook. Much time and effort have gone into the selection of a particular text, and your instructor has chosen a book that will help you to become successful in this mathematics class. Expensive textbooks can be a wise investment if you take advantage of them by reading them.

Reading a mathematics textbook is unlike reading many other types of books that you may use in your literature, history, psychology, or sociology courses. Mathematics texts are technical books that provide you with exercises to practice on. Reading a mathematics text requires slow and careful reading of each word, which takes time and effort.

Begin reading your textbook with a paper and pencil in hand. As you come across a new definition, or concept, underline it in the text and/or write it down in your notebook. Whenever you encounter an unfamiliar term, look it up and make a note of it. When you come to an example, work through it step by step. Be sure to read each word and to follow directions carefully.

Notice the helpful hints the author provides to guide you to correct solutions and prevent you from making errors. Take advantage of these pieces of expert advice.

Be sure that you understand what you are reading. Make a note of any of those things that you do not understand and ask your instructor about them. Do not hurry through the material. Learning mathematics takes time.

5.2 Exercises

Write each percent as a fraction or as a mixed number.

1. 88% **2.** 52% **3.** 7% **4.** 9% **5.** 33% **6.** 47% **7.** 55% **8.** 35%

9. 75% **10.** 25% **11.** 20% **12.** 40% **13.** 14.5% **14.** 18.5% **15.** 17.6% **16.** 78.4%

17. 64.8% **18.** 12.2% **19.** 71.25% **20.** 38.75% **21.** 176% **22.** 228% **23.** 340% **24.** 420%

25. 1200% **26.** 3600% **27.** $2\frac{1}{6}\%$ **28.** $7\frac{5}{6}\%$ **29.** $12\frac{3}{8}\%$ **30.** $15\frac{5}{8}\%$ **31.** $8\frac{4}{5}\%$ **32.** $9\frac{3}{5}\%$

Applications

33. In the 1996 presidential election, Ross Perot received $2\frac{2}{13}\%$ of the vote in the District of Columbia. Write this percent as a fraction.

34. During an inspection $17\frac{3}{4}\%$ of the labels printed had the wrong colors. What *fraction* of the labels was in the wrong color?

35. In the 1994 budget of the United States, approximately $4\frac{1}{11}\%$ was designated for the Department of Agriculture. Write this percent as a fraction.

36. In the 1994 budget of the United States, approximately $2\frac{6}{11}\%$ was designated for the Department of Transportation. Write this percent as a fraction.

Write each fraction or mixed number as a percent. Round to the nearest hundredth of a percent when necessary.

37. $\frac{3}{4}$ **38.** $\frac{1}{4}$ **39.** $\frac{1}{3}$ **40.** $\frac{2}{3}$ **41.** $\frac{5}{16}$ **42.** $\frac{9}{16}$ **43.** $\frac{7}{25}$ **44.** $\frac{9}{25}$

45. $\frac{11}{40}$ **46.** $\frac{13}{40}$ **47.** $\frac{5}{12}$ **48.** $\frac{8}{12}$ **49.** $\frac{18}{5}$ **50.** $\frac{7}{4}$ **51.** $2\frac{5}{6}$ **52.** $3\frac{1}{6}$

53. $4\frac{1}{8}$ **54.** $2\frac{5}{8}$ **55.** $\frac{3}{7}$ **56.** $\frac{2}{7}$ **57.** $\frac{15}{16}$ **58.** $\frac{11}{16}$ **59.** $\frac{26}{50}$ **60.** $\frac{43}{50}$

61. $\frac{47}{137}$ **62.** $\frac{53}{129}$ **63.** $\frac{316}{907}$ **64.** $\frac{427}{891}$

Applications

Round to the nearest hundredth of a percent.

65. The brain represents approximately $\frac{1}{40}$ of an average person's weight. Express this fraction as a percent.

66. Carbon represents approximately $\frac{9}{50}$ of an average person's weight. Express this fraction as a percent.

67. The U.S. mail service estimates that people do not send as many personal letters through the mail as they did 20 years ago. In fact in 1996 only 37/3000 of all mail delivered consisted of personal letters. Express this fraction as a percent.

68. During waking hours, a person blinks 9/2000 of the time. Express this fraction as a percent.

Express each fraction as a percent containing a fraction. (See Example 9.)

69. $\frac{5}{6}$ **70.** $\frac{1}{6}$ **71.** $\frac{11}{16}$ **72.** $\frac{15}{16}$ **73.** $\frac{3}{8}$ **74.** $\frac{5}{8}$ **75.** $\frac{3}{40}$ **76.** $\frac{11}{90}$

Mixed Practice

Complete the following table of equivalents. Round decimals to the nearest ten thousandth. Round percents to the nearest hundredth of a percent.

	Fraction	Decimal	Percent
77.	$\frac{5}{12}$		
79.		0.06	
81.			40%
83.		0.345	
85.	$\frac{3}{200}$		
87.	$\frac{5}{9}$		
89.			$3\frac{1}{8}\%$

	Fraction	Decimal	Percent
78.	$\frac{7}{12}$		
80.		0.08	
82.			60%
84.		0.625	
86.	$\frac{9}{200}$		
88.	$\frac{7}{9}$		
90.			$2\frac{5}{8}\%$

91. Write $28\frac{15}{16}\%$ as a fraction.

★ **92.** Write $18\frac{7}{12}\%$ as a fraction.

Change each fraction to a percent by using a proportion.

★ **93.** $\frac{123}{800}$

★ **94.** $\frac{417}{600}$

Cumulative Review Problems

Solve for n in each problem.

95. $\frac{15}{n} = \frac{8}{3}$ **96.** $\frac{32}{24} = \frac{n}{3}$ **97.** $\frac{n}{11} = \frac{32}{4}$ **98.** $\frac{1}{3} = \frac{32}{n}$

5.3A Solving Percent Problems Using an Equation

After studying this section, you will be able to:

1. *Translate a percent problem into an equation.*
2. *Solve a percent problem by solving an equation.*

MathPro Video 11 SSM

1 *Translating a Percent Problem into an Equation*

In word problems, like the ones in this section, we can translate from words to mathematical symbols and back again. After we have the mathematical symbols arranged in an *equation,* we solve the equation. When we find the values that make the equation true, we have also found the answer to our word problem.

To solve a percent problem, express it as an equation with an unknown quantity. We use the letter n to represent the number we do not know. The following table is helpful when translating from a percent problem to an equation.

Word	Mathematical Symbol
of	Any multiplication symbol: $\times$ or ( ) or $\cdot$
is	$=$
what	Any letter; for example, n
find	$n =$

In Examples 1–6 we show how to translate words into an equation. Please do **not** solve the problem. Translate into an equation only.

EXAMPLE 1 Translate into an equation.

What is 5% of 19.00?

↓ ↓ ↓ ↓ ↓

$n = 5\% \times 19.00$

Practice Problem 1 Translate into an equation. What is 26% of 35? ■

EXAMPLE 2 Translate into an equation.

142% of 35 is what?

↓ ↓ ↓ ↓ ↓

$142\% \times 35 = n$

Practice Problem 2 Translate into an equation. 155% of 20 is what? ■

EXAMPLE 3 Translate into an equation.

Find 0.6% of 400.

Notice here that the words *what is* are missing. The word *find* is equivalent to *what is*.

$$\begin{array}{cccc} \text{Find} & 0.6\% & \text{of} & 400. \\ \downarrow & \downarrow & \downarrow & \downarrow \\ n = & 0.6\% & \times & 400 \end{array}$$

Practice Problem 3 Translate into an equation. Find 0.08% of 350. ■

The unknown quantity, n, does not always stand alone in an equation.

EXAMPLE 4 Translate into an equation.

(a) 35% of what is 60? **(b)** 7.2 is 120% of what?

(a) $\begin{array}{ccccc} 35\% & \text{of} & \text{what} & \text{is} & 60? \\ \downarrow & \downarrow & \downarrow & \downarrow & \downarrow \\ 35\% & \times & n & = & 60 \end{array}$ **(b)** $\begin{array}{ccccc} 7.2 & \text{is} & 120\% & \text{of} & \text{what?} \\ \downarrow & \downarrow & \downarrow & \downarrow & \downarrow \\ 7.2 & = & 120\% & \times & n \end{array}$

Practice Problem 4 Translate into an equation.

(a) 58% of what is 400? **(b)** 9.1 is 135% of what? ■

EXAMPLE 5 Translate into an equation.

$$\begin{array}{ccccc} \underbrace{\text{What percent}} & \text{of} & 50 & \text{is} & 10? \\ \downarrow & \downarrow & \downarrow & \downarrow & \downarrow \\ n & \times & 50 & = & 10 \end{array}$$

We see here that the words *what percent* are represented by the letter n.

Practice Problem 5 Translate into an equation. What percent of 250 is 36? ■

EXAMPLE 6 Translate into an equation.

(a) 30 is what percent of 16? **(b)** What percent of 3000 is 2.6?

(a) $\begin{array}{ccccc} 30 & \text{is} & \underbrace{\text{what percent}} & \text{of} & 16? \\ \downarrow & \downarrow & \downarrow & \downarrow & \downarrow \\ 30 & = & n & \times & 16 \end{array}$ **(b)** $\begin{array}{ccccc} \underbrace{\text{What percent}} & \text{of} & 3000 & \text{is} & 2.6? \\ \downarrow & \downarrow & \downarrow & \downarrow & \downarrow \\ n & \times & 3000 & = & 2.6 \end{array}$

Practice Problem 6 Translate into an equation.

(a) 50 is what percent of 20? **(b)** What percent of 2000 is 4.5? ■

2 Solve a Percent Problem by Solving an Equation

The percent problems we have translated are of three types. Consider the equation $60 = 20\% \times 300$. This problem has the form

$$\text{amount} = \text{percent} \times \text{base}$$

Any of these quantities—amount, percent, or base—may be unknown.

1. When *we do not know the amount,* we have an equation like

$$n = 20\% \times 300$$

2. When *we do not know the base,* we have an equation like

$$60 = 20\% \times n$$

3. When *we do not know the percent,* we have an equation like

$$60 = n \times 300$$

We will study each type separately. It is not necessary to memorize the three types, but it is helpful to look carefully at the examples we give of each. In each example, do the computation in a way that is easiest for you. This may be using a pencil and paper, using a calculator, or, in some cases, doing the problem mentally.

Solving Percent Problems When We Do Not Know the Amount

EXAMPLE 7 What is 45% of 590?

$n = 45\% \times 590$	*Translate into an equation.*
$n = (0.45)(590)$	*Change the percent to decimal form and multiply.*
$n = 265.5$	*Multiply* 0.45×590.

Practice Problem 7 What is 82% of 350? ■

EXAMPLE 8

(a) 3.8% of 600 is what? **(b)** Find 160% of 500.

(a) 3.8% of 600 is what?

$3.8\% \times 600 = n$	*Translate into an equation.*
$(0.038)(600) = n$	*Change the percent to decimal form and multiply.*
$22.8 = n$	*Multiply* 0.038×600.

(b) Find 160% of 500. *When you translate, remember that the word* find *is equivalent to* what is.

$n = 160\% \times 500$	
$n = (1.60)(500)$	*Change the percent to decimal form and multiply.*
$n = 800$	*Multiply* 1.6×500.

Practice Problem 8

(a) 5.2% of 800 is what? **(b)** Find 230% of 400. ■

> **CALCULATOR**
>
> **Percent of a Number**
>
> You can use a calculator to find 12% of 48. Enter
>
> 12 [%] [×] 48 [=]
>
> The display should read
>
> [5.76]
>
> If your calculator *does not* have a percent key, use the keystrokes
>
> 0.12 [×] 48 [=]
>
> *Try*
>
> What is 54% of 450?

EXAMPLE 9 When Rick bought a new Dodge Neon, he had to pay a sales tax of 5% on the cost of the car, which was \$12,000. What was the sales tax?

This problem is asking

$$\begin{array}{ccccc} \text{What} & \text{is} & 5\% & \text{of} & \$12{,}000? \\ \downarrow & \downarrow & \downarrow & \downarrow & \downarrow \\ n & = & 5\% & \times & \$12{,}000 \end{array}$$

$$n = 0.05 \times 12{,}000$$

$$n = \$600$$

The sales tax was \$600.

Practice Problem 9 When Oprah bought an airplane ticket, she had to pay a tax of 8% on the cost of the ticket, which was \$350. What was the tax? ■

Solving Percent Problems When We Do Not Know the Base

If a number is multiplied by the letter n, this can be indicated by a multiplication sign, parentheses, a dot, or by placing the number in front of the letter. Thus $3 \times n = 3(n) = 3 \cdot n = 3n$. In this section we use equations like $3n = 9$ and $0.5n = 20$. To solve these equations we use the procedures developed in Chapter 4. We divide each side by the number multiplied by n.

EXAMPLE 10 90% of what is 342?

$$\begin{array}{ccccc} 90\% & \text{of} & \text{what} & \text{is} & 342? \\ \downarrow & \downarrow & \downarrow & \downarrow & \downarrow \\ 90\% & \times & n & = & 342 \end{array}$$ *Translate into an equation.*

$0.9n = 342$ *Change the percent to a decimal and multiply by n. Recall that* $0.90 = 0.9$.

$\dfrac{0.9n}{0.9} = \dfrac{342}{0.9}$ *Divide each side of the equation by* 0.9.

$n = 380$ *Divide* $342 \div 0.9$.

Practice Problem 10 45% of what is 162? ■

EXAMPLE 11 12 is 0.6% of what?

$$\begin{array}{ccccc} 12 & \text{is} & 0.6\% & \text{of} & \text{what?} \\ \downarrow & \downarrow & \downarrow & \downarrow & \downarrow \\ 12 & = & 0.6\% & \times & n \end{array}$$ *Translate into an equation.*

$12 = 0.006n$ *Change* 0.6% *to a decimal and multiply by n.*

$\dfrac{12}{0.006} = \dfrac{0.006n}{0.006}$ *Divide each side of the equation by* 0.006.

$2000 = n$ *Divide* $12 \div 0.006$.

Practice Problem 11 32 is 0.4% of what? ■

EXAMPLE 12 Dave and Elsie went out to dinner. They gave the waiter a tip that was 15% of the total bill. The waiter received \$6. What was the total bill (not including the tip)?

This problem is asking

$$\text{15\% of what is \$6?}$$

$$\downarrow \quad \downarrow \quad \downarrow \quad \downarrow \quad \downarrow$$

$$15\% \times n = 6$$

$$0.15n = 6$$

$$\frac{0.15n}{0.15} = \frac{6}{0.15}$$

$$n = 40$$

The total bill for the meal (not including the tip) was \$40.

Practice Problem 12 The coach of the university baseball team said that 30% of the players on his team are left-handed. Six people on the team are left-handed. How many people are on the team? ■

Solving Percent Problems When We Do Not Know the Percent

In solving these problems, we notice that there is no % symbol in the problem. The percent is what we are trying to find. Therefore, our answer for this type of problem will always have a percent symbol.

EXAMPLE 13 What percent of 5000 is 3.8?

$$\underbrace{\text{What percent}} \text{ of } 5000 \text{ is } 3.8?$$

$$\downarrow \quad \downarrow \quad \downarrow \quad \downarrow \quad \downarrow$$

$$n \times 5000 = 3.8$$ *Translate into an equation.*

$$5000n = 3.8$$ *Multiplication is commutative. $n \times 5000 = 5000 \times n$.*

$$\frac{5000n}{5000} = \frac{3.8}{5000}$$ *Divide each side by 5000.*

$$n = 0.00076$$ *Divide $3.8 \div 5000$.*

$$n = 0.076\%$$ *Express the decimal as a percent.*

Practice Problem 13 What percent of 9000 is 4.5? ■

EXAMPLE 14 90 is what percent of 20?

$$\underset{\downarrow}{90}\ \underset{\downarrow}{\text{is}}\ \underbrace{\text{what percent}}_{\downarrow}\ \underset{\downarrow}{\text{of}}\ \underset{\downarrow}{20?}$$

$$90 = n \times 20$$ *Translate into an equation.*

$$90 = 20n$$ *Multiplication is commutative.* $n \times 20 = 20 \times n$.

$$\frac{90}{20} = \frac{20n}{20}$$ *Divide each side by* 20.

$$4.5 = n$$ *Divide* $90 \div 20$.

$$450\% = n$$ *Express the decimal as a percent.*

CALCULATOR

Finding the Percent

You can use a calculator to find a missing percent. What percent of 95 is 19?

1. Enter as a fraction.

is [÷] of

19 [÷] 95

2. Change to a percent.

19 [÷] 95 [×] 100 [=]

The display should read

[20]

This means 20%.

Try

What percent of 625 is 250?

Practice Problem 14 198 is what percent of 33?

EXAMPLE 15 In a recent basketball game for the Chicago Bulls, Scotty Pippin made 10 of his 24 shots. What percent of his shots did he make? (Round your answer to the nearest tenth of a percent.)

This is equivalent to

$$\underset{\downarrow}{10}\ \underset{\downarrow}{\text{is}}\ \underbrace{\text{what percent}}_{\downarrow}\ \underset{\downarrow}{\text{of}}\ \underset{\downarrow}{24?}$$

$$10 = n \times 24$$

$$10 = 24n$$

$$\frac{10}{24} = \frac{24n}{24}$$

$$0.41666\ldots = n$$

To the nearest tenth of a percent we have

$$41.7\% = n$$

He made 41.7% of his shots at this game.

Practice Problem 15 In a basketball game for the Houston Rockets, Charles Barkley made 5 of his 16 shots. What percent of his shots did he make? (Round your answer to the nearest tenth of a percent.) ■

5.3A Exercises

Translate into a mathematical equation. Use the letter n *for the unknown quantity. Do* **not** *solve.*

1. What is 38% of 500?

2. What is 46% of 700?

3. 50% of what is 7?

4. 75% of what is 43?

5. 17 is what percent of 85?

6. 24 is what percent of 144?

7. Find 128% of 4000.

8. Find 320% of 350.

9. What percent of 400 is 15?

10. What percent of 600 is 424?

11. 136 is 145% of what?

12. 200 is 160% of what?

Solve each percent problem for which we do not know the amount.

13. What is 60% of 250?

14. What is 80% of 175?

15. Find 152% of 600.

16. Find 136% of 500.

Applications

17. Clarice rented a new Ford Escort for one week. Her bill before the 6% tax on rental cars was $106.50. What was the amount of her tax?

18. The Center City check-cashing service charges a fee that is 7% of the amount of the check they cash. How much is the fee for cashing a check for $240?

Solve each percent problem for which we do not know the base.

19. 26% of what is 312?

20. 32% of what is 448?

21. 52 is 4% of what?

22. 36 is 6% of what?

Applications

23. In Australia, all general sales (except for food) have a hidden tax of 22% built into the final price. Walter is planning to purchase a camera while in Australia. He wants to know the "before tax price" that the dealer is charging before he adds on the "hidden tax" of $33. Can you determine the amount of the "before tax price?"

24. Jocelyn is teaching her golden retriever puppy how to fetch. Jack, her puppy, fetched the ball 72% of the time today. Jack fetched the ball 54 times today. How many times did Jocelyn actually throw the ball for Jack?

Solve each percent problem for which we do not know the percent.

25. What percent of 30 is 18?

26. What percent of 40 is 26?

27. 56 is what percent of 200?

28. 45 is what percent of 300?

Applications

29. Simon inspected 280 waterbed mattresses. He found that 14 were defective. What percent of the waterbed mattresses were defective?

30. John paid $12 tax when he bought a mountain bike for $300. What percent tax did he pay?

Mixed Practice

Solve each percent problem. All the types are represented.

31. 18% of 280 is what?

32. 12% of 260 is what?

33. 150% of what is 102?

34. 150% of what is 138?

35. 84 is what percent of 700?

36. 72 is what percent of 900?

37. Find 0.4% of 820.

38. Find 0.3% of 540.

39. What percent of 35 is 22.4?

40. What percent of 45 is 16.2?

41. 89 is 20% of what?

42. 42 is 40% of what?

43. 42 is what percent of 120?

44. 26 is what percent of 160?

45. What is 16.5% of 240?

46. What is 18.5% of 360?

47. Find 0.06% of 2400.

48. Find 0.03% of 3200.

49. What is 38.5% of 2345?

50. What is 0.00345% of 567,000?

51. 3458 is what percent of 5832?

52. 16.8% of what is 5994.24?

Applications

53. The special effects office at Magic Movie Studio fabricated 80 minimodels of office buildings for the disaster sequence of a new film. The director of photography discovered that on camera, only 68 of them really looked authentic. What percent of the models were acceptable?

54. An Olympic equestrian rider practiced jumping over a water hazard. In 400 attempts, she and her horse touched the water 15 times. What percent of her jump attempts were not perfect?

55. At Brookstone State College 44% of the freshman class is employed part-time. There are 1260 freshman this year. How many of them are employed part-time? Round your answer to the nearest whole number.

56. A survey discovered that 28% of all registered nurses working at Woodview General Hospital are vegetarians. If there are 430 registered nurses working at Woodview, how many of them don't eat meat? Round your answer to the nearest whole number.

57. The swim team at Stonybrook College has gone on to the state championships 24 times over the years. If that translates to 60% of the time in which the team has qualified for the finals how many years has the swim team qualified for the finals?

58. North Shore Community College found that 60% of its graduates go on for further education. Last year 570 of the graduates went on for further education. How many students graduated from the college last year?

59. Find 12% of 30% of $1600.

60. Find 90% of 15% of 2700.

Cumulative Review Problems

Multiply.

61. $\begin{array}{r} 1.36 \\ \times\ 1.8 \\ \hline \end{array}$

62. $\begin{array}{r} 2.04 \\ \times\ 7.3 \\ \hline \end{array}$

Divide.

63. $0.06\overline{)170.04}$

64. $0.08\overline{)233.36}$

5.3B Solving Percent Problems Using a Proportion

After studying this section, you will be able to:

1. *Identify the parts of the percent proportion.*
2. *Use the percent proportion to solve percent problems.*

MathPro Video 12 SSM

1 *Identifying the Parts of the Percent Proportion*

In Section 5.3A we showed you how to use an equation to solve a percent problem. Some students find it easier to use proportions to solve percent problems. We will show you how to use proportions in this section. The two methods work equally well. Using percent proportions allows you to see another of the many uses of proportions that we studied in Chapter 4.

Consider the following relationship.

$$\frac{19}{25} = 76\%$$

This can be written as

$$\frac{19}{25} = \frac{76}{100}$$

In general, we can write this relationship using the percent proportion

$$\frac{\text{amount}}{\text{base}} = \frac{\text{percent number}}{100}$$

To use this equation effectively, we need to find the amount, base, and percent number in a word problem. The easiest of these three parts to find is the percent number. We use the letter p (a variable) to represent the percent number.

EXAMPLE 1 Identify the percent number p in each of the following.

(a) Find 16% of 370. **(b)** 28% of what is 25?

(c) What percent of 18 is 4.5?

(a) Find 16% of 370.
The value of p is 16.

(b) 28% of what number is 25?
The value of p is 28.

(c) $\underbrace{\text{What percent}}_{p}$ of 18 is 4.5?

We let p represent the unknown percent number.

Practice Problem 1 Identify the percent number p.

(a) Find 83% of 460. **(b)** 18% of what number is 90?

(c) What percent of 64 is 8? ■

We use the letter b to represent the base number. The base is the entire quantity or the total involved. The number that is the base usually appears after the word *of*. The amount, which we represent by the letter a, is the part being compared to the whole.

EXAMPLE 2 Identify the base b and the amount a.

(a) 20% of 320 is 64. **(b)** 12 is 60% of what?

(a) 20% of 320 is 64

320 ↑ The base is the entire quantity. It follows the word *of*. $b = 320$

64 ↑ The amount is the part compared to the whole. Here $a = 64$.

(b) 12 is 60% of what?

12 ↑ The amount 12 is the part of the base. $a = 12$

what? ↑ The base is unknown. We represent the base by the variable b.

Practice Problem 2 Identify the base b and the amount a.

(a) 30% of 52 is 15.6. **(b)** 170 is 85% of what? ■

When identifying p, b, and a in a problem, it is easiest to identify p and b first. The remaining quantity or variable is a.

EXAMPLE 3 Find p, b, and a.

(a) What is 52% of 300? **(b)** What percent of 30 is 18?

The value of p is 52. ↓

(a) What is 52% of 300?

What ↑ The amount is unknown. We let a = the amount.

300? ↑ The base usually follows the word *of*. Here $b = 300$.

(b) The value of p is not known. We let p represent the unknown percent. ↓

What percent of 30 is 18?

30 ↑ The base usually follows the word *of*. Here $b = 30$.

18? ↑ The amount is 18. $a = 18$.

Practice Problem 3 Find p, b, and a.

(a) What is 18% of 240? **(b)** What percent of 64 is 4? ■

2 Using the Percent Proportion to Solve Percent Problems

When we solve the percent proportion, we will have enough information to state the numerical value for two of the three variables a, b, p in the equation

$$\frac{a}{b} = \frac{p}{100}$$

We first identify those two values, and then substitute those values into the equation. Then we will use the skills that we acquired for solving proportions in Chapter 4 to find the value we do not know.

EXAMPLE 4 What is 16% of 380?

The percent $p = 16$. The base $b = 380$. The amount is unknown. We use the variable a. Thus

$$\frac{a}{b} = \frac{p}{100} \quad \text{becomes} \quad \frac{a}{380} = \frac{16}{100}$$

If we reduce the fraction on the right-hand side, we have

$$\frac{a}{380} = \frac{4}{25}$$

$$25a = (4)(380) \qquad \textit{Cross-multiply.}$$

$$25a = 1520 \qquad \textit{Simplify.}$$

$$\frac{25a}{25} = \frac{1520}{25} \qquad \textit{Divide each side by 25.}$$

$$a = 60.8 \qquad \textit{Divide } 1520 \div 25.$$

Thus 16% of 380 is 60.8.

Practice Problem 4 What is 32% of 550? ■

EXAMPLE 5 Find 260% of 40.

The percent $p = 260$. The number that is the base usually appears after the word *of*. The base $b = 40$. The amount is unknown. We use the variable a. Thus

$$\frac{a}{b} = \frac{p}{100} \quad \text{becomes} \quad \frac{a}{40} = \frac{260}{100}$$

If we reduce the fraction on the right-hand side, we have

$$\frac{a}{40} = \frac{13}{5}$$

$$5a = (40)(13) \qquad \textit{Cross-multiply.}$$

$$5a = 520 \qquad \textit{Simplify.}$$

$$\frac{5a}{5} = \frac{520}{5} \qquad \textit{Divide each side of the equation by 5.}$$

$$a = 104$$

Thus 260% of 40 is 104.

Practice Problem 5 Find 340% of 70. ■

EXAMPLE 6 85% of what is 221?

The percent $p = 85$. The base is unknown. We use the variable b. The amount a is 221. Thus

$$\frac{a}{b} = \frac{p}{100} \quad \text{becomes} \quad \frac{221}{b} = \frac{85}{100}$$

If we reduce the fraction on the right-hand side, we have

$$\frac{221}{b} = \frac{17}{20}$$

$$(221)(20) = 17b \qquad \textit{Cross-multiply.}$$

$$4420 = 17b \qquad \textit{Simplify.}$$

$$\frac{4420}{17} = \frac{17b}{17} \qquad \textit{Divide each side by } 17.$$

$$260 = b. \qquad \textit{Divide } 4420 \div 17.$$

Thus 85% of 260 is 221.

Practice Problem 6 68% of what is 476? ■

EXAMPLE 7 53 is 0.2% of what?

The percent $p = 0.2$. The base is unknown. We use the variable b. The amount $a = 53$. Thus

$$\frac{a}{b} = \frac{p}{100} \quad \text{becomes} \quad \frac{53}{b} = \frac{0.2}{100}$$

When we cross-multiply, we obtain

$$(53)(100) = 0.2b$$

$$5300 = 0.2b$$

$$\frac{5300}{0.2} = \frac{0.2b}{0.2}$$

$$26{,}500 = b$$

Thus 53 is 0.2% of 26,500.

Practice Problem 7 216 is 0.3% of what? ■

EXAMPLE 8 What percent of 4000 is 160?

The percent is unknown. We use the variable p. The base $b = 4000$. The amount $a = 160$. Thus

$$\frac{a}{b} = \frac{p}{100} \quad \text{becomes} \quad \frac{160}{4000} = \frac{p}{100}$$

If we reduce the fraction on the left-hand side, we have

$$\frac{1}{25} = \frac{p}{100}$$

$$100 = 25p \qquad \textit{Cross-multiply.}$$

$$\frac{100}{25} = \frac{25p}{25} \qquad \textit{Divide each side by 25.}$$

$$4 = p \qquad \textit{Divide } 100 \div 25.$$

Thus 4% of 4000 is 160.

Practice Problem 8 What percent of 3500 is 105? ■

EXAMPLE 9 19 is what percent of 95?

The percent is unknown. We use the variable p. The base $b = 95$. The amount $a = 19$. Thus

$$\frac{a}{b} = \frac{p}{100} \quad \text{becomes} \quad \frac{19}{95} = \frac{p}{100}$$

Cross-multiplying, we have

$$(19)(100) = 95p$$

$$1900 = 95p$$

$$\frac{1900}{95} = \frac{95p}{95} \qquad \textit{Divide each side of the equation by 95.}$$

$$20 = p$$

Thus 19 is 20% of 95.

Practice Problem 9 42 is what percent of 140? ■

5.3B Exercises

Identify p, b, and a. Do not solve for the unknown.

	p	b	a
1. 75% of 660 is 495.	______	______	______
2. 65% of 820 is 532.	______	______	______
3. What is 42% of 400?	______	______	______
4. What is 56% of 600?	______	______	______
5. 49% of what is 2450?	______	______	______
6. 38% of what is 2280?	______	______	______
7. 30 is what percent of 50?	______	______	______
8. 50 is what percent of 250?	______	______	______
9. What percent of 25 is 10?	______	______	______
10. What percent of 24 is 6?	______	______	______
11. 400 is 160% of what?	______	______	______
12. 900 is 225% of what?	______	______	______

Solve each problem by using the percent proportion

$$\frac{a}{b} = \frac{p}{100}$$

In problems 13–18 the amount a is not known.

13. 24% of 300 is what?

14. 54% of 500 is what?

15. Find 250% of 30.

16. Find 320% of 60.

17. 0.7% of 8000 is what?

18. 0.8% of 9000 is what?

In problems 19–24 the base b is not known.

19. 45 is 60% of what?

20. 96 is 80% of what?

21. 150% of what is 90?

22. 125% of what is 75?

23. 3000 is 0.5% of what?

24. 6000 is 0.4% of what?

In problems 25–30 the percent p is not known.

25. 70 is what percent of 280?

26. 90 is what percent of 450?

27. What percent of 140 is 3.5?

28. What percent of 170 is 3.4?

29. What percent of $5000 is $90?

30. What percent of $4000 is $64?

Mixed Practice

31. 26% of 350 is what?

32. 56% of 650 is what?

33. 180% of what is 540?

34. 160% of what is 320?

35. 82 is what percent of 500?

36. 75 is what percent of 600?

37. Find 0.7% of 520.

38. Find 0.4% of 650.

39. What percent of 66 is 16.5?

40. What percent of 49 is 34.3?

41. 68 is 40% of what?

42. 52 is 40% of what?

43. 94.6 is what percent of 220?

44. 85.8 is what percent of 260?

45. What is 12.5% of 380?

46. What is 20.5% of 320?

47. Find 0.05% of 5600.

48. Find 0.04% of 8700.

Solve each percent problem. Round to the nearest hundredth.

49. What percent of 4550 is 720?

50. What is 16.25% of 65,250?

51. 4.25% of 256.75 is what number?

52. 2760 is 5.5% of what number?

★ **53.** What is $19\frac{1}{4}\%$ of 798?

★ **54.** $140\frac{1}{2}\%$ of what number is 10,397?

★ **55.** Find 18% of 20% of $3300.
(*Hint:* First find 20% of $3300.)

★ **56.** Find 42% of 16% of $5500.
(*Hint:* First find 16% of $5500.)

Cumulative Review Problems

Simplify.

57. $\frac{4}{5} + \frac{8}{9}$

58. $\frac{7}{13} - \frac{1}{2}$

59. $\left(2\frac{4}{5}\right)\left(1\frac{1}{2}\right)$

60. $1\frac{2}{5} \div \frac{3}{4}$

5.4 Solving Applied Percent Problems

MathPro Video 12 SSM

After studying this section, you will be able to:

1. *Solve general applied percent problems.*
2. *Solve commission problems.*
3. *Solve problems when percents are added.*
4. *Solve percent of increase or decrease problems.*
5. *Solve discount problems.*
6. *Solve simple interest problems.*

1 *Solving General Applied Percent Problems*

In Sections 5.3A and 5.3B we learned the three types of applied percent problems. Some problems ask you to find a percent of a number. Some problems give you an amount and a percent and ask you to find the base (or whole). Other problems give an amount and a base and ask you to find the percent. We will now see how the three types of percent problems occur in real life.

EXAMPLE 1 Of all the computers manufactured last month, an inspector found 18 that were defective. This is 2.5% of all the computers manufactured last month. How many computers were manufactured last month?

Method A Translating to an equation

The problem is equivalent to: 2.5% of the number of computers is 18.

Let $n =$ the number of computers.

$$\underbrace{2.5\%}_{2.5\%}\ \underbrace{\text{of}}_{\times}\ \text{the}\ \underbrace{\text{number of computers}}_{n}\ \underbrace{\text{is}}_{=}\ \underbrace{18}_{18}$$

$$0.025n = 18$$

$$\frac{0.025n}{0.025} = \frac{18}{0.025}$$

$$n = 720$$

720 computers were manufactured last month.

Method B Use the Percent Proportion $\dfrac{a}{b} = \dfrac{p}{100}$.

The percent $p = 2.5$. The base is unknown. We will use the variable b. The amount $a = 18$. Thus

$$\frac{a}{b} = \frac{p}{100} \quad \text{becomes} \quad \frac{18}{b} = \frac{2.5}{100}$$

Using cross-multiplication, we have

$$(18)(100) = 2.5b$$

$$1800 = 2.5b$$

$$\frac{1800}{2.5} = \frac{2.5b}{2.5}$$

$$720 = b$$

720 computers were manufactured last month.

By either Method A or Method B, we obtain the same number of computers, 720.

Substitute 720 into the original problem to check.

2.5% of 720 computers are defective.

$$(0.025)(720) = 18 \quad ✓$$

Practice Problem 1 4800 people, or 12% of all passengers holding tickets for American Airlines flights in one month, did not show up for their flight. How many people held tickets that month? ■

EXAMPLE 2 How much sales tax will you pay on a color television priced at \$299.00 if the sales tax is 5%?

Method A Translating to an equation

The problem is equivalent to:

What is 5% of \$299?

$$\downarrow \quad \downarrow \quad \downarrow \quad \downarrow \quad \downarrow$$

$$n = 5\% \times 299$$

$$n = (0.05)(299)$$

$$n = 14.95$$

The tax is \$14.95.

Method B Using the percent proportion $\dfrac{a}{b} = \dfrac{p}{100}$

The percent $p = 5$. The base $b = 299$. The amount is unknown. We use the variable a. Thus

$$\frac{a}{b} = \frac{p}{100} \quad \text{becomes} \quad \frac{a}{299} = \frac{5}{100}$$

If we reduce the fraction on the right-hand side, we have

$$\frac{a}{299} = \frac{1}{20}$$

We then cross-multiply to obtain

$$20a = 299$$

$$\frac{20a}{20} = \frac{299}{20}$$

$$a = 14.95$$

The tax is \$14.95.

Thus, by either method, the amount of the sales tax is \$14.95.

Let's see if our answer is reasonable. Is 5% of \$299 really \$14.95? If we round each figure to one significant digit, we have 5% of 300 = 15. Since 15 is quite close to our value of \$14.95, our answer seems reasonable.

Practice Problem 2 A salesperson rented a hotel room for \$62.30 per night. The tax in her state is 8%. What tax does she pay for one night at the hotel? ■

EXAMPLE 3 A failing student attended class 39 times out of the 45 times the class met last semester. What percent of the classes did he attend? Round your answer to the nearest tenth of a percent.

Method A Translating to an equation

This problem is equivalent to:

$$\underset{39}{\underset{\downarrow}{39}}\ \underset{=}{\underset{\downarrow}{\text{is}}}\ \underset{n}{\underset{\downarrow}{\underbrace{\text{what percent}}}}\ \underset{\times}{\underset{\downarrow}{\text{of}}}\ \underset{45}{\underset{\downarrow}{45?}}$$

$$39 = 45n$$

$$\frac{39}{45} = \frac{45n}{45}$$

$$0.8666\ldots = n$$

To the nearest tenth of a percent we have $n = 86.7\%$.

Method B Using the percent proportion $\dfrac{a}{b} = \dfrac{p}{100}$

The percent is unknown. We use the variable p. The base b is 45. The amount a is 39. Thus

$$\frac{a}{b} = \frac{p}{100} \quad \text{becomes} \quad \frac{39}{45} = \frac{p}{100}$$

When we cross-multiply, we get

$$(39)(100) = 45p$$

$$3900 = 45p$$

$$\frac{3900}{45} = \frac{45p}{45}$$

$$86.666\ldots = p$$

To the nearest tenth, the answer is 86.7%.

By using either method, we discover that the failing student attended approximately 86.7% of the classes.

Verify by estimating that the answer is reasonable.

Practice Problem 3 Of the 130 flights at Orange County Airport yesterday only 105 of them were on time. What percent of the flights were on time? (Round your answer to the nearest tenth of a percent.) ■

Now you have some experience using the three types of percent problems in real-life applications. You can use either Method A or Method B to solve applied percent problems. In the following sections we will present more percent applications. We will not list all the steps of Method A or Method B. Most students will find after a careful study of Examples 1–3 that they do not need to write out all the steps of Method A or Method B when solving applied percent problems.

2 *Solving Commission Problems*

If you work as a salesperson, your earnings may be in part or in total a certain percentage of the sales you make. The amount of money you get that is a percentage of the value of your sales is called your **commission.** It is calculated by multiplying the percentage (called the **commission rate**) by the value of the sales.

Commission = commission rate × value of sales

EXAMPLE 4 A salesperson has a commission rate of 17%. She sells \$32,500 worth of goods in a department store in two months. What is her commission?

$$\text{Commission} = \text{commission rate} \times \text{value of sales}$$
$$\text{Commission} = 17\% \times \$32{,}500$$
$$= 0.17 \times 32{,}500$$
$$= 5525$$

Her commission is \$5525.00.

Does this answer seem reasonable? Check by estimating.

Practice Problem 4 A real estate salesperson earns a commission rate of 6% when he sells a \$156,000 home. What is his commission? ■

3 *Solving Problems When Percents Are Added*

Percents can be added if the base (whole) is the same. For example, 50% of your salary added to 20% of your salary = 70% of your salary. 100% of your cost added to 15% of your cost = 115% of your cost. Problems like this are often called *markup problems.* If we add 15% of the cost of an item to the original cost, the markup is 15%. We will add percents in some applied situations.

EXAMPLE 5 Walter and Mary Ann are going out to a restaurant. They have a limit of \$63.25 to spend for the evening. They want to tip the waitress 15% of the cost of the meal. How much money can they afford to spend on the meal itself?

Let n = the cost of the meal. 15% of the cost = the amount of the tip. We want to add the percents of the meal.

$$\boxed{\begin{array}{c}\text{Cost of}\\ \text{meal } n\end{array}} + \boxed{\begin{array}{c}\text{tip of 15\%}\\ \text{of the cost}\end{array}} = \boxed{\$63.25}$$

$$100\% \text{ of } n + 15\% \text{ of } n = \$63.25$$

Note that 100% of n added to 15% of n is 115% of n.

$$115\% \text{ of } n = \$63.25$$
$$1.15 \times n = 63.25$$
$$\frac{1.15 \times n}{1.15} = \frac{63.25}{1.15} \quad \textit{Divide both sides by } 1.15.$$

They can spend up to \$55.00 on the meal itself.

Does this answer seem reasonable?

Practice Problem 5 Sue and Sam have \$46.00 to spend at a restaurant, including a 15% tip. How much can they spend on the meal itself? ■

4 Solving Percent of Increase or Decrease Problems

We sometimes need to find the percent by which a number increases or decreases. If a car costs \$7000 and the price decreases \$1750, we say that the percent of decrease is $\frac{1750}{7000} = 0.25 = 25\%$.

$$\text{Percent of decrease} = \frac{\text{amount of decrease}}{\text{original amount}}$$

Similarly, if a population of 12,000 people increases by 1920 people, we say that the percent of increase is $\frac{1920}{12{,}000} = 0.16 = 16\%$.

$$\text{Percent of increase} = \frac{\text{amount of increase}}{\text{original amount}}$$

Note that for these types of problems the base is always the *original amount.*

EXAMPLE 6 The population of Center City increased from 50,000 to 59,500. What was the percent of increase?

Amount of increase

$$\begin{array}{r} 59{,}500 \\ -\ 50{,}000 \\ \hline 9{,}500 \end{array}$$

$$\text{Percent of increase} = \frac{\text{amount of increase}}{\text{original amount}} = \frac{9500}{50{,}000}$$

$$= 0.19 = 19\%$$

The percent of increase is 19%.

Practice Problem 6 A new car is sold for \$15,000. A year later its price had decreased to \$10,500. What is the percent of decrease? ■

5 Solving Discount Problems

Frequently, we see signs urging us to buy during a sale when the list price is discounted by a certain percent.

The amount of a discount is a product of the discount rate and the list price.

$$\text{Discount} = \text{discount rate} \times \text{list price}$$

EXAMPLE 7 Jeff purchased a compact disc player on sale at 35% discount. The list price was \$430.00.

(a) What was the amount of the discount?

(b) How much did Jeff pay for the CD player?

(a) Discount = discount rate × list price

$$\begin{aligned} &= 35\% \times 430 \\ &= 0.35 \times 430 \\ &= 150.5 \end{aligned}$$

The discount is \$150.50.

(b) We subtract the discount from the list price to get the selling price.

$$\begin{array}{rl} \$430.00 & \textit{list price} \\ -\ \$150.50 & \textit{discount} \\ \hline \$279.50 & \textit{selling price} \end{array}$$

Jeff paid \$279.50 for the CD player.

Practice Problem 7 Betty bought a car that lists for \$13,600 at a 7% discount.

(a) What was the discount? **(b)** What did she pay for the car? ■

6 *Solving Simple Interest Problems*

Interest is money paid for the use of money. If you deposit money in a bank, the bank uses that money and pays you interest. If you borrow money, you pay the bank interest for the use of that money. The *principal* is the amount deposited or borrowed. Interest is usually expressed as a percent rate of the principal. The interest rate is assumed to be per year, unless otherwise stated. The formula used in business to compute simple interest is

$$\text{Interest} = \text{principal} \times \text{rate} \times \text{time}$$
$$I = P \times R \times T$$

If the interest rate is *per year,* the time T *must* be in *years.*

EXAMPLE 8 Find the interest on a loan of \$7500 borrowed at 13% for one year.

$$I = P \times R \times T$$

P = principal = \$7500 $\quad R$ = rate = 13% $\quad T$ = time = 1 year

$$I = 7500 \times 13\% \times 1 = 7500 \times 0.13 = 975$$

The interest is \$975.00.

Practice Problem 8 Find the interest on a loan of \$5600 borrowed at 12% for one year. ■

Our formula is based on a yearly interest rate. Time periods of more than one year or a fractional part of a year are sometimes needed.

EXAMPLE 9 Find the interest on a loan of \$2500 that is borrowed at 9% for

(a) 3 years **(b)** 3 months

(a)

$$I = P \times R \times T$$

$P = \$2500 \quad R = 9\% \quad T = 3$ years

$$I = 2500 \times 0.09 \times 3 = 225 \times 3 = 675$$

The interest for 3 years is \$675.

(b) Now 3 months $= \frac{1}{4}$ year. The period must be in years to use the formula.

Since $T = \frac{1}{4}$ year, we have

$$I = 2500 \times 0.09 \times \frac{1}{4}$$
$$= 225 \times \frac{1}{4}$$
$$= \frac{225}{4} = 56.25$$

The interest for $\frac{1}{4}$ year is \$56.25.

Practice Problem 9 Find the interest on a loan of \$1800 that is borrowed at 11% for **(a)** 4 years **(b)** 6 months ■

CALCULATOR

Interest

You can use a calculator to find simple interest. Find the interest on \$450 invested at 6.5% for 15 months. Notice the time is in months. Since the interest formula is in years, $I = P \times R \times T$, you need to change 15 months to years by dividing 15 by 12. Enter

15 [÷] 12 [=]

Display

[1.25]

Leave this on the display and multiply as follows:

1.25 [×] 450 [×]

6.5 [%] [=]

The display should read

[36.5625]

which would round to \$36.56.

Try

(a) \$9516 invested at 12% for 30 months

(b) \$593 borrowed at 8% for 5 months

5.4 Exercises

Applications

Problems 1–22 present the three types of percent problems. They are similar to Examples 1–3. Take the time to master problems 1–22 before going on to the next ones. Round your answer to the nearest hundredth when necessary.

1. No graphite was found in 4500 pencils shipped to Sureway School Supplies. This was 2.5% of the total number of pencils received by Sureway. How many pencils in total were in the order?

2. A high-jumper on the track and field team hit the bar 58 times last week. This means that he did not succeed in 29% of his jump attempts. How many total attempts did he have last week?

3. How much tax will Colleen pay on a surround-sound system that costs $1885.00 if the tax is 5%.

4. In a high school in London, 714 out of 1020 students speak two languages fluently. What percent of students speak two languages?

5. A mechanic runs a car through a computer diagnostic check for its 60,000-mile check-up. 76 out of 95 car parts did not need any adjustment. What percentage did the mechanic need to adjust?

6. 40 out of 65 people in a room can "curl" their tongues. What percent of these people can curl successfully?

7. Scott's phone bill averages $48.60 per month. This is 115% of what his average monthly bill was last year. What was his average monthly phone bill last year?

8. Renata now earns $9.50 per hour. This is 125% of what she earned last year. What did she earn per hour last year?

9. An island in the South Pacific imposes a 7% room tax. If a room in a luxury hotel costs $92 per night, what is the tax?

10. The caterer estimates that 85% of the people at a banquet will request coffee. If the banquet has 340 guests, how many people are expected to request coffee?

11. The Incredible Chocolate Chip Company has discovered that 36 out of 400 chocolate chip cookies do not contain enough chocolate chips. What percent of the chocolate chip cookies do not have enough chips?

12. Every day this year, Sam ordered either cappuccino or espresso from the coffee bar downstairs. He had 78 espressos and 287 cappuccinos. What percent of the coffees were espressos?

13. In a survey taken by the Department of Agriculture of 450 people, 74% admitted that they hated brussels sprouts. How many people in this survey did not care for this vegetable?

14. At the backstage entrace to the Hollywood Bowl Concert Arena, 85% of the people who came to the door after a recent concert requested autographs. If 510 people asked for autographs, how many people came backstage?

15. Dean bought a new mountain bike. The sales tax in his state is 4%, he paid \$9.60 in tax. What was the price of the mountain bike before the tax?

16. Maha received a raise of 8% of her monthly salary. The amount of her raise was \$48.16 per month. What was her monthly salary before the raise?

17. Peter and Judy together earn \$5060 per month. Their mortgage payment is \$1265 per month. What percentage of their household income goes toward paying the mortgage?

18. Jackie puts aside \$65.50 per week for her monthly car payment. She earns \$327.50 per week. What percent of her income is set aside for car payments.

19. The Children's Wish Charity raised 75% of its funds from sporting promotions. Last year the charity received \$7,200,000 from its sporting promotions. What was the charity's total income last year?

20. Shannon paid \$8400 in federal and state income taxes as a lab technician, which amounted to 28% of her annual income. What was her income last year?

21. In a bin of 21,000 M&Ms, 6.5% are the color blue. How many M&Ms are blue?

22. Out of 26,400 daily telephone calls made in Meridian, 85% of the calls were made between relatives. How many phone calls were made by people calling relatives?

Problems 23–46 include commissions, percents that are added, percent of increase or decrease, discounts, and simple interest. Solve each problem. Round each answer to the nearest hundredth when necessary.

23. Laserlight Corporation offers its outside salespeople 3% commission on any industrial laser applications sold in North America. If Trevor sold \$2,500,000 worth of lasers last year, how much commission did he make?

24. Davisville Dodge pays its sales personnel a commission of 25% of the dealer profit on each car. The dealer made a profit of \$18,500 on the cars Linda sold last month. What was her commission last month?

25. In Flagstaff, Arizona, 0.9% of all babies walk before they reach the age of 11 months. If 24,000 babies were born in Flagstaff in the last twenty years, how many of them will walk before they reach the age of 11 months?

26. At a Boston Celtics basketball game scheduled for 8:30 P.M., 0.8% of the spectators were children under 12 years old. If 28,000 showed up for the game, how many children under the age of 12 were in the stands?

27. Beachfront property is very expensive. A real estate agent in South Carolina earned \$26,000 in commissions. The property she sold was worth \$650,000. What was her commission rate?

28. Adam sold encyclopedias last summer to raise tuition money. He sold \$83,500 worth of encyclopedias and was paid \$5010 in commissions. What commission rate did he earn?

29. Manuel is taking Cynthia out to dinner. He has $66 to spend. If he wants to tip the server 20%, how much can be afford to spend on their meal?

30. Belinda asked Martin out to dinner. She has $47.50 to spend. She wants to tip the waitress 15% of the cost of their meal. How much money can she afford to spend on their meal itself?

31. The Democratic National Committee has a budget of $33,000,000 to spend on the inauguration of the new president. 15% of the costs will be paid to personnel, 12% of the costs will go toward food, and 10% will go to decorations. How much money will go for personnel, food, and decorations? How much will be left over to cover security, facility rental, and all other expenses?

32. A major research facility has developed an experimental drug to treat Alzheimer's disease. Twenty percent of the research costs was paid to the staff. Sixteen percent of the research costs was paid to rent the building where the research was conducted. The company has spent $6,000,000 in research on this new drug. How much was paid to cover the cost of staff and rental of the building?

33. Bryce bought a new VCR for $236.00. The tax in Bryce's city is 6%.
(a) What is the sales tax?
(b) What is the final price of the VCR?

34. Finn bought a used Honda Accord for $10,500. The tax in his state is 8%.
(a) What is the sales tax?
(b) What is the final price of the Honda Accord?

35. The daycare enrollment in the town of Harvey increased from 5550 to 7600 children per day. What was the percent of increase?

36. The cost of a certain Chevrolet Camaro increased from $16,000 last year to $17,280 this year. What was the percent of increase?

37. A stereo originally priced at $985 was sold for $394. What was the percent of decrease?

38. Tay weighed 285 pounds two years ago. After careful supervision at a weight loss center, he reduced his weight to 171 pounds. What was the percent of decrease in his weight?

39. Susie bought her first Harley-Davidson motorcycle. The list price was $15,990, but the dealer gave her a discount of 8%.
(a) What was the discount?
(b) How much did she pay for the motorcycle?

40. Jane bought a leather sofa that had been used as the floor model. The list price of the sofa was $1250, but the dealer gave her a discount of 35%.
(a) What was the discount?
(b) How much did she pay for the sofa?

41. A Compaq laptop computer listed for $2125. This week, Lisa heard that the manufacturer is offering a rebate of 12%.
(a) What is the rebate?
(b) How much will Lisa pay for the computer?

42. Mark and Julie wanted to buy a prefabricated log cabin to put on their property in the Colorado Rockies. The price of the kit is listed at $24,000. At the after-holiday cabin sale, a discount of 14% was offered.
(a) What was the discount?
(b) How much did they pay for the carbin?

43. Manuel took out an education loan of $2500 at an interest rate of 8% for one year. He then paid back the loan.

(a) How much interest did he pay?

(b) How much did it cost to pay off the loan totally?

44. Dina received a car loan so that she could purchase a used Ford Taurus. Her loan was for $4500 at an interest rate of 7 percent for one year. She then paid the loan back in full, with interest.

(a) How much interest did she pay?

(b) How much did it cost to pay off the loan totally?

45. Adam deposited $3700 in his savings account for one year. His savings account earns 3.2% interest annually. He did not add any more money within the year, and at the end of that time, he withdrew all funds.

(a) How much interest did he earn?

(b) How much money did he withdraw from the bank?

46. Nikki had $1258 outstanding on her MasterCard, which charges 2% monthly interest. At the end of this month Nikki paid off the loan.

(a) How much interest did Nikki pay for one month?

(b) How much did it cost to pay off the loan totally?

In problems 47–52 round your answer to the nearest cent.

47. How much sales tax would you pay to purchase a new Mazda 626 that cost $18,456.82 if the sales tax rate is 4.6%?

48. The Hartling family purchased a new living room set. The list price was $1249.95. However, they got a discount of 29%. How much did they pay for the new living room set?

49. Paolo's credit card statement has a balance of $698.44. The statement said that the *monthly* rate is 1.4575%. How much interest was he charged for the month? How much would it take to pay off the total balance with interest?

50. Juan traveled 22,437 miles in his car last year. He is a salesperson, and 68% of his mileage is for business purposes. How many miles did he travel on business last year? If he can deduct 31¢ per mile for his income tax return how much can he deduct for business mileage?

51. The new Nascar concept store bought two video games so that the customers can pretend that they are really driving Nascar racing cars. The games listed for $9346.80 each. The video game dealer gave the store a 12% discount on the list price. The store had to pay a 4% sales tax on the purchase price of the video games. How much did it cost the store to buy the two games? Round the answer to the nearest cent.

52. Pterodactyl Computers sold 100 CD-ROMs for $280.00. FedEx charged $68.00 for overnight rush delivery to the wholesaler. The wholesaler increased the price by 30% over his total cost and sold them to a retail computer store. The retailer increased the price by 16%. One customer bought all 100 CD-ROMs. The clerk received a 2% tip from the customer for delivering the CD-ROMs to his office. Including the tip, how much did the customer pay? Round your answer to the nearest cent.

Cumulative Review Problems

53. Round to the nearest thousand. 1,698,481

54. Round to the nearest hundred. 2,452,399

55. Round to the nearest hundredth. 1.63474

56. Round to the nearest thousandth. 0.793468

57. Round to the nearest ten thousandth. 0.055613

58. Round to the nearest ten thousandth. 0.079152

Putting Your Skills to Work

PERCENTAGES OF BEVERAGES CONSUMED IN THE UNITED STATES

Did you know that in 1970 in the United States coffee was consumed at the rate of 33.4 gallons per person? In contrast, during that same year only 1.3 gallons of skim milk were consumed for every person in the United States. The following table shows some interesting trends in the drinking of nonalcoholic beverages in this country.

U.S. consumption of selected beverages per person measured in gallons per year

Type	1970	1975	1980	1985	1990	1995
Whole milk	25.5	21.1	17.0	14.3	10.5	9.2
Lowfat milk	4.4	7.1	9.2	10.9	12.6	12.7
Skim milk	1.3	1.3	1.3	1.5	2.6	3.3
Tea	6.8	7.5	7.3	7.1	6.8	7.1
Coffee	33.4	31.4	26.7	27.4	27.0	26.2
Bottled water	NA	NA	2.4	4.5	8.0	9.4
Soft drinks	24.3	28.2	35.1	35.7	43.7	46.9
Selected juices	NA	6.6	7.2	7.7	6.9	8.8

Source: U. S. Department of Agriculture   NA = Not Available

Use the above chart to answer the following questions. Round to the nearest tenth.

PROBLEMS FOR INDIVIDUAL STUDY AND INVESTIGATION

1. What has been the percent of increase in the drinking of skim milk from 1970 to 1995? To what do you attribute this increase?

2. What has been the percent of decrease in the drinking of coffee from 1970 to 1995? To what do you attribute this decrease?

PROBLEMS FOR GROUP INVESTIGATION AND ANALYSIS

Together with some other members of your class see if you can determine the following.

3. The consumption of whole milk during the period 1970 to 1995 decreased while the consumption of lowfat milk and skim milk increased. During this 25-year period did the consumption of all types of milk actually increase or decrease? By what percent?

4. From 1975 to 1995 the approximate population of the United States grew from 214,800,000 to 259,709,000. What was the percent of increase in the population? What was the percent of decrease in the use of tea during this same time? What can we conclude based on these calculations?

INTERNET CONNECTIONS: Go to ``http://www.prenhall.com/tobey'' to be connected

Site: Figures and Facts: Eating (Weight Watchers) or a related site

This site gives information about the eating and drinking habits of Americans. Answer the following questions.

5. Use the table of Major Frozen Foods Markets Annual Per Capita Spending to tell how much more a typical French person spends on frozen food than a typical Swiss person. Give your answer as a percent.

6. Use the table of Per Capita Snack Consumption to find the average price (in dollars per pound) paid for potato chips, tortilla chips, and pretzels, respectively.

Chapter Organizer

Topic	Procedure	Examples
Converting a decimal to a percent, p. 287.	1. Move the decimal point two places to the right. 2. Add the percent sign.	$0.19 = 19\%$ $0.516 = 51.6\%$ $0.04 = 4\%$ $1.53 = 153\%$ $0.006 = 0.6\%$
Converting a fraction with a denominator of 100 to a percent, p. 284.	1. Use the numerator only. 2. Add the percent sign.	$\frac{29}{100} = 29\%$ $\quad \frac{5.6}{100} = 5.6\%$ $\frac{3}{100} = 3\%$ $\quad \frac{7\frac{1}{3}}{100} = 7\frac{1}{3}\%$ $\frac{231}{100} = 231\%$
Changing a fraction (whose denominator is not 100) to a percent, p. 292.	1. Divide the numerator by the denominator and obtain a decimal. 2. Change the decimal to a percent.	$\frac{13}{50} = 0.26 = 26\%$ $\frac{1}{20} = 0.05 = 5\%$ $\frac{3}{800} = 0.00375 = 0.375\%$ $\frac{312}{200} = 1.56 = 156\%$
Changing a mixed number to a percent, p. 290.	1. Change the mixed number to decimal form. 2. Transform the decimal to an equivalent percent.	$3\frac{1}{4} = 3.25 = 325\%$ $\quad 6\frac{4}{5} = 6.8 = 680\%$ $7\frac{3}{8} = 7.375 = 737.5\%$
Changing a percent to a decimal, p. 286.	1. Move the decimal point two places to the left. 2. Drop the percent sign.	$49\% = 0.49$ $2\% = 0.02$ $0.5\% = 0.005$ $196\% = 1.96$ $1.36\% = 0.0136$
Changing a percent to a fraction, p. 290.	1. If the percent contains a decimal, first remove the % sign by moving the decimal point two places to the left. Then write the decimal fraction, and reduce the fraction if possible. 2. If the percent does not contain a decimal, change the percent to a fraction with a denominator of 100. Reduce the fraction if possible. 3. If the numerator contains a fraction, change the numerator to an improper fraction. Next simplify by the "invert and multiply" rule. Then reduce the fraction if possible.	$5.8\% = 0.058$ $= \frac{58}{1000} = \frac{29}{500}$ $2.72\% = 0.0272$ $= \frac{272}{10{,}000} = \frac{17}{625}$ $25\% = \frac{25}{100} = \frac{1}{4}$ $38\% = \frac{38}{100} = \frac{19}{50}$ $130\% = \frac{130}{100} = \frac{13}{10}$ $7\frac{1}{8}\% = \frac{7\frac{1}{8}}{100}$ $= 7\frac{1}{8} \div \frac{100}{1}$ $= \frac{57}{8} \times \frac{1}{100} = \frac{57}{800}$

Chapter Organizer

Topic	Procedure	Examples
Solving percent problems by translating to equations, p. 299.	1. Translate by replacing "of" by $\times$ "is" by $=$ "what" by n "find" by $n =$ 2. Solve the resulting equation.	**(a)** What is 3% of 56? $n = 3\% \times 56$ $n = (0.03)(56)$ $= 1.68$ **(b)** 16% of what is 208? $16\% \times n = 208$ $0.16n = 208$ $\frac{0.16n}{0.16} = \frac{208}{0.16}$ $n = 1300$ **(c)** What percent of 70 is 30? $n \times 70 = 30$ $70n = 30$ $\frac{70n}{70} = \frac{30}{70}$ $n = 0.4285714\ldots$ n is approximately 42.86%.
Solving percent problems by using proportions, p. 307.	1. Identify the parts of the percent proportion. a = the amount b = the base (the whole; it usually appears after the word "of") p = the percent number 2. Write the percent proportion $\frac{a}{b} = \frac{p}{100}$ using the values obtained in step 1 and solve.	**(a)** What is 28% of 420? The percent p = 28. The base b = 420. The amount a is unknown. We use the variable a. $\frac{a}{b} = \frac{p}{100}$ becomes $\frac{a}{420} = \frac{28}{100}$ If we reduce the fraction on the right-hand side, we have $\frac{a}{420} = \frac{7}{25}$ $25a = (7)(420)$ $25a = 2940$ $\frac{25a}{25} = \frac{2940}{25}$ $a = 117.6$ **(b)** 64% of what is 320? The percent p = 64. The base is unknown. We use the variable b. The amount a = 320. $\frac{a}{b} = \frac{p}{100}$ $\frac{320}{b} = \frac{64}{100}$ If we reduce the fraction on the right-hand side, we have $\frac{320}{b} = \frac{16}{25}$ $(320)(25) = 16b$ $8000 = 16b$ $\frac{8000}{16} = \frac{16b}{16}$ $500 = b$
Solving commission problems, p. 317.	Commission = commission rate $\times$ value of sales	A housewares salesperson gets a 16% commission on sales he makes. How much commission does he earn if he sells \$12,000 in housewares? Commission = (0.16)(12,000) = \$1920

Chapter Organizer

Topic	Procedure	Examples
Percent of increase or decrease problems, p. 318.	Percent of increase or decrease = amount of increase or decrease ÷ base	A car that costs \$16,500 now cost only \$15,000 last year. What is the percent of increase? $16{,}500 - 15{,}000 = 1{,}500$ increase $\frac{1500}{15{,}000} = 0.10$ Rate of increase = 10%
Solving discount problems, p. 318.	Discount = discount rate × list price	Carla purchased a color TV set that lists for \$350 at an 18% discount. **(a)** How much was the discount? **(b)** How much did she pay for the TV set? **(a)** Discount = (0.18)(350) = \$63 **(b)** 350 − 63 = 287 She paid \$287 for the color TV set.
Solving simple interest problems, p. 319.	Interest = principal × rate × time $I = P \times R \times T$	Hector borrowed \$3000 for 4 years at a simple interest rate of 12%. How much interest did he owe after 4 years? $I = P \times R \times T$ $I = (3000)(0.12)(4)$ $= (360)(4)$ $= 1440$ Hector owed \$1440 in interest.

Chapter 5 Review Problems

5.1 *Write as a percent. Round to the nearest hundredth of a percent when necessary.*

1. 0.87 **2.** 0.59 **3.** 0.276 **4.** 0.329 **5.** 0.0713 **6.** 0.0608

7. 2.52 **8.** 4.37 **9.** 1.036 **10.** 1.052 **11.** 0.006 **12.** 0.002

13. 0.0029 **14.** 0.00534 **15.** $\frac{72}{100}$ **16.** $\frac{61}{100}$ **17.** $\frac{19.5}{100}$ **18.** $\frac{21.6}{100}$

19. $\frac{0.24}{100}$ **20.** $\frac{0.98}{100}$ **21.** $\frac{4\frac{1}{12}}{100}$ **22.** $\frac{3\frac{5}{12}}{100}$ **23.** $\frac{317}{100}$ **24.** $\frac{225}{100}$

5.2 *Change each fraction to a percent. Round to the nearest hundredth of a percent when necessary.*

25. $\frac{16}{25}$ **26.** $\frac{22}{25}$ **27.** $\frac{18}{20}$ **28.** $\frac{14}{20}$ **29.** $\frac{5}{11}$ **30.** $\frac{4}{9}$ **31.** $2\frac{1}{4}$

32. $3\frac{3}{4}$ **33.** $4\frac{3}{7}$ **34.** $5\frac{5}{9}$ **35.** $\frac{152}{80}$ **36.** $\frac{165}{90}$ **37.** $\frac{3}{800}$ **38.** $\frac{5}{800}$

Change each percent to decimal form.

39. 0.2% **40.** 0.7% **41.** 21.9% **42.** 43.1% **43.** 166% **44.** 139% **45.** $32\frac{1}{8}\%$ **46.** $26\frac{3}{8}\%$

Change each percent to fractional form.

47. 82% **48.** 36% **49.** 185% **50.** 225% **51.** 16.4%

52. 30.5% **53.** $31\frac{1}{4}\%$ **54.** $43\frac{3}{4}\%$ **55.** 0.05% **56.** 0.06%

Complete the following chart.

	Fraction	Decimal	Percent
57.	$\frac{3}{5}$		
58.	$\frac{7}{8}$		
59.			37.5%
60.			56.25%
61.		0.008	
62.		0.45	

5.3 *Solve each percent problem. Round to the nearest hundredth when necessary.*

63. What is 96% of 300?

64. What is 85% of 400?

65. 3 is 20% of what number?

66. 90 is 40% of what number?

67. 50 is what percent of 125?

68. 70 is what percent of 175?

69. Find 125% of 46.

70. Find 118% of 60.

71. 92% of what number is 147.2?

72. 68% of what number is 95.2?

73. What percent of 70 is 14?

74. What percent of 60 is 28?

75. 0.5% of 2600 is what number?

76. 0.8% of 3500 is what number?

77. What percent of 28 is 130?

78. What percent of 37 is 120?

5.4 *Solve each applied percent problem. Round your answer to the nearest hundredth when necessary.*

79. Professor Wonson found that 34% of his class is left-handed. He has 150 students in his class. How many are left-handed?

80. A Vermont truck dealer found that 64% of all the trucks he sold were four-wheel drive. If he sold 150 trucks, how many had four-wheel drive?

81. Six computers in a shipment of 144 computers were found to be defective. What percent were defective?

82. Rafael obtained 16 incorrect answers on a test containing 102 questions. What percent of his answers were incorrect?

83. Today Yvonne's car has 61% of the value that it had 2 years ago. Today it is worth $6832. What was it worth 2 years ago?

84. A charity organization spent 12% of its budget for administrative expenses. It spent $9624 on administrative expenses. What was the total budget?

85. In a test of car tires, three of 20 tires failed the test. What percent failed the test?

86. Moorehouse Industries received 600 applications and hired 45 of the applicants. What percent of the applicants obtained a job?

87. Roberta earns a commission at the rate of 7.5%. Last month she sold $16,000 worth of goods. How much commission did she make last month?

88. Gary purchased a car for $18,600. The sales tax in his state is 3%. What did he pay in sales tax?

89. The average temperature in Winchester during the last 10 years has been 45.8°F. In the previous 10 years the average temperature was 44°F. What percent of increase is this?

90. Sally invested $6000 in mutual funds earning 11% simple interest. How much interest will she earn in

(a) Six months?

(b) Two years?

91. Irene purchased a dining room set at a 20% discount. The list price was $1595.

(a) What was the discount?

(b) What did she pay for the set?

92. Joan and Michael budget 38% of their income for housing. They spend $684 per month for housing. What is their monthly income?

Chapter 5 Test

Write as a percent. Round to the nearest hundredth of a percent when necessary.

1. 0.42

2. 0.01

3. 0.006

4. 0.139

5. 2.18

6. $\frac{71}{100}$

7. $\frac{2.7}{100}$

8. $\frac{3\frac{1}{7}}{100}$

Change each fraction to a percent. Round to the nearest hundredth of a percent when necessary.

9. $\frac{19}{40}$

10. $\frac{180}{450}$

11. $\frac{105}{75}$

12. $1\frac{3}{4}$

Write each decimal as a percent.

13. 0.1713

14. 3.024

Write each percent as a fraction in simplified form.

15. 152%

16. $7\frac{3}{4}\%$

1. ______
2. ______
3. ______
4. ______
5. ______
6. ______
7. ______
8. ______
9. ______
10. ______
11. ______
12. ______
13. ______
14. ______
15. ______
16. ______

17.	
18.	
19.	
20.	
21.	
22.	
23.	
24.	
25.	
26. (a)	
(b)	
27.	
28.	
29.	
30. (a)	
(b)	

Solve each percent problem. Round to the nearest hundredth if necessary.

17. What is 17% of 157?

18. 33.8 is 26% of what number?

19. What percent of 72 is 40?

20. Find 0.8% of 25,000.

21. 16% of what number is 800?

22. 92 is what percent of 200?

23. 132% of 530 is what number?

24. What percent is 8 of 350?

Solve each applied percent problem. Round to the nearest hundredth if necessary.

25. A real estate agent sells a house for $152,300. She gets a commission of 4% on the sale. What is her commission?

26. Julia and Charles bought a new dishwasher at a 33% discount. The list price was $457.
(a) What was the discount?
(b) How much did they pay for the dishwasher?

27. An inspector found that 75 out of 84 parts were not defective. What percent of the parts were not defective?

28. At Cedars University the number of hours a three-credit college course meets in a semester is 51 hours. Twenty years ago a three-credit course met for 56 hours. What is the percent of decrease?

29. A total of 5160 people voted in the city election. This was 43% of the registered voters. How many registered voters are in the city?

30. Wanda borrowed $3000 at a 16% simple rate of interest.
(a) How much interest did she pay in six months?
(b) How much interest did she pay in two years?

Cumulative Test for Chapters 1–5

Approximately one-half of this test is based on Chapter 5 material. The remainder is based on material covered in Chapters 1–4.

Do each problem. Simplify your answer.

1. Add.
$$\begin{array}{r} 38 \\ 196 \\ +\ 2007 \\ \hline \end{array}$$

2. Subtract.
$$\begin{array}{r} 23{,}007 \\ -\ 14{,}563 \\ \hline \end{array}$$

3. Multiply.
$$\begin{array}{r} 126 \\ \times\ 42 \\ \hline \end{array}$$

4. Divide. $36\overline{)3204}$

5. Add. $2\frac{1}{4} + 3\frac{1}{3}$

6. Subtract. $\frac{11}{12} - \frac{5}{6}$

7. Multiply. $3\frac{17}{36} \times \frac{21}{25}$

8. Divide. $\frac{5}{12} \div 1\frac{3}{4}$

9. Round to the nearest tenth. 5731.652

10. Add.
$$\begin{array}{r} 5.6 \\ 3.21 \\ 18.3 \\ +\ 7.008 \\ \hline \end{array}$$

11. Multiply.
$$\begin{array}{r} 5.62 \\ \times\ 0.3 \\ \hline \end{array}$$

12. Divide. $1.4\overline{)0.5152}$

13. Write as a rate in simplest form. 78 pounds to 130 square feet

14. Is this proportion true or false?
$$\frac{20}{25} = \frac{300}{375}$$

15. Solve the proportion. $\frac{8}{2.5} = \frac{n}{7.5}$

16. A college has a ratio of 3 faculty for every 19 students. The student body presently has 4263 students. How many faculty are there? Round to the nearest whole number.

1. ______________

2. ______________

3. ______________

4. ______________

5. ______________

6. ______________

7. ______________

8. ______________

9. ______________

10. ______________

11. ______________

12. ______________

13. ______________

14. ______________

15. ______________

16. ______________

17. ______

18. ______

19. ______

20. ______

21. ______

22. ______

23. ______

24. ______

25. ______

26. ______

27. ______

28. ______

29. ______

30. ______

In Problems 17–30 round to the nearest hundredth when necessary.

Write as a percent.

17. 0.023

18. $\dfrac{46.8}{100}$

19. 1.98

20. $\dfrac{3}{80}$

Write as a decimal.

21. 243%

22. $6\frac{3}{4}\%$

23. What percent of 214 is 38?

24. Find 1.7% of 6740.

25. 219 is 73% of what number?

26. 114% of 630 is what number?

27. Alice bought a new car. She got a 7% discount. The car listed for \$9000. How much did she pay for the car?

28. A total of 896 freshmen were admitted to King Frederich College. Freshmen make up 28% of the student body. How big is the student body?

29. The air pollution level in Centerville is 8.86. Ten years ago it was 7.96. What is the percent of increase of the air pollution level?

30. Fred borrowed \$1600 for two years. He was charged simple interest at a rate of 11%. How much interest did he pay?

CHAPTER 6

Measurement

You might hear a person mentioning that their new computer has a 2.1-GB hard drive and has 16K of RAM. The Pentium chip that powers the computer runs at 150 MHz. Just what do all those letters mean? What is being measured? Can you solve a problem with these kind of units? Turn to the Putting Your Skills to Work on page 364.

Pretest Chapter 6

1. ______ 17. ______
2. ______ 18. ______
3. ______ 19. ______
4. ______ 20. ______
5. ______ 21. ______
6. ______ 22. ______
7. ______ 23. ______
8. ______ 24. ______
9. ______ 25. ______
10. ______ 26. ______
11. ______ 27. ______
12. ______ 28. (a) ______
13. ______ (b) ______
14. ______ 29. ______
15. ______ 30. ______
16. ______

If you are familiar with the topics in this chapter, take this test now. Check your answers with those in the back of the book. If an answer was wrong or you couldn't do a problem, study the appropriate section of the chapter.

If you are not familiar with the topics in this chapter, don't take this test now. Instead, study the examples, work the practice problems, and then take the test.

This test will help you identify which concepts you have mastered and which you need to study further.

Section 6.1

Convert. When necessary, express your answer as a decimal rounded to the nearest hundredth.

1. 17 ft = ____ in. **2.** 7 gal = ____ pt **3.** 2 mi = ____ yd

4. 3.2 tons = ____ lb **5.** 22 min = ____ sec **6.** 6 gal = ____ qt

Section 6.2

Perform each conversion.

7. 5.32 km = ____ m **8.** 46.8 m = ____ cm **9.** 986 mm = ____ cm

10. 27 mm = ____ m **11.** 5296 mm = ____ cm **12.** 123 m = ____ km

Convert to meters and add.

13. 1.2 km + 192 m + 984 m **14.** 3862 cm + 9342 mm + 46.3 m

Section 6.3

Perform each conversion.

15. 3.82 L = ____ mL **16.** 3162 g = ____ kg **17.** 56.3 kg = ____ t

18. 4.8 kL = ____ L **19.** 568 mg = ____ g **20.** 8.9 L = ____ cm^3

Section 6.4

Perform each conversion. Round to the nearest hundredth when necessary.

21. 12 cm = ____ in. **22.** 4.2 ft = ____ m **23.** 96 km = ____ mi

24. 482 gal = ____ L **25.** 1.4 oz = ____ g **26.** 47 kg = ____ lb

Section 6.5

Solve each problem. Round your answer to the nearest hundredth when necessary.

27. Find the perimeter of the triangle at left. Express your answer in **feet.**

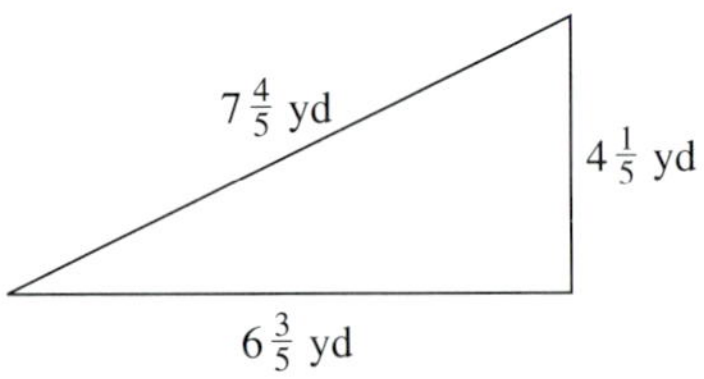

28. The radio reported the temperature today at 35°C. The record high temperature for this day is 98°F.

(a) What was the Fahrenheit temperature today?

(b) Did the temperature today set a new record?

29. Juanita traveled in Mexico for 2 hours at 95 kilometers per hour. She had to travel a distance of 130 miles. How far does she still need to travel? (Express your answer in **miles.**)

30. A pump is running at 1.5 quarts per minute. What is this rate in gallons per hour?

6.1 American Units

After studying this section, you will be able to:

1. *Identify the basic unit equivalencies in the American system.*
2. *Convert from one unit of measure to another.*

MathPro Video 12 SSM

1 *Identifying the Basic Unit Equivalencies in the American System*

We often ask questions about measurements. How far is it to work? How much does this bottle hold? What is the weight of this box? How long will it be until exams? To answer these questions we need to agree on a unit of measure for each type of measurement.

At present there are two main systems of measurement, each with its own set of units: the metric system and the American system. Nearly all countries in the world use the metric system. In the United States, however, except in science, most measurements are made in American units.

The United States is using the metric system more and more frequently, however, and may eventually convert to metric units as the standard. But for now we need to be familiar with both systems. We cover American units in this section.

One of the most familiar measuring devices is a ruler that measures lengths as great as 1 foot. It is divided into 12 inches; that is,

1 foot = 12 inches

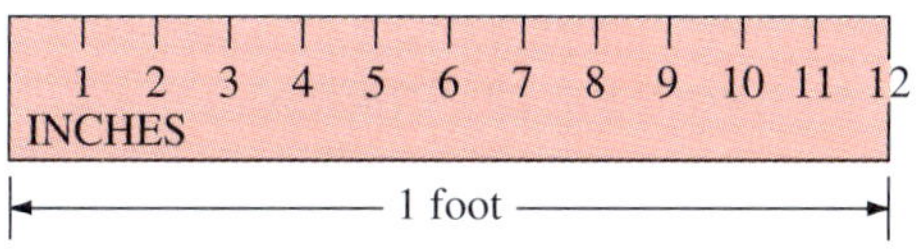

There are several other important relationships you need to know. Your instructor may require you to memorize the following facts.

Length

12 inches = 1 foot
3 feet = 1 yard
5280 feet = 1 mile
1760 yards = 1 mile

Time

60 seconds = 1 minute
60 minutes = 1 hour
24 hours = 1 day
7 days = 1 week

Note that time is measured in the same units in both the metric and the American system.

Weight

16 ounces = 1 pound
2000 pounds = 1 ton

Volume

2 cups = 1 pint
2 pints = 1 quart
4 quarts = 1 gallon

We can choose to measure an object—say, a bridge—using a small unit (an inch), a larger unit (a foot), or a still larger unit (a mile). We may say that the bridge spans 7920 inches, 660 feet, or an eighth of a mile. Although we probably would not choose to express our measurement in inches because this is not a convenient measurement to work with for an object as long as a bridge, the bridge length is the same whatever unit of measurement we use.

Notice that the smaller the measuring unit, the larger the number of those units in the final measurement. The inch is the smallest unit in our example, and the inch measurement has the greatest number of units (7920). The mile is the largest unit, and it has the smallest number of units (an eighth equals 0.125). Whatever measuring system you use, and whatever you measure (length, volume, and so on), the smaller the unit you use, the greater the number of those units.

After studying the values in the length, time, weight, and volume tables, see if you can quickly do Example 1.

EXAMPLE 1 Answer rapidly the following questions.

(a) How many inches in a foot? **(b)** How many yards in a mile?
(c) How many seconds in a minute? **(d)** How many hours in a day?
(e) How many pounds in ton? **(f)** How many cups in a pint?

(a) 12 **(b)** 1760 **(c)** 60 **(d)** 24 **(e)** 2000 **(f)** 2

Practice Problem 1 Answer rapidly the following questions.

(a) How many feet in a yard? **(b)** How many feet in a mile?
(c) How many minutes in an hour? **(d)** How many days in a week?
(e) How many ounces in a pound? **(f)** How many pints in a quart?
(g) How many quarts in a gallon? ■

2 *Converting One Measurement to Another*

To convert or change one measurement to another, we simply multiply by 1 since multiplying by 1 does not change the value of a quantity. For example, to convert 180 inches to feet, we look for a name for 1 that has inches and feet.

$$1 = \frac{1 \text{ foot}}{12 \text{ inches}}$$

$$180 \text{ inches} \times \frac{1 \text{ foot}}{12 \text{ inches}} = \frac{180 \text{ feet}}{12} = 15 \text{ feet}$$

Notice that when we multiplied, the inches divided out. We are left with the unit feet.

What name for 1 should we choose if we want to change from feet to inches? Convert 4 feet to inches.

$$1 = \frac{12 \text{ inches}}{1 \text{ foot}}$$

$$4 \text{ feet} \times \frac{12 \text{ inches}}{1 \text{ foot}} = \frac{48 \text{ inches}}{1} = 48 \text{ inches}$$

Notice that in this example feet divided out and we were left with the desired unit, inches.

When choosing a fractional name for 1, the unit we want to change to should be in the *numerator*. The unit we start with should be in the *denominator*. This unit will cancel out.

EXAMPLE 2 Convert each measurement.

(a) 8800 yards to miles **(b)** 26 feet to inches

(a) $8800 \cancel{\text{yards}} \times \dfrac{1 \text{ mile}}{1760 \cancel{\text{yards}}} = \dfrac{8800}{1760} \text{ miles} = 5 \text{ miles}$

(b) $26 \cancel{\text{feet}} \times \dfrac{12 \text{ inches}}{1 \cancel{\text{foot}}} = 26 \times 12 \text{ inches} = 312 \text{ inches}$

Practice Problem 2 Convert each measurement.

(a) 15,840 feet to miles **(b)** 17 feet to inches ■

Some conversions involve fractions or decimals. Whether you want to measure the area of a living room or the dimensions of a piece of property, it is helpful to be able to make conversions like these.

EXAMPLE 3 Convert each measurement.

(a) 26.48 miles to yards **(b)** $3\frac{2}{3}$ feet to yards

(a) $26.48 \cancel{\text{miles}} \times \dfrac{1760 \text{ yards}}{1 \cancel{\text{mile}}} = 46{,}604.8 \text{ yards}$

(b) $3\frac{2}{3} \cancel{\text{feet}} \times \dfrac{1 \text{ yard}}{3 \cancel{\text{feet}}} = \dfrac{11}{3} \times \dfrac{1}{3} \text{ yard} = \dfrac{11}{9} \text{ yards} = 1\frac{2}{9} \text{ yards}$

Practice Problem 3 Convert each measurement.

(a) 18.93 miles to feet **(b)** $16\frac{1}{2}$ inches to yards ■

EXAMPLE 4 Lynda's new car weighs 2.43 tons. How many pounds is that?

$$2.43 \cancel{\text{tons}} \times \frac{2000 \text{ pounds}}{1 \cancel{\text{ton}}} = 4860 \text{ pounds}$$

Practice Problem 4 A package weighs 760.5 pounds. How many ounces does it weigh? ■

EXAMPLE 5 The chemistry lab has 34 quarts of weak hydrochloric acid. How many gallons of this acid are in the lab? (Express your answer as a decimal.)

$$34 \cancel{\text{quarts}} \times \frac{1 \text{ gallon}}{4 \cancel{\text{quarts}}} = \frac{34}{4} \text{ gallons} = 8.5 \text{ gallons}$$

Practice Problem 5 19 pints of milk is the same as how many quarts? (Express your answer as a decimal.) ■

EXAMPLE 6 A computer printout shows that a particular job took 144 seconds. How many minutes is that? (Express your answer as a decimal.)

$$144 \text{ seconds} \times \frac{1 \text{ minute}}{60 \text{ seconds}} = \frac{144}{60} \text{ minutes} = 2.4 \text{ minutes}$$

Practice Problem 6 Joe's time card read "Hours worked today: 7.2." How many minutes are in 7.2 hours? ■

EXAMPLE 7 A window is 4 feet 5 inches wide. How many inches is that?

4 feet 5 inches means 4 feet and 5 inches. Change the 4 feet to inches and add the 5 inches.

$$4 \text{ feet} \times \frac{12 \text{ inches}}{1 \text{ foot}} = 48 \text{ inches}$$

$$48 \text{ inches} + 5 \text{ inches} = 53 \text{ inches}$$

The window is 53 inches wide.

Practice Problem 7 A bag of potatoes weighs 7 pounds 12 ounces. How many ounces is that? ■

EXAMPLE 8 The all-night Charlotte garage charges \$1.50 per hour for parking both day and night. A businessman left his car there for $2\frac{1}{4}$ days. How much was he charged?

1. *Understand the problem.*
   Here it might help to look at a simpler problem. If the businessman had left his car for 2 hours, we would multiply.

   $$\text{The rule means "per"} \rightarrow \frac{1.50 \text{ dollars}}{1 \text{ hour}} \times 2 \text{ hours} = 3.00 \text{ dollars or } \$3.00$$

   Thus, if the businessman leaves his car for 2 hours, he would be charged \$3. We see that we need to multiply by the number of hours the car was in the garage to solve the problem.

   Since the original problem gave the time in days, not hours, we will need to change the days to hours. Remember, the garage charges \$1.50 per *hour*.

2. *Solve and state the answer.*
   Now that we know that the way to solve the problem is to multiply hours, we will begin. To make our calculations easier we will write $2\frac{1}{4}$ as 2.25. Change days to hours.

   $$2.25 \text{ days} \times \frac{24 \text{ hours}}{1 \text{ day}} \times \frac{1.50 \text{ dollars}}{1 \text{ hour}} = 81 \text{ dollars or } \$81$$

   The businessman was charged \$81.

3. *Check.*

Is our answer in the desired unit? Yes. The answer is in dollars and we would expect it to be in dollars. ✓

You may want to redo the calculation or use a calculator to check. The check is up to you.

Practice Problem 8 A businesswoman parked her car at a garage for $1\frac{3}{4}$ days. The garage charges \$1.50 per hour. How much did she pay to park the car? ■

ALTERNATIVE METHOD USING PROPORTIONS How did people first come up with the idea of "multiplying by a fractional name for 1"? What mathematical principles are involved here? Actually, this is the same as solving a proportion. Consider Example 5, where we changed 34 quarts to 8.5 gallons by multiplying.

$$34 \cancel{\text{quarts}} \times \frac{1 \text{ gallon}}{4 \cancel{\text{quarts}}} = \frac{34}{4} \text{ gallons} = 8.5 \text{ gallons}$$

What we were actually doing is setting up the proportion

1 gallon is to 4 quarts as n gallons is to 34 quarts

and solving.

$$\frac{1 \text{ gallon}}{4 \text{ quarts}} = \frac{n \text{ gallons}}{34 \text{ quarts}}$$

Cross-multiply.

$$1 \text{ gallon} \times 34 \text{ quarts} = 4 \text{ quarts} \times n \text{ gallons}$$

Divide both sides of the equation by 4 quarts.

$$\frac{1 \text{ gallon} \times 34 \cancel{\text{quarts}}}{4 \cancel{\text{quarts}}} = \frac{\cancel{4} \cancel{\text{quarts}} \times n \text{ gallons}}{\cancel{4} \cancel{\text{quarts}}}$$

$$1 \text{ gallon} \times \frac{34}{4} = n \text{ gallons}$$

$$8.5 \text{ gallons} = n \text{ gallons}$$

Thus the number of gallons is 8.5. Using proportions takes a little longer, so multiplying by a fractional name for 1 is the more popular method. For more about using proportions in converting units, see Exercises 6.1, problems 67 and 68.

6.1 Exercises

From memory, write the equivalent values.

1. 1 foot = _____ inches

2. 1 yard = _____ feet

3. _____ yards = 1 mile

4. _____ feet = 1 mile

5. 1 ton = _____ pounds

6. 1 pound = _____ ounces

7. _____ quarts = 1 gallon

8. _____ cups = 1 pint

9. 1 quart = _____ pints

10. 1 day = _____ hours

11. _____ seconds = 1 minute

12. _____ minutes = 1 hour

Convert. When necessary, express your answer as a decimal.

13. 15 feet = _____ yards

14. 21 feet = _____ yards

15. 84 inches = _____ feet

16. 180 inches = _____ feet

17. 10,560 feet = _____ miles

18. 5280 yards = _____ miles

19. 7 miles = _____ yards

20. 6 miles = _____ feet

21. 12 feet = _____ inches

22. 16 feet = _____ inches

23. 41 yards = _____ feet

24. 36 yards = _____ feet

25. 75 inches = _____ feet

26. 87 inches = _____ feet

27. 192 ounces = _____ pounds

28. 144 ounces = _____ pounds

29. 13 tons = _____ pounds

30. 17 tons = _____ pounds

31. 2.25 pounds = _____ ounces

32. 4.25 pounds = _____ ounces

33. 7 gallons = _____ quarts

34. 5 gallons = _____ quarts

35. 18 pints = _____ quarts

36. 24 pints = _____ quarts

37. 31 pints = _____ cups

38. 27 pints = =_____ cups

39. 8 gallons = _____ pints

40. 6 gallons = _____ pints

41. 12 weeks = _____ days

42. 7 weeks = _____ days

43. 660 minutes = _____ hours

44. 780 minutes = _____ hours

45. 11 days = _____ hours

46. 7 days = _____ hours

47. 70 minutes = _____ seconds

48. 50 minutes = _____ seconds

49. 18 hours = _____ seconds

50. 15 hours = _____ seconds

51. 180 seconds = _____ minutes

52. 3462.56 feet = _____ inches

53. 6755 yards = _____ feet

54. 65.62 pounds = _____ ounces

Applications

55. A stockbroker left his car in an all-night garage for $2\frac{1}{2}$ days. The garage charges \$2.25 per hour. How much did he pay for parking?

56. When Barry went overseas on business, he left his car in the long-term parking lot for $3\frac{1}{2}$ days. Long-term parking costs \$2.15 per hour. How much did Barry pay when he came back from his trip?

57. Judy is making a wild mushroom sauce for pasta tonight with a large group of friends. She bought 26 ounces of wild mushrooms at \$6.00 per pound. How much were the mushrooms?

58. Kurt is trying to eat a lower-fat diet than usual. He finds a store that sells 1% ground white-meat turkey breast at \$4.00 per pound. He buys one packet weighting 18 ounces and another weighing 22 ounces. How much did he pay?

59. A physically challenged athlete proves that he is not as disabled as the general public thinks by rolling his special three-wheel racing chair 37.7 miles in one day. How many feet is that?

60. Mount Whitney in California is approximately 2.745 miles high. How many feet is that? Round your answer to the nearest 10 feet.

61. In 1994, Sue Caivun of China broke the women's world record for the pole vault when she defied gravity and found herself clearing the bar at 13 feet $5\frac{3}{4}$ inches. How many inches high was her pole vault?

62. Jan Zelezny of the Czech Republic threw his javelin 318 feet 10 inches to break the world's record in 1993. How many inches did his javelin fly?

63. A multimillion-dollar shopping complex in Italy is using nothing but marble for the floors and walls of all public areas. The lobby alone is constructed of 267,905,993 pounds of blue-gray marble. How many tons would that be? Round to the nearest ton.

64. Last week, the Red Cross and the United Nations airlifted 12,560,950 pounds of food to hungry children. How many tons of food were distributed? Round to the nearest ton.

65. A space probe traveled for 3 years to get to the planet Saturn. How many hours did the trip take? (Assume 365 days in 1 year.)

66. A seedling peach tree grew for 7 years until it produced its first fruit. How many hours would that be if you assume that there are 365 days in a year?

There are 6080 feet in a nautical mile and 5280 feet in a land mile. On sea and in the air, distance and speed are often measured in nautical miles. The ratio of regular miles to nautical miles is 38 to 33.

67. A windjammer sailboat was used as a floating school. During one school year, the ship traveled 12,800 nautical miles. What would be the equivalent in land miles? Round to the nearest whole mile.

68. A merchant ship took some tourists on board for extra income. The ship took them the equivalent of 850 land miles. How many nautical miles did they travel? Round to the nearest whole mile.

6.2 Metric Measurements: Length

MathPro Video 13 SSM

After studying this section, you will be able to:

1. *Understand prefixes in metric units.*
2. *Convert from one metric unit of length to another.*

1 *Understanding Prefixes in Metric Units*

The metric system of measurement is used in most industrialized nations of the world. As we move toward a global economy, it is important to be familiar with the metric system. The metric system is designed for ease in calculating and in converting from one unit to another.

In the metric system, the basic unit of length measurement is the meter. A meter is just slightly longer than a yard. To be more precise, the meter is approximately 39.37 inches long.

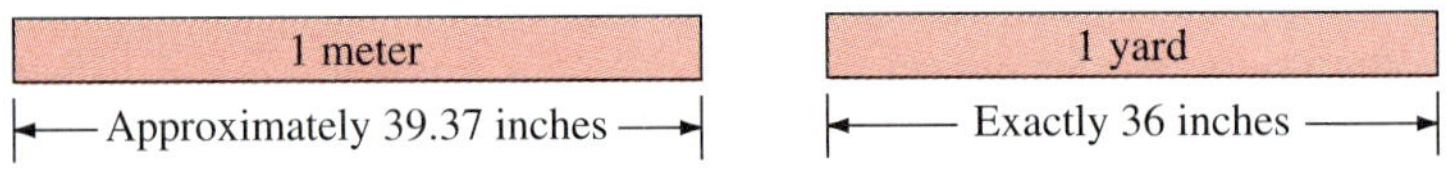

Units that are larger or smaller than the meter are based on the meter and powers of 10. For example, the unit *deka*meter is *ten* meters. The unit *deci*meter is *one-tenth* of a meter. The prefix *deka* means 10. What does the prefix *deci* mean? All the prefixes in the metric system are names for multiples of 10. A list of metric prefixes and their meaning follows.

Prefix	Meaning
kilo-	thousand
hecto-	hundred
deka-	ten
deci-	tenth
centi-	hundredth
milli-	thousandth

The most commonly used prefixes are *kilo-*, *centi-*, and *milli-*. *Kilo-* means thousand, so a *kilo*meter is a thousand meters. Similarly, *centi-* means one hundredth, so a *centi*meter is a one hundredth of a meter. And *milli-* means one thousandth, so a *milli*meter is one thousandth of a meter.

The kilometer is used to measure distances much larger than the meter. How far did you travel in a car? The centimeter is used to measure shorter lengths. What are the dimensions of this textbook? The millimeter is used to measure very small lengths. What is the width of the lead in a pencil?

EXAMPLE 1 Write the prefix that means **(a)** thousand **(b)** tenth

(a) The prefix *kilo-* is used for thousand.

(b) The prefix *deci-* is used for tenth.

Practice Problem 1 Write the prefix that means **(a)** ten **(b)** thousandth ■

2 *Converting from One Metric Unit of Length to Another*

How do we convert from one metric unit to another? For example, how do we change 5 kilometers into an equivalent number of meters?

Recall from Chapter 3 that when we multiply by 10 we move the decimal point one place to the right. When we divide by 10 we move the decimal point one place to the left. Let's see how we use that idea to change from one metric unit to another.

Changing from Larger Metric Units to Smaller Ones

When you change from one metric prefix to another moving to the **right** on this prefix chart, move the decimal point to the **right** the same number of places.

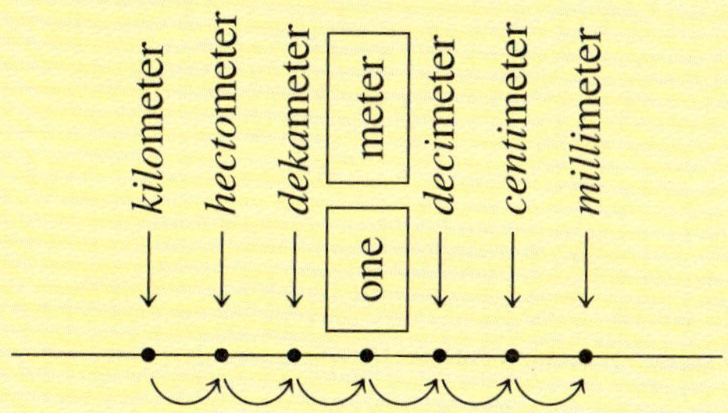

Thus 1 meter = 100 centimeters because we move two places to the right on the chart of prefixes and we also move the decimal point (1.00) two places to the right.

Now let us examine the four most commonly used metric measurements of length and their abbreviations.

Commonly Used Metric Lengths

1 kilometer (km) = 1000 meters

1 meter (m) (the basic unit of length in the metric system)

1 centimeter (cm) = 0.01 meter

1 millimeter (mm) = 0.001 meter

Now let us see how we can change a measurement stated in a larger unit to an equivalent measurement stated in smaller units.

EXAMPLE 2

(a) Change 5 kilometers to meters. **(b)** Change 20 meters to centimeters.

(a) From *kilometer* to meter we move three places to the right on the prefix chart, so we move the decimal point three places to the right.

5 kilometers = 5.000 meters (move three places) = 5000 meters

(b) To go from meters to *centimeters,* we move two places to the right on the prefix chart. Thus we move the decimal point two places to the right.

20 meters = 20.00 centimeters (move two places) = 2000 centimeters

Practice Problem 2

(a) Change 4 meters to centimeters.
(b) Change 30 centimeters to millimeters. ■

Changing from Smaller Metric Units to Larger Ones

When you change from one metric prefix to another moving to the **left** on this prefix chart, move the decimal point to the **left** the same number of places.

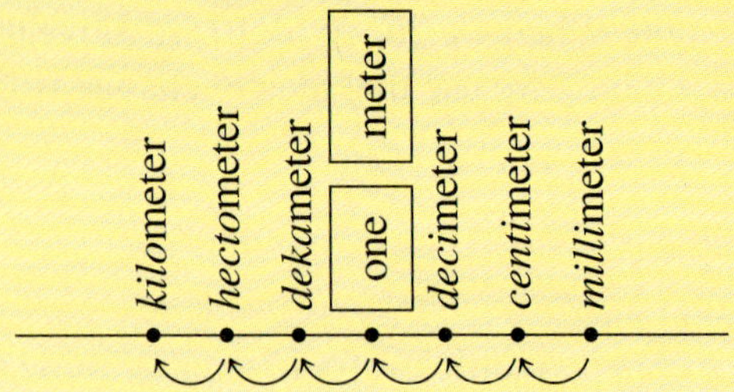

EXAMPLE 3

(a) Change 7 centimeters to meters.

(b) Change 56 millimeters to kilometers.

(a) To go from *centi*meters to meters, we move *two* places to the left on the prefix chart. Thus we move the decimal point two places to the left.

7 centimeters = 0.07 meter (move two places to the left)
= 0.07 meter

(b) To go from *milli*meters to *kilo*meters, we move six places to the left on the prefix chart. Thus we move the decimal point six places to the left.

56 millimeters = 0.000056. kilometer (move six places to the left)
= 0.000056 kilometer

Practice Problem 3

(a) Change 3 *milli*meters to meters.

(b) Change 47 *centi*meters to *kilo*meters. ■

Thinking Metric

Long distances are customarily measured in kilometers. A kilometer is about 0.62 mile. It takes about 1.6 kilometers to make a mile. The following drawing shows the relationship between the kilometer and the mile.

1 kilometer

1 mile (about 1.6 kilometers)

Many small distances are measured in centimeters. A centimeter is about 0.394 inch. It takes about 2.54 centimeters to make an inch. You can get a good idea of their size by looking at a ruler marked in both inches and centimeters.

1 centimeter

1 inch (about 2.54 centimeters)

Try to visualize how many centimeters long or wide this book is. It is 21 centimeters wide and 27.5 centimeters long.

A millimeter is a very small unit of measurement, often used in manufacturing.

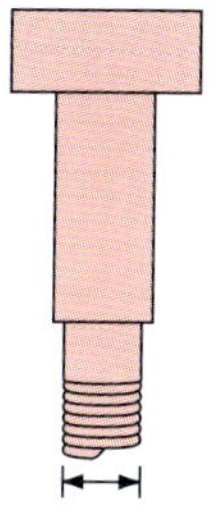

The threaded end of a bolt may be 6 mm wide.

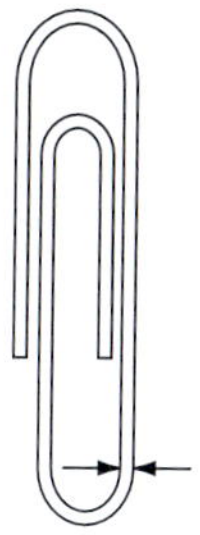

A paper clip is made of wire 0.8 mm thick.

Now that you have an understanding of the size of these metric units, let's try to select the most convenient metric unit for measuring the length of an object.

EXAMPLE 4 Bob measured the width of a doorway in his house. He wrote down "73." But what unit of measurement did he use?

(a) 73 kilometers **(b)** 73 meters **(c)** 73 centimeters

The most reasonable choice is (c), 73 centimeters. The other two units would be much too long. A meter is close to a yard, and a doorway would not be 73 yards wide! A kilometer is even larger than a meter.

Practice Problem 4 Joan measured the length of her car. She wrote down 3.8. Which unit of measurement did she use?

(a) 3.8 kilometers **(b)** 3.8 meters **(c)** 3.8 centimeters ■

The most frequent metric conversions in length are done between kilometers, meters, centimeters, and millimeters. Their abbreviations are km, m, cm, and mm, and you should be able to use them correctly.

EXAMPLE 5 Convert each measurement.

(a) 364 m to km **(b)** 982 cm to m **(c)** 16 mm to cm **(d)** 5.2 m to mm

In each of the first three cases, we move the decimal point to the left because we are going from a smaller unit to a larger unit.

(a) 364 m = 0.364 km (three places to left)

= 0.364 km

(b) 982 cm = 9.82 m (two places to left)

= 9.82 m

(c) 16 mm = 1.6 cm (one place to left)

= 1.6 cm

Here we need to move the decimal point to the right because we are going from a larger unit to a smaller unit.

(d) 5.2 m = 5200. mm (three places to right)

= 5200 mm

Practice Problem 5 Convert. **(a)** 1527 m to km **(b)** 375 cm to m **(c)** 128 mm to cm **(d)** 46 m to mm ■

The other metric units of length are the hectometer, the dekameter, and the decimeter. These are not used very frequently, but it is good to understand how their lengths relate to the basic unit, the meter. A complete list of the metric lengths we have discussed appears in the following table.

Metric Lengths with Abbreviations

1 kilometer (km) = 1000 meters

1 hectometer (hm) = 100 meters

1 dekameter (dam) = 10 meters

1 meter (m)

1 decimeter (dm) = 0.1 meter

1 centimeter (cm) = 0.01 meter

1 millimeter (mm) = 0.001 meter

EXAMPLE 6 Convert.

(a) 426 decimeters to kilometers **(b)** 9.47 hectometers to meters

(a) We are converting from a smaller unit, dm, to a larger one, km. Therefore, there will be fewer kilometers than decimeters (426 will "get smaller"). We move the decimal point four places to the left.

426 dm = 0.0426 km (four places to left)

= 0.0426 km

(b) We are converting from a larger unit, hm, to a smaller one, m. Therefore, there will be more meters than hectometers (9.47 will "get larger"). We move the decimal point two places to the right.

9.47 hm = 947. m (two places to right)

= 947 m

Practice Problem 6 Convert.

(a) 389 millimeters to dekameters **(b)** 0.48 hectometer to centimeters ■

When several metric measurements are to be added, we change them to a convenient common unit.

EXAMPLE 7 Add. 125 m + 1.8 km + 793 m

First we change the kilometer measurement to a measurement in meters.

1.8 km = 1800 m

Then we add.

$$\begin{array}{r} 125 \text{ m} \\ 1800 \text{ m} \\ +\ \ 793 \text{ m} \\ \hline 2718 \text{ m} \end{array}$$

Practice Problem 7 Add. 782 cm + 2 m + 537 m ■

SIDELIGHT: Extremely Large Metric Distances Is the biggest length in the metric system a kilometer? Is the smallest length a millimeter? No. The system extends to very large units and very small ones. Usually, only scientists use these units.

1 gigameter = 1,000,000,000 meters
1 megameter = 1,000,000 meters
1 kilometer = 1000 meters
1 meter
1 millimeter = 0.001 meter
1 micrometer = 0.000001 meter
1 nanometer = 0.000000001 meter

For example, 26 megameters equals a length of 26,000,000 meters. A length of 31 micrometers equals a length of 0.000031 meter.

SIDELIGHT: Metric Measurements for Computers A **byte** is the amount of computer memory needed to store one alphanumeric character. When referring to computers you may hear the following words: **kilobytes, megabytes,** and **gigabytes.** The following chart may help you:

1 gigabyte (GB) = one billion bytes =	1,000,000,000 bytes
1 megabyte (MB) = one million bytes* =	1,000,000 bytes
1 kilobyte (KB) = one thousand bytes† =	1000 bytes

*Sometimes in computer science 1 megabyte is considered to be 1,048,576 bytes.
†Sometimes in computer science 1 kilobyte is considered to be 1024 bytes.

We will consider these measures of computer storage in the Putting Your Skills to Work on page 364.

A Challenge for You

Complete the following:

1. 18 megameters = ________ kilometers

2. 26 millimeters = ________ micrometers

3. 17 nanometers = ________ millimeter

4. 38 meters = ________ megameter

5. 1.2 gigabytes = ________ bytes

6. 528 megabytes = ________ bytes

7. 78.9 kilobytes = ________ bytes

8. 24.9 gigabytes = ________ bytes

6.2 Exercises

Verbal and Writing Skills

Write the prefix that means

1. Hundred
2. Hundredth
3. Tenth
4. Thousandth
5. Thousand
6. Ten

The following conversions involve metric units that are very commonly used. You should be able to perform each conversion without any notes and without consulting your text.

7. 37 centimeters = ____ millimeters
8. 44 centimeters = ____ millimeters
9. 3.6 kilometers = ____ meters
10. 7.6 kilometers = ____ meters
11. 328 millimeters = ____ meter
12. 508 millimeters = ____ meter
13. 56.3 centimeters = ____ meter
14. 96.4 centimeters = ____ meter
15. 2 kilometers = ________ centimeters
16. 7 kilometers = ________ centimeters
17. 78,000 millimeters = ____ kilometer
18. 96,000 millimeters = ____ kilometer
19. 0.5386 kilometer = ________ millimeters
20. 0.3279 kilometer = ________ millimeters

Abbreviations are used in the following. Fill in the blanks with the correct value.

21. 35 mm = ____ cm = ____ m
22. 83 mm = ____ cm = ____ m
23. 3582 mm = ____ m = ________ km
24. 7812 mm = ____ m = ________ km
25. 0.32 cm = ____ m = ________ km
26. 0.81 cm = ____ m = ________ km

Applications

27. Peter measured the length of his dining room table. Choose the most reasonable measurement.
(a) 2 m **(b)** 2 km **(c)** 2 cm

28. Wally measured the length of his college identification card. It was which of the following?
(a) 10 km **(b)** 10 m **(c)** 10 cm

29. Eddie measured the width of a compact disc case. Choose the most appropriate measurement.
(a) 80.5 km **(b)** 80.5 m **(c)** 80.5 mm

30. The Botanical Gardens are located in the center of the city. If it takes approximately 10 minutes to walk directly from the east entrance to the west entrance, which would be the most likely measurement of the width of the Gardens?
(a) 1.4 km **(b)** 1.4 m **(c)** 1.4 mm

The following conversions involve some of the metric units that are not used extensively. You should be able to perform each conversion, but it is not necessary to do it from memory.

31. 3 kilometers = ____ hectometers
32. 7 hectometers = ____ dekameters
33. 27 meters = ____ decimeters
34. 35 decimeters = ____ centimeters
35. 198 millimeters = ____ decimeters
36. 2.93 decimeters = ________ hectometer
37. 48.2 meters = ____ hectometer
38. 435 hectometers = ____ kilometers
39. 0.5236 hectometer = ____ centimeters
40. 0.7175 dekameter = ____ millimeters

Change to a convenient unit of measure and add.

41. 243 m + 2.7 km + 312 m
42. 845 m + 5.79 km + 701 m
43. 5.2 cm + 361 cm + 968 mm
44. 4.8 cm + 607 cm + 834 mm
45. 82 m + 471 cm + 0.32 km
46. 46 m + 986 cm + 0.884 km

47. The outside casing of a stereo cabinet is built of plywood 0.95 centimeters thick attached to plastic 1.35 centimeters thick and a piece of mahogany veneer, 2.464 millimeters thick. How thick is the stereo casing?

48. A plywood board is 2.2 centimeters thick. A layer of tar paper is 3.42 millimeters thick. A layer of false brick siding is 2.7 centimeters thick. A house wall consists of these three layers. How thick is the wall?

Mixed Practice

Write true *or* false *for each statement.*

44. 4.8 cm + 607 cm + 834 mm **45.** 82 m + 471 cm + 0.32 km **46.** 46 m + 986 cm + 0.884 km

13. 56.3 centimeters = _____ meter

14. 96.4 centimeters = _____ meter

49. 1 meter = 0.01 centimeter

50. 1 kilometer = 0.001 meter

51. 10 centimeters = 100 millimeters

52. 10 millimeters = 1 centimeter

53. An airport runway might be 2 kilometers long.

54. A man might be 2 meters tall.

55. A page of this book is about 2 centimeters thick.

56. The glass in a drinking glass is usually about 2 millimeters thick.

57. A kilometer is shorter than a mile.

58. A meter is longer than a yard.

Applications

59. The longest runway is at Edwards Air Force Base, California. It is 1,192,000 centimeters in length.
(a) How many kilometers long is the runway?
(b) How many meters long is the runway?

60. The world's longest train run is on the Trans-Siberian line in Russia, from Moscow to Nakhodka on the Sea of Japan. The length of the run, which makes 97 stops, measures 94,380,000 centimeters. The run takes 8 days, 4 hours, and 25 minutes.
(a) How many meters is the run?
(b) How many kilometers is the run?

61. The highest railroad line in the world is a track on the Morococha branch of the Peruvian State Railways at La Cima. The track is 4818 meters high.
(a) How many centimeters high is the track?
(b) How many kilometers high is the track?

62. The break-up of the former Soviet Union has delayed the completion of the Rogunskaya dam, which will be 335 meters high.
(a) How many kilometers high is the dam supposed to be?
(b) How many centimeters high is the dam supposed to be?

63. A prism with sides measuring 0.004 millimeters (0.0001 inches), barely visible to the naked eye, was created at the National Institute of Standards and Technology in Boulder, Colorado, in 1989. How many kilometers wide is each side of the prism?

64. Type D263 glass, made by Deutsche Spezialglas AG, of Grunenplan, Germany for use in electronic and medical equipment, has a minimum thickness of 0.025 millimeters, and a maximum thickness of 0.035 millimeters. What would be the measurement, in meters of the minimum and maximum thickness of the glass.

6.3 Metric Measurements: Volume and Weight

MathPro Video 13 SSM

After studying this section, you will be able to:

1 *Convert between metric units of volume.*
2 *Convert between metric units of weight.*

1 *Converting Between Metric Units of Volume*

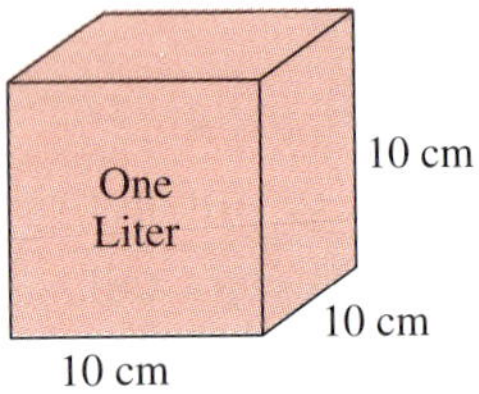

As products are distributed worldwide, more and more of them are being sold in metric units. Soft drinks come in 1-, 2-, or 3-liter bottles. Labels often print the amount in both American units and metric units. Get used to looking at these labels to gain a sense of the size of metric units of volume.

The basic metric unit for volume is the liter. A liter is defined as the volume of a box 10 cm × 10 cm × 10 cm or 1000 cm^3. A cubic centimeter may be written as cc, so we sometimes see 1000 cc = 1 liter. A liter is slightly larger than a quart; 1 liter of liquid is 1.057 quarts of that liquid.

The most common metric units of volume are the milliliter, the liter, and the kiloliter. Often a capital letter L is used as an abbreviation for *liter.*

> 1 kiloliter (kL) = 1000 liters
>
> 1 liter (L)
>
> 1 milliliter (mL) = 0.001 liter

We know that 1000 cc = 1 liter. Dividing each side of that equation by 1000, we get 1 cc = 1 mL.

> When you change one metric prefix to another, move the decimal point in the same direction and the appropriate number of places. Use the prefix chart as a guide.
>
> 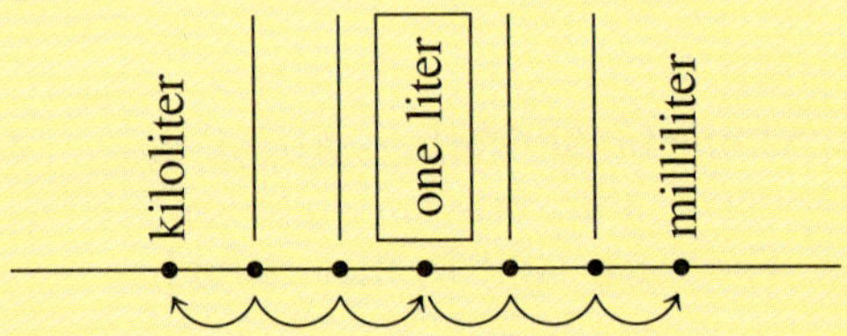
>

The prefixes for liter follow the pattern we have seen for meter. The *kilo-* is three places to the left of the liter, and the *milli-* is three places to the right.

EXAMPLE 1 Convert.

(a) 3 L = ____ mL **(b)** 24 kL = ____ L **(c)** 0.084 L = ____ mL

The prefix *milli-* is three places to the right. We move the decimal point three places to the right.

(a) 3 L = 3000 mL **(b)** 24 kL = 24,000 L **(c)** 0.084 L = 84 mL

Practice Problem 1 Convert.

(a) 5 L = ____ mL **(b)** 84 kL = ____ L **(c)** 0.732 L = ____ mL ■

EXAMPLE 2 Convert.

(a) 26.4 mL = ____ L **(b)** 5982 mL = ____ L **(c)** 6.7 L = ____ kL

The unit L is three places to the left. We move the decimal three places to the left.

(a) 26.4 mL = 0.0264 L **(b)** 5982 mL = 5.982 L **(c)** 6.7 L = 0.0067 kL

Practice Problem 2 Convert.

(a) 15.8 mL = ____ L **(b)** 12,340 mL = ____ L **(c)** 86.3 L = ____ kL ■

The cubic centimeter is often used in medicine. Recall 1 mL = 1 cm^3 or 1 cc.

EXAMPLE 3 Convert.

(a) 26 mL = ____ cm^3 **(b)** 0.82 L = ____ cm^3

A milliliter and a cubic centimeter are equivalent.

(a) 26 mL = 26 cm^3

We use the same rule to convert liters to cubic centimeters as we do to convert liters to milliliters.

(b) 0.82 L = 820 cm^3

Practice Problem 3 Convert.

(a) 396 mL = ____ cm^3 **(b)** 0.096 L = ____ cm^3 ■

EXAMPLE 4 A special cleaning fluid to rinse test tubes in the chemistry lab costs \$40.00 per liter. What is the cost per milliliter?

Since the milliliter is only $\frac{1}{1000}$ of a liter, its cost is only $\frac{1}{1000}$ of \$40.00.

$$\frac{\$40.00}{1000} = \$0.040$$

Thus each milliliter costs \$0.04, which is quite expensive.

Practice Problem 4 A purified acid costs \$110 per liter. What does it cost per milliliter? ■

2 *Converting Between Metric Units of Weight*

In the metric system the basic unit of weight is the gram. A *gram* is the weight of the water in a box that is 1 centimeter on each side. To get an idea of how small a gram is, we note that two small paper clips weigh about 1 gram. A gram is only about 0.035 ounce.

One kilogram is 1000 times larger than a gram. A kilogram weighs about 2.2 pounds. The most common measures of weight in the metric system are shown in the following chart.

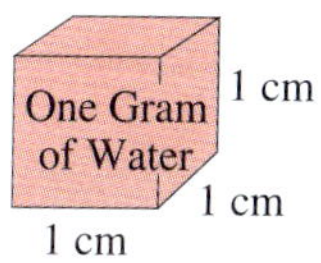

Common Metric Weight Measurements

1 metric ton (t) = 1,000,000 grams

1 kilogram (kg) = 1000 grams

1 gram (g)

1 milligram (mg) = 0.001 gram

When you change one metric prefix to another, move the decimal point in the same direction and the appropriate number of places. Use the prefix chart as a guide.

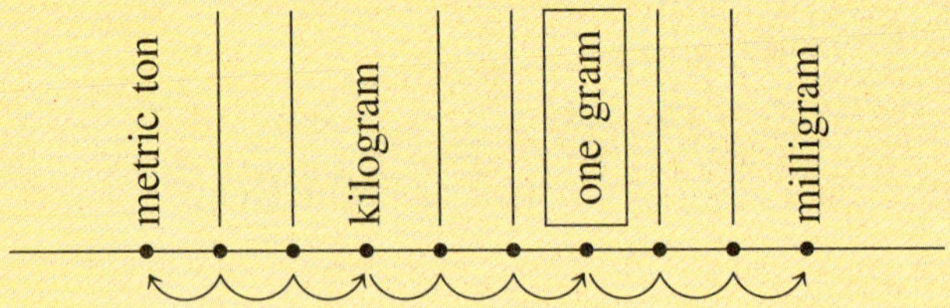

As we convert from a larger unit to the next smaller unit on the chart (for example, from kg to g), we move three places to the right. This moves the decimal point in the measurement three places to the right (4 kg = 4000 g).

EXAMPLE 5 Convert.

(a) 2 t = ____ kg **(b)** 0.42 kg = ____ g **(c)** 19.63 g = ____ mg

(a) 2 t = **2000 kg** **(b)** 0.42 kg = **420 g**

(c) 19.63 g = **19,630 mg** ■

Practice Problem 5 Convert.

(a) 3.2 t = ____ kg **(b)** 7.08 kg = ____ g **(c)** 0.526 g = ____ mg ■

As we convert from a smaller unit to the next larger unit on the chart (for example, from g to kg), we move three places to the left. This moves the decimal point in the measurement three places to the left (6000 g = 6 kg).

EXAMPLE 6 Convert.

(a) 283 kg = ____ t **(b)** 14.628 g = ____ kg **(c)** 7.98 mg = ____ g

(a) 283 kg = 0.283 t **(b)** 14.628 g = 0.014628 kg

(c) 7.98 mg = 0.00798 mg

Practice Problem 6 Convert.

(a) 59 kg = ____ t **(b)** 6.152 g = ____ kg **(c)** 28.3 mg = ____ g ■

EXAMPLE 7 If a chemical costs \$0.03 per gram, what will it cost per kilogram?

Since there are 1000 grams in a kilogram, a chemical that costs \$0.03 per gram would cost 1000 times as much per kilogram.

$$1000 \times \$0.03 = \$30.00$$

The chemical would cost \$30.00 per kilogram.

Practice Problem 7 If coffee costs \$10.00 per kilogram, what will it cost per gram? ■

EXAMPLE 8 Select the most reasonable weight for Tammy's Toyota.

(a) 820 t **(b)** 820 g **(c)** 820 kg **(d)** 820 mg

The most reasonable answer is **(c)** 820 kg. The other weight values are much too large or much too small. Since a kilogram is slightly more than 2 pounds, we see that this weight, 820 kg, more closely tells the weight of a car.

Practice Problem 8 Select the most reasonable weight for Hank, starting linebacker for the college football team.

(a) 120 kg **(b)** 120 g **(c)** 120 mg ■

SIDELIGHT When dealing with very small particles or atomic elements, scientists sometimes use units smaller than a gram.

Small Weight Measurements

1 milligram = 0.001 gram

1 microgram = 0.000001 gram

1 nanogram = 0.000000001 gram

1 picogram = 0.000000000001 gram

We could make the following conversions.

2.6 picograms = 0.0026 nanogram

29.7 micrograms = 0.0297 milligram

58 nanograms = 58,000 picograms

58 nanograms = 0.058 microgram

See Exercises 6.3, problems 75 and 76.

6.3 Exercises

Verbal and Writing Skills

Write the metric unit that represents each of the following.

1. One thousand liters

2. One thousandth of a liter

3. One thousandth of a gram

4. One thousand grams

5. 1,000,000 grams

6. 1000 kilograms

Perform each conversion.

7. 64 kL = _____ L

8. 37 kL = _____ L

9. 4.7 L = _____ mL

10. 8.6 L = _____ mL

11. 18.9 mL = _____ L

12. 31.5 mL = _____ L

13. 752 L = _____ kL

14. 493 L = _____ kL

15. 2.43 kL = __________ mL

16. 1.76 kL = __________ mL

17. 82 mL = _____ cm^3

18. 152 mL = _____ cm^3

19. 5261 mL = __________ kL

20. 28,156 mL = __________ kL

21. 74 L = _____ cm^3

22. 122 L = __________ cm^3

23. 162 g = _____ kg

24. 294 g = _____ kg

25. 35 mg = _____ g

26. 13 mg = _____ g

27. 6328 mg = _____ g

28. 986 mg = _____ g

29. 2.92 kg = _____ g

30. 14.6 kg = _____ g

31. 17 t = _____ kg

32. 36 t = _____ kg

33. 0.32 g = __________ kg

34. 0.78 g = __________ kg

35. 7896 g = __________ t

36. 12,315 g = __________ t

37. 5.9 kg = __________ mg

38. 0.83 kg = __________ mg

Fill in the blanks with the correct values.

39. 7 mL = _____ L = __________ kL

40. 18 mL = _____ L = __________ kL

41. 128 cm^3 = _____ L = __________ kL

42. 199.8 cm^3 = _____ L = __________ kL

43. 522 mg = _____ g = __________ kg

44. 49 mg = _____ g = __________ kg

45. 3607 g = _____ kg = __________ t

46. 7183 g = _____ kg = __________ t

47. Alice bought a jar of apple juice at the store. Choose the most reasonable measurement for its contents.
(a) 0.32 kL
(b) 0.32 L
(c) 0.32 mL

48. A nurse gave an injection of insulin to a diabetic patient. Choose the most reasonable measurement for the dose.
(a) 4 kL
(b) 4 L
(c) 4 mL

49. A shipowner purchased a new oil tanker. Choose the most reasonable measurement for its weight.
(a) 4000 t
(b) 4000 kg
(c) 4000 g

50. Robert bought a new psychology textbook. Choose the most reasonable measurement for its weight.
(a) 0.49 t
(b) 0.49 kg
(c) 0.49 g

Find the convenient unit of measure and add.

51. 83 L + 822 mL + 30.1 L

52. 152 L + 473 mL + 77.3 L

53. 5 t + 3.82 t + 983 kg

54. 7.8 t + 669 kg + 5.23 t

55. 24 mg + 136 mg + 0.26 kg

56. 78 mg + 221 mg + 0.14 kg

Mixed Practice *Write* true *or* false *for each statement.*

57. 1 milliliter = 0.001 liter

58. 1 kiloliter = 1000 liters

59. Milk can be purchased in kiloliter jugs at the food store.

60. Small amounts of medicine are often measured in liters.

61. 1 kilogram = 1000 grams

62. 1 gram = 1000 milligrams

63. 1 metric ton = 100 grams

64. 0.1 gram = 10 milligrams

65. A nickel coin weighs about 5 grams.

66. A convenient size for a family purchase is 2 kilograms of ground beef.

Applications *Convert the units.*

67. A recently developed medication to prevent rejection in heart transplant patients COELS, $358,000 per liter. How much would one milliliter cost?

68. Australia is using a form of natural gas fuel for newly developed cars. If a rancher uses 3.50 kiloliters per year for natural gas and the price is $0.39 per liter, how much will the rancher spend per year?

69. A physician in Mexico is striving to perfect a new anticancer drug. At the moment, he is using a chemical solution that costs $95.50 per gram. How much would it cost to buy 6 kilograms of the solution?

70. A pharmaceutical firm developed a new vaccine that costs $6.00 per milliliter to produce. How much will it cost the firm to produce 1 liter of the vaccine?

71. A very rare essence of an almost extinct flower found in the Amazon Jungle of South America is extracted by a biogenetic company trying to copy and synthesize it. The company estimates that if the procedure is successful, the product will cost the company $850 per milliliter to produce. How much will it cost the company to produce 0.4 liter of the engineered essence?

72. In a recent year the price of gold was $9.40 per gram. At that price how much would a kilogram of gold cost?

73. A government rocket is carrying cargo into space for a private company. The charge for the freight shipment is $22,450 per kilogram of the special cargo, and the cargo weighs 0.45 metric ton. How much will the private company have to pay.

74. The individual coal-cutting record is 45.4 metric tons per person in one shift, (6 hours), by five Soviet miners. If the coal is sold for $2.00 per kilogram, what is the value of the coal?

To Think About *Complete the following.*

★ **75.** 5632 picograms = ________ micrograms

★ **76.** 0.076182 milligram = ________ nanograms

Cumulative Review Questions

77. 14 out of 70 is what percent?

78. What is 23% of 250?

79. What is 1.7% of $18,900?

80. A salesperson earns a commission of 8%. She sold furniture worth $8960. How much commission will she earn?

6.4 Conversion of Units (Optional)

MathPro Video 13 SSM

After studying this section, you will be able to:

1. *Convert units of length, volume, or weight between the metric and American systems.*
2. *Convert temperature readings between Fahrenheit and Celsius degrees.*

1 *Converting Units of Length, Volume, or Weight Between the Metric and American Systems*

So far we've seen how to convert units when working *within* either the American or the metric system. Many people, however, work in *both* the metric and the American systems. If you study such fields as chemistry, electromechanical technology, business, X-ray technology, nursing, or computers, you will probably need to convert measurements between the two systems. We learn that skill in this section.

To convert between American units and metric units, it is helpful to have equivalent values. The most commonly used equivalents are listed below. Most of these equivalents are approximate.

EQUIVALENT MEASURES

	American to Metric	Metric to American
Units of length	1 mile = 1.61 kilometers	1 kilometer = 0.62 mile
	1 yard = 0.914 meter	1 meter = 3.28 feet
	1 foot = 0.305 meter	1 meter = 1.09 yards
	1 inch = 2.54 centimeters	1 centimeter = 0.394 inch
Units of volume	1 gallon = 3.79 liter	1 liter = 0.264 gallon
	1 quart = 0.946 liter	1 liter = 1.06 quarts
Units of weight	1 pound = 0.454 kilogram	1 kilogram = 2.2 pounds
	1 ounce = 28.35 gram	1 gram = 0.0353 ounce

Remember that to convert from one unit to the other you multiply by a fraction that is equivalent to 1. Create a fraction from the equivalent measures table so that the unit in the denominator cancels the unit you are changing.

To change 5 miles to kilometers, we look in the table and find that 1 mile = 1.61 kilometers. We will use the fraction

$$\frac{1.61 \text{ kilometers}}{1 \text{ mile}}$$

We want to have 1 mile in the denominator.

$$5 \cancel{\text{miles}} \times \frac{1.61 \text{ kilometers}}{1 \cancel{\text{mile}}} = 5 \times 1.61 \text{ kilometers} = 8.05 \text{ kilometers}$$

Thus 5 miles = 8.05 kilometers.

EXAMPLE 1

(a) Convert 3 feet to meters. **(b)** Convert 6 inches to centimeters.

(a) $3 \cancel{\text{feet}} \times \dfrac{0.305 \text{ meter}}{1 \cancel{\text{foot}}} = 0.915 \text{ meter}$

(b) $6 \cancel{\text{inches}} \times \dfrac{2.54 \text{ centimeters}}{1 \cancel{\text{inch}}} = 15.24 \text{ centimeters}$

Practice Problem 1

(a) Convert 7 feet to meters. **(b)** Convert 4 inches to centimeters. ■

Unit abbreviations are quite common, so we will use them for the remainder of this section. We list them here for your reference.

American Measure (Alphabetical Order)	Standard Abbreviation
feet	ft
gallon	gal
inch	in.
mile	mi
ounce	oz
pound	lb
quart	qt
yard	yd

Metric Measure	Standard Abbreviation
centimeter	cm
gram	g
kilogram	kg
kilometer	km
liter	L
meter	m
millimeter	mm

EXAMPLE 2

(a) Convert 26 m to yd. **(b)** Convert 1.9 km to mi. **(c)** Convert 14 gal to L.
(d) Convert 2.5 L to qt. **(e)** Convert 5.6 lb to kg. **(f)** Convert 152 g to oz.

(a) $26 \cancel{\text{m}} \times \dfrac{1.09 \text{ yd}}{1 \cancel{\text{m}}} = 28.34 \text{ yd}$ **(b)** $1.9 \cancel{\text{km}} \times \dfrac{0.62 \text{ mi}}{1 \cancel{\text{km}}} = 1.178 \text{ mi}$

(c) $14 \cancel{\text{gal}} \times \dfrac{3.79 \text{ L}}{1 \cancel{\text{gal}}} = 53.06 \text{ L}$ **(d)** $2.5 \cancel{\text{L}} \times \dfrac{1.06 \text{ qt}}{1 \cancel{\text{L}}} = 2.65 \text{ qt}$

(e) $5.6 \cancel{\text{lb}} \times \dfrac{0.454 \text{ kg}}{1 \cancel{\text{lb}}} = 2.5424 \text{ kg}$ **(f)** $152 \cancel{\text{g}} \times \dfrac{0.0353 \text{ oz}}{1 \cancel{\text{g}}} = 5.3656 \text{ oz}$

Practice Problem 2

(a) Convert 17 m to yd. **(b)** Convert 29.6 km to mi.
(c) Convert 26 gal to L. **(d)** Convert 6.2 L to qt.
(e) Convert 16 lb to kg. **(f)** Convert 280 g to oz. ■

Some conversions require more than one step.

EXAMPLE 3 Convert 235 cm to ft. Round your answer to the nearest hundredth of a foot.

Our first fraction converts centimeters to inches. Our second fraction converts inches to feet.

$$235 \text{ cm} \times \frac{0.394 \text{ in.}}{1 \text{ cm}} \times \frac{1 \text{ ft}}{12 \text{ in.}} = \frac{92.59}{12} \text{ ft}$$

$$= 7.72 \text{ ft (rounded to the nearest hundredth)}$$

Practice Problem 3 Convert 180 cm to ft. ■

The same rules can be followed for a rate such as 50 miles per hour.

EXAMPLE 4 Convert 100 km/hr to mi/hr.

$$\frac{100 \text{ km}}{\text{hr}} \times \frac{0.62 \text{ mi}}{1 \text{ km}} = 62 \text{ mi/hr}$$

Thus 100 km/hr is approximately equal to 62 mi/hr.

Practice Problem 4 Convert 88 km/hr to mi/hr. (Round to the nearest hundredth.) ■

EXAMPLE 5 A camera film that is 35 mm wide is how many inches wide?

We first convert from millimeters to centimeters by moving the decimal point in the number 35 one place to the left.

$$35 \text{ mm} = 3.5 \text{ cm}$$

Then we convert to inches using a unit fraction.

$$3.5 \text{ cm} \times \frac{0.394 \text{ in.}}{1 \text{ cm}} = 1.379 \text{ in.}$$

Practice Problem 5 The city police use a 9-mm automatic pistol. If the pistol fires a bullet 9 mm wide, how many inches wide is it? (Round to the nearest hundredth.) ■

SIDELIGHT Suppose we consider a rectangle that measures 2 yards wide by 4 yards long. The area would be 2 yards × 4 yards = 8 square yards. How could you change 8 square yards to square meters? Suppose that we look at 1 square yard. Each side is 1 yard long, which is equivalent to 0.9144 meter.

$$\text{Area} = 1 \text{ yard} \times 1 \text{ yard} = 0.9144 \text{ meter} \times 0.9144 \text{ meter}$$

$$\text{Area} = 1 \text{ square yard} = 0.8361 \text{ square meter}$$

Thus $1 \text{ yd}^2 = 0.8361 \text{ m}^2$. Therefore

$$8 \text{ yd}^2 \times \frac{0.8361 \text{ m}^2}{1 \text{ yd}^2} = 6.6888 \text{ m}^2$$

8 square yards = 6.6888 square meters.

2 *Converting Temperature Readings Between Fahrenheit and Celsius Degrees*

In the metric system temperature is measured on the Celsius scale. Water boils at 100° (100°C) and freezes at 0° (0°C) on the Celsius scale. In the Fahrenheit system water boils at 212° (212°F) and freezes at 32° (32°F).

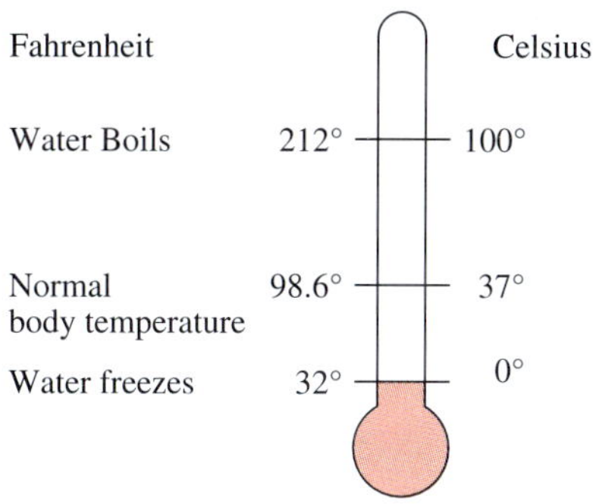

To convert Celsius to Fahrenheit, we can use the formula

$$F = 1.8 \times C + 32$$

EXAMPLE 6 When the temperature is 35°C, what is the Fahrenheit reading?

$$\begin{aligned} F &= 1.8 \times C + 32 \\ &= 1.8 \times 35 + 32 \\ &= 63 + 32 \\ &= 95 \end{aligned}$$

The temperature is 95°F.

Practice Problem 6 Convert 20°C to Fahrenheit temperature. ■

To convert Fahrenheit temperature to Celsius, we can use the formula

$$C = \frac{5 \times F - 160}{9}$$

where F is the number of Fahrenheit degrees and C is the number of Celsius degrees.

EXAMPLE 7 When the temperature is 176°F, what is the Celsius reading?

$$\begin{aligned} C &= \frac{5 \times F - 160}{9} \\ &= \frac{5 \times 176 - 160}{9} \\ &= \frac{880 - 160}{9} = \frac{720}{9} = 80 \end{aligned}$$

The temperature is 80°C.

Practice Problem 7 Convert 181°F to Celsius temperature. ■

CALCULATOR

Converting Temperature

You can use your calculator to convert temperature readings between Fahrenheit and Celsius. To convert 30°C to Fahrenheit temperature enter:

1.8 [×] 30 [+] 32 [=]

Display:

[86]

The temperature is 86°F.

To convert 82.4°F to Celsius temperature enter:

5 [×] 82.4 [−] 160 [=] [÷] 9 [=]

Display:

[28]

The temperature is 28°C.

6.4 Exercises

Perform each conversion. Round the answer to the nearest hundredth when necessary.

1. 7 ft to m

2. 11 ft to m

3. 9 in. to cm

4. 13 in. to cm

5. 14 m to yd

6. 18 m to yd

7. 26.5 m to yd

8. 29.3 m to yd

9. 15 km to mi

10. 12 km to mi

11. 24 yd to m

12. 31 yd to m

13. 82 mi to km

14. 68 mi to km

15. 25 m to ft

16. 35 m to ft

17. 17.5 cm to in.

18. 19.6 cm to in.

19. 200 m to yd

20. 350 m to yd

21. 5 gal to L

22. 5 gal to L

23. 280 gal to L

24. 140 gal to L

25. 23 qt to L

26. 28 qt to L

27. 19 L to gal

28. 15 L to gal

29. 4.5 L to qt

30. 6.5 L to qt

31. 32 lb to kg

32. 27 lb to kg

33. 7 oz to g

34. 9 oz to g

35. 16 kg to lb

36. 14 kg to lb

37. 126 g to oz

38. 186 g to oz

39. 166 cm to ft

40. 142 cm to ft

41. 16.5 ft to cm

42. 19.5 ft to cm

43. 50 km/hr to mi/hr

44. 60 km/hr to mi/hr

45. 60 mi/hr to km/hr

46. 40 mi/hr to km/hr

47. A wire that is 13 mm wide is how many inches wide?

48. A bolt that is 7-mm wide is how many inches wide?

49. 40°C to Fahrenheit

50. 60°C to Fahrenheit

51. 85°C to Fahrenheit

52. 105°C to Fahrenheit

53. 12°C to Fahrenheit

54. 21°C to Fahrenheit

55. 68°F to Celsius

56. 131°F to Celsius

57. 168°F to Celsius

58. 112°F to Celsius

59. 86°F to Celsius

60. 98°F to Celsius

Applications

Solve. Round answers to the nearest hundredth when necessary.

61. Mr. and Mrs. Weston have traveled 67 miles on a boat cruise from Seattle, Washington, to Victoria Island, Vancouver, B.C., Canada. They have 36 kilometers until their rendezvous point with another boat. How many kilometers in total will they have traveled?

62. Marcia is traveling from Ixtapa to Zihuatenejo in Mexico. She drives from the center of town to the cliff that makes the descent down to the beaches. The odometer on her American car shows that the first part of her short trip has taken 4 miles. The sign carved into a wooden post says Zihuatenejo 14 KILOMETERS. How many kilometers in total will she have traveled when she arrives at the beach?

63. Pierre had a Jeep imported into France. During a trip from Paris to Lyon, he used 38 liters of gas. The tank, which he filled before starting the trip, holds 15 gallons of gas. How many liters of gas were left in the tank when he arrived, if he had a full tank to start with?

64. A surgeon is irrigating an abdominal cavity after a cancerous growth is removed. There is a supply of 3 gallons of distilled water in the operating room. The surgeon uses a total of 7 liters of the water during the procedure. How many liters of water are left over after the operation?

65. One of the heaviest males documented in medical records weighed 635 kg in 1978. What would have been his weight in pounds?

66. The average weight for a 7-year-old girl is 22.2 kilograms. What is the average weight in pounds?

67. In the Australian summer, Ayers Rock in Central Australia is visited by tourists who climb beginning at 4 o'clock in the morning, when the temperature is 19° Celsius, because after 7 o'clock in the morning, the temperature can reach 45°C and people could die of dehydration while climbing. What would be equivalent Fahrenheit temperatures?

68. The holiday turkey in Buenos Aires, Argentina, was roasted at 200° Celsius for 4 hours, (20 minutes per pound). What would have been the temperature in Fahrenheit in Joplin, Missouri?

69. Metallic tungsten, or wolfram (W), melts at 6188° Fahrenheit. At what Celsius temperature would metallic tungsten melt?

70. On top of Old Smoky, water boils at 205°F. What Celsius temperature is this?

Round your answer to the nearest thousandth.

71. A pathologist found 0.768 ounce of lead in the liver of a child who died of lead poisoning. How many grams of toxic lead were in the child's liver?

72. A prospector in the Yukon, in Alaska, found a gold nugget that weighed 2.552 ounces while panning in a river. How many grams were in the nugget?

Round your answer to four decimal places.

73. 28 square inches = ? square centimeters

74. 36 square meters = ? square yards

75. Change 48 feet per second to miles per hour.

76. Change 55 miles per hour to feet per second.

Cumulative Review Problems

Do the operations in the correct order.

77. $2^3 \times 6 - 4 + 3$

78. $5 + 2 - 3 + 5 \times 3^2$

79. $2^2 + 3^2 + 4^3 + 2 \times 7$

80. $5^2 + 4^2 + 3^2 + 3 \times 8$

Putting Your Skills to Work

THE MATHEMATICS OF COMPUTER CAPACITY

As we mentioned on page 349 a **byte** is the amount of computer memory needed to store one alphanumeric character. One kilobyte (KB or K) is usually defined as one thousand bytes. One megabyte (MB) is usually defined as one million bytes while one gigabyte (GB) is one billion bytes.

PROBLEMS FOR INDIVIDUAL INVESTIGATION

1. Sharon purchased a new computer with a 1.2-GB hard drive. She installed a version of Windows that uses 16.5 MB of hard drive space. She has stored several personal programs on her computer that use 867 K of hard drive space. What is the maximum amount of space still available on her hard drive?

Speed on a computer is often measured in hertz. One **hertz** is a unit of frequency of one cycle per second. One megahertz (MHz) is one million cycles per second. One gigahertz (GHz) is one billion cycles per second. Although different types of tasks on a computer are measured in different ways, it is generally true that the ''speed'' at which the computer can do a task is measured by megahertz. Theoretically, a 100-MHz computer is twice as fast as a 50-MHz computer.

PROBLEMS FOR GROUP INVESTIGATION AND COOPERATIVE LEARNING

Together with other members of your class, determine the following. Round to the nearest thousandth.

2. On a 40-MHz computer, it took 3.2 seconds for Mary Lou to have her 20-page paper scanned for spelling errors. Approximately how long will it take her using the same software on a 150-MHz computer to do the same task?

3. Engineers and scientists are now researching the possibility of a 1-GHz personal computer in the near future. Theoretically, how long would it take such a computer to scan Mary Lou's paper for spelling errors?

INTERNET CONNECTIONS: Go to ''http://www.prenhall.com/tobey'' to be connected

Site: Windows Station Configurator (Vektron International) or a related site

This site shows some of the options currently available from a computer manufacturer. Answer the following questions.

4. If you choose the largest available hard drive, how much more disk space will you have than you would with the smallest hard drive? Give your answer as a percent.

5. The processor determines the speed of the computer. Suppose it would take 24 seconds for the fastest available computer to complete an operation. How long would the same operation take on the slowest available computer?

6.5 Applied Problems

After studying this section, you will be able to:

1 *Solve applied problems involving metric and American units.*

MathPro Video 14 SSM

1 *Solving Applied Problems Involving Metric and American Units*

Once again, we will be using the blueprint for solving applied problems that we used in Sections 1.8, 2.9, and 3.7. The following ideas will be helpful.

Read the problem carefully. Draw a picture if this helps you. Fill in the Mathematics Blueprint for Problem Solving. Decide what conversion of units (if any) is needed to help you solve the problem. Perform the calculations. State the answer and the unit of measure. Estimate the answer. Compare the exact answer with the estimate. See if your answer is reasonable.

EXAMPLE 1 Find the perimeter of the triangle on the right. Express the answer in *feet.*

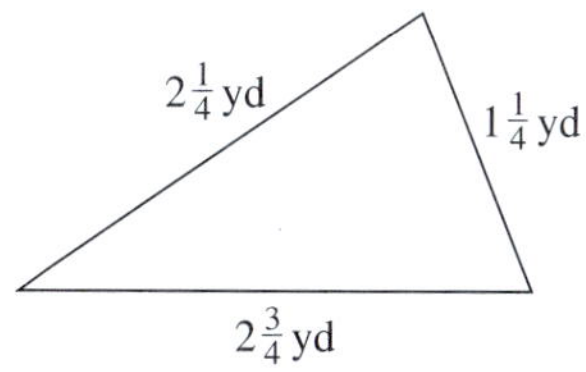

1. *Understand the problem.*
   The perimeter is the sum of the lengths of the sides. We are asked to express the answer in feet. Remember to convert the yards to feet.
2. *Solve and state the answer.*
   Add the three sides.

$$\begin{array}{r} 2\frac{1}{4} \text{ yards} \\ 1\frac{1}{4} \text{ yards} \\ + \; 2\frac{3}{4} \text{ yards} \\ \hline 5\frac{5}{4} \text{ yards} \end{array} = 6\frac{1}{4} \text{ yards}$$

Convert $6\frac{1}{4}$ yards to feet using the fact that 1 yard = 3 feet. To make the calculation easier, we will change $6\frac{1}{4}$ to $\frac{25}{4}$.

$$6\frac{1}{4} \cancel{\text{yards}} \times \frac{3 \text{ feet}}{1 \cancel{\text{yard}}} = \frac{25}{4} \cancel{\text{yards}} \times \frac{3 \text{ feet}}{1 \cancel{\text{yard}}} = \frac{75}{4} \text{ feet} = 18\frac{3}{4} \text{ feet}$$

The perimeter of the triangle is $18\frac{3}{4}$ feet.

3. *Check.*
   We will check by estimating the answer.

$$2\frac{1}{4} \text{ yd} \approx 2 \text{ yd} \qquad 1\frac{1}{4} \text{ yd} \approx 1 \text{ yd} \qquad 2\frac{3}{4} \text{ yd} \approx 3 \text{ yd}$$

Now we add the three sides, using our estimated values.

$$2 + 1 + 3 = 6 \text{ yards} \qquad 6 \text{ yards} = 18 \text{ feet}$$

Our estimated answer, 18 feet, is close to our calculated answer, $18\frac{3}{4}$ feet. Thus our answer seems reasonable. ✓

Practice Problem 1 Find the perimeter of the rectangle on the right. Express the answer in *feet.* ■

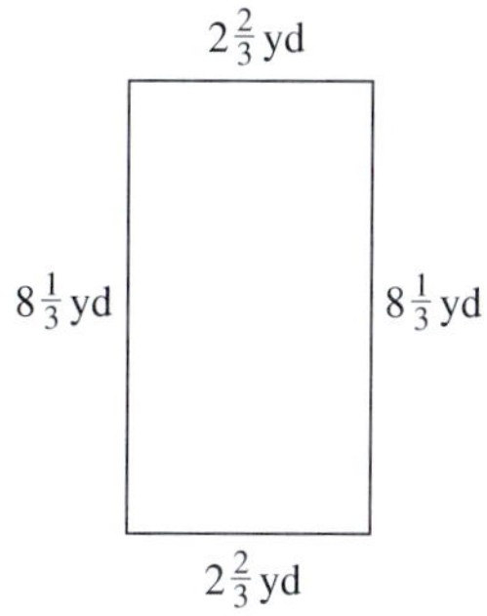

We use the Mathematics Blueprint for Problem Solving when the solution to a problem is not obvious.

EXAMPLE 2 How many 210-*liter* gasoline barrels can be filled from a tank of 5.04 *kiloliters* of gasoline?

1. *Understand the problem.*

MATHEMATICS BLUEPRINT FOR PROBLEM SOLVING			
Gather the Facts	**What Am I Asked to Do?**	**How Do I Proceed?**	**Key Points to Remember**
We have 5.04 kiloliters of gasoline. We are going to divide the gasoline into smaller barrels that hold 210 liters each.	Find out how many of these smaller 210-liter barrels can be filled.	We need to get all measurements in the same units. We choose to convert 5.04 kiloliters to liters. Then we divide that result by 210 to find out how many barrels will be filled.	To convert 5.04 kiloliters to liters, we move the decimal point 3 places to the right.

2. *Solve and state the answer.*
First we convert 5.04 kiloliters to liters.

$$5.04 \text{ kiloliters} = 5040 \text{ liters}$$

Now we find out how many barrels will be filled. How many 210-liter barrels will 5040 liters fill? Visualize fitting 210-liter barrels into a big barrel that holds 5040 liters.

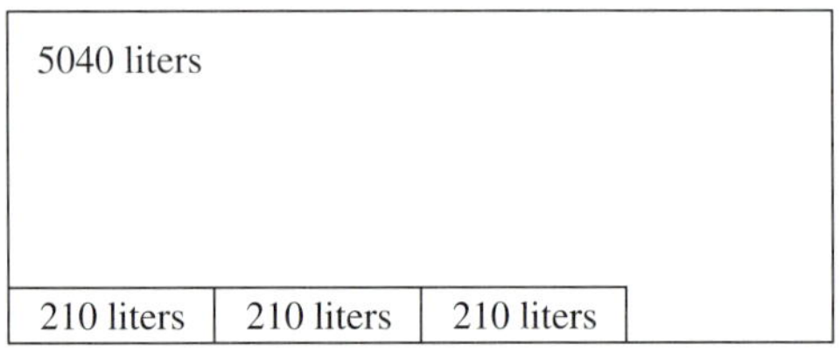

We need to divide

$$\frac{5040 \text{ liters}}{210 \text{ liters}} = 24$$

Thus we can fill 24 of the 210-liter barrels.

3. *Check.*
Estimate each value. 5.04 kiloliters is approximately 5 kiloliters or 5000 liters. 210-liter barrels hold approximately 200 liters. How many times does 200 fit into 5000?

$$\frac{5000}{200} = 25$$

We estimate that 25 barrels can be filled. This is very close to our calculated value of 24 barrels. Thus our answer is reasonable. ✓

Practice Problem 2 A lab assistant must use 18.06 liters of solution to fill 42 jars. How many milliliters of the solution will go into each jar? ■

6.5 Exercises

Applications

Solve each.

1. Find the perimeter of the triangle. Express your answer in *yards.*

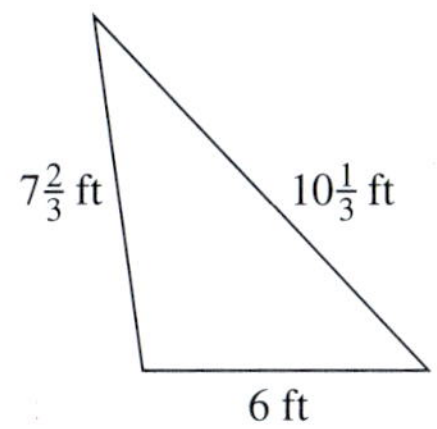

2. Find the perimeter of the triangle. Express your answer in *yards.*

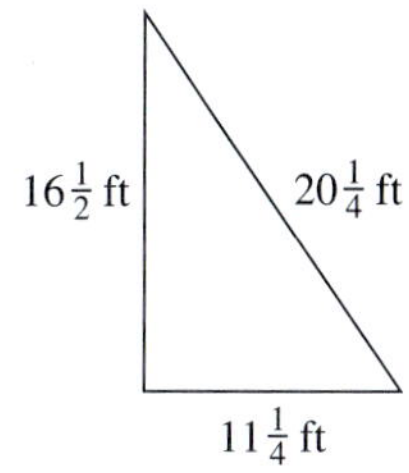

3. The triangular giraffe enclosure at the zoo is being refenced. One side is 86 yards long and a second side is 77 yards long. If the zoo has 522 *feet* of fencing, how much will be left over for the third side? Express your answer in *yards.*

4. A farmer has 600 feet of fencing to fence in a triangular region. One side is to be 75 yards and a second 65 yards. How much fencing will the farmer have left for the third side? Express your answer in *yards.*

5. A rectangular picture window measures 87 centimeters × 152 centimeters. Window insulation is applied along all four sides. The insulation costs $7.00 per meter. What will it cost to insulate the window?

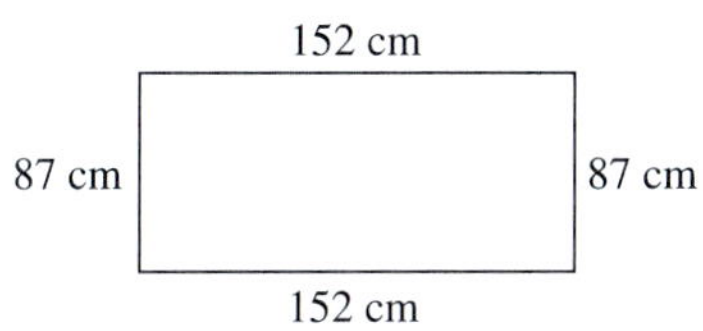

6. A rectangular doorway measures 90 centimeters × 200 centimeters. Weatherstripping is applied on the top and the two sides. The weatherstripping costs $6.00 per meter. What does it cost to weatherstrip the door?

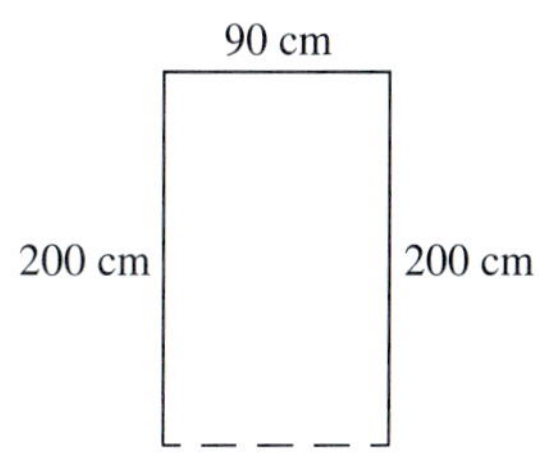

7. A beverage company is filling new bottles of strawberry-kiwi fruit juice to be sent to a test market of college students. The packaging division has to put 62 liters into 80 bottles. How many milliliters of juice will be in each bottle?

8. A lab assistant must use 28 *liters* of solvent to fill 40 jars. How many *milliliters* of solvent will be in each jar?

9. A tack supply company, which makes saddles, fittings, bridles, and other equipment for horses and riders, has a length of braided leather 12.4 meters long. The leather must be cut into 4 equal pieces. How many *centimeters* long will each piece be?

10. A stretch of road 1.863 kilometers long has 230 parking spaces of equal length painted in white on the pavement. How many meters long is each parking space?

11. A hallway outside the kindergarten classrooms in Aalborg, Denmark, has one wall that is 67 meters long, one fifth of the wall is dedicated to paintings made by the children. How many centimeters long is the art portion of the wall?

12. An antenna 186 *meters* tall is on the roof of a building. The top $\frac{1}{8}$ of the antenna has a special rubberized insulation. How many *centimeters* long is the portion of the antenna with this insulation?

13. How many 24-*milliliter* samples can be obtained from 3 *liters* of a chemical solution?

14. A gourmet balsamic vinegar factory is filling 200-liter oak casks from 8 kiloliters of balsamic vinegar. How many oak casks can be filled?

15. A bottle of superglue states on its label that it should not be used at temperatures above 85°F. The sign on the display in the store reads NOT TO BE USED AT TEMPERATURES ABOVE 27°C. What is the discrepancy, in degrees Fahrenheit between the two suggested temperatures?

16. The temperature in Dakar, Senegal, today is 33°C. The temperature on this date last year was 94°F. What is the difference, in degrees Fahrenheit, between the temperature in Dakar today and one year ago?

17. A Swedish flight attendant is heating passenger meals for the new American Airline he is working for. He is used to heating meals at 180°C. His co-worker tells him that the food must be heated at 350°F. What is the difference in temperature in degrees Fahrenheit between the two temperatures? Which temperature is hotter?

18. A French cook wishes to bake potatoes in an oven at 195°C. The American assistant set the oven at 400°F. By how many degrees Fahrenheit was the oven temperature off? Was the oven too hot or too cold?

19. Diana and Louisa are traveling in Switzerland at 95 kilometers/hour for 6.2 hours. The total trip will be 560 miles long. How many miles do they still need to travel? Round your answer to the nearest mile.

20. Alex is traveling from Niagara Falls, New York, to Toronto, Canada. He has been driving at 100 kilometers/hour for 4.8 hours. His total trip will be 415 miles. How many miles does Alex have left to travel? Round your answer to the nearest mile.

21. Sharon and James Hanson traveled from Arizona to Acapulco, Mexico. The last day of their trip they traveled 520 miles and it took them 8 hours. The maximum speed limit is 110 kilometers/hour.

(a) How many kilometers per hour did they average on the last day of the trip?

(b) Did they break the speed limit?

22. A small corporate jet travels at 600 kilometers/hour for 1.5 hours. The pilot said the plane will be on time if it travels at 350 miles per hour.

(a) What was the jet's speed in miles per hour?

(b) Will it arrive on time?

23. A very popular snack bar by the community pool sells quite a bit of soda from the poolside soda fountain. On an average day, 8 pints per minute of various soft drinks are sold. How many gallons per hour of soda are dispensed?

24. A swimming pool is filled at the rate of 6 pints per minute. How many gallons per hour is this?

25. 3.4 tons of bananas are off-loaded from a ship that has just arrived from Costa Rica. The port taxes imported fruit at $0.015 per pound. What is the tax on the entire shipment?

26. Trucks in Sam's home state are taxed each year at $0.03 per pound. Sam's empty truck weighs 1.8 tons. What is his annual tax?

27. A box of raisin bran contains 14 ounces of cereal. Of this, 11 ounces are bran flakes and the rest is raisins. How many *grams* of raisins are in the box?

28. A can of peaches contains 16 ounces. Fred discovered that 5 ounces were syrup and the rest was fruit. How many *grams* of fruit are in the can?

★ **29.** Maria bought 18 quarts of motor oil for $1.39 per quart. Her uncle from Mexico is visiting this summer. He said his car will use 12 liters of oil while driving this summer. **(a)** How many extra quarts of oil did Maria buy? **(b)** How much did this extra oil cost her?

★ **30.** Carlos is mixing a fruit spray to retard the damage caused by beetles. He has 16 quarts of spray concentrate available. The old recipe he used in Mexico called for 11 liters of spray concentrate. **(a)** How much extra spray concentrate in quarts does he have? **(b)** If the spray costs $2.89 per quart, how much will it cost him to prepare the old recipe?

31. Jackson's small motorcycle gets 56 kilometers per liter. If he drives 392 kilometers to Quebec City, Canada, and gas is $0.78 per liter,

(a) How much would the gas used for the trip cost?

(b) What would be the number of *miles per gallon* Jackson gets on his motorcycle?

32. Ryan's Volvo station wagon runs on diesel fuel. He gets 20 miles per gallon on the highway. If he drives a distance of 275 kilometers to Mexico City, and diesel fuel costs $1.30 per gallon, how much does the fuel used for the trip cost him? Round your answer to the nearest cent. How many kilometers per liter does he get?

Cumulative Review Problems

Solve for n.

33. $\frac{n}{16} = \frac{2}{50}$

34. $\frac{1.7}{n} = \frac{51}{6}$

35. Justin is going to India. While reading his atlas he sees that Bombay is 6 inches from where he would like to travel. The scale shows that 3 inches represent 7.75 miles on the ground. If he travels by train on a straight track, how far will the train go to take him to his destination?

36. Thompson has a scale model of a famous fishing schooner. Each 2 centimeters on the model represents an actual length of 7.5 yards. The model is 11 centimeters long. How long is the famous fishing schooner?

Chapter Organizer

Topic	Procedure	Examples
Changing from one American unit to another, p. 338.	**1.** Find the equality statement that relates what you want to find and what you know. **2.** Form a unit fraction. The denominator will contain the units you now have. **3.** Simplify and multiply by the unit fraction.	Convert 210 inches to feet. **1.** Use 12 inches = 1 foot. **2.** Unit fraction = $\dfrac{1 \text{ foot}}{12 \text{ inches}}$. **3.** $210 \cancel{\text{in.}} \times \dfrac{1 \text{ ft}}{12 \cancel{\text{in.}}} = \dfrac{210}{12} \text{ ft}$ $= 17.5 \text{ ft}$ Convert 86 yards to feet. **1.** Use 1 yard = 3 feet. **2.** Unit fraction = $\dfrac{3 \text{ feet}}{1 \text{ yard}}$. **3.** $86 \cancel{\text{yd}} \times \dfrac{3 \text{ ft}}{1 \cancel{\text{yd}}} = 86 \times 3 \text{ ft} = 258 \text{ ft}$
Changing from one metric unit to another, pp. 345, 352, 353.	When you change from one prefix to another moving to the *left,* move the decimal point to the *left* the same number of places. kilo- $= 10^3$ hecto- $= 10^2$ deka- $= 10^1$ one unit $= 1$ deci- $= 0.1$ centi- $= 0.01$ milli- $= 0.001$ When you change from one prefix to another moving to the *right,* move the decimal point to the *right* the same number of places.	Change 7.2 meters to kilometers. **1.** Move three decimal places to the left. $0_{\wedge}007.2$ **2.** 7.2 m = 0.0072 km. Change 196 centimeters to meters. **1.** Move two places to the left. $1_{\wedge}96.$ **2.** 196 cm = 1.96 m. Change 17.3 liters to milliliters. **1.** Move three decimal places to the right. $17.300_{\wedge}$ **2.** 17.3 L = 17,300 mL.
Changing from American units to metric units, p. 358.	**1.** From the list of equivalent measures, pick an equality statement that begins with the unit you now have. 1 mi = 1.61 km 1 yd = 0.914 m 1 ft = 0.305 m 1 in. = 2.54 cm 1 gal = 3.79 L 1 qt = 0.946 L 1 lb = 0.454 kg 1 oz = 28.35 g **2.** Multiply by a unit fraction.	Convert 7 gallons to liters. **1.** 1 gal = 3.79 L. **2.** $7 \cancel{\text{gal}} \times \dfrac{3.79 \text{ L}}{1 \cancel{\text{gal}}} = 26.53 \text{ L}$ Convert 18 pounds to kilograms. **1.** 1 lb = 0.454 kg. **2.** $18 \cancel{\text{lb}} \times \dfrac{0.454 \text{ kg}}{1 \cancel{\text{lb}}} = 8.172 \text{ kg}$

Chapter Organizer		
Topic	**Procedure**	**Examples**
Changing from metric units to American units, p. 358.	1. From the list of equivalent measures, pick an equality statement that begins with the unit you now have and ends with the unit you want. 1 km = 0.62 mi 1 m = 3.28 ft 1 m = 1.09 yd 1 cm = 0.394 in. 1 L = 0.264 gal 1 L = 1.06 qt 1 kg = 2.2 lb 1 g = 0.0353 oz 2. Multiply by a unit fraction.	Convert 605 grams to ounces. 1. 1 g = 0.0353 oz. 2. $605\,\cancel{g} \times \dfrac{0.0353 \text{ oz}}{1\,\cancel{g}} = 21.3565 \text{ oz}$ Convert 80 km/hr to mi/hr. 1. 1 km = 0.62 mi. 2. $80 \dfrac{\cancel{km}}{hr} \times \dfrac{0.62 \text{ mi}}{1\,\cancel{km}} = 49.6 \text{ mi/hr}$
Changing from Celsius to Fahrenheit temperature, p. 361.	1. To convert Celsius to Fahrenheit, we use the formula $F = 1.8 \times C + 32$ 2. We replace C by the Celsius temperature. 3. We calculate to find the Fahrenheit temperature.	Convert 65°C to Fahrenheit. 1. $F = 1.8 \times C + 32$ 2. $F = 1.8 \times 65 + 32$ 3. $F = 117 + 32 = 149$ The temperature is 149°F.
Changing from Fahrenheit to Celsius temperature, p. 361.	1. To convert Fahrenheit to Celsius, we use the formula $C = \dfrac{5 \times F - 160}{9}$ 2. We replace F by the Fahrenheit temperature. 3. We calculate to find the Celsius temperature.	Convert 50°F to Celsius. 1. $C = \dfrac{5 \times F - 160}{9}$ 2. $C = \dfrac{5 \times 50 - 160}{9}$ 3. $C = \dfrac{250 - 160}{9} = \dfrac{90}{9} = 10$ The temperature is 10°C.

Chapter 6 Review Problems

6.1 *Convert. When necessary, express your answer as a decimal. Round to the nearest hundredth.*

1. 27 ft = ____ yd
2. 33 ft = ____ yd
3. 3 mi = ____ yd
4. 4 mi = ____ yd
5. 90 in. = ____ ft
6. 78 in. = ____ ft
7. 15,840 ft = ____ mi
8. 10,560 ft = ____ mi
9. 4 tons = ____ lb
10. 7 tons = ____ lb
11. 92 oz = ____ lb
12. 100 oz = ____ lb
13. 15 gal = ____ qt
14. 21 gal = ____ qt
15. 31 pt = ____ qt
16. 27 pt = ____ qt
17. 2160 sec = ____ min
18. 1380 sec = ____ min
19. 14 hr = ____ min
20. 11 hr = ____ min

6.2 *Convert. Do not round your answer.*

21. 56 cm = ____ mm
22. 29 cm = ____ mm
23. 1763 mm = ____ cm
24. 2598 mm = ____ cm
25. 9.2 m = ____ cm
26. 7.4 m = ____ cm
27. 5 km = ____ m
28. 7 km = ____ m
29. 285 m = ____ km
30. 473 m = ____ km

Change all units to meters and add.

31. 6.2 m + 121 cm + 0.52 m
32. 9.8 m + 673 cm + 0.48 m
33. 0.024 km + 1.8 m + 983 cm
34. 0.078 km + 5.5 m + 609 cm

6.3 *Convert. Do not round your answer.*

35. 17 kL = ____ L
36. 23 kL = ____ L
37. 59 mL = ____ L
38. 77 mL = ____ L
39. 196 kg = ________ g
40. 721 kg = ________ g
41. 778 mg = ____ g
42. 459 mg = ____ g
43. 125 kg = ____ t
44. 705 kg = ____ t
45. 76 kg = ____ g
46. 41 kg = ____ g
47. 765 cm^3 = ____ mL
48. 423 cm^3 = ____ mL
49. 2.43 L = ____ cm^3
50. 1.93 L = ____ cm^3

6.4 *Perform each conversion. Round your answer to the nearest hundredth.*

51. 42 kg = ____ lb
52. 33 kg = ____ lb
53. 15 ft = ____ m
54. 9 ft = ____ m
55. 13 oz = ____ g
56. 17 oz = ____ g
57. 1.8 ft = ____ cm
58. 1.3 ft = ____ cm
59. 14 cm = ____ in.
60. 18 cm = ____ in.
61. 20 lb = ____ kg
62. 30 lb = ____ kg
63. 12 yd = ____ m
64. 14 yd = ____ m
65. 80 km/hr = ____ mi/hr
66. 70 km/hr = ____ mi/hr
67. 15°C = ____ °F
68. 25°C = ____ °F
69. 221°F = ____ °C
70. 185°F = ____ °C
71. 32°F = ____ °C

6.5 *Solve each problem. Round your answer to the nearest hundredth when necessary.*

72. A roof is layered with 1.76-cm plywood, 4.32-mm tar paper, and 0.93-cm shingles. How thick is the roof?

73. Find the perimeter of the triangle.
(a) Express your answer in feet.
(b) Express your answer in inches.

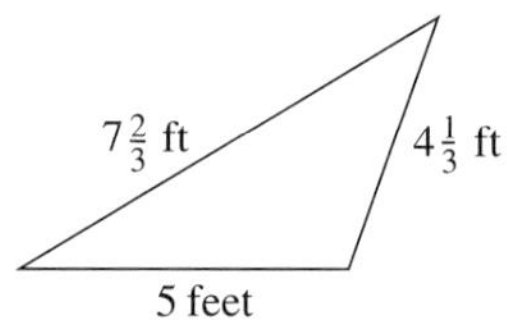

74. Find the perimeter of the rectangle.
(a) Express your answer in meters.
(b) Express your answer in kilometers.

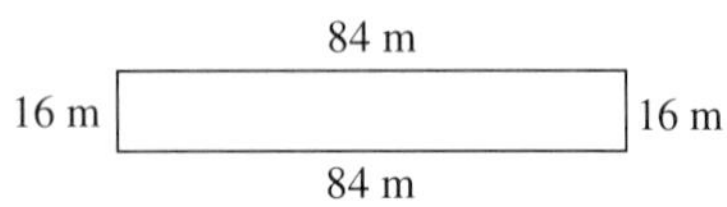

75. The unit price on a box of cereal was $0.14 per ounce. The net weight was 450 grams. How much did the cereal cost?

76. Keshia traveled at 90 km/hr for 3 hours. She needs to travel 200 miles. How much farther does she need to travel?

77. The width of a house in Mexico is 12.6 meters. Kim arrived with enough wooden trim for a house that is 43 feet wide. Did she have enough? How many feet extra or how short was this amount?

78. A chemist has a solution of 4 *liters*. She needs to place it into 24 equal-sized smaller jars. How many *milliliters* will be contained in each jar?

79. A shipment of parts for Europe weighs 2 tons. The cost for overseas shipping is $0.32 per kilogram. How much will it cost to send the shipment overseas?

80. A flagpole is 19 *meters* long. The bottom $\frac{1}{5}$ of it is coated with a special water seal before being placed in the ground. How many *centimeters* long is the portion that has the water seal?

81. A German cook wishes to bake a cake at 185° Celsius. The oven is set at 390° Fahrenheit. By how many degrees Fahrenheit is the oven temperature different from what is desired? Is the oven too hot or not hot enough?

Chapter 6 Test

Convert. Express your answer as a decimal rounded to the nearest hundredth when necessary.

1. 1.6 tons = ________ lb

2. 19 ft = ________ in.

3. 21 gal = ________ qt

4. 36,960 ft = ________ mi

5. 3 cups = ________ qt

6. 1800 sec = ________ min

Perform each conversion. Do not round your answer.

7. 27.3 cm = ________ m

8. 9.2 km = ________ m

9. 46 mm = ________ cm

10. 9.88 cm = ________ m

11. 12.7 m = ________ cm

12. 0.936 cm = ________ mm

13. 46 L = ________ kL

14. 127 L = ________ mL

15. 28.9 mg = ________ g

16. 983 g = ________ kg

17. 0.92 L = ________ mL

18. 9.42 g = ________ mg

Perform each conversion. Round to the nearest hundredth when necessary.

19. 42 mi = ________ km

20. 1.78 yd = ________ m

21. 9 cm = ________ in.

22. 38 L = ________ gal

23. 7.3 kg = ________ lb

24. 3 oz = ________ g

1. ________	**13.** ________
2. ________	**14.** ________
3. ________	**15.** ________
4. ________	**16.** ________
5. ________	**17.** ________
6. ________	**18.** ________
7. ________	**19.** ________
8. ________	**20.** ________
9. ________	**21.** ________
10. ________	**22.** ________
11. ________	**23.** ________
12. ________	**24.** ________

25. (a) ______

(b) ______

26. (a) ______

(b) ______

27. ______

28. (a) ______

(b) ______

Solve each problem. Round your answer to the nearest hundredth when necessary.

25. A rectangular picture frame measures 3 m × 7 m.

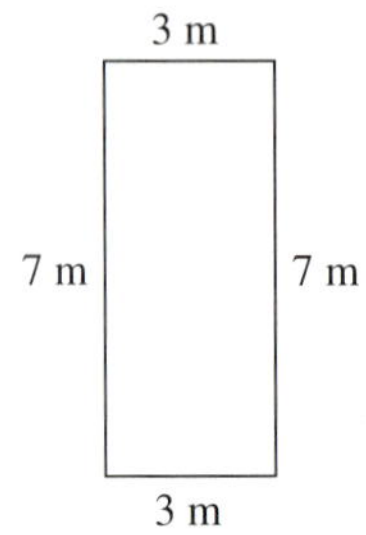

(a) What is the perimeter of the picture frame in meters?

(b) What is the perimeter of the picture frame in yards?

26. The temperature is 80°F today. Kristen's computer contains a warning not to operate above 35°C.

(a) How many degrees Fahrenheit are there between the two temperatures?

(b) Can she use her computer today?

27. A pump is running at 5.5 quarts per minute. How many gallons per hour is this?

28. The speed limit on a Canadian road is 100 km/hr.

(a) How far can Samuel travel at this speed limit in 3 hours?

(b) If Samuel has to travel 200 miles, how much farther will he need to go after 3 hours of driving at 100 km/hr?

Cumulative Test for Chapters 1–6

Approximately one half of this test is based on Chapter 6 material. The remainder is based on material covered in Chapters 1–5.

Do each problem. Simplify your answer.

1. Subtract. $\begin{array}{r} 9824 \\ -\ 3796 \\ \hline \end{array}$

2. Multiply. $\begin{array}{r} 608 \\ \times\ 305 \\ \hline \end{array}$

3. Divide. $28\overline{)1932}$

4. Add. $\frac{1}{7} + \frac{3}{14} + \frac{2}{21}$

5. Subtract. $3\frac{1}{8} - 1\frac{3}{4}$

6. Is this proportion true or false?
$$\frac{21}{35} = \frac{12}{20}$$

7. Solve the proportion. $\frac{0.4}{n} = \frac{2}{30}$

8. A piece of wire 6.5 centimeters long weighs 68 grams. What will a 20-centimeter length of the same wire weigh? (Round to the nearest hundredth.)

9. What percent of 66 is 165?

10. Find 18% of 360.

11. 0.5% of what number is 100?

Convert. Express your answer as a decimal rounded to the nearest hundredth when necessary.

12. 38 qt = ________ gal

13. 2.5 tons = ________ lb

14. 7 pt = ________ qt

15. 25 feet = ________ in.

1. ________

2. ________

3. ________

4. ________

5. ________

6. ________

7. ________

8. ________

9. ________

10. ________

11. ________

12. ________

13. ________

14. ________

15. ________

Perform each conversion. Do not round your answer.

16. 3.7 km = _____ m **17.** 62.8 g = _____ kg **18.** 0.79 L = _____ mL

19. 5 cm = _____ m **20.** 42 lb = _____ oz

Perform each conversion. Round to the nearest hundredth when necessary.

21. 28 gal = _____ L **22.** 96 lb = _____ kg

23. 7.87 m = _____ ft **24.** 9 mi = _____ km

25. Find the perimeter in *meters* of this triangle.

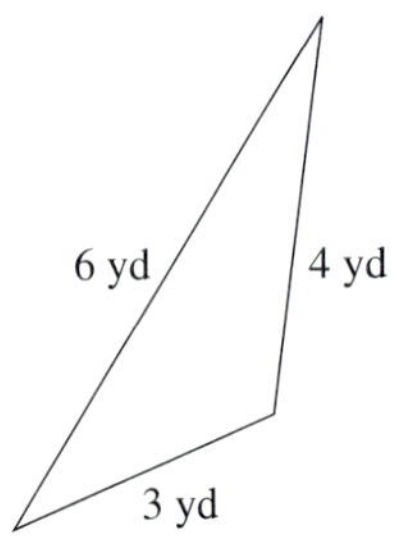

26. Change 15°C to Fahrenheit temperature. Now find the difference between 15°C and 15°F. Which figure represents the higher temperature?

27. Ricardo traveled on a Mexican highway at 100 km/hr for $1\frac{1}{2}$ hours. He needs to travel a total distance of 100 miles. How far does he still need to travel? (Express your answer in miles.)

28. Two metal sheets are 0.72 centimeter thick and 0.98 centimeter thick, respectively. An insulating foil is placed between them that is 0.38 millimeter thick. When the three layers are placed together, what is the total thickness? Write the exact answer.

16. ________

17. ________

18. ________

19. ________

20. ________

21. ________

22. ________

23. ________

24. ________

25. ________

26. ________

27. ________

28. ________

CHAPTER 8

Statistics

As you look at trucks speeding down the highway do you ever wonder how many years these trucks have been on the road? Or do you wonder how many miles these trucks have been driven? As we build newer and more modern trucks, is the nation's fleet of trucks getting younger or getting older? If you would like to use your knowledge of statistics and graphs to help you answer these questions, please turn to the Putting Your Skills to Work on page 477.

Pretest Chapter 8

1. ______

2. ______

3. ______

4. ______

5. ______

6. ______

7. ______

8. ______

9. ______

10. ______

11. ______

If you are familiar with the topics in this chapter, take this test now. Check your answers with those in the back of the book. If an answer was wrong or you couldn't do a problem, study the appropriate section of the chapter.

If you are not familiar with the topics in this chapter, don't take this test now. Instead, study the examples, work the practice problems, and then take the test.

This test will help you identify which concepts you have mastered and which you need to study further.

Section 8.1

The 5000 students on campus were recorded by age. The circle graph depicts their distribution by age.

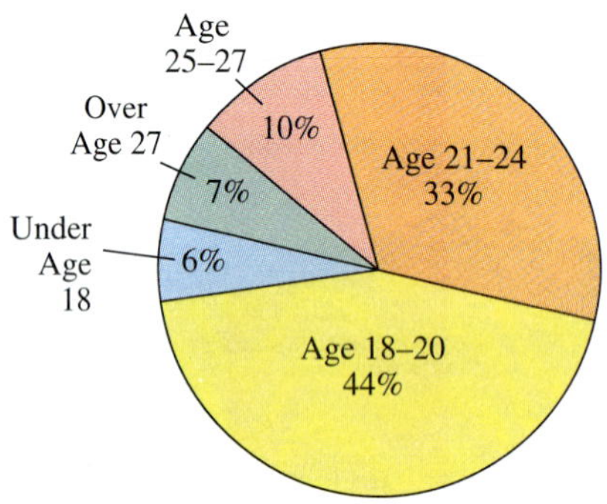

1. What age group has the smallest percent of the student body?

2. What percent of the students are between 21 and 27?

3. What percent of the students are 25 or older?

4. If 5000 students are at the university, *how many students* are between 21 and 24 years of age?

5. *How many students* are over age 27?

Section 8.2

The double-bar graph indicates the number of people unemployed in Pacerville during each quarter of 1993 and 1994.

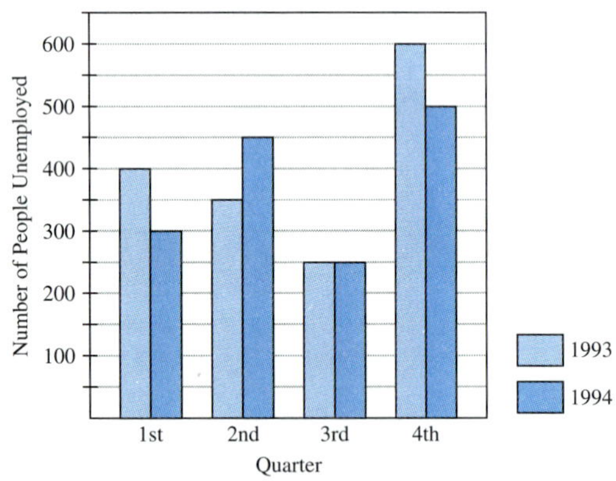

6. How many people were unemployed in Pacerville during the first quarter of 1994?

7. How many people were unemployed in Pacerville during the fourth quarter of 1993?

8. When were the fewest number of people unemployed in Pacerville?

9. When were the greatest number of people unemployed in Pacerville?

10. How many more people were unemployed in the fourth quarter of 1993 than the fourth quarter of 1994?

11. How many more people were unemployed during the second quarter of 1994 than the second quarter of 1993?

The line graph indicates sales and production of color television sets by a major manufacturer during the specified months.

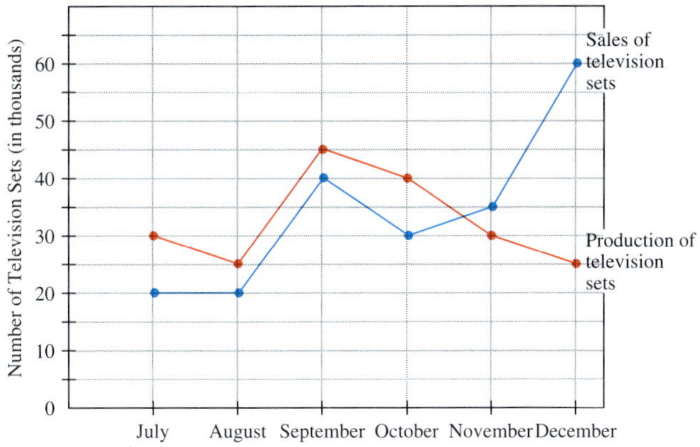

12. During what months was the production of television sets the lowest?

13. During what month was the sales of television sets the highest?

14. What was the first month in which the production of television sets was lower than the sales of television sets?

15. How many television sets were sold in August?

16. How many television sets were produced in November?

Section 8.3

The histogram tells us the number of miles a car was driven before the car was discarded or sold to a junk dealer.

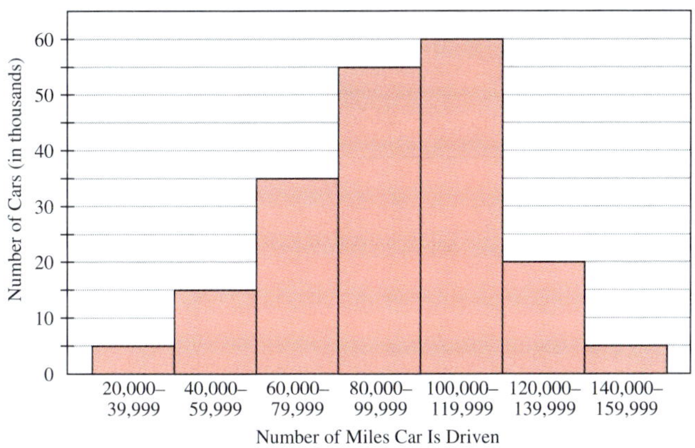

17. How many discarded or junked cars had been driven 80,000 to 99,999 mi?

18. How many discarded or junked cars had been driven 100,000 to 119,999 mi?

19. How many discarded or junked cars had been driven 120,000 mi or more?

20. How many discarded or junked cars had been driven less than 60,000 mi?

Section 8.4

A typist produced the following number of pages.

Mon.	Tues.	Wed.	Thurs.	Fri.
40	42	44	27	32

21. Find the *mean* number of pages typed per day.

22. Find the *median* number of pages typed per day.

12. ________

13. ________

14. ________

15. ________

16. ________

17. ________

18. ________

19. ________

20. ________

21. ________

22. ________

8.1 Circle Graphs

MathPro Video 17 SSM

After studying this section, you will be able to:

1. *Read a circle graph with numerical values.*
2. *Read a circle graph with percentage values.*

1 *Reading a Circle Graph with Numerical Values*

Statistics is that branch of mathematics that collects and studies data. Once the data are collected, they must be organized so that the information is easily readable. We use graphs to give a visual representation of the data that is easy to read. Graphs appeal to the eye. Their visual nature allows them to communicate information about the complicated relationships among statistical data. For this reason, newspapers often use graphs to help their readers quickly grasp information.

Circle graphs are especially helpful for showing the relationship of parts to a whole. The entire circle represents 100%; the pie-shaped pieces represent the subcategories. The following circle graph divides the 10,000 students at Westline College into five categories.

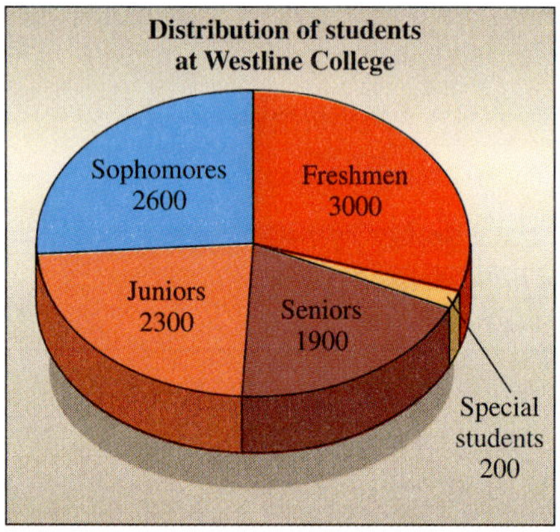

EXAMPLE 1 What is the largest category of students?

The largest pie-shaped section of the circle is labeled ''Freshmen.'' Thus the largest category is *freshmen students.*

Practice Problem 1 What is the smallest category of students? ■

EXAMPLE 2 How many students are either sophomores or juniors?

There are 2600 sophomores and 2300 juniors. If we add these two numbers, we have $2600 + 2300 = 4900$. Thus we see that there are *4900 students who are either sophomores or juniors.*

Practice Problem 2 How many students are either freshmen or special students? ■

EXAMPLE 3 What is the ratio of freshmen to seniors?

Number of freshmen $\longrightarrow$ 3000
Number of seniors $\longrightarrow$ 1900

Thus $\dfrac{3000}{1900} = \dfrac{30}{19}$

The ratio of freshmen to seniors is $\dfrac{30}{19}$.

Practice Problem 3 What is the ratio of freshmen to sophomores? ■

EXAMPLE 4 What is the ratio of seniors to the total number of students?

There are 1900 seniors. We find the total of all the students by adding the number of students in each section of the graph. There are 10,000 students. The ratio of seniors to the total number of students is

$$\frac{1900}{10,000} = \frac{19}{100}$$

Practice Problem 4 What is the ratio of freshmen to the total number of students? ■

2 *Reading a Circle Graph with Percentage Values*

Together, the Great Lakes form the largest body of fresh water in the world. The total area of these five lakes is about 290,000 mi^2. The percentage of this total area taken up by each of the Great Lakes is shown in the circle graph at the right.

Percentage of area occupied by each of the Great Lakes

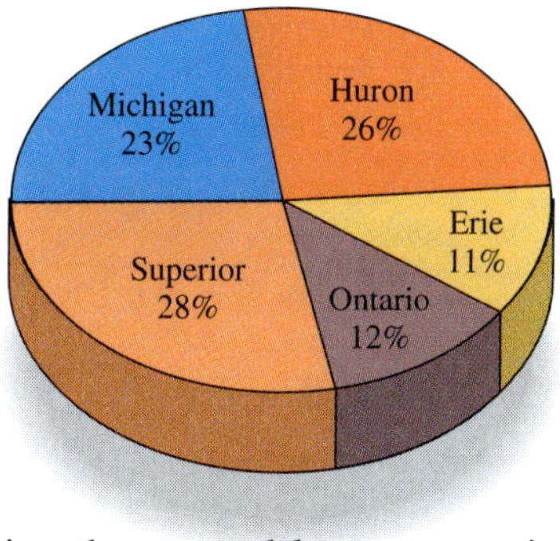

EXAMPLE 5 Which of the Great Lakes occupies the second-largest area in square miles?

The largest percent corresponds to the biggest area, which is occupied by Lake Superior. The second-largest percent corresponds to Lake Huron's area. Therefore, *Lake Huron has the second-largest area in square miles.*

Practice Problem 5 Which of the Great Lakes occupies the third-largest area in square miles? ■

EXAMPLE 6 What percent of the total area is occupied either by Lake Erie or Lake Ontario?

If we add 11% for Lake Erie and 12% for Lake Ontario, we get

$$11\% + 12\% = 23\%$$

Thus 23% of the area is occupied by either Lake Erie or Lake Ontario.

Practice Problem 6 What percent of the total area is occupied by either Lake Superior or Lake Michigan? ■

EXAMPLE 7 How many of the total 290,000 mi^2 are occupied by Lake Michigan?

Remember that we multiply percent times base to obtain amount. Here 23% of 290,000 is what is occupied by Lake Michigan.

$$(0.23)(290{,}000) = n$$
$$= 66{,}700$$

Thus about 66,700 mi^2 are occupied by Lake Michigan.

Practice Problem 7 How many of the total 290,000 mi^2 are occupied by Lake Superior? ■

Sometimes a circle graph is used to investigate the distribution of one part of a larger group. For example, there were 690,000 bachelor's degrees awarded in the United States in 1993 to students majoring in the six most popular subject areas: Business, Social Sciences, Engineering, Health Sciences, Psychology, and English. This circle graph shows how the degrees in these six subject areas are distributed.

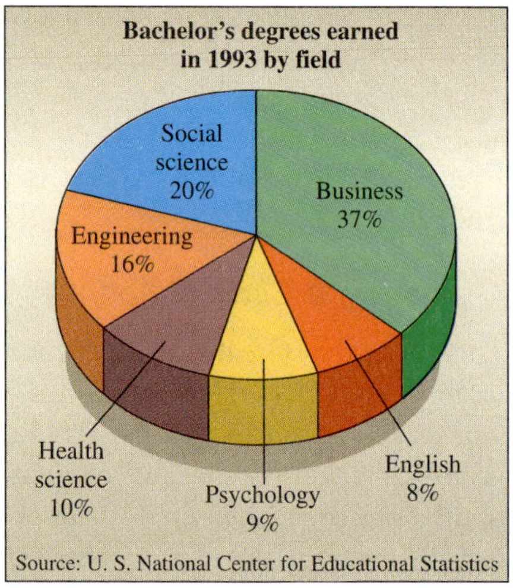

EXAMPLE 8

(a) What percent of the bachelor's degrees represented in this circle graph are in the fields of psychology or business?

(b) Of the approximately 690,000 degrees awarded in these six fields, how many are awarded in the field of Engineering?

(a) We add 37% to 9% to obtain 46%. Thus 46% of the bachelor's degrees represented by this circle are in the fields of psychology or business.

(b) We take 16% of the 690,000 people who obtained degrees in these six areas. Thus we have (0.16)(690,000) = 110,400. There were 110,400 degrees in Engineering awarded in 1993.

Practice Problem 8

(a) What percent of the bachelor's degrees represented in this circle graph are in the fields of Health Science or English?

(b) How many bachelor's degrees in Social Science were awarded in 1993? ■

8.1 Exercises

The following circle graph displays Bob and Linda McDonald's monthly $2000 family budget. Use the circle graph to answer questions 1–10.

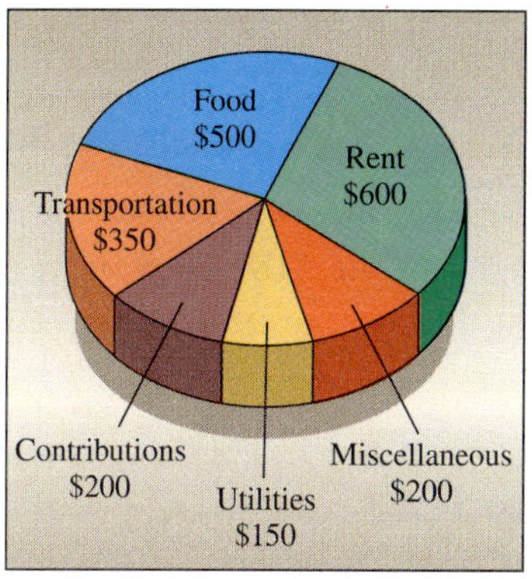

1. What category takes the largest amount of the budget?

2. What category takes the least amount of the budget?

3. How much money is allotted each month for contributions?

4. How much money is allotted each month for transportation?

5. How much money in total is allotted each month for utilities and transportation?

6. How much money in total is allotted for either food or rent?

7. What is the ratio of money spent for food to the amount of money spent for transportation?

8. What is the ratio of money spent for rent to the amount of money spent for utilities?

9. What is the ratio of money spent for rent to the total amount of the monthly budget?

10. What is the ratio of money spent for food to the total amount of the monthly budget?

A major league pitcher has thrown 650 pitches during the first part of the baseball season. The following circle graph shows the results of his pitches. Use the circle graph to answer questions 11–20.

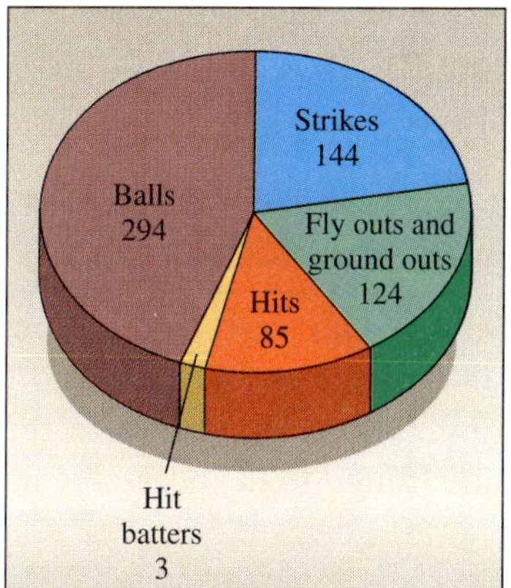

Results of 650 pitches

11. What category had the least number of pitches?

12. What category had the second-highest number of pitches?

13. How many pitches were balls?

14. How many pitches were strikes?

15. How many pitches were either hits or balls?

16. How many pitches were either strikes or fly outs and ground outs?

17. What is the ratio of the number of strikes to the total number of pitches?

18. What is the ratio of the number of balls to the total number of pitches?

19. What is the ratio of the number of balls to the number of strikes?

20. What is the ratio of the number of fly outs and ground outs to the number of hits?

In 1992 there were approximately 118,000 women physicians in the United States. The following circle graph divides them by age. Use the circle graph to answer questions 21–26.

21. What percent of the women physicians are between the ages of 45 and 54?

22. What percent of the women physicians are between the ages of 35 and 44?

23. What percent of the women physicians are under the age of 45?

24. What percent of the women physicians are over the age of 45?

25. How many of the 118,000 women physicians are between the ages of 35 and 44?

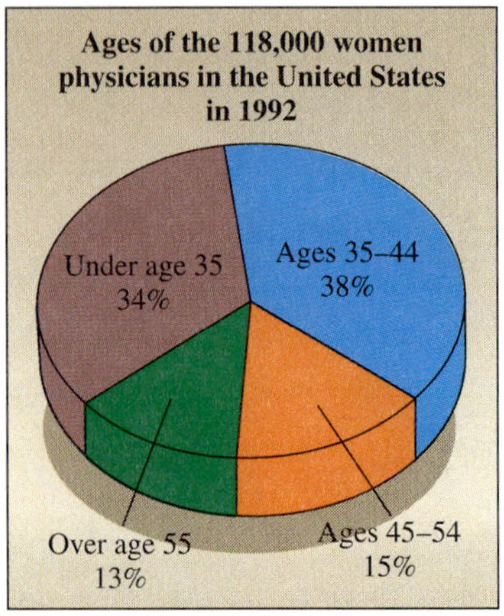

26. How many of the 118,000 women physicians are over age 55?

Researchers estimate that the religious faith distribution of the 5,000,000,000 people in the world in 1988 was approximately that displayed in the following circle graph. Use the graph to answer questions 27–32.

27. Approximately how many of the 5,000,000,000 people are Christians?

28. Approximately how many of the 5,000,000,000 people are Moslems?

29. What percent of the world's population is either Moslem or nonreligious?

30. What percent of the world's population is either Hindu or Buddhist?

31. What percent of the world's population is *not* Moslem?

32. What percent of the world's population is *not* Christian?

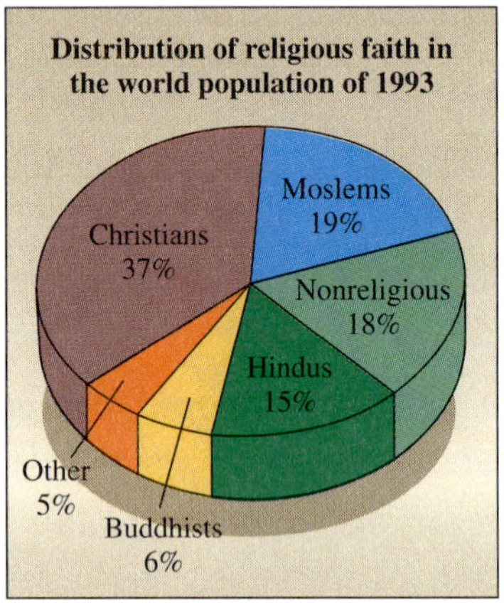

33. In 1993 it was estimated that there were 75,000,000 Anglicans in the world. What percent of the 5,000,000,000 people in the world were Anglican? Round to the nearest tenth of a percent.

34. The circle contains a sector labeled Christians. What percent of the Christian sector represents Anglicans? Round to the nearest tenth of a percent.

Cumulative Review Problems

35. Find the area of a triangle with base 6 in. and height 14 in.

36. Find the area of a parallelogram with base of 17 in. and height 12 in.

37. How many gallons of paint will it take to cover the four sides of a barn. Two sides measure 7 yd by 12 yd and two sides measure 7 yd by 20 yd. Assume a gallon of paint covers 28 square yards.

38. A circular reflector has a radius of 8 cm. How many grams of silver will it take to cover the reflector if each gram will cover 64 sq cm? Assume that the area of the reflector is covered on one side only. (Use $\pi \approx 3.14$).

8.2 Bar Graphs and Line Graphs

After studying this section, you will be able to:

MathPro Video 18 SSM

1. *Read and interpret a bar graph.*
2. *Read and interpret a double-bar graph.*
3. *Read and interpret a line graph.*
4. *Read and interpret a comparison line graph.*

1 *Reading and Interpreting a Bar Graph*

Bar graphs are helpful for seeing changes over a period of time. Bar graphs or line graphs are especially helpful when the same type of data is repeatedly studied. The following bar graph shows the population of California from 1940 to 1990.

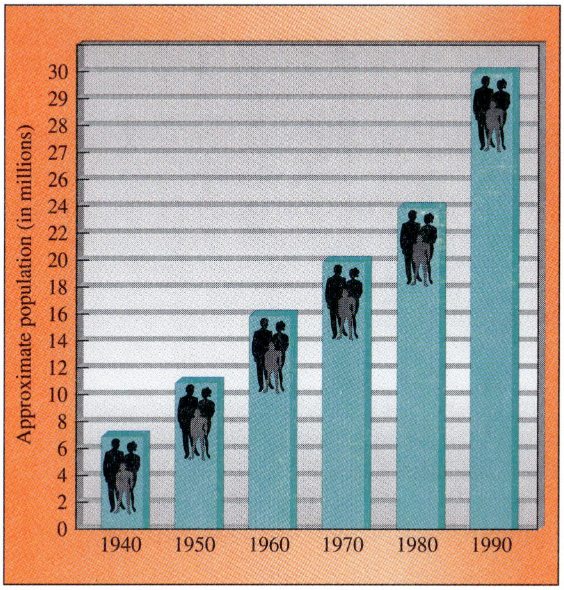

EXAMPLE 1 What was the approximate population of California in 1970?

The bar for 1970 rises to 20. This represents 20 million; thus the approximate population is 20,000,000.

Practice Problem 1 What was the approximate population of California in 1980? ■

EXAMPLE 2 What was the increase in population from 1980 to 1990?

The bar for 1980 rises to 24. Thus the approximate population is 24,000,000. The bar for 1990 rises to 30. Thus the approximate population is 30,000,000. To find the increase in population from 1980 to 1990, we subtract.

$$30{,}000{,}000 - 24{,}000{,}000 = 6{,}000{,}000$$

Practice Problem 2 What was the increase in population from 1940 to 1960? ■

2 *Reading and Interpreting a Double-Bar Graph*

Double-bar graphs are useful for making comparisons. For example, when a company is analyzing its sales, it may want to compare different years or different quarters. The following double-bar graph illustrates the sales of new cars at a local Ford dealership for two different years, 1995 and 1996. The sales are recorded for each quarter of the year.

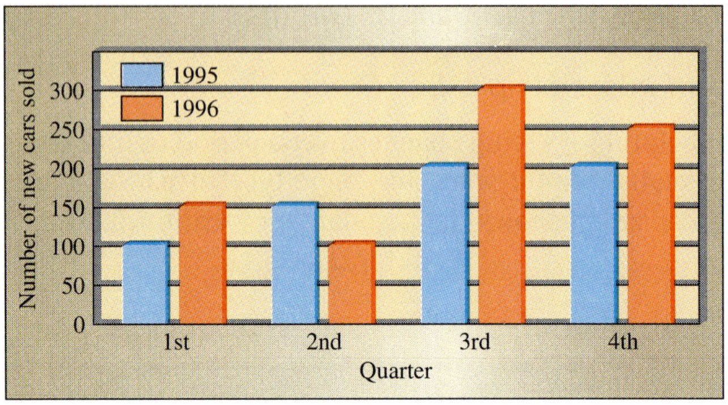

EXAMPLE 3 How many cars were sold in the second quarter of 1995?

The bar rises to 150 for the second quarter of 1995. Therefore, 150 cars were sold.

Practice Problem 3 How many cars were sold in the fourth quarter of 1996? ■

EXAMPLE 4 How many more cars were sold in the third quarter of 1996 than in the third quarter of 1995?

From the double-bar graph, we see that 300 cars were sold in the third quarter of 1996 and that 200 cars were sold in the third quarter of 1995.

$$\begin{array}{r} 300 \\ -\ 200 \\ \hline 100 \end{array}$$

Thus 100 more cars were sold.

Practice Problem 4 How many fewer cars were sold in the second quarter of 1996 than in the second quarter of 1995? ■

3 *Reading and Interpreting a Line Graph*

A line graph is useful for showing trends over a period of time. In a line graph only a few points are actually plotted from measured values. The points are then connected by straight lines to show a trend. The intervening values between points may not lie exactly on the line. The following line graph shows the number of customers per month coming into a restaurant in a vacation community.

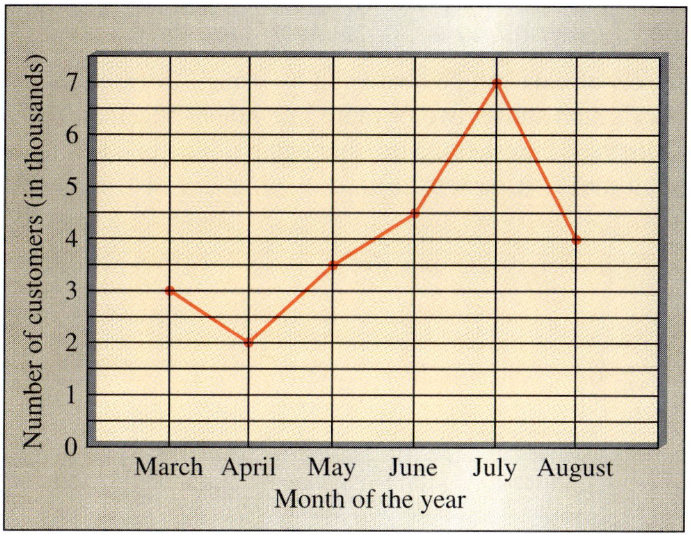

EXAMPLE 5 In which month did the fewest number of customers come into the restaurant?

The lowest point on the graph occurs in the month of April. *Thus the fewest number of customers came in April.*

Practice Problem 5 In which month did the greatest number of customers come into the restaurant? ■

EXAMPLE 6

(a) Approximately how many customers per month came during the month of June?

(b) From May to June did the number of customers increase or decrease?

(a) Notice that the dot is halfway between 4 and 5. This represents a value halfway between 4000 and 5000 customers. Thus we would estimate *4500 customers* came during the month of June.

(b) Between May and June the line goes up, so *the number of customers increased.*

Practice Problem 6

(a) Approximately how many customers per month came during the month of May?

(b) From March to April did the number of customers increase or decrease? ■

EXAMPLE 7 Between what two months is the *increase* in attendance the largest?

The line from June to July goes upward at the steepest angle. This represents the largest increase. (You can check this by reading the numbers from the left axis.) Thus the greatest increase in attendance is *between June and July.*

Practice Problem 7 Between what two months did the biggest *decrease* occur? ■

4 *Reading and Interpreting a Comparison Line Graph*

Two or more sets of data can be compared by using a *comparison line graph.* A comparison line graph shows two or more line graphs together. A different style for each line distinguishes them. Note that using a blue line and a red line in the following graph makes it easy to read.

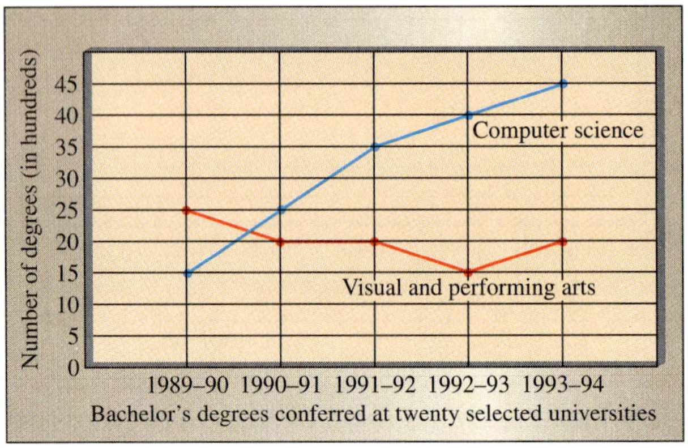

EXAMPLE 8 How many bachelor's degrees in computer science were awarded in the academic year 1991–92?

Because the dot corresponding to 1991–92 is opposite 35 and the scale is in hundreds, we have $35 \times 100 = 3500$. Thus *3500 degrees were awarded in computer science in 1991–92.*

Practice Problem 8 How many bachelor's degrees in visual and performing arts were awarded in the academic year 1992–93? ■

EXAMPLE 9 In what academic year were more degrees awarded in the visual and performing arts than in computer science?

The only year when more bachelor's degrees were awarded in the visual and performing arts was *the academic year 1989–90.*

Practice Problem 9 What was the first academic year in which more degrees were awarded in computer science than in the visual and performing arts? ■

8.2 Exercises

The bar graph below shows the approximate population of Texas from 1940 to 1990. Use the graph to answer questions 1–6.

1. What was the approximate population in 1960?
2. What was the approximate population in 1950?
3. What was the approximate population in 1980?
4. What was the approximate population in 1990?
5. According to this bar graph, between what years did the population of Texas increase by the smallest amount?
6. According to this bar graph, between what years did the population of Texas increase by the largest amount?

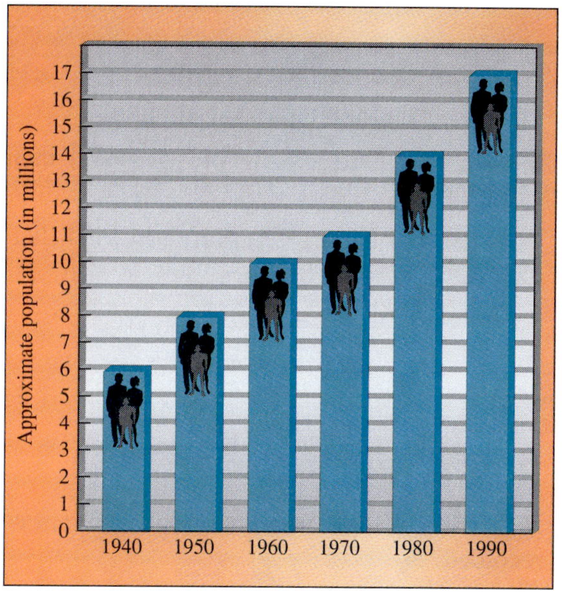

The following double-bar graph illustrates the number of students at 10 selected universities at 20-year intervals. The student population is divided into men and women. Use the graph to answer questions 7–14.

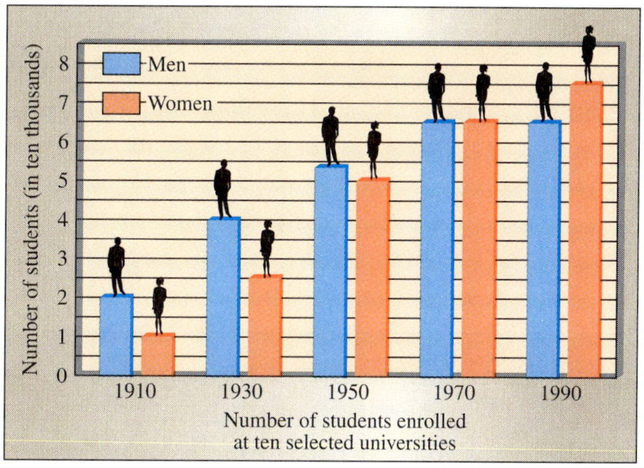

7. How many women students were enrolled in 1950?
8. How many men students were enrolled in 1930?
9. In what year was the number of women students enrolled greater than the number of men students enrolled?
10. In what year was the number of women students enrolled equal to the number of men students enrolled?
11. How many more women students were enrolled in 1950 than in 1930?
12. How many more men students were enrolled in 1970 than in 1950?
13. In what 20-year period did the greatest increase in the enrollment of women take place?
14. In what 20-year period did the greatest increase in the enrollment of men take place?

The following line graph shows Wentworth Construction Company's profits during the last six years. Use the graph to answer questions 15–20.

15. What was the profit in 1993?

16. What was the profit in 1989?

17. What year had the lowest profit?

18. What year had the highest profit?

19. How much greater was the profit in 1991 than in 1990?

20. How much greater was the profit in 1994 than in 1989?

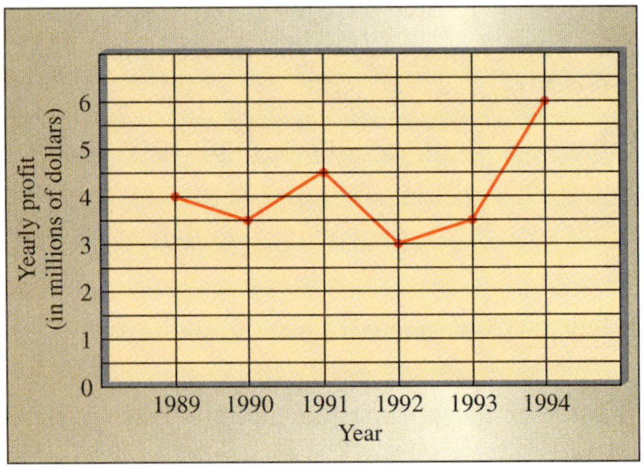

The following comparison line graph indicates the rainfall for the last six months of two different years in Springfield. Use the graph to answer questions 21–26.

21. In September 1993, how many inches of rain were recorded?

22. In October of 1992, how many inches of rain were recorded?

23. During what months was the rainfall of 1993 less than the rainfall of 1992?

24. During what months was the rainfall of 1993 greater than the rainfall of 1992?

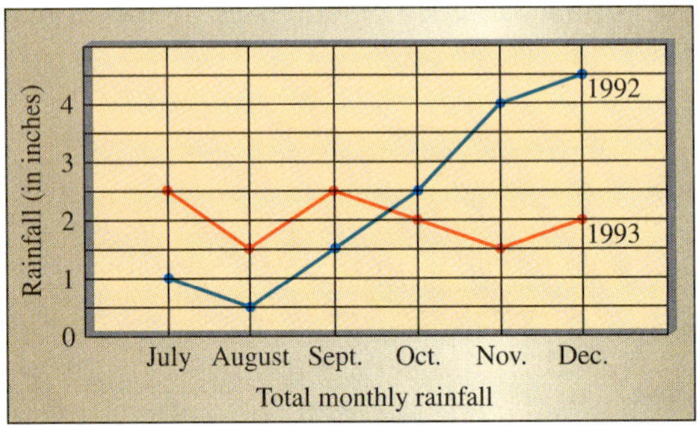

25. How many more inches of rain fell in November of 1992 than in October 1992?

26. How many more inches of rain fell in September of 1992 than in August 1992?

To Think About

The table below shows the number of pizzas sold at a pizza place near a college campus.

Number of Pizzas Sold by Alfredo's Pizza Parlor	300	400	100	200	600
Month of the Year	Jan.	Feb.	Mar.	Apr.	May

27. Use the graph paper on the right to construct a line graph of the information in the table. Let the vertical scale (height) represent the number of pizzas sold. Let the horizontal scale (width) represent the month.

28. Is the biggest change on the graph an increase or a decrease in the number of pizzas sold per month?

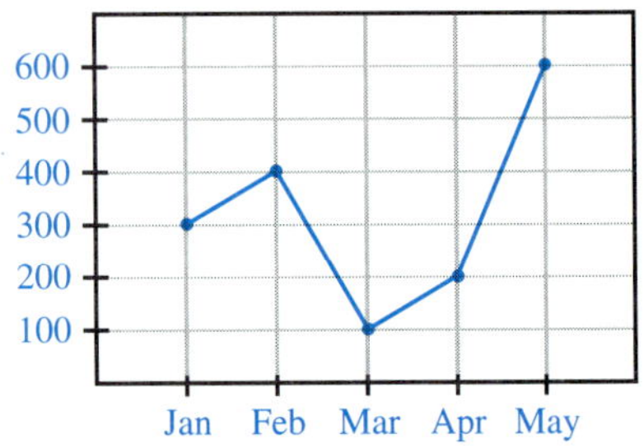

Cumulative Review Problems

Do each calculation in the proper order.

29. $7 \times 6 + 3 - 5 \times 2$

30. $(5 + 6)^2 - 18 \div 9 \times 3$

Putting Your Skills to Work

THE AGING OF AMERICA'S TRUCKS

Does it seem like the trucks on our nation's roads are getting older? Statistics kept on truck registrations show that more and more trucks are staying on the road for a greater number of years.

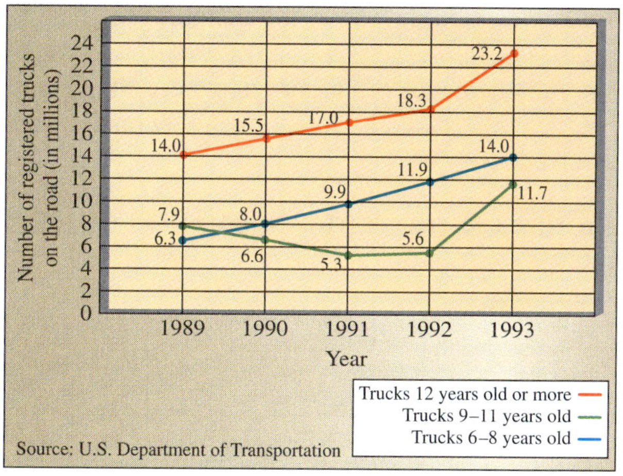

Based on the graph above please determine the following.

PROBLEMS FOR INDIVIDUAL INVESTIGATION AND ANALYSIS

1. During what year were more registered trucks on the road that were 9–11 years old than there were trucks that were 6–8 years old?

2. Which age category saw the greatest increase in the number of registered trucks during the period 1989 to 1993?

PROBLEMS FOR GROUP INVESTIGATION AND COOPERATIVE LEARNING

Together with some other members of your class determine the answers to the following.

3. From 1989 to 1993 the mean age of a registered truck increased from 7.9 years to 8.6 years. If there were 52.8 million trucks on the road in 1989, how many million trucks would be expected in 1993 due to this increase in age of the trucks?

4. There were 53.2 million trucks registered in 1989 and 82.5 million trucks registered in 1993. What percent of the registered trucks in 1989 were 6 years old or older? What percent of the registered trucks in 1993 were 6 years old or older? What can you conclude from these statistics?

INTERNET CONNECTIONS: Go to `http://www.prenhall.com/tobey` to be connected

Site: Truck Inventory and Use Survey (Bureau of Transportation Statistics/U.S. Department of Transportation) or a related site. *Note: This site requires the Adobe Acrobat Reader (which is free) to view the documents.*

View the Truck Inventory and Use Survey for your state. This document includes many data tables relating to the trucks in your state. Locate the information in Table 1 that gives the percent of trucks of varying ages over the past 20 years.

5. What percentage of the trucks used in 1987 were over 4 years old? in 1992?

6. Which age category saw the greatest increase in the percent of trucks in use from 1982 to 1987?

8.3 Histograms

MathPro Video 18 SSM

After studying this section, you will be able to:

1 *Understand and interpret a histogram.*
2 *Construct a histogram from raw data.*

1 *Understanding and Interpreting a Histogram*

In business or in higher education you are often asked to take data and organize them in some way. This section shows you the technique for making a *histogram*—a type of bar graph.

Suppose that a mathematics professor announced the results of a class test. The 40 students in the class scored between 50 and 99 on the test. The results are displayed on this chart.

SCORES ON THE TEST	CLASS FREQUENCY
50–59	4
60–69	6
70–79	16
80–89	8
90–99	6

The results in the table can be organized in a special type of bar graph known as a ***histogram.*** In a histogram the width of each bar is the same. The width represents the range of scores on the test. This is called a ***class interval.*** The height of each bar gives the class frequency of each class interval. The ***class frequency*** is the number of times a score occurs in a particular class interval.

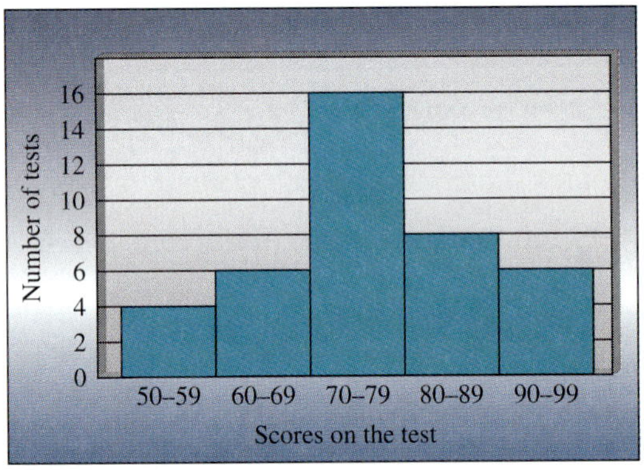

EXAMPLE 1 How many students scored a B on the test if the professor considers a test score of 80–89 as a B?

Since the 80–89 bar rises to a height of 8, there were *eight students* who scored a B on the test.

Practice Problem 1 How many students scored a D on the test if the professor considers a test score of 60–69 as a D? ■

EXAMPLE 2 How many students scored less than 80 on the test?

From the histogram, we see that there are three different bar heights to be included. Four tests were scored 50–59, six tests were scored 60–69, and 16 tests were scored 70–79. When we combine 4 + 6 + 16 = 26, we can see that *26 students scored less than 80* on the test.

Practice Problem 2 How many students scored greater than 69 on the test? ■

The following histogram tells us about the length of life of 110 new light bulbs tested at a research center. The number of hours the bulbs lasted is indicated on the horizontal scale. The frequency of bulbs lasting that long is indicated on the vertical scale.

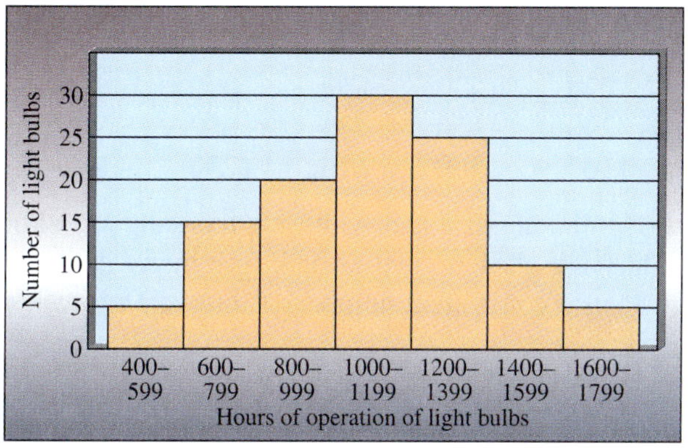

EXAMPLE 3 How many light bulbs lasted between 1400 and 1599 hours?

The bar with a range of 1400–1599 hours rises to 10. Thus 10 light bulbs lasted that long.

Practice Problem 3 How many light bulbs lasted between 800 and 999 hours? ■

EXAMPLE 4 How many light bulbs lasted fewer than 1000 hours?

We see that there are three different bar heights to be included. Five bulbs lasted 400–599 hours, 15 bulbs lasted 600–799 hours, and 20 bulbs lasted 800–999 hours. We add 5 + 15 + 20 = 40. Thus 40 light bulbs lasted less than 1000 hours.

Practice Problem 4 How many light bulbs lasted more than 1199 hours? ■

2 *Constructing a Histogram from Raw Data*

To construct a histogram, we start with *raw data,* data that have not yet been organized or interpreted. We perform the following steps.

> **1.** Select data class intervals of equal width for the data.
>
> **2.** Make a table with class intervals and a *tally* (count) of how many numbers occur in each interval. Add up the tally to find the class frequency for each class interval.
>
> **3.** Draw the histogram.

First we will practice making the table. Later we will use the table to draw the histogram.

EXAMPLE 5 Complete a table to determine the frequency of each class interval for the following data. Each number represents the number of kilowatt-hours of electricity used in a home during a one-month period.

770	520	850	900	1100
1200	1150	730	680	900
1160	590	670	1230	980

1. We select class intervals of equal width for the data. We choose intervals of 200. We might have chosen smaller or larger intervals, but we choose 200 because it gives us a convenient number of intervals to work with, as we will see.

2. We make a table. We write down the class intervals, then count (tally) how many numbers occur within each interval. Then we write the total. This is the class frequency.

KILOWATT-HOURS USED CLASS INTERVAL	TALLY	FREQUENCY
500–699	||||	4
700–899	|||	3
900–1099	|||	3
1100–1299	~~||||~~	5

Practice Problem 5 Each number below represents the weight in pounds of a new car.

2250	1760	2000	2100	1900
1640	1820	2300	2210	2390
2150	1930	2060	2350	1890

Complete the following table to determine the frequency of each class interval for the data above.

WEIGHT IN POUNDS CLASS INTERVAL	TALLY	FREQUENCY
1600–1799		
1800–1999		
2000–2199		
2200–2399		

■

EXAMPLE 6 Draw a histogram from the table in Example 5.

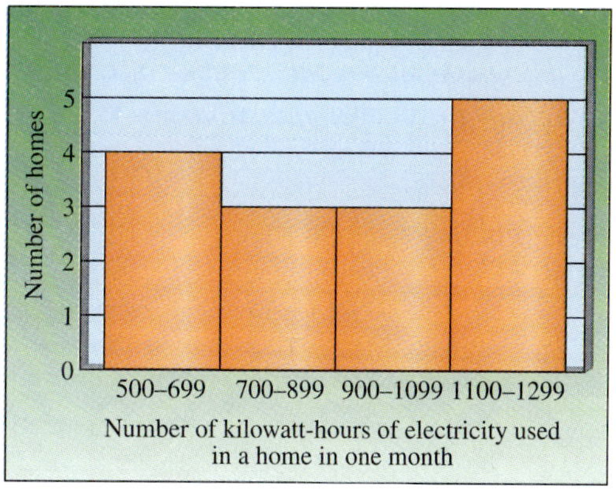

Practice Problem 6 Draw a histogram using the data from the table in Practice Problem 5. ■

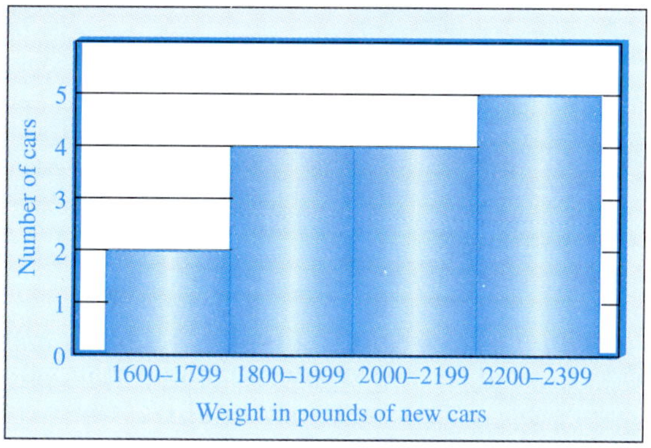

EXAMPLE 7 Draw a histogram from the following table of data concerning books that were purchased in 1994 in the United States.

Age Category of People Purchasing Books	Approximate Number of Books Purchased in the U.S. in 1994
Under 25	90,000,000
25–34	350,000,000
35–44	580,000,000
45–54	450,000,000
55–64	310,000,000
65 and Older	350,000,000

Source: U.S. Bureau of Economic Analysis

Practice Problem 7 Based on the histogram, between what two age categories is the greatest difference in the number of books purchased? ■

8.3 Exercises

A total of 190 rental cars were studied. The following histogram shows the number of rental cars fitting a certain rating classification in miles per gallon. Use the histogram to answer questions 1–8.

1. How many cars achieve between 28 and 30.9 mi per gallon?
2. How many cars achieve between 19 and 21.9 mi per gallon?
3. How many cars achieve between 25 and 27.9 mi per gallon?
4. How many cars achieve between 16 and 18.9 mi per gallon?
5. How many cars achieve more than 24.9 mi per gallon?
6. How many cars achieve less than 22 mi per gallon?
7. How many cars achieve between 19 and 27.9 mi per gallon?
8. How many cars achieve between 16 and 24.9 mi per gallon?

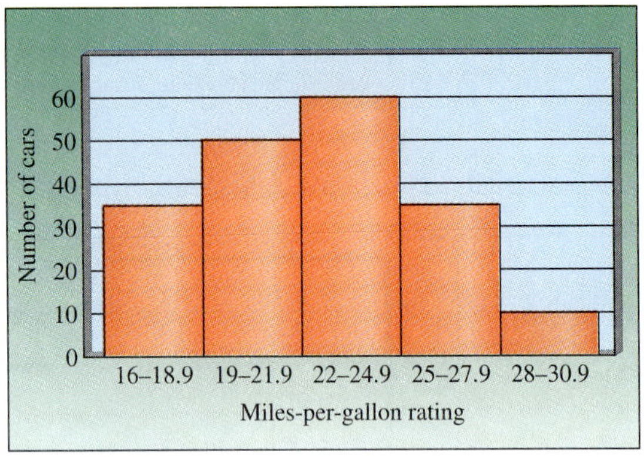

In an air-pollution study a large factory was studied for 47 operating days. Each day the number of tons of sulfur oxides emitted was measured. The following histogram shows the number of days that emissions of sulfur oxide reached a certain class interval. Use the histogram to answer questions 9–18.

9. For how many days did the factory emit between 13 and 16.9 tons of sulfur oxides?
10. For how many days did the factory emit between 17 and 20.9 tons of sulfur oxides?
11. For how many days did the factory emit between 25 and 28.9 tons of sulfur oxides?
12. For how many days did the factory emit between 21 and 24.9 tons of sulfur oxides?
13. For how many days did the factory emit less than 25 tons of sulfur oxides?
14. For how many days did the factory emit more than 16.9 tons of sulfur oxides?
15. For how many days did the factory emit between 13 and 24.9 tons of sulfur oxides?
16. For how many days did the factory emit between 17 and 28.9 tons of sulfur oxides?

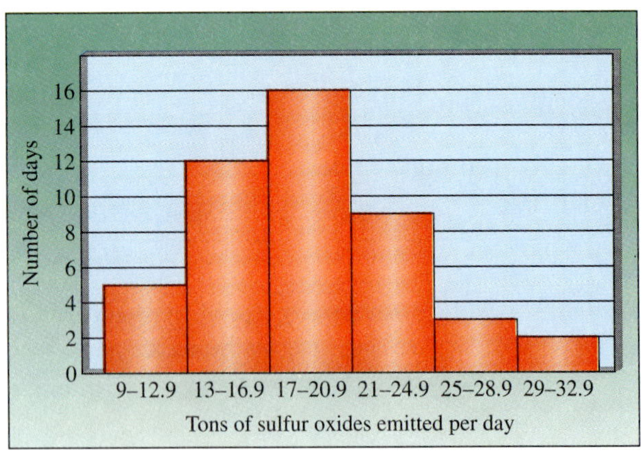

17. On what percent of the 47 operating days were 17 or more tons of sulfur oxides emitted? Round to the nearest tenth of a percent.
18. On what percent of the 47 operating days were less than 21 tons of sulfur oxides emitted? Round to the nearest tenth of a percent.

Complete a table to determine the frequency of each class interval of the following data. Each number represents the highest temperature in degrees Fahrenheit for the day in Boston for February.

23°	26°	30°	18°	42°	17°	19°
51°	42°	38°	36°	12°	18°	14°
20°	24°	26°	30°	18°	17°	16°
35°	38°	40°	33°	19°	22°	26°

	TEMPERATURE CLASS INTERVAL	TALLY	FREQUENCY
19.	12°–16°	____	____
20.	17°–21°	____	____
21.	22°–26°	____	____
22.	27°–31°	____	____
23.	32°–36°	____	____
24.	37°–41°	____	____
25.	42°–46°	____	____
26.	47°–51°	____	____

27. Construct a histogram using the table prepared in problems 19–26.

Complete a table to determine the frequency of each class interval of the following data. Each number represents the cost of a prescription purchased by the Lin family this year.

\$28.50	\$16.00	\$32.90	\$46.20	\$ 9.85
\$27.30	\$16.00	\$41.95	\$36.00	\$24.20
\$ 7.65	\$ 8.95	\$ 4.50	\$11.35	\$ 7.75
\$12.30	\$21.85	\$46.20	\$15.50	\$ 4.50

	PURCHASE PRICE CLASS INTERVAL	TALLY	FREQUENCY
28.	\$ 4.00–\$ 9.99	____	____
29.	\$10.00–\$15.99	____	____
30.	\$16.00–\$21.99	____	____
31.	\$22.00–\$27.99	____	____
32.	\$28.00–\$33.99	____	____
33.	\$34.00–\$39.99	____	____
34.	\$40.00–\$45.99	____	____
35.	\$46.00–\$51.99	____	____

36. Construct a histogram using the table constructed in problems 28–35.

Cumulative Review Problems

37. Solve for n. $\frac{126}{n} = \frac{36}{17}$

38. Solve for n. $\frac{n}{18} = \frac{3.5}{9}$

39. Ben and Trish Hale are going to have a Texas barbeque next week. According to Trish's grandmother, the recipe calls for 3 pounds of chicken for every 5 people. The Hales are expecting 20 people. How much chicken should they buy?

40. Tim and Judy Newitt worked as scientists on Mount Washington last year. They found that every 23 in. of snow corresponded to 2 in. of water. During the month of January they measured 150 in. of snow at the mountain weather observatory. How many inches of water does this correspond to? Round your answer to the nearest tenth of an inch.

8.4 Mean and Median

MathPro Video 18 SSM

After studying this section, you will be able to:

1 *Find the mean of a set of numbers.*
2 *Find the median value of a set of numbers.*

1 *Finding the Mean of a Set of Numbers*

We often want to know the "middle value" of a group of numbers. In this section we learn that, in statistics, there is more than one way of describing this middle value: there is the *mean* of the group of numbers, and there is the *median* of the group of numbers. In some situations it's more helpful to look at the mean, and in others it's more helpful to look at the median. We'll learn to tell which situations lend themselves to one or the other.

The **mean** of a set of values is the sum of the values divided by the number of values. The mean is often called the **average.**

EXAMPLE 1 Find the average or mean test score of a student who has test grades of 71, 83, 87, 99, 80, and 90.

We take the sum of the six tests and divide the sum by 6.

$$\begin{array}{l} \text{Sum of tests} \longrightarrow \\ \text{Number of tests} \longrightarrow \end{array} \frac{71 + 83 + 87 + 99 + 80 + 90}{6} = \frac{510}{6} = 85$$

The mean is 85.

Practice Problem 1 Find the average or mean of the following test scores: 88, 77, 84, 97, 89. ■

The mean value is often rounded to a certain decimal-place accuracy.

EXAMPLE 2 Carl recorded the mi/gal achieved by his car for the last two months. His results were:

Week	1	2	3	4	5	6	7	8
Miles per Gallon	26	24	28	29	27	25	24	23

What is the mean mi/gal figure for the last 8 weeks? Round your answer to the nearest mile per gallon.

$$\begin{array}{l} \text{Sum of values} \longrightarrow \\ \text{Number of values} \longrightarrow \end{array} \frac{26 + 24 + 28 + 29 + 27 + 25 + 24 + 23}{8}$$

$$= \frac{206}{8} = 26 \quad \text{rounded to the nearest whole number}$$

Mean mi/gal rating is 26.

Practice Problem 2 Bob and Wally kept records of their phone bill for the last six months. Their bills were \$39.20, \$43.50, \$81.90, \$34.20, \$51.70, and \$48.10. Find the mean monthly bill. Round your answer to the nearest cent. ■

2 *Finding the Median Value of a Set of Numbers*

If a set of numbers is arranged in order from smallest to largest, the **median** is that value that has the same number of values above it as below it.

EXAMPLE 3 Find the median value of the following package weights: 6 lb, 7 lb, 9 lb, 12 lb, 13 lb, 16 lb, 17 lb.

The numbers are arranged in order.

$$\underbrace{6, 7, 9}_{\text{three numbers}} \qquad \underset{\substack{\uparrow \\ \text{middle number}}}{12} \qquad \underbrace{13, 16, 17}_{\text{three numbers}}$$

There are three numbers smaller than 12 and three numbers larger than 12. Thus 12 is the median.

Practice Problem 3 Find the median value of the following lengths of telephone calls: 2 minutes, 3 minutes, 7 minutes, 12 minutes, 13 minutes, 15 minutes, 18 minutes, 20 minutes, 21 minutes. ■

If the numbers are not arranged in order, then the first step is to put them in order.

EXAMPLE 4 Find the median cost of the following costs for a microwave oven: \$100, \$60, \$120, \$200, \$190, \$130, \$320, \$290, \$180.

We must arrange the numbers in order from smallest to largest (or largest to smallest).

$$\underbrace{\$60, \$100, \$120, \$130}_{\text{four numbers}} \qquad \underset{\substack{\uparrow \\ \text{middle number}}}{\$180} \qquad \underbrace{\$190, \$200, \$290, \$320}_{\text{four numbers}}$$

Thus \$180 is the median cost.

Practice Problem 4 Find the median weekly salary of the following salaries: \$320, \$150, \$400, \$600, \$290, \$180, \$450. ■

If a list of numbers contains an even number of items, then of course there is no one middle number. In this situation we obtain the median by taking the average of the two middle numbers.

EXAMPLE 5 Find the median of the following numbers: 13, 16, 18, 26, 31, 33, 38, 39.

$$\underbrace{13, 16, 18}_{\text{three numbers}} \qquad \underset{\substack{\uparrow \\ \text{two middle} \\ \text{numbers}}}{26, 31} \qquad \underbrace{33, 38, 39}_{\text{three numbers}}$$

The average (mean) of 26 and 31 is

$$\frac{26 + 31}{2} = \frac{57}{2} = 28.5$$

Thus the median value is 28.5.

Practice Problem 5 Find the median value of the following numbers: 88, 90, 100, 105, 118, 126. ■

SIDELIGHT When would someone want to use the mean, and when would someone want to use the median? Which is more helpful? The mean, or average, is used more frequently. It is most helpful when the data are distributed fairly evenly, that is, when no one value is "much larger" or "much smaller" than the rest. For example, a company had employees with annual salaries of \$9000, \$11,000, \$14,000, \$15,000, \$17,000, and \$20,000. All the salaries fall within a fairly limited range. The mean salary

$$\frac{9000 + 11{,}000 + 14{,}000 + 15{,}000 + 17{,}000 + 20{,}000}{6} = \$14{,}333.33$$

gives us a reasonable idea of the "average" salary. However, suppose the company had six employees with salaries of \$9000, \$11,000, \$14,000, \$15,000, \$17,000, and \$90,000. Talking about the mean salary, which is \$26,000, is deceptive. No one earns a salary very close to the mean salary. The "average" worker in that company does not earn around \$26,000. In this case, the median value is more appropriate. Here the median is \$14,500. Some problems of this type are included in Exercises 8.4, problems 35 and 36.

EXPLORING MATHEMATICAL TOPICS: THE MODE

Another value that is sometimes used to describe a set of data is the mode. The **mode** of a set of data is the number or numbers that occur most often. For example, if a student had test scores of 89, 94, 96, 89, and 90, we would say that the mode is 89. If two values occur most often, we say that the data have two modes (or are bimodal). For example, if the ages of students in a calculus class were 33, 27, 28, 28, 21, 19, 18, 25, 26, and 33, we would say that the modes were 28 and 33. The mode is not used as frequently in statistics as the mean and the median. We will practice finding the mode in Exercises 37–42.

8.4 Exercises

In problems 1–14, find the mean. Round to the nearest tenth when necessary.

1. A student received grades of 89, 92, 83, 96, and 99 on math quizzes.

2. A student received grades of 77, 88, 90, 92, 83, and 84 on six history quizzes.

3. The Windy City Passport Photo Center received the following number of telephone calls over the last six days: 23, 45, 63, 34, 21, and 42?

4. The local Hertz rental car office received the following number of inquiries over the last seven days: 34, 57, 61, 22, 43, 80, and 39.

5. The last five houses built in town sold for the following prices: $89,000, $93,000, $62,000, $102,000, $89,000.

6. Luis priced a sofa at six local stores. The prices were $499, $359, $600, $450, $529, $629.

7. Sam watched television last week for the following number of hours:

Mon.	Tues.	Wed.	Thurs.	Fri.	Sat.	Sun.
8	3	2	5	6.5	7.5	3

8. Steve's car got the following miles per gallon results during the last 6 months:

Jan.	Feb.	Mar.	Apr.	May	June
23	22	25	28	29	30

9. The population on the island of Guam has increased significantly over the last 30 years. Find an approximate value for the mean population for this 30-year period from the following population chart:

1960 Population	1970 Population	1980 Population	1990 Population
67,000	86,000	107,000	134,000

10. Desmond has taken four three-credit courses at college each of four semesters. His semester grade averages are:

Semester 1	Semester 2	Semester 3	Semester 4
2.90	2.66	2.52	2.48

11. The captain of the college baseball team achieved the following results:

	Game 1	Game 2	Game 3	Game 4	Game 5
Hits	0	2	3	2	2
Times at Bat	5	4	6	5	4

Find his batting average by dividing his total number of hits by the total times at bat.

12. The captain of the college bowling team had the following results after practice:

	Practice 1	Practice 2	Practice 3	Practice 4
Score (Pins)	541	561	840	422
Number of Games	3	3	4	2

Find her bowling average by dividing the total number of pins scored by the total number of games.

13. Frank and Wally traveled to the West Coast during the summer. The number of miles they drove and the number of gallons of gas they used are recorded below.

	Day 1	Day 2	Day 3	Day 4
Miles Driven	276	350	391	336
Gallons of Gas	12	14	17	14

Find the average miles per gallon achieved by the car on the trip by dividing the total number of miles driven by the total number of gallons used.

14. Cindy and Andrea traveled to Boston this fall. The number of miles they drove and the number of gallons of gas they used are recorded below.

	Day 1	Day 2	Day 3	Day 4
Miles Driven	260	375	408	416
Gallons of Gas	10	15	17	16

Find the average miles per gallon achieved by the car on the trip by dividing the total number of miles driven by the total number of gallons used.

In problems 15–30, find the median value.

15. 22, 36, 45, 47, 48, 50, 58

16. 37, 39, 46, 53, 57, 60, 63

17. 1052, 968, 1023, 999, 865, 1152

18. 1400, 1329, 1200, 1386, 1427, 1350

19. 0.52, 0.69, 0.71, 0.34, 0.58

20. 0.26, 0.12, 0.35, 0.43, 0.28

21. The annual salaries of the employees of a local cable television office are \$17,000, \$11,600, \$23,500, \$15,700, \$26,700, \$31,500.

22. The costs of six cars recently purchased by the Weston Company were \$18,270, \$11,300, \$16,400, \$9100, \$12,450, \$13,800.

23. The number of minutes spent on the phone per day by each of eight San Diego teenagers living on the same block is 40 minutes, 108 minutes, 62 minutes, 12 minutes, 24 minutes, and 31 minutes.

24. The ages of 10 people swimming laps at the YMCA pool each morning: 60, 18, 24, 36, 39, 32, 70, 12, 15, and 85.

25. The phone bill for Dr. Price's cellular phone over the last seven months: \$109, \$207, \$420, \$218, \$97, \$330, and \$185.

26. The price of the same compact disc sold at several different music stores or by mail-order: \$15.99, \$11.99, \$5.99, \$12.99, \$14.99, \$9.99, \$13.99, \$7.99, and \$10.99.

27. The numbers of potential actors who tried out for the school play at Hamilton-Wenham Regional High School over the last ten years: 36, 48, 44, 64, 60, 71, 22, 36, 53, and 37.

28. The number of injuries during the high school football season for the Badgers over the last eight years: 10, 17, 14, 29, 30, 19, 25, and 21.

Find the mean. Round to the nearest cent when necessary.

29. The salaries of eight small business owners in Big Rapids: \$30,000, \$74,500, \$47,890, \$89,000, \$57,645, \$78,090, \$110,370, and \$65,800.

30. The prices of nine laptop computers with a Pentium chip: \$5679, \$6902, \$1530, \$2738, \$2999, \$4105, \$3655, \$5980, and \$4430.

Find the median value.

31. 2576, 8764, 3700, 5000, 7200, 4700, 9365, 1987

32. 15.276, 21.375, 18.90, 29.2, 14.77, 19.02

Find the mean.

33. Oil is imported into the United States in great quantities. The following chart is provided by the U.S. Energy Information Administration.

Year	Amount of Oil Imported (in millions of barrels)
1991	2151
1992	2110
1993	2220
1994	2578
1995	2643

(a) Find the mean number of barrels of oil imported into the United States per year.

(b) If one barrel contains 42 gallons of oil, find the mean number of gallons of oil imported into the United States per year.

34. The capacity of the United States to produce electricity continues to increase as the demand for electricity continues to grow. The following chart is provided by the U.S. Energy Information Administration.

Year	Summer Capacity (in millions of kilowatts)
1990	685.1
1991	690.9
1992	695.4
1993	694.3
1994	703.0

(a) Find the mean capacity of the United States to produce electricity in the summer measured in number of kilowatts.

(b) A kilowatt is 1000 watts. Find the mean capacity to produce electricity in the summer measured in watts.

To Think About

35. A local travel office has 10 employees. Their monthly salaries are $1500, $1700, $1650, $1300, $1440, $1580, $1820, $1380, $2900, $6300.
(a) Find the mean.
(b) Find the median.
(c) Which of these numbers best represents ''what the average person earns''? Why?

36. A college track star in California ran the 100 meter event in eight track meets. Her times were 11.7 seconds, 11.6 seconds, 12.0 seconds, 12.1 seconds, 11.9 seconds, 18 seconds, 11.5 seconds, 12.4 seconds.
(a) Find the mean.
(b) Find the median.
(c) Which of these numbers best represents ''her average running time''? Why?

Exploring Mathematical Topics: The Mode

Find the mode for each of the following.

37. 60, 65, 68, 60, 72, 59, 80

38. 86, 84, 82, 87, 84, 88, 90

39. 121, 150, 116, 150, 121, 181, 117, 123

40. 144, 143, 140, 141, 149, 144, 141, 150

41. The last six bicycles sold at the Skol Bike shop cost: $249, $649, $439, $259, $269, $249

42. The last six color televisions sold at the local Circuit City cost: $315, $430, $515, $330, $430, $615

Chapter Organizer

Topic	Procedure	Examples
Circle graphs, p. 466.	The percentage of the 200 police officers within a given age range is illustrated. The following circle graph describes the age of the 200 men and women of the Glover City police force. 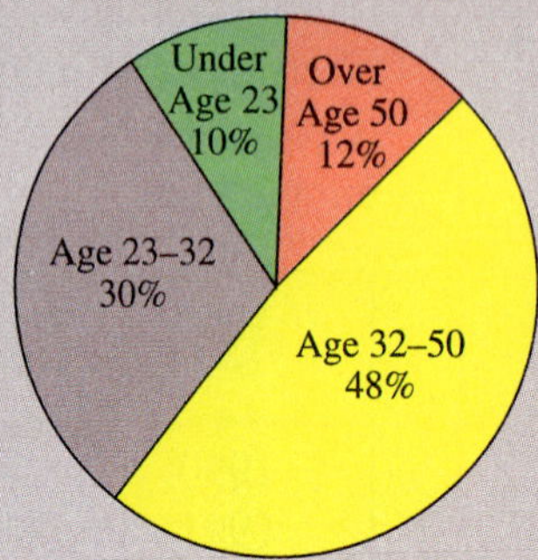 	**1.** What percent of the police force is between 23 and 32 years old? 30% **2.** How many men and women in the police force are over 50 years old? 12% of 200 = (0.12)(200) = 24 people
Bar graphs and double-bar graphs, p. 471.	The following double-bar graph illustrates the sales of color television sets by a major store chain for 1989 and 1990 in three regions of the country. 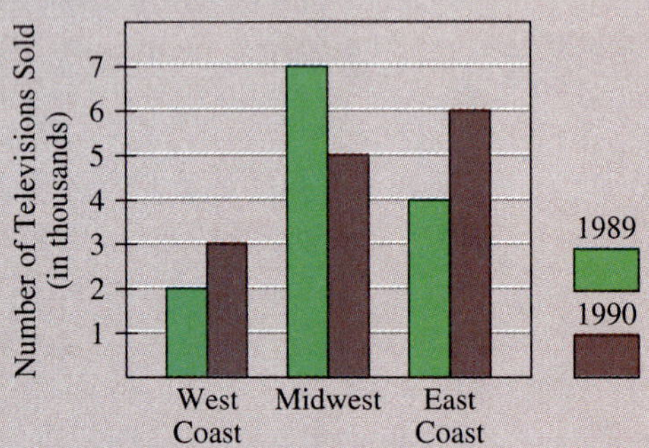 	**1.** How many color television sets were sold by the chain on the East Coast in 1990? 6000 sets **2.** How many *more* color television sets were sold in 1990 than in 1989 on the West Coast? 3000 sets were sold in 1990; 2000 sets were sold in 1989. 3000 − 2000 1000 sets more in 1990
Line graphs and comparison line graphs, pp. 472–74.	The following line graph indicates the number of visitors to Wetlands State Park during a four-month period in 1989 and 1990. 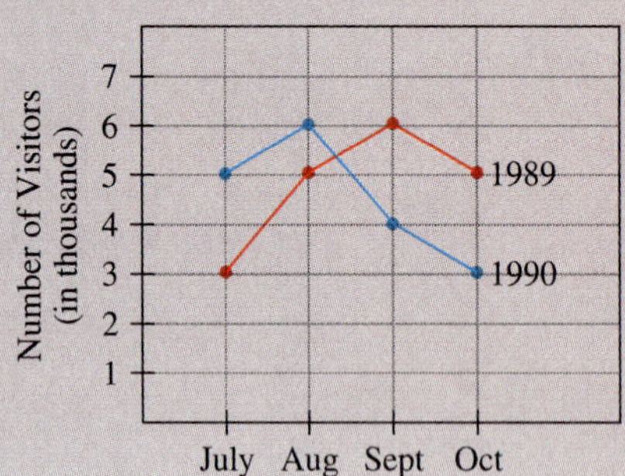 	**1.** How many visitors came to the park in July of 1989? 3000 visitors **2.** In what months were there more visitors in 1989 than in 1990? September and October **3.** The sharpest *decrease* in attendance took place between what two months? Between August 1990 and September 1990

Chapter Organizer

Topic	Procedure	Examples
Histograms, p. 478.	The following histogram indicates the number of quizzes in a math class that were scored within each interval on a 15-point quiz. 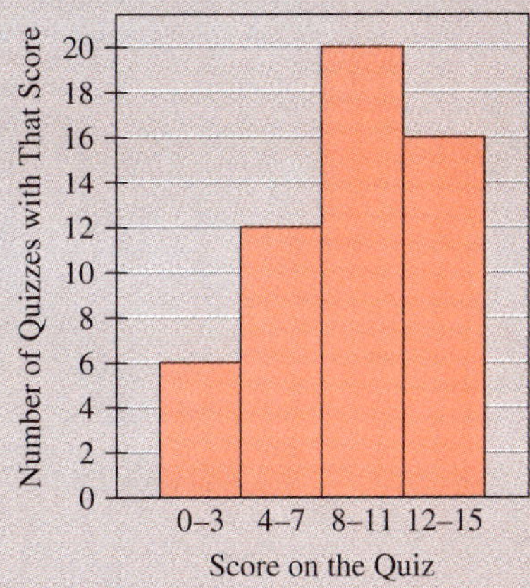	**1.** How many quizzes had a score between 8 and 11? 20 quizzes **2.** How many quizzes had a score of less than 8? $12 + 6 = 18$ quizzes
Finding the mean, p. 484.	The *mean* of a set of values is the sum of the values divided by the number of values. The mean is often called the *average.*	**1.** Find the mean of 19, 13, 15, 25, and 18. $\frac{19 + 13 + 15 + 25 + 18}{5} = \frac{90}{5} = 18$ The mean is 18.
Finding the median, p. 485.	**1.** Arrange the numbers in order from smallest to largest. **2.** If there is an odd number of values, the middle value is the median. **3.** If there is an even number of values, the average of the two middle values is the median.	**1.** Find the median of 19, 29, 36, 15, and 20. First we arrange in order from smallest to largest: 15, 19, 20, 29, 36. $\underbrace{15, 19}_{\text{two numbers}}$ 20 ↑ middle number $\underbrace{29, 36}_{\text{two numbers}}$ The median is 20. **2.** Find the median of 67, 28, 92, 37, 81, and 75. First we arrange in order from smallest to largest: 28, 37, 67, 75, 81, 92. There is an even number of values. 28, 37 67, 75 ↑ two middle numbers 81, 92 $\frac{67 + 75}{2} = \frac{142}{2} = 71$ The median is 71.

Chapter 8 Review Problems

A student found that there were a total of 120 personal computers owned by students in the dormitory. The following circle graph displays the distribution of manufacturers of these computers. Use the graph to answer questions 1–8.

1. How many personal computers were manufactured by IBM?

2. How many personal computers were manufactured by Apple?

3. How many personal computers were manufactured by Leading Edge or Compaq?

4. How many personal computers were manufactured by Hewlett-Packard or miscellaneous companies?

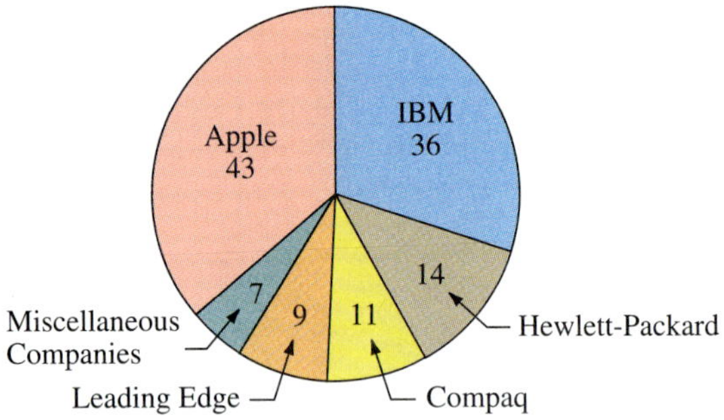

5. What is the *ratio* of the number of computers manufactured by IBM to the number of computers manufactured by Leading Edge?

6. What is the *ratio* of the number of computers manufactured by Hewlett-Packard to the number of computers manufactured by miscellaneous companies?

7. What *percent* of the 120 computers is manufactured by Leading Edge?

8. What *percent* of the computers is manufactured by IBM?

Nancy and Wally Worzowski's family monthly budget of $2400 is displayed in the following circle graph. Use the graph to answer questions 9–16.

9. What percent of the budget is allotted for transportation?

10. What percent of the budget is allotted for savings?

11. What percent of the budget is used up by the food and rent categories?

12. What percent of the budget is used up by the transportation, utilities, and savings categories?

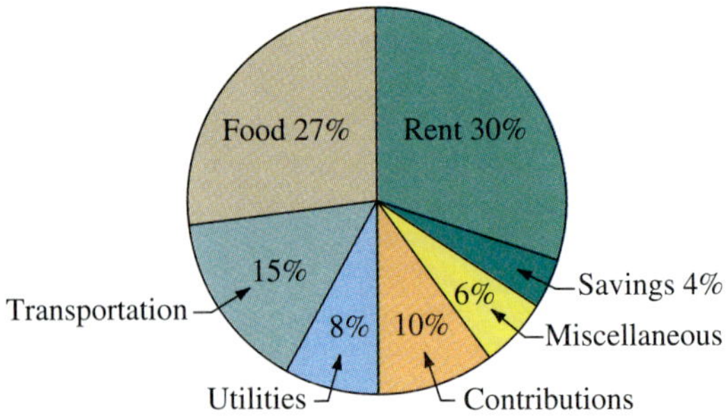

13. Of the total $2400, how much money per month is budgeted for utilities?

14. Of the total $2400, how much money per month is budgeted for transportation?

15. Of the total $2400, how much money per month is budgeted for the rent and savings categories?

16. Of the total $2400, how much money per month is budgeted for the transportation and food categories?

The following double-bar graph illustrates the number of customers at Reid's Steak House for each quarter for the years 1993 and 1994. Use the graph to answer questions 17–24.

17. How many customers came to the restaurant in the second quarter of 1993?

18. How many customers came to the restaurant in the third quarter of 1994?

19. When did the restaurant have the greatest number of customers?

20. When did the restaurant have the fewest number of customers?

21. By how much did the number of customers increase from the first quarter of 1993 to the first quarter of 1994?

22. By how much did the number of customers increase from the fourth quarter of 1993 to the fourth quarter of 1994?

23. During the third quarter (July, August, September) there was road construction in front of the restaurant in both 1993 and 1994. Does the graph indicate the possibility that this might have caused a drop in the number of customers?

24. In the second quarter (April, May, June) the owner spent more on advertising in 1993 than in 1994. Does the graph suggest that this change might have caused a drop in the number of customers?

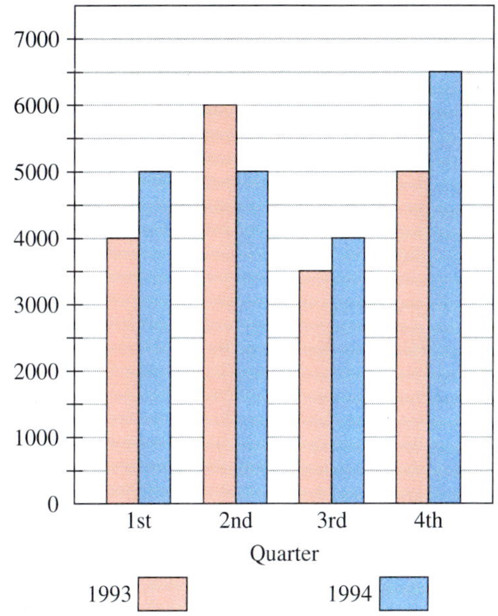

The following line graph shows the number of graduates of Williamston University during the last 6 years. Use the graph to answer questions 25–32.

25. How many Williamston University students graduated in 1993?

26. How many Williamston University students graduated in 1992?

27. How many Williamston University students graduated in 1994?

28. How many Williamston University students graduated in 1991?

29. How many more Williamston University students graduated in 1990 than in 1989?

30. How many more Williamston University students graduated in 1992 than in 1991?

31. Between what two years did the number of graduates decline?

32. Between what two years did the number of graduates increase by the greatest amount?

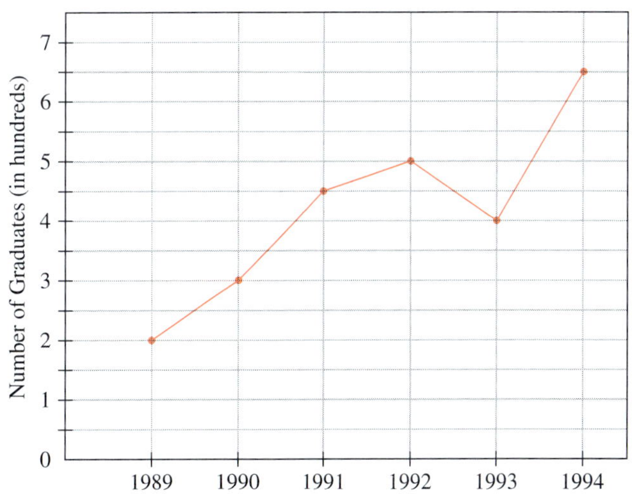

The following comparison line graph shows the number of ice cream cones purchased at the Junction Ice Cream Stand during a five-month period in 1991 and in 1992. Use this graph to answer questions 33–40.

33. How many ice cream cones were purchased in July 1992?

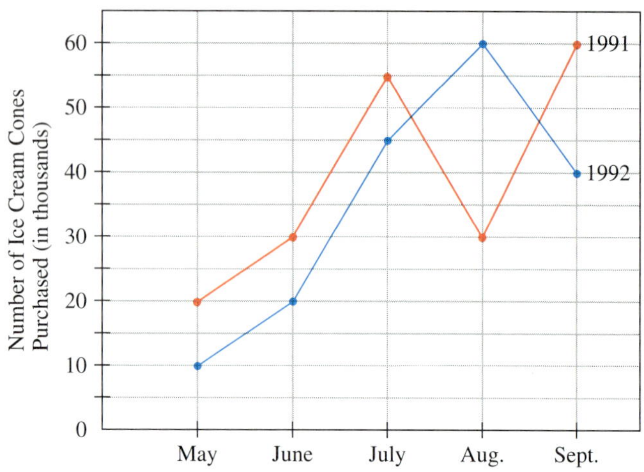

34. How many ice cream cones were purchased in August 1991?

35. How many more ice cream cones were purchased in May 1991 than in May 1992?

36. How many more ice cream cones were purchased in August 1992 than in August 1991?

37. How many more ice cream cones were purchased in July 1991 than in June 1991?

38. How many more ice cream cones were purchased in September 1991 than in August 1991?

39. At the location of the ice cream stand, August 1991 was cold and rainy. The months of May, June, July, and September of 1991 were warm and sunny. What trend on the graph do you think is dependent on the weather?

40. July and August of 1992 were warm and very sunny, while May and June of 1992 were cloudy at the location of the ice cream stand. What trend on the graph do you think is dependent on the weather?

The state highway department made a survey of the bridges over all of its county and state highways. The number of bridges of each age interval is displayed in the following histogram. Use the histogram to answer questions 41–46.

41. How many bridges are between 40 and 59 years old?

42. How many bridges are between 60 and 79 years old?

43. The greatest number of bridges in the state are between ____ and ____ years old.

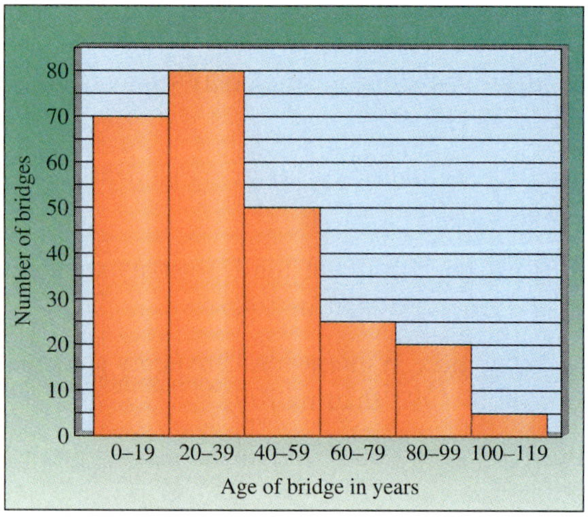

44. The highway commissioner has ordered an immediate inspection of all bridges older than 79 years old. How many bridges will be inspected?

45. All bridges less than 40 years old have had a recent inspection by the highway department. How many bridges were inspected recently?

46. How many more bridges in the state are in the category 60–79 years old than are in the category 80–99 years old?

Complete a table to determine the frequency of each class interval of the following data. Each number tells how many defective televisions were identified daily in the production of 400 new color television sets by a major manufacturer during the last 28 days.

3	5	8	2	0	1	7
13	6	3	4	1	8	12
0	2	16	5	7	13	14
12	10	17	5	4	0	3

	Number of Defects Class Interval	Tally	Frequency
47.	0–3		
48.	4–7		
49.	8–11		
50.	12–15		
51.	16–19		

52. Construct a histogram using the table prepared in problems 47–51.

53. Based on the data of problems 47–51, how often were between 0 and 7 defective television sets identified in the production?

Find the mean value (average) for each of the following items in problems 54–59.

54. The last seven-day maximum temperature readings in Los Angeles in July were: 86°, 83°, 88°, 95°, 97°, 100°, 81°.

55. The amount of groceries purchased by the Michael Stallard family each week for the last seven weeks was \$87, \$105, \$89, \$120, \$139, \$160, \$98.

56. The number of college textbooks purchased by each of eight men living at Jenkins House during his four years of college was 76, 20, 91, 57, 42, 21, 75, 82.

57. The number of women students enrolled in the school of engineering at Westwood University during each of the last 10 years was 151, 140, 148, 156, 183, 201, 205, 228, 231, 237.

58. The number of cars parked in the Central City garage for each of the last five days was 1327, 1561, 1429, 1307, 1481.

59. The number of employees throughout the nation employed by Freedom Rent a Car for the last six years was 882, 913, 1017, 1592, 1778, 1936.

Find the median value for each of the following in problems 60–65.

60. The scores on a recent mathematics exam: 69, 57, 100, 87, 93, 65, 77, 82, 88.

61. The number of students taking Abnormal Psychology for the fall semester for the last nine years at Elmson College: 77, 83, 91, 104, 87, 58, 79, 81, 88.

62. The costs of eight trucks purchased by the Highway Department: \$28,500, \$29,300, \$21,690, \$35,000, \$37,000, \$43,600, \$45,300, \$38,600.

63. The costs of 10 houses recently purchased at Stillwater: \$98,000, \$150,000, \$120,000, \$139,000, \$170,000, \$156,000, \$135,000, \$144,000, \$154,000, \$126,000.

64. The number of cups of coffee consumed by each of the students of the 7:00 A.M. Biology III class during the last semester: 38, 19, 22, 4, 0, 1, 5, 9, 18, 36, 43, 27, 21, 19, 25, 20.

65. The number of deliveries made each day by the Northfield House of Pizza: 21, 16, 0, 3, 19, 24, 13, 18, 9, 31, 36, 25, 28, 14, 15, 26.

66. The scores on eight tests taken by Wong Yin in calculus last semester were 96, 98, 88, 100, 31, 89, 94, 98. Which is a better measure of his usual score, the *mean* or the *median?* Why?

67. The ten salespersons of People's Dodge sold the following number of cars last month: 13, 16, 8, 4, 5, 19, 15, 18, 39, 12. Which is a better measure of the usual sales of these salespersons, the *mean* or the *median?* Why?

Chapter 8 Test

A state Highway Safety Commission recently reported the results of inspecting 300,000 automobiles. The following circle graph depicts the percent of automobiles that passed and the percent that had one or more safety violations. Use this graph to answer questions 1–5.

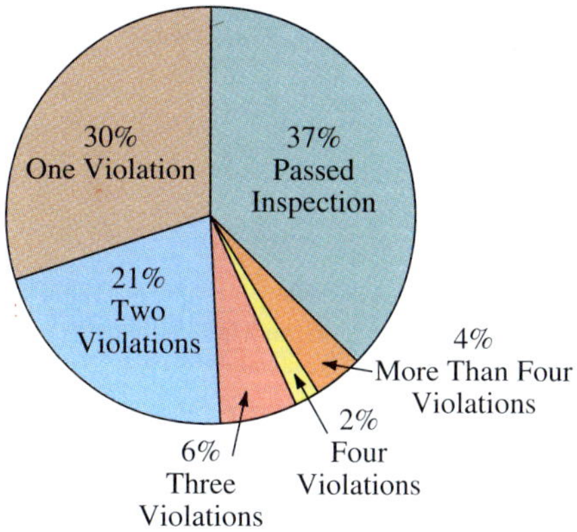

1. What percent of the automobiles passed inspection?
2. What percent of the automobiles had two safety violations?
3. What percent of the automobiles had more than two safety violations?
4. If 300,000 automobiles were inspected, *how many* of them had one safety violation?
5. If 300,000 automobiles were inspected, *how many* of them had either two violations or three violations?

The following double-bar graph indicates the number of cars sold at Boley's Chrysler during each quarter of 1992 and 1993. Use the graph to answer questions 6–11.

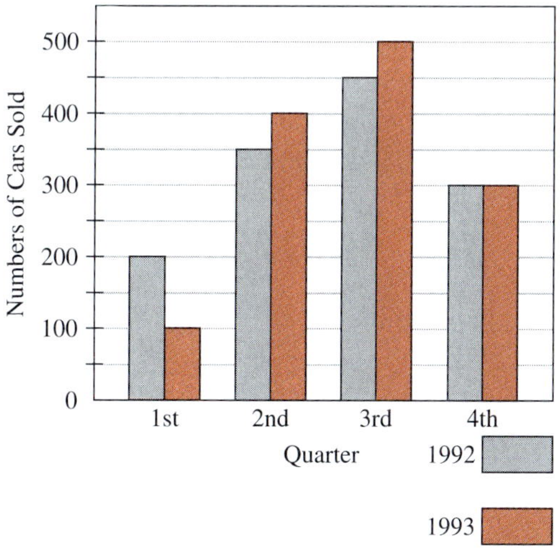

6. How many cars were sold in the second quarter of 1992?
7. How many cars were sold in the third quarter of 1993?
8. When was the greatest number of cars sold?
9. During which quarter were more cars sold in 1992 than in 1993?
10. How many more cars were sold in the second quarter of 1993 than in the second quarter of 1992?
11. How many more cars were sold in the third quarter of 1992 than in the fourth quarter of 1992?

1. ______
2. ______
3. ______
4. ______
5. ______
6. ______
7. ______
8. ______
9. ______
10. ______
11. ______

12. ______

13. ______

14. ______

15. ______

16. ______

17. ______

18. ______

19. ______

20. ______

21. ______

22. ______

A research study by 10 midwestern universities produced the following line graph. Use the graph to answer questions 12–16.

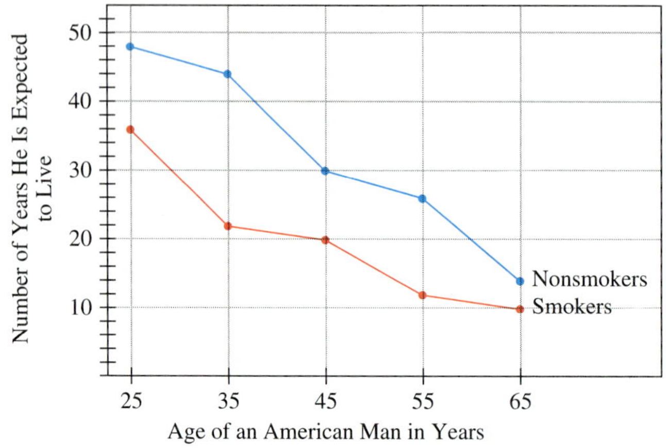

12. Approximately how many more years is a 45-year-old American man expected to live if he smokes?

13. Approximately how many more years is a 55-year-old American man expected to live if he does not smoke?

14. According to this graph, approximately how much longer is a 25-year-old nonsmoker expected to live than a 25-year-old smoker?

15. According to this graph, at what age is the difference between the life expectancy of a smoker and a nonsmoker the greatest?

16. According to this graph, at what age is the difference between the life expectancy of a smoker and a nonsmoker the smallest?

The following histogram was prepared by a consumer research group. Use the histogram to answer questions 17–20.

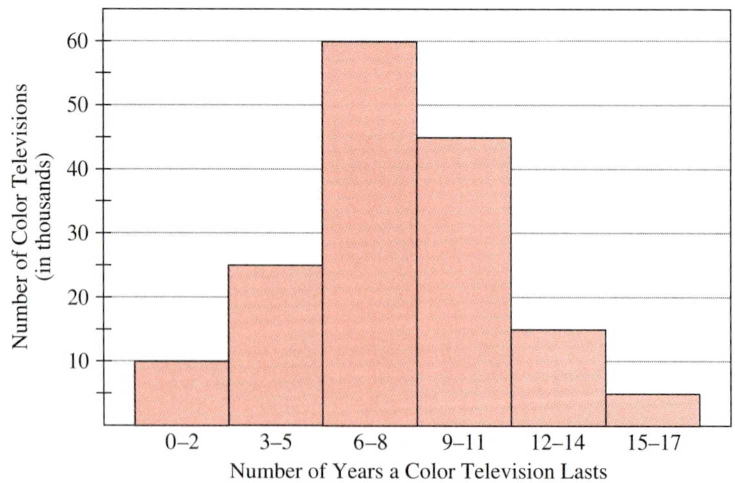

17. How many color television sets lasted 6–8 years?

18. How many color television sets lasted 3–5 years?

19. How many color television sets lasted more than 11 years?

20. How many color television sets lasted 9–14 years?

A chemistry student had the following scores on eight quizzes in her chemistry class: 19, 16, 15, 12, 18, 17, 14, 10

21. Find the *mean* quiz score.

22. Find the *median* quiz score.

Cumulative Test for Chapters 1–8

Approximately one-half of this test is based on Chapter 8 material. The remainder is based on material covered in Chapters 1–7.

1. Add.
$1376 + 2804 + 9003 + 7642$

2. Multiply. 3004×26

3. Subtract. $7\frac{1}{5} - 3\frac{3}{8}$

4. Divide. $10\frac{4}{5} \div 3\frac{1}{2}$

5. Round to the nearest hundredth.
2864.3719

6. Subtract. $\begin{array}{r} 200.58 \\ -\ 127.93 \\ \hline \end{array}$

7. Divide. $52.0056 \div 0.72$

8. Find n. $\frac{7}{n} = \frac{35}{3}$

9. Of every 2030 cars manufactured, three have major engine defects. If the total number of these cars manufactured was 26,390, approximately how many would have major engine defects?

10. What is 1.3% of 25?

11. 72% of what number is 252?

12. Convert 198 cm to m.

13. Convert 18 yd to ft.

14. Find the area of a circle with radius of 3 in. Round your answer to the nearest tenth.

15. Find the perimeter of a square with a side of 17 in.

The 12,000-member student body of Mason University consists of five groups: freshmen, sophomores, juniors, seniors, and graduate students. The distribution by category is displayed on the following graph.

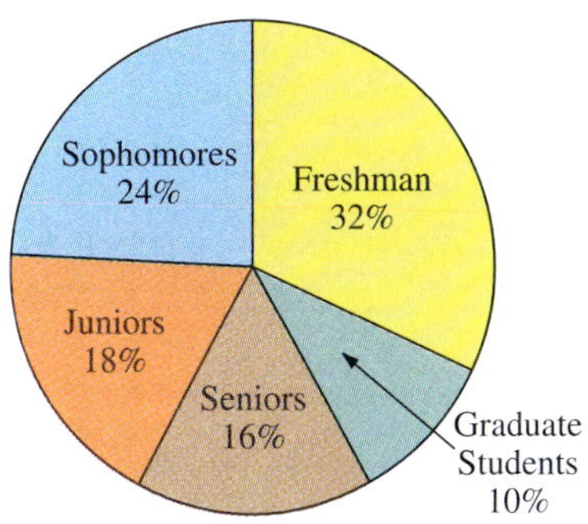

16. What *percent* are either juniors or seniors?

17. How *many* of the 12,000 students are freshmen?

1. ______
2. ______
3. ______
4. ______
5. ______
6. ______
7. ______
8. ______
9. ______
10. ______
11. ______
12. ______
13. ______
14. ______
15. ______
16. ______
17. ______

18. ____

19. ____

20. ____

21. ____

22. ____

23. ____

24. ____

25. ____

The following double-bar graph indicates the quarterly profits for Dedalon Corporation for 1993 and 1994.

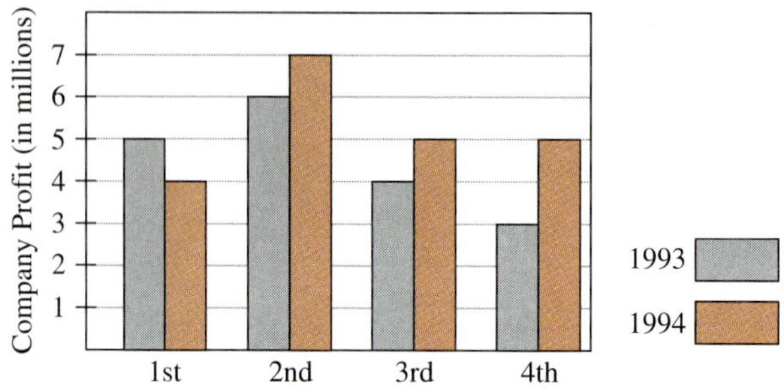

18. What was the quarterly profit for Dedalon Corporation in the fourth quarter of 1993?

19. How much greater was the profit of Dedalon Corporation in the second quarter of 1994 than in the second quarter of 1993?

The following comparison line graph depicts the annual rainfall in Dixville compared to the annual rainfall in Weston for five specific years.

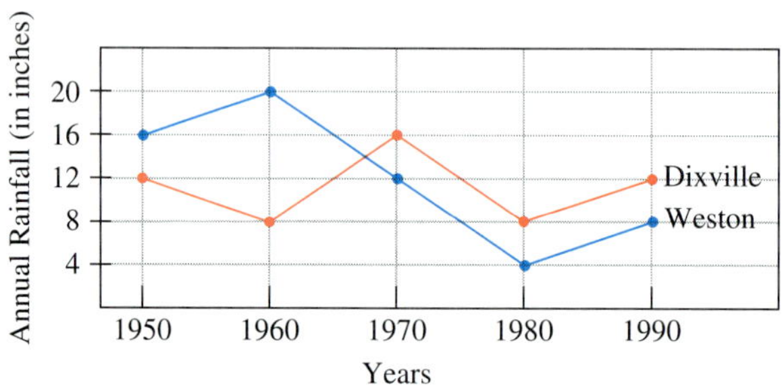

20. How many inches of rain fell in Dixville in 1970?

21. In what years was the annual rainfall in Weston greater than the annual rainfall in Dixville?

The following histogram depicts the number of students in the math course Basic Mathematics who fall into various age groups.

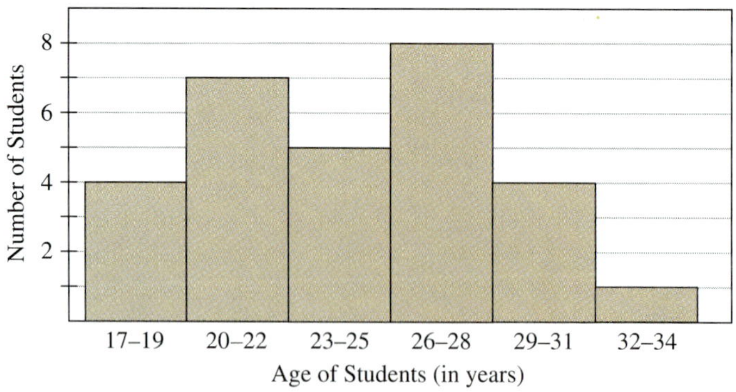

22. How many students are 26–28 years of age?

23. How many students are under 26 years of age?

The following are the hourly wages of six employees of the Hamilton House of Pizza: \$5.00, \$4.50, \$3.95, \$4.90, \$7.00, \$13.65.

24. Find the *mean* hourly wage.

25. Find the *median* hourly wage.

Glossary

Absolute value of a number (9.1) The absolute value of a number is the distance between that number and zero on the number line. When we find the absolute value of a number, we use the $| \ |$ notation. To illustrate, $|-4| = 4$, $|6| = 6$, $|-20 - 3| = |-23| = 23$, $|0| = 0$.

Addends (1.2) When two or more numbers are added, the numbers being added are called addends. In the problem $3 + 4 = 7$, the numbers 3 and 4 are both addends.

Algebraic expression (10.6) An algebraic expression consists of variables, numerals, and operation signs.

Altitude of a triangle (7.3) The height of a triangle.

Amount of a percent equation (5.3A) The product we obtain when we multiply a percent times a number. In the equation $75 = 50\% \times 150$, the amount is 75.

Angle (7.3) An angle is formed whenever two lines meet.

Area (7.1) The measure of the surface inside a geometric figure. Area is measured in square units, such as square feet.

Associative property of addition (1.2) The property that tells us that when three numbers are added, it does not matter which two numbers are added first. An example of the associative property is $5 + (1 + 2) = (5 + 1) + 2$. Whether we add $1 + 2$ first and then add 5 to that, or add $5 + 1$ first and then add that result to 2, we will obtain the same result.

Associative property of multiplication (1.4) The property that tells us that when we multiply three numbers, it does not matter which two numbers we group together first to multiply; the result will be the same. An example of the associative property of multiplication is $2 \times (5 \times 3) = (2 \times 5) \times 3$.

Base (1.6) The number that is to be repeatedly multiplied in exponent form. When we write $16 = 2^4$, the number 2 is the base.

Base of a percent equation (5.3A) The quantity we take a percent of. In the equation $80 = 20\% \times 400$, the base is 400.

Billion (1.1) The number 1,000,000,000.

Borrowing (1.3) The renaming of a number in order to facilitate subtraction. When we subtract $42 - 28$, we rename 42 as 3 tens plus 12. This represents 3 tens and 12 ones. This renaming is called borrowing.

Box (7.7) A three-dimensional object whose every side is a rectangle. Another name for a box is a *rectangular solid.*

Building fraction property (2.6) For whole numbers a, b, c where neither b nor c equals zero

$$\frac{a}{b} = \frac{a}{b} \times 1 = \frac{a}{b} \times \frac{c}{c} = \frac{a \times c}{b \times c}$$

Building up a fraction (2.6) To make one fraction into an equivalent fraction by making the denominator and numerator larger numbers. For example, the fraction $\frac{3}{4}$ can be built up to the fraction $\frac{30}{40}$.

Caret (3.5) A symbol $_\wedge$ used to indicate the new location of a decimal point when performing division of decimal fractions.

Celsius temperature (6.4) A temperature scale in which water boils at 100 degrees (100°C) and freezes at 0 degrees (0°C). To convert Celsius temperature to Fahrenheit, we use the formula $F = 1.8 \times C + 32$.

Center of a circle (7.6) The point in the middle of a circle from which all points on the circle are an equal distance.

Centimeter (6.2) A unit of length commonly used in the metric system to measure small distances. 1 centimeter = 0.01 meter.

Circle (7.6) A figure for which all points are at an equal distance from a given point.

Circumference of a circle (7.6) The distance around the rim of a circle.

Commission (5.4) The amount of money a salesperson is paid that is a percentage of the value of the sales made by that salesperson. The commission is obtained by multiplying the commission rate times the value of the sales. If a salesman sells \$120,000 of insurance and his commission rate is 0.5%, then his commission is $0.5\% \times 120{,}000 = \$600.00$.

Common denominator (2.7) Two fractions have a common denominator if the same number appears in the denominator of each fraction. $\frac{3}{7}$ and $\frac{1}{7}$ have a common denominator of 7.

Commutative property of addition (1.2) The property that tells us that the order in which two numbers are added does not change the sum. An example of the commutative property of addition is $3 + 6 = 6 + 3$.

Commutative property of multiplication (1.4) The property that tells us that the order in which two numbers are multiplied does not change the value of the answer. An example of the commutative property of multiplication is $7 \times 3 = 3 \times 7$.

Composite number (2.2) A composite number is a whole number greater than 1 that can be divided by whole numbers other than itself. The number 6 is a composite number since it can be divided exactly by 2 and 3 (as well as by 1 and 6).

Cone (7.7) A three-dimensional object shaped like an ice-cream cone or the sharpened end of a pencil.

Cross-multiplying (4.3) If you have a proportion such as $\frac{n}{5} = \frac{12}{15}$, then to cross-multiply, you form products to have $n \times 15 = 5 \times 12$.

Cubic centimeter (6.3) A metric measurement of volume equal to 1 milliliter.

Cup (6.1) One of the smallest units of volume in the American system. 2 cups = 1 pint.

Cylinder (7.7) A three-dimensional object shaped like a tin can.

Debit (1.2) A debit in banking is the removing of money from an account. If you had a savings account and took $300 out of it on Wednesday, we would say that you had a debit of $300 from your account. Often a bank will add a service charge to your account and use the word *debit* to mean that it has removed money from your account to cover the charge.

Decimal fraction (3.1) A fraction whose denominator is a power of 10.

Decimal places (3.4) The number of digits to the right of the decimal point in a decimal fraction. The number 1.234 has three decimal places, while the number 0.129845 has six decimal places. A whole number such as 42 is considered to have zero decimal places.

Decimal point (3.1) The period that is used when writing a decimal fraction. In the number 5.346, the period between the 5 and the 3 is the decimal point. It separates the whole number from the fractional part that is less than 1.

Decimal system (1.1) Our number system is called the decimal system or base 10 system because the value of numbers written in our system is based on tens and ones.

Decimeter (6.2) A unit of length not commonly used in the metric system. 1 decimeter = 0.1 meter.

Degree (7.3) A unit used to measure an angle. A degree is $\frac{1}{360}$ of a complete revolution. An angle of 32 degrees is written as 32°.

Dekameter (6.2) A unit of length not commonly used in the metric system. 1 dekameter = 10 meters.

Denominator (2.1) The number on the bottom of a fraction. In the fraction $\frac{2}{9}$ the denominator is 9.

Deposit (1.2) A deposit in banking is the placing of money in an account. If you had a checking account and on Tuesday you placed $124 into that account, we would say that you made a deposit of $124.

Diameter of a circle (7.6) A line segment across the circle that passes through the center of the circle. The diameter of a circle is equal to twice the radius of the circle.

Difference (1.3) The result of performing a subtraction. In the problem $9 - 2 = 7$ the number 7 is the difference.

Digits (1.1) The symbols 0, 1, 2, 3, 4, 5, 6, 7, 8, and 9 are called digits.

Discount (5.4) The amount of reduction in a price. The discount is a product of the discount rate times the list price. If the list price of a television is $430.00 and it has a discount rate of 35%, then the amount of discount is 35% × $430.00 = $150.50. The price would be reduced by $150.50.

Distributive property of multiplication over addition (1.4) The property illustrated by $5 \times (4 + 3) = (5 \times 4) + (5 \times 3)$. In general, for any numbers $a, b, c,$ it is true that $a(b + c) = a \times b + a \times c$.

Dividend (1.5) The number that is being divided by another. In the problem $14 \div 7 = 2$, the number 14 is the dividend.

Divisor (1.5) The number that you divide into another number. In the problem $30 \div 5 = 6$, the number 5 is the divisor.

Earned run average (4.4) A ratio formed by finding the number of runs a pitcher would give up in a

9-inning game. If a pitcher has an earned run average of 2, it means that, on the average, he gives up two runs for every nine innings he pitches.

Equal fractions (2.2) Fractions that represent the same number. The fractions $\frac{3}{4}$ and $\frac{6}{8}$ are equal fractions.

Equality test of fractions (2.2) Two fractions $\frac{a}{b}$ and $\frac{c}{d}$ are equal if the product $a \times d = b \times c$. In this case, A, B, C, D are whole numbers and $B, D \neq 0$.

Equations (10.3) Mathematical statements with variables that say that two expressions are equal, such as $x + 3 = -8$ and $2s + 5s = 34 - 4s$.

Equilateral triangle (7.3) A triangle with three equal sides.

Equivalent equations (10.3) Equations that have the same solution.

Equivalent fractions (2.2) Two fractions that are equal.

Expanded notation for a number (1.1) A number is written in expanded notation if it is written as a sum of hundreds, tens, ones, etc. The expanded notation for 763 is 700 + 60 + 3.

Exponent (1.6) The number that indicates the number of times a factor occurs. When we write $8 = 2^3$, the number 3 is the exponent.

Factors (1.4) Each of the numbers that are multiplied. In the problem $8 \times 9 = 72$, the numbers 8 and 9 are factors.

Fahrenheit temperature (6.4) A temperature scale in which water boils at 212 degrees (212°F) and freezes at 32 degrees (32°F). To convert Fahrenheit temperature to Celsius, we use the formula $C = \frac{5 \times F - 160}{9}$.

Foot (6.1) American system unit of length. 3 feet = 1 yard.

Fundamental theorem of arithmetic (2.2) Every composite number has a unique product of prime numbers.

Gallon (6.1) A unit of volume in the American system. 4 quarts = 1 gallon.

Gigameter (6.2) A metric unit of length equal to 1,000,000,000 meters.

Gram (6.3) The basic unit of weight in the metric system. A gram is defined as the weight of the water in a box that is 1 centimeter on each side. 1 gram = 1000 milligrams. 1 gram = 0.001 kilograms.

Hectometer (6.2) A unit of length not commonly used in the metric system. 1 hectometer = 100 meters.

Height (7.2) The distance between two parallel sides in a four-sided figure such as a parallelogram or a trapezoid.

Height of a cone (7.7) The distance from the vertex of a cone to the base of the cone.

Height of a pyramid (7.7) The distance from the point on a pyramid to the base of the pyramid.

Height of a triangle (7.3) The distance of a line drawn from a vertex perpendicular to the other side, or an extension of the other side, of the triangle. This is sometimes called the *altitude of a triangle*.

Hexagon (7.2) A six-sided figure.

Hypotenuse (7.5) The side opposite the right angle in a right triangle. The hypotenuse is always the longest side of a right triangle.

Improper fraction (2.3) A fraction in which the numerator is greater than or equal to the denominator. The fractions $\frac{34}{29}$, $\frac{8}{7}$, and $\frac{6}{6}$ are all improper fractions.

Inch (6.1) The smallest unit of length in the American system. 12 inches = 1 foot.

Inequality symbol (3.2) The symbol that is used to indicate if a number is greater than another number or less than another number. Since 5 is greater than 3, we would write this with a "greater than" symbol as follows: $5 > 3$. The statement 7 is less than 12 would be written as follows: $7 < 12$.

Interest (5.4) The money that is paid for the use of money. If you deposit money in a bank, the bank uses that money and pays you interest. If you borrow money, you pay the bank interest for the use of that money. Simple interest is determined by the formula $I = P \times R \times T$. Compound interest is usually determined by a table, a calculator, or a computer.

Invert a fraction (2.5) To invert a fraction is to interchange the numerator and the denominator. If we invert $\frac{5}{9}$, we obtain the fraction $\frac{9}{5}$. To invert a fraction is sometimes referred to as *to take the reciprocal of a fraction*.

Irreducible (2.2) A fraction that cannot be reduced (simplified) is called irreducible.

Isosceles triangle (7.3) A triangle with two sides equal.

Kilogram (6.3) The most commonly used metric unit of weight. 1 kilogram = 1000 grams.

Kiloliter (6.3) The metric unit of volume normally used to measure large volumes. 1 kiloliter = 1000 liters.

Kilometer (6.2) The unit of length commonly used in the metric system to measure large distances. 1 kilometer = 1000 meters.

Least common denominator (LCD) (2.6) The least common denominator (LCD) of two or more fractions is the smallest number that can be divided without remainder by each fraction's denominator. The LCD of $\frac{1}{3}$ and $\frac{1}{4}$ is 12. The LCD of $\frac{5}{6}$ and $\frac{4}{15}$ is 30.

Legs of a right triangle (7.5) The two shortest sides of a right triangle.

Length of a rectangle (7.1) Each of the longer sides of a rectangle.

Like terms (10.1) Like terms have identical variables with identical exponents. $-5x$ and $3x$ are like terms. $-7xyz$ and $-12xyz$ are like terms.

Line segment (7.3) A portion of a straight line that has a beginning and an end.

Liter (6.3) The standard metric measurement of volume. 1 liter = 1000 milliliters. 1 liter = 0.001 kiloliter.

Mean (8.4) The mean of a set of values is the sum of the values divided by the number of values. The mean of the numbers 10, 11, 14, 15 is 12.5. In everyday language, when people use the word "average," they are usually referring to the mean.

Median (8.4) If a set of numbers is arranged in order from smallest to largest, the median is that value that has the same number of values above it as below it. The median of the numbers 3, 7, 8 is 7. If the list contains an even number of items, we obtain the median by finding the mean of the two middle numbers. The median of the numbers 5, 6, 10, 11 is 8.

Megameter (6.2) A metric unit of length equal to 1,000,000 meters.

Meter (6.2) The basic unit of length in the metric system. 1 meter = 1000 millimeters. 1 meter = 0.001 kilometer.

Metric ton (6.3) A metric unit of measurement for very heavy weights. 1 metric ton = 1,000,000 grams.

Microgram (6.3) A unit of weight equal to 0.000001 gram.

Micrometer (6.2) A metric unit of length equal to 0.000001 meter.

Mile (6.1) Largest unit of length in the American system. 5280 feet = 1 mile, 1760 yards = 1 mile.

Milligram (6.3) A metric unit of weight used for very, very small objects. 1 milligram = 0.001 gram.

Milliliter (6.3) The metric unit of volume normally used to measure small volumes. 1 milliliter = 0.001 liter.

Millimeter (6.2) A unit of length commonly used in the metric system to measure very small distances. 1 millimeter = 0.001 meter.

Million (1.1) The number 1,000,000.

Minuend (1.3) The number being subtracted from in a subtraction problem. In the problem $8 - 5 = 3$, the number 8 is the minuend.

Mixed number (2.3) A number created by the sum of a whole number greater than 1 and a proper fraction. The numbers $4\frac{5}{6}$ and $1\frac{1}{8}$ are both mixed numbers. Mixed numbers are sometimes referred to as *mixed fractions*.

Mode (8.4) The mode of a set of data is the number or numbers that occur most often.

Multiplicand (1.4) The first factor in a multiplication problem. In the problem $7 \times 2 = 14$, the number 7 is the multiplicand.

Multiplier (1.4) The second factor in a multiplication problem. In the problem $6 \times 3 = 18$, the number 3 is the multiplier.

Nanogram (6.3) A unit of weight equal to 0.000000001 gram.

Nanometer (6.2) A metric unit of length equal to 0.000000001 meter.

Negative numbers (9.1) All of the numbers to the left of zero on the number line. The numbers -1.5, -16, -200.5, -4500 are all negative numbers. All negative numbers are written with a negative sign in front of the digits.

Number line (1.7) A line on which numbers are placed in order from smallest to largest.

Numerator (2.1) The number on the top of a fraction. In the fraction $\frac{3}{7}$ the numerator is 3.

Numerical coefficients (10.1) The numbers in front of the variables in one or more terms. If we look at $-3xy + 12w$, we find that the numerical coefficient of the xy term is -3 while the numerical coefficient of the w term is 12.

Octagon (7.2) An eight-sided figure.

Odometer (1.8) A device on an automobile that displays how many miles the car has been driven since it was first put into operation.

Opposite of a signed number (9.2) The opposite of a signed number is a number that has the same absolute value. The opposite of -5 is 5. The opposite of 7 is -7.

Order of operations (1.6) An agreed-upon procedure to do a problem with several arithmetic operations in the proper order.

Ounce (6.1) Smallest unit of weight in the American system. 16 ounces = 1 pound.

Overtime (2.9) The pay earned by a person if he works more than a certain number of hours per week. In most jobs that pay by the hour, a person will earn $1\frac{1}{2}$ times as much per hour for every hour beyond 40 hours worked in one workweek. For example, Carlos earns $6.00 per hour for the first 40 hours in a week and overtime for each additional hour. He would earn $9.00 per hour for all hours he worked in that week beyond 40 hours.

Parallel lines (7.2) Two straight lines that are always the same distance apart.

Parallelogram (7.2) A four-sided figure with both pairs of opposite sides parallel.

Parentheses (1.4) One of several symbols used in mathematics to indicate multiplication. For example, (3)(5) means 3 multiplied by 5. A parentheses is also used as a grouping symbol.

Percent (5.1) The word *percent* means per one hundred. For example, 14 percent means $\frac{14}{100}$.

Percent of decrease (5.4) The percent that something decreases is determined by dividing the amount of decrease by the original amount. If a tape deck sold for $300 and its price was decreased by $60, the percent of decrease would be $\frac{60}{300} = 0.20 = 20\%$.

Percent of increase (5.4) The percent that something increases is determined by dividing the amount of increase by the original amount. If the population of a town was 5000 people and the population increased by 500 people, the percent of increase would be $\frac{500}{5000} = 0.10 = 10\%$.

Percent proportion (5.3B) The percent proportion is the equation $\frac{a}{b} = \frac{p}{100}$ where a is the amount, b is the base, and p is the percent number.

Percent symbol (5.1) A symbol that is used to indicate percent. To indicate 23 percent, we write 23%.

Perfect square (7.4) When a whole number is multiplied by itself, the number that is obtained is a perfect square. The numbers 1, 4, 9, 16, 25, 36, 49, 64, 81, and 100 are all perfect squares.

Perimeter (7.1) The distance around a figure.

Perpendicular lines (7.1) Lines that meet at an angle of 90 degrees.

Pi (7.6) Pi is an irrational number that we obtain if we divide the circumference of a circle by the diameter of a circle. It is represented by the symbol π. Accurate to eleven decimal places, the value of pi is given by 3.14159265359. For most work in this textbook, the value of 3.14 is used to approximate the value of pi.

Picogram (6.3) A unit of weight equal to 0.000000000001 gram.

Pint (6.1) Unit of volume in the American system. 2 pints = 1 quart.

Placeholder (1.1) The use of a digit to indicate a place. Zero is a placeholder in our number system. It holds a position and shows that there is no other digit in that place.

Place-value system (1.1) Our number system is called a place-value system because the placement of the digits tells the value of the number. If we use the digits 5 and 4 to write the number 54, the result is different than if we placed them in opposite order and wrote 45.

Positive numbers (9.1) All of the numbers to the right of zero on the number line. The numbers 5, 6.2, 124.186, 5000 are all positive numbers. A positive number such as +5 is usually written without the positive sign.

Pound (6.1) Basic unit of weight in the American system. 2000 pounds = 1 ton. 16 ounces = 1 pound.

Power of 10 (1.4) Whole numbers that begin with 1 and end in one or more zeros are called powers of 10. The numbers 10, 100, 1000, etc., are all powers of 10.

Prime factors (2.2) Factors that are prime numbers. If we write 15 as a product of prime factors, we have $15 = 5 \times 3$.

Prime number (2.2) A prime number is a whole number greater than 1 that can only be divided by 1 and itself. The first fifteen prime numbers are 2, 3, 5, 7, 11, 13, 17, 19, 23, 29, 31, 37, 41, 43, 47. The list of prime numbers goes on indefinitely.

Principal (5.4) The amount of money deposited or borrowed on which interest is computed. In the simple interest formula $I = P \times R \times T$, the P stands for the principal. (The other letters are I = interest, R = interest rate, and T = the amount of time.)

Product (1.4) The answer in a multiplication problem. In the problem $3 \times 4 = 12$ the number 12 is the product.

Proper fraction (2.3) A fraction in which the numerator is less than the denominator. The fractions $\frac{3}{4}$ and $\frac{15}{16}$ are proper fractions.

Proportion (4.2) A statement that two ratios or two rates are equal. The statement $\frac{3}{4} = \frac{15}{20}$ is a proportion. The statement $\frac{5}{7} = \frac{7}{9}$ is false, and is therefore not a proportion.

Pyramid (7.7) A three-dimensional object made up of a geometric figure for a base and triangular sides that meet at a point. Some pyramids are shaped like the great pyramids of Egypt.

Pythagorean Theorem (7.5) A statement that for any right triangle the square of the hypotenuse equals the sum of the squares of the two legs of the triangle.

Quadrilateral (7.2) A four-sided geometric figure.

Quadrillion (1.1) The number 1,000,000,000,000,000.

Quart (6.1) Unit of volume in the American system. 4 quarts = 1 gallon.

Quotient (1.5) The answer after performing a division problem. In the problem $60 \div 6 = 10$ the number 10 is the quotient.

Radius of a circle (7.6) A line segment from the center of a circle to any point on the circle. The radius of a circle is equal to one-half of the diameter of the circle.

Rate (4.1) A rate compares two quantities that have different units. Examples of rates are \$5.00 an hour and 13 pounds for every 2 inches. In fraction form, these two rates would be written as $\frac{\$5.00}{1\text{ hour}}$ and $\frac{13\text{ pounds}}{2\text{ inches}}$.

Ratio (4.1) A ratio is a comparison of two quantities that have the same units. To compare 2 to 3, we can express the ratio in three ways: the ratio of 2 to 3; 2:3; or the fraction $\frac{2}{3}$.

Ratio in simplest form (4.1) A ratio is in simplest form when the two numbers do not have a common factor.

Rectangle (7.1) A four-sided figure that has four right angles.

Reduced fraction (2.2) A fraction for which the numerator and denominator have no common factor other than 1. The fraction $\frac{5}{7}$ is a reduced fraction. The fraction $\frac{15}{21}$ is not a reduced fraction because both numerator and denominator have a common factor of 3.

Regular hexagon (7.2) A six-sided figure with all sides equal.

Regular octagon (7.2) An eight-sided figure with all sides equal.

Remainder (1.5) When two numbers do not divide exactly, a part is left over. This part is called the remainder. For example, $13 \div 2 = 6$ with 1 left over; the 1 is the remainder.

Repeating decimals (3.6) Decimals that have a digit or a group of digits that repeat. The decimals 0.33333333333 . . . and 1.234234234234 . . . are repeating decimals. The pattern of repeating continues indefinitely. Repeating decimals can be written in a form with a bar over the repeating digit(s). Thus the decimals above could be written as $0.\overline{3}$ and $1.\overline{234}$.

Right angle (7.1) and (7.3) An angle that measures 90 degrees.

Right triangle (7.3) A triangle with one 90-degree angle.

Rounding (1.7) The process of writing a number in an approximate form for convenience. The number 9756 rounded to the nearest hundred is 9800.

Sales tax (5.4) The amount of tax on a purchase. The sales tax for any item is a product of the sales tax rate times the purchase price. If an item is purchased for \$12.00 and the sales tax rate is 5%, the sales tax is $5\% \times 12.00 = \$0.60$.

Scientific notation (9.5) A positive number is written in scientific notation if it is in the form $a \times 10^n$ where a is a number greater than or equal to 1, but less than 10, and n is an integer. If we write 5678 in scientific notation, we have 5.678×10^3. If we write 0.00825 in scientific notation, we have 8.25×10^{-3}.

Semicircle (7.6) One-half of a circle. The semicircle usually includes the diameter of a circle connected to one-half of the circumference of the circle.

Sides of an angle (7.3) The two line segments that meet to form an angle.

Signed numbers (9.1) All of the numbers on a number line. Numbers like -33, 2, 5, -4.2, 18.678, -8.432 are all signed numbers. A negative number always has a negative sign in front of the digits. A positive number such as $+3$ is usually written without the positive sign in front of it.

Similar triangles (7.8) Two triangles that have the same shape but are not necessarily the same size. The corresponding angles of similar triangles are equal. The corresponding sides of similar triangles have the same ratio.

Simple interest (5.4) The interest determined by the formula $I = P \times R \times T$ where I = the interest obtained, P = the principal or the amount borrowed or invested, R = the interest rate (usually on an annual basis), and T = the number of time periods (usually years).

Solution of an equation (10.3) A number is a solution of an equation if replacing the variable by the number makes the equation always true. The solution of $x - 5 = -20$ is the number -15.

Sphere (7.7) A three-dimensional object shaped like a perfectly round ball.

Square (7.1) A rectangle with all four sides equal.

Square root (7.4) The square root of a number is one of two identical factors of that number. The square root of 9 is 3. The square root of 121 is 11.

Square root sign (7.4) The symbol $\sqrt{}$. When we want to find the square root of 25, we write $\sqrt{25}$. The answer is 5.

Standard notation for a number (1.1) A number written in ordinary terms. For example, 70 + 2 in standard notation is 72.

Subtrahend (1.3) The number being subtracted. In the problem $7 - 1 = 6$, the number 1 is the subtrahend.

Sum (1.2) The result of an addition of two or more numbers. In the problem $7 + 3 + 5 = 15$, the number 15 is the sum.

Trapezoid (7.2) A four-sided figure with at least two parallel sides.

Term (10.1) A number, a variable, or a product of a number and one or more variables. $5x$, $2ab$, $-43cdef$ are three examples of terms, separated in an expression by a + sign or a − sign.

Terminating decimals (3.6) Every fraction can be written as a decimal. If the division process of dividing denominator into numerator ends with a remainder of zero, the decimal is a terminating decimal. Decimals such as 1.28, 0.007856, and 5.123 are terminating decimals.

Triangle (7.3) A three-sided figure.

Trillion (1.1) The number 1,000,000,000,000.

Unit fraction (6.1) A fraction used to change one unit to another. For example, to change 180 inches to feet, we multiply by the unit fraction $\frac{1 \text{ foot}}{12 \text{ inches}}$. Thus we have

$$180 \text{ inches} \times \frac{1 \text{ foot}}{12 \text{ inches}} = 15 \text{ feet}$$

Variable (10.1) A letter that is used to represent a number.

Vertex of an angle (7.3) The point at which two line segments meet to form an angle.

Vertex of a cone (7.7) The sharp point of a cone.

Volume (7.7) The measure of the space inside a three-dimensional object. Volume is measured in cubic units such as cubic feet.

Width of a rectangle (7.1) Each of the shorter sides of a rectangle.

Whole numbers (1.1) The whole numbers are the set of numbers 0, 1, 2, 3, 4, 5, 6, 7, 8, 9, 10, 11, 12, ... The set goes on forever. There is no largest whole number.

Word names for whole numbers (1.1) The notation for a number in which each digit is expressed by a word. To write 389 with a word name, we would write three hundred eighty-nine.

Zero (1.1) The smallest whole number. It is normally written 0.

Solutions to Practice Problems

Chapter 1

1.1 Practice Problems

1. **(a)** $3182 = 3000 + 100 + 80 + 2$
   (b) $520{,}890 = 500{,}000 + 20{,}000 + 800 + 90$
   (c) $709{,}680{,}059 = 700{,}000{,}000 + 9{,}000{,}000 + 600{,}000 + 80{,}000 + 50 + 9$
2. **(a)** 492 **(b)** 80,427
3. **(a)** 7 **(b)** 9 **(c)** 4,000
   (d) 900,000 for the first 9, 9 for the last 9
4. **(a)** Two thousand, seven hundred thirty-six
   (b) Nine hundred eighty thousand, three hundred six
   (c) Twelve million, twenty-one
5. The national debt in 1989 was two trillion, eight hundred fifty-seven billion, four hundred six million, three hundred thousand dollars.
6. **(a)** 803 **(b)** 30,229
7. **(a)** 13,000
   (b) 88,000
   (c) 10,000

1.2 Practice Problems

1. **(a)** $\begin{array}{r} 7 \\ +5 \\ \hline 12 \end{array}$ **(b)** $\begin{array}{r} 9 \\ +4 \\ \hline 13 \end{array}$ **(c)** $\begin{array}{r} 3 \\ +0 \\ \hline 3 \end{array}$

2. $$\begin{array}{rl} 7 & 7 \\ +6 & \\ +5 & 11 \\ +8 & \\ +2 & 10 \\ \hline 28 & \end{array}$$

3. $$\begin{array}{rll} 1 & & \\ +7 & & \\ +2 & 10 & 10 \\ +9 & & \\ +3 & & \\ \hline 22 & & \end{array}$$

4. $$\begin{array}{r} 8246 \\ +1702 \\ \hline 9948 \end{array}$$

5. $$\begin{array}{r} {\scriptstyle 1} \\ 56 \\ +36 \\ \hline 92 \end{array}$$

6. $$\begin{array}{r} {\scriptstyle 2\,1} \\ 789 \\ 63 \\ +297 \\ \hline 1149 \end{array}$$

7. **(a)** $\begin{array}{r} {\scriptstyle 1\,1\,1} \\ 127 \\ 9876 \\ +342 \\ \hline 10{,}345 \end{array}$ **(b)** Check by adding in opposite order. $\begin{array}{r} {\scriptstyle 1\,1\,1} \\ 342 \\ 9876 \\ +127 \\ \hline 10{,}345 \end{array}$ same

8. $$\begin{array}{r} {\scriptstyle 12\ 12} \\ 18{,}316 \\ 24{,}789 \\ +22{,}965 \\ \hline 66{,}070 \end{array}$$ total women

9. $$\begin{array}{r} 1000 \\ 2000 \\ 1000 \\ +2000 \\ \hline 6000 \text{ ft} \end{array}$$

1.3 Practice Problems

1. **(a)** $\begin{array}{r} 9 \\ -6 \\ \hline 3 \end{array}$ **(b)** $\begin{array}{r} 12 \\ -5 \\ \hline 7 \end{array}$ **(c)** $\begin{array}{r} 17 \\ -8 \\ \hline 9 \end{array}$ **(d)** $\begin{array}{r} 14 \\ -0 \\ \hline 14 \end{array}$ **(e)** $\begin{array}{r} 18 \\ -9 \\ \hline 9 \end{array}$

2. $$\begin{array}{r} 7695 \\ -3481 \\ \hline 4214 \end{array}$$

3. $$\begin{array}{r} {\scriptstyle 2\,14} \\ \not{3}\not{4} \\ -16 \\ \hline 18 \end{array}$$

4. $$\begin{array}{r} {\scriptstyle 8\,13} \\ 6\not{9}\not{3} \\ -426 \\ \hline 267 \end{array}$$

5. $$\begin{array}{r} {\scriptstyle 16} \\ {\scriptstyle 8\,9\,\not{6}\,10} \\ \not{9}\not{0}\not{7}\not{0} \\ -5886 \\ \hline 3184 \end{array}$$

6. **(a)** $\begin{array}{r} 8964 \\ -985 \\ \hline 7979 \end{array}$ **(b)** $\begin{array}{r} 50{,}000 \\ -32{,}508 \\ \hline 17{,}492 \end{array}$

7. Subtraction $\begin{array}{r} 9763 \\ -5732 \\ \hline 4031 \end{array}$ Checking by Addition $\begin{array}{r} 5732 \\ +4031 \\ \hline 9763 \end{array}$ IT CHECKS

8. **(a)** $\begin{array}{r} 284{,}000 \\ -96{,}327 \\ \hline 187{,}673 \end{array}$ Checking by Addition $\begin{array}{r} 96{,}327 \\ +187{,}673 \\ \hline 284{,}000 \end{array}$ IT CHECKS

   (b) $\begin{array}{r} 8{,}526{,}024 \\ -6{,}397{,}518 \\ \hline 2{,}128{,}506 \end{array}$ Checking by Addition $\begin{array}{r} 6{,}397{,}518 \\ +2{,}128{,}506 \\ \hline 8{,}526{,}024 \end{array}$ IT CHECKS

9. **(a)** $x = 5$ **(b)** $x = 12$

10. **(a)** $\begin{array}{r} 23{,}667{,}947 \\ -14{,}227{,}799 \\ \hline 9{,}440{,}148 \end{array}$ **(b)** $\begin{array}{r} 11{,}198{,}655 \\ -9{,}579{,}677 \\ \hline 1{,}618{,}978 \end{array}$

11. **(a)** From the bar graph:

    $\begin{array}{lr} \text{1995 sales} & 114 \\ \text{1994 sales} & -78 \\ \hline \text{Sales increase} & 36 \end{array}$

    (b) From the bar graph:

    $\begin{array}{lr} \text{Springfield} & 91 \\ \text{Riverside} & -78 \\ \hline & 13 \text{ more homes} \end{array}$

    (c) $\begin{array}{lr} \text{1995 sales} & 271 \\ \text{1994 sales} & -240 \\ \hline & 31 \end{array}$ $\begin{array}{lr} \text{1996 sales} & 284 \\ \text{1995 sales} & -271 \\ \hline & 13 \end{array}$

    Therefore, the greatest increase in sales occurred from 1994 to 1995.

1.4 Practice Problems

1. **(a)** 64 **(b)** 42 **(c)** 40 **(d)** 63 **(e)** 81

2.
$$\begin{array}{r} 3021 \\ \times \quad 3 \\ \hline 9063 \end{array}$$

3.
$$\begin{array}{r} {\scriptstyle 2\ \ } \\ 43 \\ \times\ 8 \\ \hline 344 \end{array}$$

4.
$$\begin{array}{r} {\scriptstyle 5\,6\ } \\ 579 \\ \times \quad 7 \\ \hline 4053 \end{array}$$

5. **(a)** $1267 \times 10 = 12{,}670$
(b) $1267 \times 1000 = 1{,}267{,}000$
(c) $1267 \times 10{,}000 = 12{,}670{,}000$
(d) $1267 \times 1{,}000{,}000 = 1{,}267{,}000{,}000$

6. **(a)** $9 \times 6 \times 10{,}000 = 54 \times 10{,}000 = 540{,}000$
(b) $15 \times 4 \times 100 = 60 \times 100 = 6{,}000$
(c) $21 \times 8 \times 1000 = 168 \times 1000 = 168{,}000$

7.
$$\begin{array}{r} 323 \\ \times\ 32 \\ \hline 646 \\ 969\ \\ \hline 10{,}336 \end{array}$$

8.
$$\begin{array}{r} 385 \\ \times\ 69 \\ \hline 3465 \\ 2310\ \\ \hline 26{,}565 \end{array}$$

9.
$$\begin{array}{r} 34 \\ \times 20 \\ \hline 680 \end{array}$$

10.
$$\begin{array}{r} 130 \\ \times\ 50 \\ \hline 6500 \end{array}$$

11.
$$\begin{array}{r} 923 \\ \times 675 \\ \hline 4615 \\ 6461\ \\ 5538\ \ \\ \hline 623{,}025 \end{array}$$

12. $25 \times 4 \times 17 = (25 \times 4) \times 17 = 100 \times 17 = 1700$

13.
$$\begin{aligned} 8 \times 4 \times 3 \times 25 &= 8 \times 3 \times (4 \times 25) \\ &= 24 \times 100 \\ &= 2400 \end{aligned}$$

14.
$$\begin{array}{r} 17348 \\ \times \quad 378 \\ \hline 138784 \\ 121436\ \\ 52044\ \ \\ \hline 6{,}557{,}544 \end{array}$$
The total sales of cars is about $6,557,544.

15. Area = 5 yards × 7 yards = 35 square yards.

1.5 Practice Problems

1. **(a)** $4\overline{)36}$ with quotient 9 **(b)** $5\overline{)25}$ with quotient 5 **(c)** $9\overline{)72}$ with quotient 8 **(d)** $6\overline{)30}$ with quotient 5

2. **(a)** $\dfrac{7}{1} = 7$ **(b)** $\dfrac{8}{8} = 1$ **(c)** $\dfrac{0}{5} = 0$

(d) $\dfrac{12}{0}$ Cannot be done **(e)** $\dfrac{0}{0}$ Cannot be done

3.
$$\begin{array}{r} 7\text{ R3} \\ 6\overline{)45} \\ \underline{42} \\ 3 \end{array} \qquad \text{Check.} \quad \begin{array}{r} 6 \\ \times\ 7 \\ \hline 42 \\ +\ 3 \\ \hline 45 \end{array}$$

4.
$$\begin{array}{r} 21\text{ R3} \\ 6\overline{)129} \\ \underline{12}\ \\ 9 \\ \underline{6} \\ 3 \end{array} \qquad \text{Check.} \quad \begin{array}{r} 21 \\ \times \quad 6 \\ \hline 126 \\ +\quad 3\text{ R} \\ \hline 129 \end{array}$$

5.
$$\begin{array}{r} 529\text{ R5} \\ 8\overline{)4237} \\ \underline{40}\ \ \\ 23\ \\ \underline{16}\ \\ 77 \\ \underline{72} \\ 5 \end{array}$$

6.
$$\begin{array}{r} 7\text{ R5} \\ 32\overline{)229} \\ \underline{224} \\ 5 \end{array}$$

7.
$$\begin{array}{r} 1278\text{ R9} \\ 33\overline{)42183} \\ \underline{33}\ \ \ \\ 91\ \ \\ \underline{66}\ \ \\ 258\ \\ \underline{231}\ \\ 273 \\ \underline{264} \\ 9 \end{array}$$

8.
$$\begin{array}{r} 25\text{ R27} \\ 128\overline{)3227} \\ \underline{256}\ \\ 667 \\ \underline{640} \\ 27 \end{array}$$

9.
$$\begin{array}{r} 16{,}852 \\ 7\overline{)117{,}964} \end{array} \qquad \text{Check.} \quad \begin{array}{r} 16{,}852 \\ \times \qquad 7 \\ \hline 117{,}964 \end{array}$$
The cost of one car is $16,852.

10. $\begin{array}{r} 367 \\ 14\overline{)5138} \end{array}$ The average speed was 367 mph.

1.6 Practice Problems

1. **(a)** $12 \times 12 \times 12 \times 12 = 12^4$. 12 is the base, 4 is the exponent.
(b) $2 \times 2 \times 2 \times 2 \times 2 \times 2 = 2^6$. 2 is the base, 6 is the exponent.

2. **(a)** $12^2 = 12 \times 12 = 144$
(b) $6^3 = 6 \times 6 \times 6 = 216$
(c) $2^6 = 2 \times 2 \times 2 \times 2 \times 2 \times 2 = 64$
(d) $1^{10} = 1 \times 1 \times 1 \times 1 \times 1 \times 1 \times 1 \times 1 \times 1 \times 1 = 1$

3. **(a)** $(7)(7)(7) + (8)(8) = 343 + 64 = 407$
(b) $(9)(9) + 1 = 81 + 1 = 82$
(c) $5^4 + 5 = (5)(5)(5)(5) + 5 = 625 + 5 = 630$

4.
$$\begin{aligned} 7 + 4^3 \times 3 &= 7 + 64 \times 3 && \text{Exponents} \\ &= 7 + 192 && \text{Multiply} \\ &= 199 && \text{Add} \end{aligned}$$

5.
$$\begin{aligned} 7 - 3 \times 4 \div 2 + 5 &= 7 - 12 \div 2 + 5 && \text{Multiplying} \\ &= 7 - 6 + 5 && \text{Dividing} \\ &= 1 + 5 && \text{Subtracting} \\ &= 6 && \text{Adding} \end{aligned}$$

6. $4^3 - 2 + 3^2$

$= 4 \times 4 \times 4 - 2 + 3 \times 3$	Evaluating exponents.
$= 64 - 2 + 9$	$4^3 = 64$ and $3^2 = 9$.
$= 62 + 9$	Subtracting
$= 71$	Adding

7. $(17 + 7) \div 6 \times 2 + 7 \times 3 - 4$

$= 24 \div 6 \times 2 + 7 \times 3 - 4$	Combining inside parenthesis
$= 4 \times 2 + 7 \times 3 - 4$	Dividing
$= 8 + 7 \times 3 - 4$	Multiplying
$= 8 + 21 - 4$	Multiplying
$= 29 - 4$	Adding
$= 25$	Subtracting

8. $5^2 - 6 \div 2 + 3^4 + 7 \times (12 - 10)$

$= 5^2 - 6 \div 2 + 3^4 + 7 \times 2$	Parenthesis
$= 25 - 6 \div 2 + 81 + 7 \times 2$	Exponents
$= 25 - 3 + 81 + 7 \times 2$	Divide
$= 25 - 3 + 81 + 14$	Multiply
$= 22 + 81 + 14$	Subtract
$= 117$	Add

1.7 Practice Problems

1.

65,528	Locate the thousands round-off place.
65,(5)28	The first digit to the right is 5 or more. We will increase the thousands digit by 1.
66,000	All digits to the right of thousands are replaced by zero.

2. 17(2),963 = 170,000 to the nearest ten thousand.

3. (a) 53,282 = 53,280 to nearest ten. The digit to the right of tens was less than 5.

   (b) 164,485 = 164,000 to nearest thousand. The digit to the right of thousands was less than 5.

   (c) 1,365,273 = 1,400,000 to nearest hundred thousand. The digit to the right of hundred thousands was greater than 5.

4. (a) 935,682 = 936,000. The digit to the right of thousands was greater than 5.

   (b) 935,682 = 900,000. The digit to the right of hundred thousands was less than 5.

   (c) 935,682 = 1,000,000. The digit to the right of millions was greater than 5.

5. 9,460,000,000,000,000 = 9,500,000,000,000,000 meters to the nearest hundred trillion.

6.

Actual Sum	Estimated Sum	
3456	3000	
9876	10000	
5421	5000	
+ 1278	+ 1000	
20,031	19,000	Close to the actual sum

7.

\$697	\$700
35	40
+ 19	+ 20
	\$760

We estimate that the total cost is \$760. (The exact answer is \$751, so we can see that our answer is quite close.

8. 10,000 + 10,000 + 20,000 + 60,000 = 100,000
This is significantly different from 81,358, so we would suspect that an error has been made. In fact, Ming did make an error. The exact sum is actually 101,358!

9. 30,000,000 − 10,000,000 = 20,000,000
We estimate that 20,000,000 more people lived in California than in Florida.

10. $9000 \times 7000 = 63{,}000{,}000$
We estimate the product to be 63,000,000.

11. $40\overline{)80{,}000}$ = 2,000 Our estimate is 2,000.

12. $60\overline{)2{,}000{,}000}$ = 33,333 Our estimate is \$33,333 for one truck.

1.8 Practice Problems

Practice Problem 1

1. Understand the problem.

MATHEMATICS BLUEPRINT FOR PROBLEM SOLVING			
Gather the Facts	What Am I Asked to Do?	How Do I Proceed?	Key Points to Remember
The deductions are: \$135 \$28 \$13 \$34	Find out the total amount of deductions.	I must add the four deductions to obtain the total.	Watch out! Gross pay of \$1352 is not needed to solve the problem.

2. Solve and state the answer:
135 + 28 + 13 + 34 = 210
The total amount taken out of Diane's paycheck is \$210.
3. Check. Estimate the answer to see if it is reasonable.

Practice Problem 2

1. Understand the problem.

MATHEMATICS BLUEPRINT FOR PROBLEM SOLVING			
Gather the Facts	What Am I Asked to Do?	How Do I Proceed?	Key Points to Remember
Clinton had 1,964,842 votes. Bush had 1,876,495 votes.	Find out by how many votes Clinton beat Bush.	I must subtract the amounts.	Clinton is a Democrat and Bush is a Republican.

2. Solve and state the answer:

$$\begin{array}{r} 1{,}964{,}842 \\ -\ 1{,}876{,}495 \\ \hline 88{,}347 \end{array}$$

Clinton beat Bush by 88,347 votes.

3. Check. Estimate the answer to see if it is reasonable.

Practice Problem 3

1. Understand the problem.

MATHEMATICS BLUEPRINT FOR PROBLEM SOLVING			
Gather the Facts	What Am I Asked to Do?	How Do I Proceed?	Key Points to Remember
1 gallon is 1,024 fluid drams.	Find out how many fluid drams in 9 gallons.	I need to multiply 1024 by 9.	I must use fluid drams as my measure in my answer.

2. Solve and state the answer:

$$\frac{1024 \text{ fluid drams}}{1\, \cancel{\text{gal}}} \times 9\, \cancel{\text{gal}} = 9{,}216 \text{ fluid drams.}$$

There would be 9,216 fluid drams in 9 gallons.

3. Check. Estimate the answer to see if it is reasonable.

Practice Problem 4

1. Understand the problem.

MATHEMATICS BLUEPRINT FOR PROBLEM SOLVING			
Gather the Facts	What Am I Asked to Do?	How Do I Proceed?	Key Points to Remember
Donna bought 45 shares of stock. She paid \$1620 for them.	Find out the cost per share of stock.	I need to divide \$1620 by 45.	Use dollars as the unit in the answer.

2. Solve and state the answer:

$$\begin{array}{r} \$36 \\ 45\overline{)1620} \\ \underline{135} \\ 270 \\ \underline{270} \\ 0 \end{array}$$

Donna paid $36 per share for the stock.

3. Check. Estimate the answer to see if it is reasonable.

Practice Problem 5

1. Understand the problem.

MATHEMATICS BLUEPRINT FOR PROBLEM SOLVING			
Gather the Facts	What Am I Asked to Do?	How Do I Proceed?	Key Points to Remember
50 tables at \$200 each. 180 chairs at \$40 each. 6 carts at \$65 each.	Find the total cost.	1. Multiply the number of purchases by each price. 2. Add the total costs of each of the items.	There are three types of items involved in the total purchase.

2. Solve and state the answer:

50 tables at \$200 each = 50 × 200 = \$10,000 cost of tables
180 chairs at \$40 each = 180 × 40 = 7,200 cost of chairs
6 carts at \$65 each = 6 × 65 = 390 cost of carts
\$17,590 total cost

The total purchase was \$17,590

3. Check. Estimate the answer to see if it is reasonable.

Practice Problem 6

1. Understand the problem.

MATHEMATICS BLUEPRINT FOR PROBLEM SOLVING			
Gather the Facts	What Am I Asked to Do?	How Do I Proceed?	Key Points to Remember
Old Balance: \$498 New Deposits: \$607 \$163 Interest: \$36 Withdrawals: \$ 19 \$158 \$582 \$ 74	Find her new balance after the transactions.	(a) Add the new deposits and interest to the old balance. (b) Add the withdrawals. (c) Subtract the results from steps (a) and (b).	Deposits and interest are added and withdrawals are subtracted from savings accounts.

2. Solve and state the answer:

(a) \$498 + 607 + 163 + 36 = \$1304

(b) \$ 19 + 158 + 582 + 74 = \$833

(c) \$1304 − 833 = \$471

Her balance this month is \$471.

3. Check. Estimate the answer to see if it is reasonable.

Practice Problem 7

1. Understand the problem.

MATHEMATICS BLUEPRINT FOR PROBLEM SOLVING			
Gather the Facts	What Am I Asked to Do?	How Do I Proceed?	Key Points to Remember
Odometer at end of trip: 51,118 miles Beginning of trip: 50,698 Used on trip: 12 gallons of gas.	Find the number of miles per gallon that the car obtained on the trip.	(a) Subtract the two odometer readings. (b) Divide that number by 12.	The gas tank was full at the beginning of the trip. 12 gallons fills the tank at the end of the trip.

2. Solve and state the answer:

51,118 odometer at end of trip
− 50,698 odometer at start of trip
420 miles traveled on trip

$$\frac{420 \text{ miles}}{12 \text{ gallons of gas used}} = 12\overline{)420} = 35 \text{ miles per gallon on the trip}$$

$$\begin{array}{r} 35 \\ 12\overline{)420} \\ 36 \\ \hline 60 \\ 60 \\ \hline \end{array}$$

3. Check. Estimate the answer to see if it is reasonable.

Chapter 2

2.1 Practice Problems

1. (a) Five parts out of twelve are shaded. The fraction is $\frac{5}{12}$.

(b) Three parts out of six are shaded. The fraction is $\frac{3}{6}$.

(c) Two parts out of three are shaded. The fraction is $\frac{2}{3}$.

2. (a) $\frac{4}{5}$ of the object is shaded. 

(b) $\frac{3}{7}$ of the group is shaded. 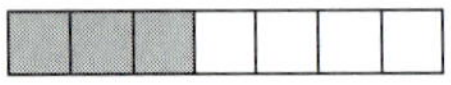

3. (a) $\frac{9}{17}$ represents 9 women out of 17.

(b) The total class is $382 + 351 = 733$.

The fractional part that is women is $\frac{351}{733}$.

(c) $\frac{7}{8}$ of a yard of material.

4. Total number of defective items $1 + 2 = 3$. Total number of items $7 + 9 = 16$. A fraction that represents the portion of the items that were defective is $\frac{3}{16}$.

2.2 Practice Problems

1. (a)
$$\begin{aligned} 18 &= 9 \times 2 \\ &= 3 \times 3 \times 2 \\ &= 3^2 \times 2 \end{aligned}$$

(b)
$$\begin{aligned} 72 &= 9 \times 8 \\ &= 3 \times 3 \times 2 \times 2 \times 2 \\ &= 3^2 \times 2^3 \end{aligned}$$

(c)
$$\begin{aligned} 400 &= 10 \times 40 \\ &= 5 \times 2 \times 5 \times 8 \\ &= 5 \times 2 \times 5 \times 2 \times 2 \times 2 \\ &= 5^2 \times 2^4 \end{aligned}$$

2. (a) $\frac{30 \div 6}{42 \div 6} = \frac{5}{7}$

(b) $\frac{60 \div 12}{132 \div 12} = \frac{5}{11}$

3. (a) $\frac{120}{135} = \frac{2 \times 2 \times 2 \times \cancel{3} \times \cancel{5}}{3 \times 3 \times \cancel{3} \times \cancel{5}} = \frac{8}{9}$

(b) $\frac{715}{880} = \frac{\cancel{5} \times \cancel{11} \times 13}{\cancel{5} \times \cancel{11} \times 2 \times 2 \times 2 \times 2} = \frac{13}{16}$

4. (a) $\frac{84}{108} \stackrel{?}{=} \frac{7}{9}$

$84 \times 9 = 108 \times 7$

$756 = 756$ Yes

(b) $\frac{3}{7} \stackrel{?}{=} \frac{79}{182}$

$3 \times 182 = 7 \times 79$

$546 = 553$ No

2.3 Practice Problems

1. (a) $4\frac{3}{7} = \frac{7 \times 4 + 3}{7} = \frac{28 + 3}{7} = \frac{31}{7}$

(b) $6\frac{2}{3} = \frac{3 \times 6 + 2}{3} = \frac{18 + 2}{3} = \frac{20}{3}$

(c) $19\frac{4}{7} = \frac{7 \times 19 + 4}{7} = \frac{133 + 4}{7} = \frac{137}{7}$

2. (a) $4\overline{)17}$ gives quotient 4; 16; remainder 1 so $\frac{17}{4} = 4\frac{1}{4}$

(b) $5\overline{)36}$ gives quotient 7; 35; remainder 1 so $\frac{36}{5} = 7\frac{1}{5}$

(c) $27\overline{)116}$ gives quotient 4; 108; remainder 8 so $\frac{116}{27} = 4\frac{8}{27}$

(d) $13\overline{)91}$ gives quotient 7; 91; remainder 0 so $\frac{91}{13} = 7$

3. $\frac{51}{15} = \frac{3 \times 17}{3 \times 5} = \frac{\overset{1}{\cancel{3}} \times 17}{\underset{1}{\cancel{3}} \times 5} = \frac{17}{5}$

4. $\frac{16}{80} = \frac{1}{5}$

Therefore, $3\frac{16}{80} = 3\frac{1}{5}$

5. $\frac{1001}{572} = 1\frac{429}{572}$

Now the fraction $\frac{429}{572} = \frac{3 \times \overset{1}{\cancel{11}} \times \overset{1}{\cancel{13}}}{2 \times 2 \times \underset{1}{\cancel{11}} \times \underset{1}{\cancel{13}}} = \frac{3}{4}$.

Thus $\frac{1001}{572} = 1\frac{429}{572} = 1\frac{3}{4}$.

2.4 Practice Problems

1. (a) $\frac{6}{7} \times \frac{3}{13} = \frac{18}{91}$ **(b)** $\frac{1}{5} \times \frac{11}{12} = \frac{11}{60}$

2. $\frac{\overset{5}{\cancel{55}}}{\underset{9}{\cancel{72}}} \times \frac{\overset{2}{\cancel{16}}}{\underset{3}{\cancel{33}}} = \frac{5}{9} \times \frac{2}{3} = \frac{10}{27}$

3. (a) $7 \times \frac{5}{13} = \frac{7}{1} \times \frac{5}{13} = \frac{35}{13}$ or $2\frac{9}{13}$

(b) $\frac{13}{16} \times 8 = \frac{13}{\underset{2}{\cancel{16}}} \times \frac{\overset{1}{\cancel{8}}}{1} = \frac{13}{2}$ or $6\frac{1}{2}$

4. $\frac{3}{\underset{1}{\cancel{8}}} \times \overset{12{,}300}{\cancel{98{,}400}} = \frac{3}{1} \times 12{,}300 = 36{,}900$

There are 36,900 ft^2 in the wetland area.

5. (a) $2\frac{1}{6} \times \frac{4}{7} = \frac{13}{\underset{3}{\cancel{6}}} \times \frac{\overset{2}{\cancel{4}}}{7} = \frac{26}{21}$ or $1\frac{5}{21}$

(b) $10\frac{2}{3} \times 13\frac{1}{2} = \frac{\overset{16}{32}}{\underset{1}{\cancel{3}}} \times \frac{\overset{9}{\cancel{27}}}{\underset{1}{\cancel{2}}} = \frac{144}{1} = 144$

(c) $\frac{3}{5} \times 1\frac{1}{3} \times \frac{5}{8} = \frac{\overset{1}{\cancel{3}}}{\underset{1}{\cancel{5}}} \times \frac{\overset{1}{\cancel{4}}}{\underset{1}{\cancel{3}}} \times \frac{\overset{1}{\cancel{5}}}{\underset{2}{\cancel{8}}} = \frac{1}{2}$

(d) $3\frac{1}{5} \times 2\frac{1}{2} = \frac{\overset{8}{\cancel{16}}}{\underset{1}{\cancel{5}}} \times \frac{\overset{1}{\cancel{5}}}{\underset{1}{\cancel{2}}} = \frac{8}{1} = 8$

6. Area $= lw = 1\frac{1}{5} \times 4\frac{5}{6} = \frac{\overset{1}{\cancel{6}}}{5} \times \frac{29}{\underset{1}{\cancel{6}}} = \frac{29}{5} = 5\frac{4}{5}\ \text{m}^2$

7. Since $8 \cdot 10 = 80$ and $9 \cdot 9 = 81$

We know that $\frac{8}{9} \cdot \frac{10}{9} = \frac{80}{81}$

Therefore, $x = \frac{10}{9}$

2.5 Practice Problems

1. (a) $\frac{7}{13} \div \frac{3}{4} = \frac{7}{13} \times \frac{4}{3} = \frac{28}{39}$

(b) $\frac{16}{35} \div \frac{24}{25} = \frac{\overset{2}{\cancel{16}}}{\underset{7}{\cancel{35}}} \times \frac{\overset{5}{\cancel{25}}}{\underset{3}{\cancel{24}}} = \frac{10}{21}$

2. (a) $\frac{3}{17} \div \frac{6}{1} = \frac{\overset{1}{\cancel{3}}}{17} \times \frac{1}{\underset{2}{\cancel{6}}} = \frac{1}{34}$

(b) $\frac{14}{1} \div \frac{7}{15} = \frac{\overset{2}{\cancel{14}}}{1} \times \frac{15}{\underset{1}{\cancel{7}}} = 30$

3. (a) $1 \div \frac{11}{13} = \frac{1}{1} \times \frac{13}{11} = \frac{13}{11}$ or $1\frac{2}{11}$

(b) $\frac{14}{17} \div 1 = \frac{14}{17} \times \frac{1}{1} = \frac{14}{17}$

(c) $\frac{3}{11} \div 0$ Division by zero cannot be done. This problem cannot be done.

(d) $0 \div \frac{9}{16} = \frac{0}{1} \times \frac{16}{9} = \frac{0}{9} = 0$

4. (a) $1\frac{1}{5} \div \frac{7}{10} = \frac{6}{\underset{1}{\cancel{5}}} \times \frac{\overset{2}{\cancel{10}}}{7} = \frac{12}{7}$ or $1\frac{5}{7}$

(b) $2\frac{1}{4} \div 1\frac{7}{8} = \frac{9}{4} \div \frac{15}{8} = \frac{\overset{3}{\cancel{9}}}{\underset{1}{\cancel{4}}} \times \frac{\overset{2}{\cancel{8}}}{\underset{5}{\cancel{15}}} = \frac{6}{5}$ or $1\frac{1}{5}$

5. (a) $\frac{5\frac{2}{3}}{7} = 5\frac{2}{3} \div 7 = \frac{17}{3} \times \frac{1}{7} = \frac{17}{21}$

(b) $\frac{1\frac{2}{5}}{2\frac{1}{3}} = 1\frac{2}{5} \div 2\frac{1}{3} = \frac{7}{5} \div \frac{7}{3} = \frac{\overset{1}{\cancel{7}}}{5} \times \frac{3}{\underset{1}{\cancel{7}}} = \frac{3}{5}$

6. $x \div \frac{3}{2} = \frac{22}{36}$

$x \cdot \frac{2}{3} = \frac{22}{36}$

$\frac{11}{12} \cdot \frac{2}{3} = \frac{22}{36}$ Thus $x = \frac{11}{12}$

7. $19\frac{1}{4} \div 14 = \frac{\overset{11}{\cancel{77}}}{4} \times \frac{1}{\underset{2}{\cancel{14}}} = \frac{11}{8}$ or $1\frac{3}{8}$

Each piece will be $1\frac{3}{8}$ inches long.

2.6 Practice Problems

1. (a) The LCD of $\frac{3}{4}$ and $\frac{11}{12}$ is 12.

12 can be divided by 4 and 12.

(b) The LCD of $\frac{1}{7}$ and $\frac{8}{35}$ is 35.

35 can be divided by 7 and 35.

2. The LCD of $\frac{3}{7}$ and $\frac{5}{6}$ is 42.

42 can be divided by 7 and 6.

3. (a) $14 = 2 \times 7$

$10 = 2 \times 5$

The LCD $= 2 \times 5 \times 7 = 70$.

(b) $15 = 3 \times 5$

$50 = 2 \times 5 \times 5$

The LCD $= 2 \times 3 \times 5 \times 5 = 150$

(c) $16 = 2 \times 2 \times 2 \times 2$

$12 = 2 \times 2 \times 3$

The LCD $= 2 \times 2 \times 2 \times 2 \times 3 = 48$

4. $49 = 7 \times 7$

$21 = 7 \times 3$

$7 = 7 \times 1$

LCD $= 7 \times 7 \times 3 = 147$.

5. (a) $\frac{3}{5} \times \frac{8}{8} = \frac{24}{40}$ **(c)** $\frac{2}{7} = \frac{2}{7} \times \frac{4}{4} = \frac{8}{28}$

(b) $\frac{7}{11} \times \frac{4}{4} = \frac{28}{44}$ $\frac{3}{4} = \frac{3}{4} \times \frac{7}{7} = \frac{21}{8}$

6. (a) The LCD is 60 because it can be divided by 20 and 15.

(b) $\frac{3}{20} \times \frac{3}{3} = \frac{9}{60}$

$\frac{11}{15} \times \frac{4}{4} = \frac{44}{60}$

2.7 Practice Problems

1. $\frac{3}{17} + \frac{12}{17} = \frac{15}{17}$

2. (a) $\frac{1}{12} + \frac{5}{12} = \frac{6}{12} = \frac{1}{2}$

(b) $\frac{13}{15} + \frac{7}{15} = \frac{20}{15} = \frac{4}{3}$ or $1\frac{1}{3}$

3. (a) $\frac{5}{19} - \frac{2}{19} = \frac{3}{19}$ **(b)** $\frac{21}{25} - \frac{6}{25} = \frac{15}{25} = \frac{3}{5}$

4.

$$\begin{array}{r} \frac{2}{15} = \frac{2}{15} \\ \frac{2}{5} \times \frac{3}{3} = \frac{6}{15} \\ \hline \frac{8}{15} \end{array}$$

5. LCD = 48 $\frac{5}{12} \times \frac{4}{4} = \frac{20}{48}$ $\frac{5}{16} \times \frac{3}{3} = \frac{15}{48}$

$\frac{5}{12} + \frac{5}{16} = \frac{20}{48} + \frac{15}{48} = \frac{35}{48}$

6. LCD = 48

$$\begin{array}{r} \frac{3}{16} \times \frac{3}{3} = \frac{9}{48} \\ \frac{1}{8} \times \frac{6}{6} = \frac{6}{48} \\ \frac{1}{12} \times \frac{4}{4} = \frac{4}{48} \\ \hline \frac{19}{48} \end{array}$$

7. LCD = 96 $\frac{9}{48} \times \frac{2}{2} = \frac{18}{96}$

$\frac{5}{32} \times \frac{3}{3} = \frac{15}{96}$ $\frac{9}{48} - \frac{5}{32} = \frac{18}{96} - \frac{15}{96} = \frac{3}{96} = \frac{1}{32}$

8.

$$\begin{array}{r} \frac{9}{10} \times \frac{2}{2} = \frac{18}{20} \\ -\frac{1}{4} \times \frac{5}{5} = -\frac{5}{20} \\ \hline \frac{13}{20} \end{array}$$

There is $\frac{13}{20}$ gallon left.

9. The LCD of $\frac{3}{10}$ and $\frac{23}{25}$ is 50.

$\frac{3}{10} \times \frac{5}{5} = \frac{15}{50}$ Now rewriting: $x + \frac{15}{50} = \frac{46}{50}$

$\frac{23}{25} \times \frac{2}{2} = \frac{46}{50}$ $\frac{31}{50} + \frac{15}{50} = \frac{46}{50}$

2.8 Practice Problems

1.
$$\begin{array}{r} 5\frac{1}{12} \\ +\,9\frac{5}{12} \\ \hline 14\frac{6}{12} = 14\frac{1}{2} \end{array}$$

2.
$$\begin{array}{rcr} 6\frac{1}{4} & = & 6\frac{5}{20} \\ +\,2\frac{2}{5} & = & +\,2\frac{8}{20} \\ \hline & & 8\frac{13}{20} \end{array}$$

3. LCD = 12
$$\begin{array}{rcr} 7\boxed{\frac{1}{4}\times\frac{3}{3}} & = & 7\frac{3}{12} \\ +\,3\boxed{\frac{5}{6}\times\frac{2}{2}} & & +\,3\frac{10}{12} \\ \hline & & 10\frac{13}{12} = 10 + 1\frac{1}{12} = 11\frac{1}{12} \end{array}$$

4. LCD = 12
$$\begin{array}{rcr} 12\frac{5}{6} & = & 12\frac{10}{12} \\ -\,7\frac{5}{12} & = & -\,7\frac{5}{12} \\ \hline & & 5\frac{5}{12} \end{array}$$

5a. LCD = 24
$$\begin{array}{rcr} 9\boxed{\frac{1}{8}\times\frac{3}{3}} & = & 9\frac{3}{24} = 8\frac{27}{24} \\ -\,3\boxed{\frac{2}{3}\times\frac{8}{8}} & = & -\,3\frac{16}{24} \quad -\,3\frac{16}{24} \\ \hline & & 5\frac{11}{24} \end{array}$$

Borrow 1 from 9:
$9\frac{3}{24} = 8 + 1 + \frac{3}{24} = 8\frac{27}{24}$

b.
$$\begin{array}{rcr} 18 & = & 17\frac{18}{18} \\ -\,6\frac{7}{18} & = & -\,6\frac{7}{18} \\ \hline & & 11\frac{11}{18} \end{array}$$

6.
$$\begin{array}{rcrcr} 6\frac{1}{4} & = & 6\frac{3}{12} & = & 5\frac{15}{12} \\ -\,4\frac{2}{3} & = & -\,4\frac{8}{12} & = & -\,4\frac{8}{12} \\ \hline & & & & 1\frac{7}{12} \end{array}$$

They had $1\frac{7}{12}$ gallons left over.

2.9 Practice Problems

Practice Problem 1

1. Understand the problem.

MATHEMATICS BLUEPRINT FOR PROBLEM SOLVING			
Gather the Facts	**What Am I Asked to Do?**	**How Do I Proceed?**	**Key Points to Remember**
Gas amounts: $18\frac{7}{10}$ gal, $15\frac{2}{5}$ gal, $14\frac{1}{2}$ gal	Find out how many gallons of gas she bought altogether.	Add the three amounts.	When adding mixed numbers, the LCD is needed for the fractions.

2. Solve and state the answer:

$$\text{LCD} = 10 \quad 18\frac{7}{10} = 18\frac{7}{10}$$
$$15\frac{2}{5} = 15\frac{4}{10}$$
$$14\frac{1}{2} = 14\frac{5}{10}$$
$$47\frac{16}{10} = 48\frac{6}{10}$$
$$= 48\frac{3}{5} \text{ gallons}$$

Total is $48\frac{3}{5}$ gallons.

3. Check. Estimate the answer to see if it is reasonable.

Practice Problem 2

1. Understand the problem.

MATHEMATICS BLUEPRINT FOR PROBLEM SOLVING			
Gather the Facts	What Am I Asked to Do?	How Do I Proceed?	Key Points to Remember
Poster: $12\frac{1}{4}$ in. Top border: $1\frac{3}{8}$ in. Bottom border: 2 in.	Find the length of the inside portion of the poster.	(a) Add the two border lengths. (b) Subtract this total from the poster length.	When adding mixed numbers, the LCD is needed for the fractions.

2. Solve and state the answer:

(a) $1\frac{3}{8} + 2 = 3\frac{3}{8}$

(b) $12\frac{1}{4} - 3\frac{3}{8} = 12\frac{2}{8} - 3\frac{3}{8} = 11\frac{10}{8} - 3\frac{3}{8} = 8\frac{7}{8}$

The inside portion is $8\frac{7}{8}$ inches.

3. Check. Estimate the answer to see if it is reasonable.

Practice Problem 3

1. Understand the problem.

MATHEMATICS BLUEPRINT FOR PROBLEM SOLVING			
Gather the Facts	What Am I Asked to Do?	How Do I Proceed?	Key Points to Remember
Regular tent uses $8\frac{1}{4}$ yards. Large tent uses $1\frac{1}{2}$ times the regular. She makes 6 regular and 15 large tents.	Find out how many yards of cloth will be needed to make the tents.	Find the amount used for regular tents, and the amount used for large tents. Then add the two.	Large tents use $1\frac{1}{2}$ times the regular amount.

2. Solve and state the answer:

We multiply $6 \times 8\frac{1}{4}$ for regular tents and $16 \times 1\frac{1}{2} \times 8\frac{1}{4}$ for large tents. Then add total yardage.

$$6 \times 8\frac{1}{4} \text{ (regular tents)} = \overset{3}{\cancel{6}} \times \frac{33}{\underset{2}{\cancel{4}}} = \frac{99}{2} = 49\frac{1}{2} \text{ yards.}$$

$$\text{(Large tents)} \quad 16 \times 1\frac{1}{2} \times 8\frac{1}{4} = \overset{2}{\cancel{\overset{8}{\cancel{16}}}} \times \frac{3}{\underset{1}{\cancel{2}}} \times \frac{33}{\underset{1}{\cancel{4}}} = \frac{198}{1} = 198 \text{ yards.}$$

Total yards for all tents $198 + 49\frac{1}{2} = 247\frac{1}{2}$ yards.

3. Check. Estimate the answer to see if it is reasonable.

Practice Problem 4

1. Understand the problem.

MATHEMATICS BLUEPRINT FOR PROBLEM SOLVING			
Gather the Facts	What Am I Asked to Do?	How Do I Proceed?	Key Points to Remember
He purchases 12-foot boards. Each shelf is $2\frac{3}{4}$ ft. He needs four shelves for each bookcase and he is making two bookcases.	(a) Find out how many boards he needs to buy. (b) Find out how many feet of shelving are actually needed. (c) Find out how many feet will be left over.	Find out how many $2\frac{3}{4}$ shelves he can get from one board. Then see how many boards he needs to make all eight shelves.	There will be three answers to this problem. Don't forget to calculate the leftover wood.

2. Solve and state the answer:

We want to know how many $2\frac{3}{4}$ ft. shelves are in a 12-ft. board.

$$12 \div 2\frac{3}{4} = \frac{12}{1} \div \frac{11}{4} = \frac{12}{1} \times \frac{4}{11} = \frac{48}{11} = 4\frac{4}{11} \text{ shelves}$$

He will get 4 shelves from each board with some left over.

(a) For two bookcases, he needs 8 shelves. He gets four shelves out of each board. $8 \div 4 = 2$. He will need two 12-ft. boards.

(b) He needs 8 shelves at $2\frac{3}{4}$ feet.

$$8 \times 2\frac{3}{4} = 8 \times \frac{11}{4} = 22 \text{ feet.}$$

He actually needs 22 feet of shelving.

(c) 24 feet of shelving bought
− 22 feet of shelving used
2 feet of shelving left over.

3. Check. Estimate the answer to see if it is reasonable.

Practice Problem 5

1. Understand the problem.

MATHEMATICS BLUEPRINT FOR PROBLEM SOLVING			
Gather the Facts	What Am I Asked to Do?	How Do I Proceed?	Key Points to Remember
Distance is $199\frac{3}{4}$ miles. He uses $8\frac{1}{2}$ gallons of gas	Find out how many miles per gallon he gets.	Divide the distance by the number of gallons.	Change mixed numbers to improper fractions before dividing.

2. Solve and state the answer:

$$199\frac{3}{4} \div 8\frac{1}{2} = \frac{799}{4} \div \frac{17}{2}$$

$$= \frac{\overset{47}{\cancel{799}}}{\underset{2}{\cancel{4}}} \times \frac{\overset{1}{\cancel{2}}}{\underset{1}{\cancel{17}}}$$

$$= \frac{47}{2} = 23\frac{1}{2}$$

He gets $23\frac{1}{2}$ miles per gallon.

3. Check. Estimate the answer to see if it is reasonable.

Chapter 3

3.1 Practice Problems

1. **(a)** 0.073 seventy-three thousandths
(b) 4.68 four and sixty-eight hundredths
(c) 0.0017 seventeen ten-thousandths
(d) 561.78 five hundred sixty-one and seventy-eight hundredths

2. Seven thousand eight hundred sixty-three and $\frac{4}{100}$

3. **(a)** $\frac{9}{10} = 0.9$ **(b)** $\frac{136}{1000} = 0.136$ **(c)** $2\frac{56}{100} = 2.56$

(d) $34\frac{86}{1000} = 34.086$

4. **(a)** $0.37 = \frac{37}{100}$

(b) $182.3 = 182\frac{3}{10}$

(c) $0.7131 = \frac{7131}{10{,}000}$

(d) $42.019 = 42\frac{19}{1000}$

5. **(a)** $8.5 = 8\frac{5}{10} = 8\frac{1}{2}$ **(b)** $0.58 = \frac{58}{100} = \frac{29}{50}$

(c) $36.25 = 36\frac{25}{100} = 36\frac{1}{4}$ **(d)** $106.013 = 106\frac{13}{1000}$

6. $\frac{2}{1{,}000{,}000{,}000} = \frac{1}{500{,}000{,}000}$ The concentration of PCBs is $\frac{1}{500{,}000{,}000}$

3.2 Practice Problems

1. Since 4 < 5

5.74 5.75 therefore, 5.74 < 5.75

2. 0.894 > 0.890, so 0.894 > 0.89

3. 2.45, 2.543, 2.46, 2.54, 2.5

It is helpful to add extra zeros and to place the decimals that begin with 2.4 in a group and the decimals that begin with 2.5 in the other.

2.450, 2.460 2.543, 2.540, 2.500

In order, we have from smallest to largest

2.450, 2.460, 2.500, 2.540, 2.543

It is OK to leave the extra terminal zeros in our answer.

4. 723.9

5. **(a)** 12.92**6**47

Since digit to right of thousandth is less than 5, we drop the digits 4 and 7.

12.926

(b) 0.00**7**892

Since digit to right of thousandths is greater than 5, we round up.

0.008

6. 15,700.0

7.

		Rounded to Nearest Dollar
Medical bills	375.50	376
Taxes	981.39	981
Retirement	980.49	980
Charity	817.65	818

3.3 Practice Problems

1. **(a)**
```
   1
  9.8
  3.6
+ 5.4
-----
 18.8
```
(b)
```
  1 1 1
  300.72
  163.75
+ 291.08
--------
  755.55
```
(c)
```
  21 1
   8.9000
  37.0560
   0.0023
+  9.4500
---------
  55.4083
```

2.
```
   1  1
  93,521.8
+  1,634.8
----------
  95,156.6 miles
```

3.
```
  $ 80.95
   133.91
   256.47
    53.08
 + 381.32
 --------
  $905.73
```

4. **(a)**
```
    7 18
    38.8
  - 26.9
  ------
    11.9
```
(b)
```
    2034.908
  - 1986.325
  ----------
      48.583
```

5. **(a)** 19.000 − 12.579 = 6.421 **(b)** 283.076 − 96.380 = 186.696

6. 87,160.1 − 82,370.9 = 4,789.2 miles between oil changes

7. 15.3 − 10.8 = 4.5 x is 4.5

3.4 Practice Problems

1. 0.09 × 0.6 = 0.054
2 decimal places
+ 1 decimal place
3 decimal places in product

2. **(a)** 5.3 × 0.9 = 4.77
1 decimal place
+ 1 decimal place
2 decimal places in product

(b) 2.831 × 0.07 = 1.9817
3 decimal places
+ 1 decimal place
4 decimal places in product

(c) 0.47 × 0.28
376
94
0.1316
2 decimal places
+ 2 decimal places
4 decimal places in product

(d) 0.436 × 18.39 = 8.01804
3 decimal places
+ 2 decimal places
5 decimal places in product

3. 0.4264 × 38 = 16.2032
4 decimal places
+ 0 decimal place
4 decimal places in product

4. Area = length × width
1.26 × 2.3 = 2.898 square millimeters

5. **(a)** 0.0561 × 10 = 0.561 Decimal point moved one place to the right.
(b) 1462.37 × 100 = 146,237 Decimal point moved two places to the right.

6. **(a)** 0.26 × 1000 = 260 Decimal point moved three places to the right. One extra zero needed.
(b) 5862.89 × 10,000 = 58,628,900 Decimal point moved four places to the right. Two extra zeros needed.

7. **(a)** $7{,}684 \times 10^4 = 76{,}840$ Decimal point moved four places to the right. One extra zero needed.
(b) $0.00073 \times 10^2 = 0.073$ Decimal point moved two places to the right.

8. 156.2 × 1000 = 156,200 meters

3.5 Practice Problems

1. **(a)** 7)1.806 = 0.258
14
40
35
56
56
0

(b) 16)0.0928 = 0.0058
80
128
128
0

2. 46)23.820 = 0.517 = 0.52 to the nearest hundredth
23 0
82
46
360
322

3. 19)$3538.75 = $186.25 each month
19
163
152
118
114
47
38
95
95

4. **(a)** 0.09‸)0.10‸08 = 1.12
9
10
9
18
18
0

(b) 0.037‸)1.702‸ = 46.
1 48
222
222
0

5. **(a)** 1.8‸)0.0‸414 = 0.023
36
54
54

(b) 0.0036‸)8.3160‸ = 2310.
72
111
108
36
36
0

6. **(a)** 3.8‸)521.6‸00 = 137.26 The answer rounded to the nearest tenth is 137.3
38
141
114
27 6
26 6
1 00
76
24 0
22 8

(b) 8.05‸)0.17‸0000 = 0.0211 The answer rounded to the nearest thousandth is 0.021
16 10
900
805
950
805

7. 28.5‸)454.4‸0 = 15.9 The truck got approximately 16 miles per gallon.
285
169 4
142 5
269 0
256 5

8. 0.12‸)0.69‸6 = 5.8 n is 5.8
60
9 6
9 6

3.6 Practice Problems

1. (a)

$$\begin{array}{r} 0.3125 \\ 16\overline{)5.0000} \\ \underline{48} \\ 20 \\ \underline{16} \\ 40 \\ \underline{32} \\ 80 \\ \underline{80} \\ 0 \end{array}$$

(b)

$$\begin{array}{r} 0.1375 \\ 80\overline{)11.0000} \\ \underline{80} \\ 300 \\ \underline{240} \\ 600 \\ \underline{560} \\ 400 \\ \underline{400} \\ 0 \end{array}$$

2. (a) $0.6363 = 0.\overline{63}$

$$\begin{array}{r} 0.6363 \\ 11\overline{)7.0000} \\ \underline{66} \\ 40 \\ \underline{33} \\ 70 \\ \underline{66} \\ 40 \\ \underline{33} \\ 7 \end{array}$$

(b) $0.533 = 0.5\overline{3}$

$$\begin{array}{r} 0.533 \\ 15\overline{)8.000} \\ \underline{75} \\ 50 \\ \underline{45} \\ 50 \\ \underline{45} \\ 5 \end{array}$$

(c) $0.295454 = 0.29\overline{54}$

$$\begin{array}{r} 0.295454 \\ 44\overline{)13.000000} \\ \underline{88} \\ 420 \\ \underline{396} \\ 240 \\ \underline{220} \\ 200 \\ \underline{176} \\ 240 \\ \underline{220} \\ 200 \\ \underline{176} \\ 24 \end{array}$$

3. (a)

$$2\frac{11}{18} = 2 + \frac{11}{18} = 2 + .6\overline{1} = 2.6\overline{1}$$

$0.611 = 0.6\overline{1}$

$$\begin{array}{r} 0.611 \\ 18\overline{)11.000} \\ \underline{108} \\ 20 \\ \underline{18} \\ 20 \\ \underline{18} \\ 2 \end{array}$$

(b) $1.03703 = 1.\overline{037}$

$$\begin{array}{r} 1.03703 \\ 27\overline{)28.00000} \\ \underline{27} \\ 100 \\ \underline{81} \\ 190 \\ \underline{189} \\ 100 \\ \underline{81} \\ 19 \end{array}$$

4. $0.7916 = 0.792$ rounded to the nearest thousandth

$$\begin{array}{r} 0.7916 \\ 24\overline{)19.0000} \\ \underline{168} \\ 220 \\ \underline{216} \\ 40 \\ \underline{24} \\ 160 \\ \underline{144} \\ 16 \end{array}$$

5.

$$\begin{aligned} 0.9 - 0.3 + 0.8 \times 0.3 &= 0.9 - 0.3 + 0.24 \\ &= 0.6 + 0.24 \\ &= 0.84 \end{aligned}$$

6.

$$\begin{aligned} 0.3 \times 0.5 + (0.4)^3 - 0.036 &= 0.3 \times 0.5 + 0.064 - 0.036 \\ &= 0.15 + 0.064 - 0.036 \\ &= 0.214 - 0.036 \\ &= 0.178 \end{aligned}$$

7. $6.56 \div (2 - 0.36) + (8.5 - 8.3)^2$

$= 6.56 \div (1.64) + (0.2)^2$ Parenthesis
$= 6.56 \div (1.64) + 0.04$ Exponents
$= 4 + 0.04$ Divide
$= 4.04$ Add

3.7 Practice Problems

Practice Problem 1

1. Understand the problem.

MATHEMATICS BLUEPRINT FOR PROBLEM SOLVING			
Gather the Facts	What Am I Asked to Do?	How Do I Proceed?	Key Points to Remember
Coat Cost: \$198.95 Tax on Coat: \$9.95 Boot Cost: \$59.95 Tax on Boots: \$3.00 Renée has \$280 to spend.	Find out how much money Renée had left after her purchases.	Add the cost of each item and the tax on each item to obtain the total purchase price. Subtract that total from \$280.	Be sure to line up the decimal points when adding and subtracting the decimal values.

2. Solve and state the answer:

Add the four amounts.

$$\begin{array}{r} \$198.95 \\ 9.95 \\ 59.95 \\ +\quad 3.00 \\ \hline \$271.85 \end{array}$$

Subtract all expenses from \$280.

$$\begin{array}{r} \$280.00 \\ -\ 271.85 \\ \hline \$\quad 8.15 \end{array}$$

Renée has \$8.15 left after her purchases.

3. Check. $200 + $10 + $60 + $3 = $273.

$$\begin{array}{r} \$280 \\ -\ 273 \\ \hline \$\ \ 7 \end{array}$$

The answer is reasonable.

Practice Problem 2

1. Understand the problem.

MATHEMATICS BLUEPRINT FOR PROBLEM SOLVING			
Gather the Facts	**What Am I Asked to Do?**	**How Do I Proceed?**	**Key Points to Remember**
She works 51 hours. She gets paid $9.36 per hour for 40 hours. She gets paid time-and-a-half for 11 hours.	Find the amount Melinda earned working 51 hours last week.	Add the earnings of 40 hours at $9.36 per hour to the earnings of 11 hours at overtime pay.	Overtime pay is time-and-a-half, which is 1.5 × $9.36.

2. Solve and state the answer:

(a) Calculate regular earnings for 40 hours.

$$\begin{array}{r} \$9.36 \\ \times\ \ 40 \\ \hline \$374.40 \end{array}$$

(b) Calculate overtime pay rate.

$$\begin{array}{r} \$9.36 \\ \times\ \ 1.5 \\ \hline \$14.04 \end{array}$$

(c) Calculate overtime earnings for 11 hours.

$$\begin{array}{r} \$\ 14.04 \\ \times\ \ \ \ 11 \\ \hline \$154.44 \end{array}$$

(d) Add the two amounts.

$$\begin{array}{rl} \$374.40 & \text{Regular earnings} \\ +\ 154.44 & \text{Overtime earnings} \\ \hline \$528.84 & \text{Total earnings} \end{array}$$

Melinda earned $528.84 last week.

3. Check. 40 × $9 = $360
10 × $14 = $140

$$\begin{array}{r} \$360 \\ +\ 140 \\ \hline \$500 \end{array}$$

The answer is reasonable.

Practice Problem 3

1. Understand the problem.

MATHEMATICS BLUEPRINT FOR PROBLEM SOLVING			
Gather the Facts	**What Am I Asked to Do?**	**How Do I Proceed?**	**Key Points to Remember**
The total amount of steak is 17.4 pounds. Each package contains 1.45 pounds. Each package costs $4.60 per pound.	(a) Find out how many packages of steak the butcher will have. (b) Find the cost of each package.	(a) Divide the total, 17.4, by the amount in each package, 1.45, to find the number of packages. (b) Multiply the cost of one pound, $4.60, by the amount in one package, 1.45.	There will be two answers to this problem.

2. Solve and state the answer:

(a)

$$\begin{array}{r} 12. \\ 1.45_\wedge\overline{)17.40_\wedge} \\ 14\ 5 \\ \hline 290 \\ 290 \\ \hline \end{array}$$

The butcher will have 12 packages of steak.

(b)

$$\begin{array}{r} \$4.60 \\ \times\ 1.45 \\ \hline \$6.67 \end{array}$$

Each package will cost $6.67.

3. Check.

(a) $1.5_\wedge\overline{)17.0_\wedge}$ with quotient $11.$

15

$\overline{2\,0}$

(b) $\$5 \times 1 = \$5$

The answers are reasonable.

Chapter 4

4.1 Practice Problems

1. **(a)** $\dfrac{36}{40} = \dfrac{9}{10}$ **(b)** $\dfrac{18}{15} = \dfrac{6}{5}$ **(c)** $\dfrac{220}{270} = \dfrac{22}{27}$

2. **(a)** $\dfrac{18}{36} = \dfrac{1}{2}$ **(b)** $\dfrac{25}{90} = \dfrac{5}{18}$

3. $\dfrac{44 \text{ dollars}}{900 \text{ tons}} = \dfrac{11 \text{ dollars}}{225 \text{ tons}}$

4. $\dfrac{212 \text{ miles}}{4 \text{ hours}} = \dfrac{53 \text{ miles}}{1 \text{ hour}}$ 53 miles/hour

5.

Selling price	$170.40
− purchase price	− 129.60
= profit	$ 40.80

She made a profit of $40.80 on 120 batteries.

$120\overline{)40.80}$ with quotient $.34$; 360; 480; 480; 0

Her profit was $0.34 per battery.

6. **(a)** $\dfrac{\$2.04}{12 \text{ ounces}} = \$0.17$ per ounce

$\dfrac{\$2.80}{20 \text{ ounces}} = \$0.14$ per ounce

(b) Fred saves $0.03 per ounce by buying the larger size.

4.2 Practice Problems

1. 6 is to 8 as 9 is to 12.

$\dfrac{6}{8} = \dfrac{9}{12}$

2. $\dfrac{2 \text{ hours}}{72 \text{ miles}} = \dfrac{3 \text{ hours}}{108 \text{ miles}}$

3. **(a)** $\dfrac{10}{18} \times \dfrac{25}{45}$ (cross products)

$10 \times 45 \stackrel{?}{=} 18 \times 25$

$450 = 450$ yes

Thus $\dfrac{10}{18} = \dfrac{25}{45}$ is a proportion.

(b) $\dfrac{42}{100} \times \dfrac{22}{55}$ (cross products)

$42 \times 55 \stackrel{?}{=} 100 \times 22$

$2310 \neq 2200$ no

Thus $\dfrac{42}{100} \neq \dfrac{22}{55}$ is *not* a proportion.

4. **(a)** $\dfrac{2.4}{3} \stackrel{?}{=} \dfrac{12}{15}$

$\dfrac{2.4}{3} \times \dfrac{12}{15}$

$3 \times 12 = 36$

↑ products equal ↓

$2.4 \times 15 = 36$

Thus $\dfrac{2.4}{3} = \dfrac{12}{15}$ *is* a proportion.

(b) $\dfrac{2\frac{1}{3}}{6} \stackrel{?}{=} \dfrac{14}{38}$

$2\dfrac{1}{3} \times 38 = \dfrac{7}{3} \times \dfrac{38}{1} = \dfrac{266}{3} = 88\dfrac{2}{3}$

$6 \times 14 = 84$

$\dfrac{2\frac{1}{3}}{6} \times \dfrac{14}{38}$ products not equal

$2\dfrac{1}{3} \times 38 = 88\dfrac{2}{3}$

Thus $\dfrac{2\frac{1}{3}}{6} \neq \dfrac{14}{38}$ is *not* a proportion.

5. **(a)** $\dfrac{1260}{7} \stackrel{?}{=} \dfrac{3530}{20}$

$1260 \times 20 \stackrel{?}{=} 7 \times 3530$

$25{,}200 \neq 24{,}710$ No

The rates are *not* equal.

(b) $\dfrac{2}{11} \stackrel{?}{=} \dfrac{16}{88}$

$2 \times 88 \stackrel{?}{=} 11 \times 16$

$176 = 176$ Yes

The rates *are* equal.

4.3 Practice Problems

1. **(a)** $5 \times n = 45$

$\dfrac{5 \times n}{5} = \dfrac{45}{5}$

$n = 9$

(b) $7 \times n = 84$

$\dfrac{7 \times n}{7} = \dfrac{84}{7}$

$n = 12$

2. **(a)** $108 = 9 \times n$

$\dfrac{108}{9} = \dfrac{9 \times n}{9}$

$12 = n$

(b) $210 = 14 \times n$

$\dfrac{210}{14} = \dfrac{14 \times n}{14}$

$15 = n$

3. **(a)** $15 \times n = 63$

$\dfrac{15 \times n}{15} = \dfrac{63}{15}$

$n = 4.2$

(b) $39.2 = 5.6 \times n$

$\dfrac{39.2}{5.6} = \dfrac{5.6 \times n}{5.6}$

$7 = n$

4. $\dfrac{24}{n} = \dfrac{3}{7}$

$24 \times 7 = n \times 3$

$168 = n \times 3$

$\dfrac{168}{3} = \dfrac{n \times 3}{3}$

$56 = n$

5. $\dfrac{176}{4} = \dfrac{286}{n}$

$176 \times n = 286 \times 4 = 1144$

$\dfrac{176 \times n}{176} = \dfrac{1144}{176}$

$n = 6.5$

6. $80 \times 6 = 5 \times n$

$480 = 5 \times n$

$\dfrac{480}{5} = \dfrac{5 \times n}{5}$

$96 = n$

n is $96

7. $264.25 \times 2 = 3.5 \times n$

$528.5 = 3.5 \times n$

$\dfrac{528.5}{3.5} = \dfrac{3.5 \times n}{3.5}$

$151 = n$

n is 151 meters

4.4 Practice Problems

1. $\dfrac{27 \text{ defective engines}}{243 \text{ engines produced}} = \dfrac{n \text{ defective engines}}{4131 \text{ engines produced}}$

$27 \times 4131 = 243 \times n$

$111{,}537 = 243 \times n$

$\dfrac{111{,}537}{243} = \dfrac{243 \times n}{243}$

$459 = n$

Thus we estimate that 459 engines would be defective.

2. $\dfrac{5 \text{ students who quit}}{19 \text{ students}} = \dfrac{n \text{ students who quit}}{7410 \text{ total students}}$

$5 \times 7410 = 19 \times n$

$37{,}050 = 19 \times n$

$\dfrac{37{,}050}{19} = \dfrac{19 \times n}{19}$

$1950 = n$

Approximately 1,950 students quit smoking.

3. $\dfrac{9 \text{ gallons of gas}}{234 \text{ miles traveled}} = \dfrac{n \text{ gallons of gas}}{312 \text{ miles traveled}}$

$9 \times 312 = 234 \times n$

$2808 = 234 \times n$

$\dfrac{2808}{234} = \dfrac{234 \times n}{234}$

$12 = n$

She would need 12 gallons of gas.

4. $\dfrac{80 \text{ rpm}}{16 \text{ mph}} = \dfrac{90 \text{ rpm}}{n \text{ mph}}$

$80 \times n = 16 \times 90$

$80 \times n = 1440$

$\dfrac{80 \times n}{80} = \dfrac{1440}{80}$

$n = 18$

Alicia will be riding 18 miles per hour.

5. $\dfrac{4050 \text{ walk-in}}{729 \text{ purchase}} = \dfrac{5500 \text{ walk-in}}{n \text{ purchase}}$

$4050 \times n = 729 \times 5500$

$4050 \times n = 4{,}009{,}500$

$\dfrac{4050 \times n}{4050} = \dfrac{4{,}009{,}500}{4050}$

$n = 990$

Tom will expect 990 people to make a purchase in his store.

6. $\dfrac{50 \text{ bears tagged in 1st sample}}{n \text{ bears in forest}} = \dfrac{4 \text{ bears tagged in 2nd sample}}{50 \text{ bears caught in 2nd sample}}$

$50 \times 50 = n \times 4$

$2500 = n \times 4$

$\dfrac{2500}{4} = \dfrac{n \times 4}{4}$

$625 = n$

We estimate that there are 625 bears in the forest.

Chapter 5

5.1 Practice Problems

1. (a) $\dfrac{51}{100} = 51\%$ (b) $\dfrac{68}{100} = 68\%$

(c) $\dfrac{7}{100} = 7\%$ (d) $\dfrac{26}{100} = 26\%$

2. (a) $\dfrac{238}{100} = 238\%$

(b) $\dfrac{505}{100} = 505\%$

(c) $\dfrac{121}{100} = 121\%$

3. (a) $\dfrac{0.5}{100} = 0.5\%$ (b) $\dfrac{0.06}{100} = 0.06\%$ (c) $\dfrac{0.003}{100} = 0.003\%$

4. (a) $47\% = .47$

(b) $2\% = 0.02$

(c) $90\% = 0.90$

5. (a) $80.6\% = 0.806$ (b) $2.5\% = 0.025$

(c) $0.29\% = 0.0029$ (d) $231\% = 2.31$

6. (a) $0.78 = 78\%$

(b) $0.02 = 2\%$

(c) $5.07 = 507\%$

(d) $0.029 = 2.9\%$

(e) $0.006 = 0.6\%$

5.2 Practice Problems

1. (a) $71\% = \dfrac{71}{100}$ (b) $25\% = \dfrac{25}{100} = \dfrac{1}{4}$ (c) $8\% = \dfrac{8}{100} = \dfrac{2}{25}$

2. (a) $8.4\% = 0.084 = \dfrac{84}{1000} = \dfrac{21}{250}$

(b) $55.25\% = .5525 = \dfrac{5525}{10{,}000} = \dfrac{221}{400}$

(c) $28.5\% = .285 = \dfrac{285}{1000} = \dfrac{57}{200}$

3. (a) $170\% = 1.70 = 1\dfrac{7}{10}$

(b) $288\% = 2.88 = 2\dfrac{88}{100} = 2\dfrac{22}{25}$

4. $7\dfrac{5}{8}\% = 7\dfrac{5}{8} \div 100$

$= \dfrac{61}{8} \times \dfrac{1}{100}$

$= \dfrac{61}{800}$

5. $21\dfrac{1}{4}\% = 21\dfrac{1}{4} \div 100 = \dfrac{85}{4} \times \dfrac{1}{100} = \dfrac{85}{400} = \dfrac{17}{80}$

6. $\dfrac{5}{8}$ $\quad 8\overline{)5.000}$ gives 0.625 $\quad 62.5\%$

7. (a) $\dfrac{3}{5} = 0.60 = 60\%$

(b) $\dfrac{21}{25} = 0.84 = 84\%$

(c) $\dfrac{7}{16} = 0.4375 = 43.75\%$

8. (a) $\dfrac{7}{9} = 0.7777\overline{7} = 0.7778 = 77.78\%$

(b) $\dfrac{19}{30} = 0.6333\overline{3} = 0.6333 = 63.33\%$

9. $\dfrac{7}{12}$ If we divide

$$\begin{array}{r} .58 \\ 12\overline{)7.00} \\ 60 \\ \hline 100 \\ 96 \\ \hline 4 \end{array}$$

we see that $\dfrac{7}{12} = 0.58\dfrac{4}{12} = 0.58\dfrac{1}{3}$.

Thus $\dfrac{7}{12} = 58\dfrac{1}{3}\%$.

10.

Fraction	Decimal	Percent
$\dfrac{23}{99}$	0.2323	23.23%
$\dfrac{129}{250}$	0.516	51.6%
$\dfrac{97}{250}$	0.388	$38\dfrac{4}{5}\%$

5.3A Practice Problems

1. What is 26% of 35?

$\downarrow \ \downarrow \ \downarrow \ \downarrow \ \downarrow$

$n = 26\% \times 35$

2. 155% of 20 is what?

$\downarrow \ \downarrow \ \downarrow \ \downarrow \ \downarrow$

$155\% \times 20 = n$

3. Find 0.08% of 350

$\downarrow \ \downarrow \ \downarrow \ \downarrow$

$n = 0.08\% \times 350$

4. (a) $58\% \times n = 400$ (b) $9.1 = 135\% \times n$

5. What percent of 250 is 36?

$$\underbrace{\text{What percent}}_{n}\ \text{of 250 is 36?} \quad\Rightarrow\quad n \times 250 = 36$$

6. **(a)** $50 = n \times 20$ **(b)** $n \times 2000 = 4.5$

7. What is 82% of 350?

$$n = 82\% \times 350$$
$$n = 0.82(350)$$
$$n = 287$$

8. **(a)** $5.2\% \times 800 = n$ **(b)** $n = 230\% \times 400$

$(.052)(800) = n$ $n = (2.30)(400)$

$41.6 = n$ $n = 920$

9. The problem asks: What is 8% of \$350?

$n = 8\% \times 350$

$n = \$28$ tax

10. $45\% \times n = 162$

$.45 \times n = 162$

$$\frac{.45 \times n}{.45} = \frac{162}{.45}$$

$n = 360$

11. $32 = 0.4\% \times n$

$32 = 0.004n$

$$\frac{32}{0.004} = \frac{0.004n}{0.004}$$

$8000 = n$

12. The problem asks: 30% of what is 6?

$30\% \times n = 6$

$.30n = 6$

$$\frac{.30n}{.30} = \frac{6}{.30}$$

$n = 20$

13. What percent of 9000 is 4.5?

$$\underbrace{\text{What percent}}_{n}\ \text{of 9000 is 4.5?} \quad\Rightarrow\quad n \times 9000 = 4.5$$

$9000n = 4.5$

$$\frac{9000n}{9000} = \frac{4.5}{9000}$$

$n = 0.0005$

$n = 0.05\%$

14. $198 = n \times 33$

$$\frac{198}{33} = \frac{n \times 33}{33}$$

$6 = n$

Now express n as a percent: 600%

15. The problem asks: 5 is what percent of 16?

$5 = n \times 16$

$$\frac{5}{16} = \frac{n \times 16}{16}$$

$0.3125 = n$

Now express n as a percent rounded to tenths: 31.3%

5.3B Practice Problems

1. **(a)** Find 83% of 460.

$p = 83$

(b) 18% of what number is 90?

$p = 18$

(c) What percent of 64 is 8?

The percent is unknown. Use the variable p.

2. **(a)** $b = 52$, $a = 15.6$ **(b)** Base $= b$, $a = 170$

3. **(a)** What is 18% of 240?

Percent $p = 18$

Base $b = 240$

Amount is unknown; use the variable a.

(b) What percent of 64 is 4?

Percent is unknown; use the variable p.

Base $b = 64$

Amount $a = 4$

4. Percent $p = 32$
Base $b = 550$
Variable a

$$\frac{a}{b} = \frac{p}{100}$$
$$\frac{a}{550} = \frac{32}{100}$$
$$100a = (550)(32)$$
$$a = 176$$

Thus 32% of 550 is 176.

5. Percent $p = 340$
Base $b = 70$
Variable a

$$\frac{a}{70} = \frac{340}{100}$$
$$100a = (70)(340)$$
$$a = 238$$

Thus 340% of 70 is 238.

6. 68% of what is 476?

Percent $p = 68$

Base is unknown; use base $= b$.

Amount $a = 476$

$$\frac{a}{b} = \frac{p}{100} \quad \text{becomes} \quad \frac{476}{b} = \frac{68}{100}$$

If we reduce the right-hand fraction, we have

$$\frac{476}{b} = \frac{17}{25}$$

$(476)(25) = 17b$ Using cross multiplication.

$11{,}900 = 17b$ Simplifying.

$\dfrac{11{,}900}{17} = \dfrac{17b}{17}$ Dividing each side by 17.

$700 = b$ Result of $11{,}900 \div 17$.

Thus 68% of 700 is 476.

7. Percent $p = 0.3$
Variable b
Amount $a = 216$

$$\frac{216}{b} = \frac{0.3}{100}$$
$$0.3b = (216)(100)$$
$$b = 72{,}000$$

Thus 216 is 0.3% of 72,000

8. Variable p
Base $b = 3500$
Amount $a = 105$

$$\frac{105}{3500} = \frac{p}{100}$$
$$\frac{21}{700} = \frac{p}{100}$$
$$700p = (21)(100)$$
$$p = 3$$

Thus 3% of 3500 is 105.

9. Variable p
Base $b = 140$
Amount $a = 42$

$$\frac{42}{140} = \frac{p}{100}$$
$$\frac{21}{70} = \frac{p}{100}$$
$$70p = (21)(100)$$
$$p = 30$$

Thus 42 is 30% of 140

5.4 Practice Problems

1. Let n = number of people with reserved airline tickets.

12% of $n = 4800$

$0.12 \times n = 4800$

$$\frac{0.12 \times n}{0.12} = \frac{4800}{0.12} \quad n = 40{,}000$$

40,000 people had airline tickets that month.

2. The problem asks: What is 8% of \$62.30?

$n = .08 \times 62.30$

$n = 4.984$

The tax is \$4.98.

3. The problem asks: 105 is what percent of 130?

$105 = n \times 130$

$$\frac{105}{130} = n$$

$0.8077 = n$

Now express n as a percent rounded to tenths: 80.8%

Thus 80.8% of the flights were on time.

4. Commission = commission rate × value of sales
Commission = 6% × \$156,000
= 0.06 × 156,000
= 9360
His commission is \$9360.

5. 100% Cost of meal + tip of 15% = \$46.00
Let n = Cost of meal
100% of n + 15% of n = \$46.00
115% of n = 46.00
$1.15 \times n = 46.00$
$n = 40.00$
They can spend \$40.00 on the meal itself.

6. 15,000
− 10,500
4,500 the amount of decrease

$$\text{Percent of decrease} = \frac{\text{amount of decrease}}{\text{original amount}} = \frac{\$4500}{\$15,000} = 0.30 = 30\%$$

The percent of decrease is 30%.

7. **(a)** 7% of \$13,600 is the discount
0.07 × 13,600 = the discount
\$952 is the discount
(b) \$13,000 list price
− 952 discount
\$12,048 amount Betty paid for the car.

8. $I = P \times R \times T$
$P = \$5600 \quad R = 12\% \quad T = 1$ year
$I = 5600 \times 12\% \times 1$
$= 5600 \times 0.12$
$= 672$
The interest is \$672 for 1 year.

9. **(a)** $I = P \times R \times T$
$I = 1800 \times 0.11 \times 4$
$I = \$792$
(b) $I = P \times R \times T$
$I = 1800 \times 0.11 \times \frac{1}{2}$
$I = \$99$

Chapter 6

6.1 Practice Problems

1. **(a)** 3 feet in 1 yard **(b)** 5280 feet in 1 mile
(c) 60 minutes in 1 hour **(d)** 7 days in 1 week
(e) 16 ounces in 1 pound **(f)** 2 pints in 1 quart
(g) 4 quarts in 1 gallon

2. **(a)** $15{,}840\ \cancel{\text{ft.}} \times \frac{1 \text{ mile}}{5280\ \cancel{\text{ft.}}} = \frac{15{,}840}{5280}$ miles = 3 miles.
(b) $17\ \cancel{\text{ft.}} \times \frac{12 \text{ inches}}{1\ \cancel{\text{ft.}}} = 17 \times 12$ inches = 204 inches.

3. **(a)** $18.93\ \cancel{\text{miles}} \times \frac{5280 \text{ feet}}{1\ \cancel{\text{mile}}} = 99{,}950.4$ feet
(b) $16\frac{1}{2}\ \cancel{\text{inches}} \times \frac{1 \text{ yard}}{36\ \cancel{\text{inches}}} = 16\frac{1}{2} \times \frac{1}{36}$ yards
$= \frac{\overset{11}{\cancel{33}}}{2} \times \frac{1}{\underset{12}{\cancel{36}}}$ yards $= \frac{11}{24}$ yards

4. $760.5\ \cancel{\text{lbs.}} \times \frac{16 \text{ oz.}}{1\ \cancel{\text{lb.}}} = 760.5 \times 16$ oz. = 12,168 oz.

5. $19\ \cancel{\text{pints}} \times \frac{1 \text{ quart}}{2\ \cancel{\text{pints}}} = \frac{19}{2}$ quarts = 9.5 quarts

6. $7.2\ \cancel{\text{hrs.}} \times \frac{60 \text{ min.}}{1\ \cancel{\text{hr.}}} = 7.2 \times 60$ min. = 432 minutes.

7. **Step 1:** $7\ \cancel{\text{lbs.}} \times \frac{16 \text{ oz.}}{1\ \cancel{\text{lb.}}} = 7 \times 16$ oz. = 112 oz.
Step 2: 112 oz. + 12 oz. = 124 ounces.
The potatoes weigh 124 ounces.

8. **Step 1:** $1\frac{3}{4}\ \cancel{\text{days}} \times \frac{24 \text{ hours}}{1\ \cancel{\text{day}}} = \frac{7}{4} \times \frac{24}{1}$ hours = 42 hrs parked.
Step 2: 42 hrs $\times \frac{1.50 \text{ dollars}}{1 \text{ hour}}$ = 63.00 dollars.
She paid \$63.00.

6.2 Practice Problems

1. **(a)** deka- **(b)** milli-
2. **(a)** 4 meters = 4.00$_\wedge$ centimeters = 400 cm.
(b) 30 centimeters = 30.0$_\wedge$ millimeters = 300 mm.
3. **(a)** 3 mm = $_\wedge$003. m = 0.003 m
(b) 47 cm = $_\wedge$00047. km. = 0.00047 km.
4. The car length would logically be choice (b) 3.8 meters. [A meter is close to a yard and 3.8 yards seems reasonable!]
5. **(a)** 1.527 km **(c)** 12.8 cm
(b) 3.75 m **(d)** 46,000 mm
6. **(a)** 389 mm = 0.0389 dekameter (four places to left)
(b) 0.48 hectometer = 4800 cm (four places to right)
7. 782 cm = 7.82 m
2 m = 2.00 m
537 m = 537.00 m
546.82 m

6.3 Practice Problems

1. **(a)** 5 L = 5000 mL
(b) 84 kL = 84,000 L
(c) 0.732 L = 732 mL
2. **(a)** 15.8 mL = 0.0158 L
(b) 12340 mL = 12.34 L
(c) 863.3 L = 0.0863 kL
3. **(a)** 396 mL = 396 cm^3
(because 1 milliliter = 1 cubic centimeter)
(b) 0.096 L = 96 cm^3
4. $\frac{\$110.00}{1000} = \$0.11$ per milliliter
5. **(a)** 3.2 t = 3200 kg
(b) 7.08 kg = 7080 g
(c) 0.526 g = 526 mg
6. **(a)** 59 kg = 0.059 t
(b) 6.152 g = 0.006152 kg
(c) 28.3 mg = 0.0283 g
7. A gram is $\frac{1}{1000}$ of a kilogram. If the coffee costs \$10.00 per kilogram, then 1 gram would be $\frac{1}{1000}$ of \$10.
$\frac{1}{1000} \times \$10 = \frac{\$10.00}{1000} = \$0.01$
The coffee costs 1¢ per gram.
8. The most reasonable answer is (a) 120 kg. [A kg is slightly more than 2 lbs.]

6.4 Practice Problems

1. **(a)** $7\ \cancel{\text{ft}} \times \frac{0.305 \text{ m}}{1\ \cancel{\text{ft}}} = 2.135$ m
(b) $4\ \cancel{\text{in.}} \times \frac{2.54 \text{ cm}}{1\ \cancel{\text{in.}}} = 10.16$ cm
2. **(a)** $17\ \cancel{\text{m}} \times \frac{1.09 \text{ yd.}}{1\ \cancel{\text{m}}} = 18.53$ yd.
(b) $29.6\ \cancel{\text{km}} \times \frac{0.62 \text{ mi.}}{1\ \cancel{\text{km}}} = 18.352$ mi.
(c) $26\ \cancel{\text{gal}} \times \frac{3.79 \text{ L}}{1\ \cancel{\text{gal}}} = 98.54$ L
(d) $6.2\ \cancel{\text{L}} \times \frac{1.06 \text{ qt.}}{1\ \cancel{\text{L}}} = 6.572$ qt.
(e) $16\ \cancel{\text{lb.}} \times \frac{0.454 \text{ kg}}{1\ \cancel{\text{lb}}} = 7.264$ kg.

(f) $280\cancel{g} \times \dfrac{0.0353 \text{ oz.}}{1\cancel{g}} = 9.884$ oz.

3. $180\cancel{cm} \times \dfrac{0.394 \cancel{in.}}{1\cancel{cm}} \times \dfrac{1 \text{ ft}}{12\cancel{in.}} = 5.91$ ft

4. $\dfrac{88\cancel{km}}{\text{hr}} \times \dfrac{0.62 \text{ mi}}{1\cancel{km}} = 54.56$ mi/hr

5. 9 mm = 0.9 cm

$0.9\cancel{cm} \times \dfrac{0.394 \text{ in.}}{\cancel{cm}} = 0.3546$ in. rounded to 0.35 in.

6. $F = 1.8 \times C + 32$
$= 1.8 \times 20 + 32$
$= 36 + 32$
$= 68$

The temperature is 68°F.

7. $C = \dfrac{5 \times 181 - 160}{9} = \dfrac{905 - 160}{9}$

$= \dfrac{745}{9} = 82.87°\text{C}$ rounded to the nearest tenth

Practice Problem 2

1. Understand the problem.

MATHEMATICS BLUEPRINT FOR PROBLEM SOLVING			
Gather the Facts	What Am I Asked to Do?	How Do I Proceed?	Key Points to Remember
He must use 18.06 liters of solution. He has 42 jars to fill.	Find out how many milliliters of a solvent will go into each jar.	We need to convert 18.06 liters to milliliters, and then divide that result by 42.	To convert 18.06 liters to milliliters we move the decimal point 3 places to the right.

2. Solve and state the answer:
(a) 18.06 L = 18,060 mL
(b) $\dfrac{18{,}060 \text{ mL}}{42 \text{ jars}} = 430$ mL/jar

Thus, 430 mL of a solvent will go into each jar.

3. Check. 18.06 L is approximately 18 L, or 18,000 mL.
$\dfrac{18{,}000}{40} = 450$ and is reasonable.

6.5 Practice Problems

Practice Problem 1

Step 1:
$$\begin{array}{r} 2\frac{2}{3}\text{ yd} \\ 8\frac{1}{3}\text{ yd} \\ 2\frac{2}{3}\text{ yd} \\ +\ 8\frac{1}{3}\text{ yd} \\ \hline 22\text{yd} \end{array}$$

Step 2: $22\cancel{yd} \times \dfrac{3 \text{ ft}}{1\cancel{yd}} = 66$ ft.

The perimeter is 66 ft.

Chapter 7

7.1 Practice Problems

1. $P = 2l + 2w$
$= (2)(6) + (2)(1.5)$
$= 12 + 3 = 15$ m

2. $P = 2l + 2w$
$= (2)(1200) + 2(860)$
$= 2400 + 1720$
$= 4120$ ft.

3. $P = 4s$
$= (4)(5.8) = 23.2$ cm

4. $P = 4 + 2 + 5 + 1 + 1 + 1 = 14$ cm.

5. $P = 4 + 4 + 5.5 + 2.5 + 1.5 + 1.5$
$= 19$ ft

Cost $= 19\cancel{ft} \times \dfrac{0.16 \text{ dollar}}{1\cancel{ft}} = \$3.04$

6. $A = lw = (29)(17) = 493 \text{ m}^2$

7. $A = (s)^2$
$= (11.8)^2$
$= 139.24 \text{ mm}^2$

8. Area of rectangle $= (18)(20) = 360 \text{ ft}^2$
Area of square $= 6^2 = 36 \text{ ft}^2$
The total area $= 396 \text{ ft}^2$

7.2 Practice Problems

1. Perimeter $= (2)(7.6) + (2)(3.5)$
$= 15.2 + 7.0 = 22.2$ cm

2. $A = bh$
$= (10.3)(1.5)$
$= 15.45 \text{ km}^2$

3. Perimeter $= (2)(4.1) + (2)(11.6)$
$= 8.2 + 23.2 = 31.4$ km

Area $= bh$
$= (4.1)(9.8) = 40.18 \text{ km}^2$

4. $P = 7 + 15 + 21 + 13 = 56$ yd.

5. $A = \dfrac{h(b + B)}{2} = \dfrac{(2.6)(1.3 + 1.9)}{2} = \dfrac{(2.6)(3.2)}{2} = \dfrac{8.32}{2} = 4.16 \text{ mm}^2$

6. (a) $A = \dfrac{h(b + B)}{2}$
$= \dfrac{140(180 + 130)}{2}$
$= 21{,}700 \text{ yd}^2$

(b) $21{,}700\cancel{yd^2} \times \dfrac{1 \text{ gallon}}{100\cancel{yd^2}} = 217$ gallons

Thus 217 gallons of sealer are needed.

7. The area of the trapezoid

$$A = \frac{(9.2)(12.6 + 19.8)}{2} = \frac{(9.2)(32.4)}{2} = \frac{298.08}{2} = 149.04 \text{ cm}^2$$

The area of the rectangle is

$A = (8.3)(12.6) = 104.58 \text{ cm}^2$

Total area $= 253.62 \text{ cm}^2$

7.3 Practice Problems

1. The sum of the angles in a triangle is 180°. The two given angles total 125° + 15° = 140°. Thus 180° − 140° = 40°. Angle A must be 40°.
2. $P = 10.5 + 10.5 + 8.5 = 29.5$ m
3. $P = 30 \quad 30 \div 3 = 10$
   The length of each side is 10 cm
4. $A = \frac{bh}{2} = \frac{(38)(13)}{2} = \frac{494}{2} = 247 \text{ m}^2$
5. $A = \frac{bh}{2} = \frac{(5)(16)}{2} = 40 \text{ cm}^2$
6. Area of rectangle $= (11)(24) = 264 \text{ cm}^2$
   Area of triangle $= \frac{(11)(7)}{2} = \frac{77}{2} = 38.5 \text{ cm}^2$
   Total area $= 302.5 \text{ cm}^2$

7.4 Practice Problems

1. (a) $\sqrt{49} = 7$ (b) $\sqrt{169} = 13$
2. $\sqrt{49} - \sqrt{4} = 7 - 2 = 5$
3. (a) Yes, because (12)(12) = 144 (b) $\sqrt{144} = 12$
4. (a) $\sqrt{3} \approx 1.732$ (b) $\sqrt{13} \approx 3.606$ (c) $\sqrt{5} \approx 2.236$
5. $\sqrt{100 \text{ in}^2} = 10$ in.
6. $\sqrt{22 \text{ m}^2} \approx 4.690$ m

7.5 Practice Problems

1. Hypotenuse $= \sqrt{(6)^2 + (8)^2} = \sqrt{36 + 64} = \sqrt{100} = 10$ m
2. Hypotenuse $= \sqrt{(12)^2 + (16)^2} = \sqrt{144 + 256} = \sqrt{400} = 20$ m
3. Hypotenuse $= \sqrt{(3)^2 + (7)^2} = \sqrt{9 + 49} = \sqrt{58}$
   Rounded to the nearest thousandth, hypotenuse ≈ 7.616 cm.
4. Leg $= \sqrt{(17)^2 - (15)^2} = \sqrt{289 - 225} = \sqrt{64} = 8$ m
5. Leg $= \sqrt{(10)^2 - (5)^2} = \sqrt{100 - 25} = \sqrt{75}$
   Rounded to the nearest thousandth, the leg ≈ 8.660 m.
6. The distance between the holes is the hypotenuse.
   Hypotenuse $= \sqrt{(5)^2 + (2)^2} = \sqrt{25 + 4} = \sqrt{29} \approx 5.385$ cm
7. Leg $= \sqrt{(30)^2 - (27)^2} = \sqrt{900 - 729} = \sqrt{171}$
   The distance to the nearest tenth is 13.1 yd.
8. (a) $y = \frac{1}{2} \times 12 = 6$ ft.
   $x = \sqrt{(12)^2 - (6)^2} = \sqrt{144 - 36} = \sqrt{108} \approx 10.4$ ft.
   (b) Hypotenuse $= \sqrt{2} \times \text{leg} \approx 1.414 \times 8 \approx 11.3$ m

7.6 Practice Problems

1. $C = \pi d = (3.14)(9) = 28.26 = 28.3$ m to nearest tenth
2. $C = 2\pi r = (2)(3.14)(14) = 87.9$ in.
3. (a) $C = \pi d = (3.14)(30) = 94.2$ in
   (b) Change 94.2 in. to ft.
   $94.2 \cancel{\text{in}} \times \frac{1 \text{ ft}}{12 \cancel{\text{in}}} = 7.85$ ft
   (c) When the wheel makes 2 revolutions, the bicycle travels 15.7 ft. [7.85 × 2]
4. $A = \pi r^2 = (3.14)(5^2) = (3.14)(25) = 78.5 \text{ km}^2$
5. $r = \frac{d}{2} = \frac{22}{2} = 11$ m
   $A = \pi r^2 = (3.14)(11)^2 = 379.9 \text{ m}^2$
   The area of the circle is 379.9 m^2
6. $r = \frac{d}{2} = \frac{10}{2} = 5$ ft.
   $A = \pi r^2 = (3.14)(5)^2 = 78.5 \text{ ft}^2$
   Change 78.5 ft^2 to yd^2
   $78.5 \cancel{\text{ft}^2} \times \frac{1 \text{ yd}^2}{9 \cancel{\text{ft}^2}} = 8.7 \text{ yd}^2$
   Find the cost: $\frac{\$12}{1 \cancel{\text{yd}^2}} \times 8.7 \cancel{\text{yd}^2} = \$104.40$
   The cost of the pool cover is \$104.40.
7. Area of square − area of circle = shaded area

s^2	−	πr^2	= shaded area
5^2	−	$(3.14)(2^2)$	
25	−	(3.14)(4)	
25.00	−	12.56	= 12.4 ft^2 shaded area (rounded to nearest tenth)

8. $r = \frac{d}{2} = \frac{8}{2} = 4$ ft.
   A semicircle $= \frac{\pi r^2}{2} = \frac{(3.14)(4)^2}{2} = 25.12 \text{ ft}^2$
   A rectangle $= lw = (12)(8) = 96 \text{ ft}^2$
   $25.12 \text{ ft}^2 + 96.00 \text{ ft}^2 = 121.12 \text{ ft}^2$ The total area is 121.1 ft^2

7.7 Practice Problems

1. $V = lwh = (6)(5)(2) = (30)(2) = 60 \text{ m}^3$
2. $V = \pi r^2 h = (3.14)(2)^2(5) = (3.14)(4)(5) = 62.8 \text{ in}^3$
3. $V = \frac{4\pi r^3}{3} = \frac{(4)(3.14)(6^3)}{3} = \frac{(4)(3.14)(6)(6)\cancel{(6)}^{2}}{\cancel{3}_{1}}$
   $= (12.56)(36)(2) = 904.32$
   $= 904.3 \text{ m}^3$ Rounded to nearest tenth

4. $V = \frac{\pi r^2 h}{3}$

$= \frac{(3.14)(5)^2(12)}{3}$

$= 314 \text{ m}^3$

5. (a) $V = \frac{Bh}{3}$

$= \frac{(6)(6)(10)}{3} = \frac{(36)(10)}{3}$

$= \frac{\overset{12}{\cancel{36}}(10)}{\underset{1}{\cancel{3}}} = 120 \text{ m}^3$

(b) $V = \frac{Bh}{3}$

$= \frac{(7)(8)(15)}{3} = \frac{(7)(8)(\overset{5}{\cancel{15}})}{\underset{1}{\cancel{3}}}$

$= (56)(5) = 280 \text{ m}^3$

7.8 Practice Problems

1. $\frac{11}{27} = \frac{15}{n}$

$11n = (27)(15)$

$11n = 405$

$\frac{11n}{11} = \frac{405}{11}$

$n = 36.\overline{81}$

$= 36.8$ meters measured to the nearest tenth.

2. a corresponds to p, b corresponds to m, c corresponds to n

3. Small triangle has a perimeter of 30 yards.

Let p = the unknown perimeter.

$\frac{12}{40} = \frac{30}{p}$

$12p = (40)(30)$

$12p = 1200$

$\frac{12p}{12} = \frac{1200}{12}$

$p = 100$

The larger triangle has a perimeter of 100 yards.

4. $\frac{h}{5} = \frac{20}{2}$

$2h = 100$

$h = 50$ ft.

5. $\frac{3}{29} = \frac{1.8}{n}$

$3n = (1.8)(29)$

$3n = 52.2$

$\frac{3n}{3} = \frac{52.2}{3}$

$n = 17.4$ The width is 17.4 meters.

7.9 Practice Problems

Practice Problem 1

1. Understand the problem.

MATHEMATICS BLUEPRINT FOR PROBLEM SOLVING			
Gather the Facts	**What Am I Asked to Do?**	**How Do I Proceed?**	**Key Points to Remember**
Mike needs to sand three rooms: 24 ft × 13 ft 12 ft × 9 ft 16 ft × 3 ft He can sand 80 ft^2 in 15 min.	Find out how long it will take him to sand all three rooms.	(a) Find the total area to be sanded. (b) Then find out how long it will take him to sand the total area.	Area = length × width. To get the total time, multiply the total area by the fraction: $\frac{15 \text{ min.}}{80 \text{ ft}^2}$

2. Solve and state the answer:

(a) $24 \times 13 = 312 \text{ ft}^2$ room 1

$12 \times 9 = 108 \text{ ft}^2$ room 2

$16 \times 3 = 48 \text{ ft}^2$ room 3

468 ft^2 total area

(b) $468 \cancel{\text{ft}^2} \times \frac{15 \text{ min}}{80 \cancel{\text{ft}^2}} = 87.75$ min. It takes Mike 88 min. to sand the rooms.

3. Check. Estimate the answer to see if it is reasonable.

Practice Problem 2

1. Understand the problem.

MATHEMATICS BLUEPRINT FOR PROBLEM SOLVING			
Gather the Facts	**What Am I Asked to Do?**	**How Do I Proceed?**	**Key Points to Remember**
The trapezoid has a height of 9 ft. The bases are 18 ft and 12 ft. The rectangular portion measures 24 ft × 15 ft. Roofing costs \$2.75 per square yard.	(a) Find the area of the roof. (b) Find the cost to install new roofing.	(a) Find the area of the entire roof. Change square feet to square yards. (b) Multiply by \$2.75.	9 square feet = 1 square yard.

2. Solve and state the answer:

(a) Area of trapezoid $= \frac{1}{2}h(b + B)$

$= \frac{1}{2}(9)(12 + 18)$

$= 135 \text{ ft}^2$

Area of rectangle $= lw$

$= (15)(24)$

$= 360 \text{ ft}^2$

Total Area $= 135 \text{ ft}^2 + 360 \text{ ft}^2 = 495 \text{ ft}^2$

Change square feet to square yards.

$495 \cancel{\text{ft}^2} \times \frac{1 \text{ yd}^2}{9 \cancel{\text{ft}^2}} = 55 \text{ yd}^2$

The area of the roof is 55 yd^2

(b) Cost $= 55 \cancel{\text{yd}^2} \times \frac{\$2.75}{1 \cancel{\text{yd}^2}} = \$151.25$

The cost to install new roofing would be \$151.25.

3. Check. Estimate the answers to see if they seem reasonable.

Chapter 8

8.1 Practice Problems

1. The smallest category of students is *special students.*
2. 3000 freshmen + 200 *special students* = 3200
   There are 3200 students who are either freshmen or *special students.*
3. There are 3000 freshmen but only 2600 sophomores. The ratio of freshmen to sophomores is $\frac{3000}{2600} = \frac{15}{13}$.
4. The ratio of freshmen to the total number of students is: $\frac{3000}{10000} = \frac{3}{10}$.
5. Lake Michigan occupies the third largest area with 23%.
6. The percent of the total area occupied by either Lake Superior or Lake Michigan is: 28% + 23% = 51%.
7. Lake Superior has 28% of the area. Now 28% of 290,000 mi^2 is $(0.28)(290{,}000) = 81{,}200 \text{ mi}^2$
8. (a) 10% + 8% = 18%
   (b) 20% of 690,000 = (0.20)(690,000) = 138,000

8.2 Practice Problems

1. The bar rises to 24. The approximate population was 24,000,000.
2. 16,000,000 − 7,000,000 = 9,000,000. The bar graph increases 9,000,000.
3. The bar rises to 250. The number of new cars sold in the fourth quarter of 1996 was 250.
4. 150 − 100 = 50. Thus, 50 fewer cars were sold.
5. The greatest number of customers came in July since the highest point of the graph occurs in July.
6. (a) 3500 (b) The number decreased.
7. The sharpest line of decrease is from July to August. The line slopes downward most steeply between those two months. Thus the biggest decrease in customers is between July and August.
8. Because the dot corresponding to 1992–93 is opposite 15 and the scale is in hundreds, we have 15 × 100 = 1500. Thus, 1500 degrees in visual and performing arts were awarded.
9. The computer science line goes above the visual and performing arts line first in 1990–91. Thus the first academic year with more degrees in computer science was 1990–91.

8.3 Practice Problems

1. The 60–69 bar rises to a height of 6. Thus six tests would have a D grade.
2. From the histogram, *16* tests were 70–79, *8* tests were 80–89, and *6* tests were 90–99. When we combine 16 + 8 + 6 = 30, we can see that 30 students scored greater than 69 on the test.
3. The 800–999 bar rises to a height of 20. Thus 20 light bulbs lasted between 800 and 999 hours.
4. From the histogram, *25* bulbs lasted 1200–1399 hours, *10* lasted 1400–1599 hours, and *5* lasted 1600–1799 hours. When we combine 25 + 10 + 5 = 40, we can see that 40 light bulbs lasted more than 1199 hours.
5. Weight in Pounds

Class Interval	Tally	Frequency
1600–1799	||	2
1800–1999	||||	4
2000–2199	||||	4
2200–2399	~~||||~~	5

6.

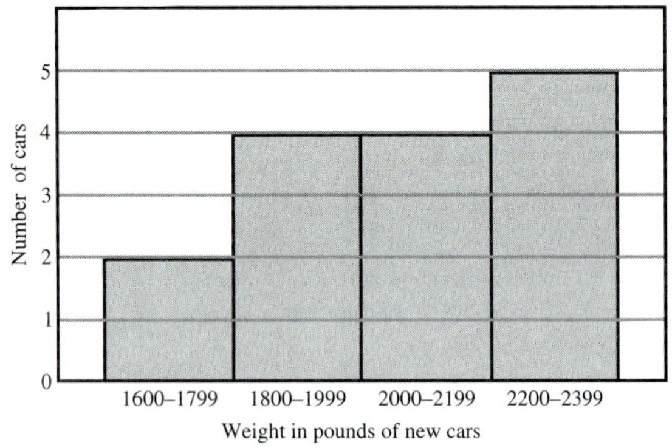

7. The greatest difference occurs between the under age 25 category and the 25–34 age category.

8.4 Practice Problems

1. $\frac{88 + 77 + 84 + 97 + 89}{5} = \frac{435}{5} = 87$ mean value
2. $\frac{\$39.20 + \$43.50 + \$81.90 + \$34.20 + \$51.70 + \$48.10}{6} = \$49.77$
   The mean monthly phone bill is \$49.77.
3. 2, 3, 7, 12, (four values) 13, (middle value) 15, 18, 20, 21 (four values)
   The median value is 13 minutes.
4. \$150, \$180, \$290, (three numbers) \$320, (middle number) \$400, \$450, \$600 (three numbers)
   Thus, \$320 is the median salary.
5. 88, 90, (two values) 100, 105, (two middle values) 118, 126 (two values)
   $\frac{100 + 105}{2} = \frac{205}{2} = 102.5$ median value

Chapter 9

9.1 Practice Problems

1. (a) $\begin{array}{r} 9 \\ +\ 14 \\ \hline 23 \end{array}$ (b) $\begin{array}{r} -7 \\ +\ -3 \\ \hline -10 \end{array}$ (c) $\begin{array}{r} -4.5 \\ +\ -1.9 \\ \hline -6.4 \end{array}$

2. (a) $\begin{array}{r} -\frac{2}{15} \\ +\ -\frac{12}{15} \\ \hline -\frac{14}{15} \end{array}$ (b) $\begin{array}{r} \frac{5}{12} = \frac{5}{12} \\ +\ \frac{1}{4} \times \frac{3}{3} = +\frac{3}{12} \\ \hline \frac{8}{12} = \frac{2}{3} \end{array}$ (c) $\begin{array}{r} -\frac{1}{6} = -\frac{7}{42} \\ +\ -\frac{2}{7} = +\ -\frac{6}{42} \\ \hline = -\frac{13}{42} \end{array}$

Selected Answers

Chapter 1

Pretest Chapter 1

1. Seventy-eight million, three hundred ten thousand, four hundred thirty-six **2.** 30,000 + 8,000 + 200 + 40 + 7 **3.** 5,064,122 **4.** 41,217,000 **5.** 2,237,000 **6.** 244 **7.** 50,570 **8.** 14,729 **9.** 2111 **10.** 76,311 **11.** 1,981,652 **12.** 108 **13.** 18,606 **14.** 6734 **15.** 331,420 **16.** 10,605 **17.** 7376 R 1 **18.** 139 **19.** 6^4 **20.** 81 **21.** 29 **22.** 52 **23.** 23 **24.** 271,000 **25.** 26,500 **26.** 60,000,000 **27.** 2600 **28.** 30,000,000,000 **29.** 1166 miles **30.** $22.00 **31.** $299 **32. (a)** 148,192 **(b)** 684,933

1.1 Exercises

1. 6000 + 700 + 30 + 1 **3.** 100,000 + 8,000 + 200 + 70 + 6 **5.** 20,000,000 + 3,000,000 + 700,000 + 60,000 + 1000 + 300 + 40 + 5 **7.** 100,000,000 + 3,000,000 + 200,000 + 60,000 + 700 + 60 + 8 **9.** 671 **11.** 9863 **13.** 76,036 **15.** 706,200 **21.** Fifty-three **23.** Eight thousand, nine hundred thirty-six **25.** Thirty-six thousand, one hundred eighteen **27.** One hundred five thousand, two hundred sixty-one **29.** Twenty-three million, five hundred sixty-one thousand, two hundred forty-eight **31.** Four billion, three hundred two million, one hundred fifty-six thousand, two hundred **33.** 375 **35.** 56,281 **37.** 100,079,826 **39.** One thousand, nine hundred sixty-five **41.** 9 million or 9,000,000 **43.** 16 million or 16,000,000 **45.** $201 billion or $201,000,000,000 **47.** $380 billion or $380,000,000,000 **49. (a)** 5 **(b)** 2 **50. (a)** 1 **(b)** 3 **51. (a)** 3 **(b)** 4 **52. (a)** 9 **(b)** 7 **53.** 613,001,033,208,003 **54.** Three quintillion, six hundred eighty-two quadrillion, nine hundred sixty-eight trillion, nine billion, nine hundred thirty-one million, nine hundred sixty thousand, seven hundred forty-seven

1.2 Exercises

3.

+	3	5	4	8	0	6	7	2	9	1
2	5	7	6	10	2	8	9	4	11	3
7	10	12	11	15	7	13	14	9	16	8
5	8	10	9	13	5	11	12	7	14	6
3	6	8	7	11	3	9	10	5	12	4
0	3	5	4	8	0	6	7	2	9	1
4	7	9	8	12	4	10	11	6	13	5
1	4	6	5	9	1	7	8	3	10	2
8	11	13	12	16	8	14	15	10	17	9
6	9	11	10	14	6	12	13	8	15	7
9	12	14	13	17	9	15	16	11	18	10

5. 23 **7.** 26 **9.** 60 **11.** 99 **13.** 4579 **15.** 13,336 **17.** 13,861 **19.** 42,739 **21.** 121 **23.** 1132 **25.** 17,909 **27.** 11,579,426 **29.** 1,135,280,240 **31.** 2,303,820 **33.** 248 **35.** 20,909 **37.** $794 **39.** $7986 **41.** 855 feet **43.** 121,100,000 square miles **45.** 2,922,339 votes **47. (a)** 1134 students **(b)** 1392 students **49. (a)** $9553 **(b)** $7319 **(c)** $13,047 **50.** 12,117,289,695 **51.** 1161 **52.** Answers may vary. A sample is: You could not add the addends in reverse order to check the addition. **53.** Answers may vary. A sample is: You could not group the addends in groups that sum to 10 to make column addition easier. **54.** Seventy-six million, two hundred eight thousand, nine hundred forty-one **55.** One hundred twenty-one million, three hundred seventy-four **56.** 8,724,396 **57.** 9,051,719

1.3 Exercises

1. 6 **3.** 8 **5.** 1 **7.** 9 **9.** 16 **11.** 9 **13.** 7 **15.** 6 **17.** 3 **19.** 9 **21.** 61 $\begin{array}{r} 37 \\ +\ 61 \\ \hline 98 \end{array}$ **23.** 12 $\begin{array}{r} 73 \\ +\ 12 \\ \hline 85 \end{array}$

25. 31 $\begin{array}{r} 95 \\ +\ 31 \\ \hline 126 \end{array}$ **27.** 625 $\begin{array}{r} 143 \\ +\ 625 \\ \hline 768 \end{array}$ **29.** 1341 $\begin{array}{r} 422 \\ +\ 1341 \\ \hline 1763 \end{array}$ **31.** 11,191 $\begin{array}{r} 13{,}205 \\ +11{,}191 \\ \hline 24{,}396 \end{array}$ **33.** 553,101 $\begin{array}{r} 433{,}201 \\ +\ 553{,}101 \\ \hline 986{,}302 \end{array}$ **35.** $\begin{array}{r} 19 \\ +\ 110 \\ \hline 129 \end{array}$ Correct

37. $\begin{array}{r} 1113 \\ +\ 5067 \\ \hline 6180 \end{array}$ Correct **39.** $\begin{array}{r} 6030 \\ -\ 5020 \\ \hline 1010 \end{array}$ Incorrect **41.** $\begin{array}{r} 98{,}763 \\ -\ 42{,}531 \\ \hline 56{,}232 \end{array}$ Incorrect **43.** 46 **45.** 37 **47.** 75 **49.** 581 **51.** 3,296 **53.** 34,092 **55.** 2,314

57. 223,116 **59.** $x = 5$ **61.** $x = 21$ **63.** $x = 27$ **65.** He got 182 more votes. **67.** 6,984,914 **69.** He received $248. **71.** 961,235 people **73.** 3,686,434 people **75.** 320,317 people **77.** 29 homes **79.** 80 homes **81.** Between 1994 and 1995 **83.** Willow Creek and Irving **85.** If a and b represent the same number. For example, if $a = 10$ and $b = 10$. **86.** It is true for all a and b if $c = 0$. For example, if $a = 5$, $b = 3$, and $c = 0$. **87.** 21,222,415,864 **88.** 1,909,111,121,312 **89.** 8,466,084 **90.** Two hundred ninety-six thousand, three hundred eight **91.** 168 **92.** 1,119,534

1.4 Exercises

3.

×	6	2	3	8	0	5	7	9	1	4
5	30	10	15	40	0	25	35	45	5	20
7	42	14	21	56	0	35	49	63	7	28
1	6	2	3	8	0	5	7	9	1	4
0	0	0	0	0	0	0	0	0	0	0
6	36	12	18	48	0	30	42	54	6	24
2	12	4	6	16	0	10	14	18	2	8
3	18	6	9	24	0	15	21	27	3	12
8	48	16	24	64	0	40	56	72	8	32
4	24	8	12	32	0	20	28	36	4	16
9	54	18	27	72	0	45	63	81	9	36

5. 96 **7.** 208 **9.** 18,306 **11.** 303,612 **13.** 70 **15.** 522 **17.** 1630 **19.** 7609 **21.** 100,208 **23.** 942,808 **25.** 1560 **27.** 2,715,800 **29.** 482,000 **31.** 3,162,500,000 **33.** 8460 **35.** 63,600 **37.** 56,000,000 **39.** 31,682 **41.** 7884 **43.** 5696 **45.** 15,175 **47.** 41,537 **49.** 69,312 **51.** 148,567 **53.** 823,823 **55.** 41,830 **57.** 89,496 **59.** 217,980 **61.** 1,435,083 **63.** 4,097,115 **65.** 10,400 **67.** 9210 **69.** 70 **71.** 308 **73.** 700 **75.** 360 **77.** 0 **79.** 3216 square feet **81.** $12,075 **83.** $1200 **85.** $3192 **87.** 612 miles **89.** 1020 hamburgers **91.** $5,224,000,000 **93.** 54,000 **94.** 8 **95.** 8 **96.** 9 **97.** 7 **98.** No, it would not always be true. In our number system 62 = 60 + 2. But in roman numerals IV ≠ I + V. The digit system in roman numerals involves subtraction. Thus (XII) × (IV) ≠ (XII × I) + (XII × V). **99.** Yes. $5 \times (8 - 3) = 5 \times 8 - 5 \times 3$; $a \times (b - c) = (a \times b) - (a \times c)$ **100.** 6,756 **101.** 1249 **102.** $139 **103.** $59

1.5 Exercises

3. 7 **5.** 3 **7.** 5 **9.** 4 **11.** 3 **13.** 7 **15.** 9 **17.** 9 **19.** 7 **21.** 4 **23.** 7 **25.** 0 **27.** 9 **29.** 1 **31.** 5 R 2 **33.** 9 R 4 **35.** 25 R 3 **37.** 23 R 4 **39.** 32 **41.** 37 **43.** 322 R 1 **45.** 127 R 1 **47.** 753 **49.** 1357 R 4 **51.** 1757 R 5 **53.** 2478 R 3 **55.** 5 R 5 **57.** 5 R 7 **59.** 7 **61.** 160 R 10 **63.** 48 R 12 **65.** 615 R 11 **67.** 210 R 8 **69.** 202 R 7 **71.** 4 R 4 **73.** 7 R 26 **75.** 27 **77.** 730 pounds of feed per horse **79.** $245,192 per hour **81.** $46,179 **83.** 34 minutes **85.** a and b must represent the same number. For example, if $a = 12$, then $b = 12$. **87.** 5504 **88.** 1,038,490 **89.** 406,195 **90.** 66,844

Putting Your Skills to Work

1. $10,552 **3.** $18,760

1.6 Exercises

7. 6^4 **9.** 5^3 **11.** 8^4 **13.** 9^1 **15.** 16 **17.** 16 **19.** 216 **21.** 10,000 **23.** 1 **25.** 64 **27.** 81 **29.** 225 **31.** 343 **33.** 256 **35.** 1 **37.** 625 **39.** 1,000,000 **41.** 1024 **43.** 9 **45.** 2401 **47.** 82 **49.** 225 **51.** 520 **53.** $30 - 3 + 10$; $27 + 10 = 37$ **55.** $2 + 72 \div 2$; $2 + 36 = 38$ **57.** $100 \div 25 + 3$; $4 + 3 = 7$ **59.** $7 \times 9 + 4 - 8$; $63 + 4 - 8 = 59$ **61.** 304 **63.** $20 \div 20 = 1$ **65.** $950 \div 5 = 190$ **67.** $64 - 20 = 44$ **69.** $9 + 16 \div 4 = 9 + 4 = 13$ **71.** $42 - 4 \div 4 = 42 - 1 = 41$ **73.** $100 - 9 \times 4 = 100 - 36 = 64$ **75.** 56 **77.** $20 \times 3 \times 2 \div 2$; $120 \div 2 = 60$ **79.** $144 - 0$; 144 **81.** 49 **83.** $5 + 3 - 1$; 7 **85.** 61 **87.** $12 \div 2 \times 3 - 16$; $18 - 16 = 2$ **89.** $9 \times 6 \div 9 + 4 \times 3$; $6 + 12 = 18$ **91.** $4(4)^3 \div 8 = 4(64) \div 8 = 32$ **93.** $1200 - 8(3) \div 6 = 1200 - 4 = 1196$ **95.** $12 - 10 + 8 = 10$ **97.** $15 - 6 + 10$; 19 **99.** $15 + 21 - 20$; 16 **101.** 100,000 + 50,000 + 6000 + 300 + 10 + 2 **102.** 200,765,909 **103.** Two hundred sixty-one million, seven hundred sixty-three thousand, two **104.** 14,528 **105.** 1156 **106.** 90,000 **107.** 612 minutes; approximately 36,700 seconds

1.7 Exercises

3. 80 **5.** 70 **7.** 90 **9.** 530 **11.** 4240 **13.** 400 **15.** 2800 **17.** 7700 **19.** 8000 **21.** 1000 **23.** 28,000 **25.** 800,000 **27.** 15,000,000 stars **29.** **(a)** 143,000 **(b)** 143,400 **31.** **(a)** 3,700,000 square miles or 9,600,000 square kilometers **(b)** 3,710,000 square miles or 9,600,000 square kilometers

33. $\begin{array}{r} 500 \\ 200 \\ +\ 900 \\ \hline 1600 \end{array}$ **35.** $\begin{array}{r} 30 \\ 80 \\ 60 \\ +\ 30 \\ \hline 200 \end{array}$ **37.** $\begin{array}{r} 100,000 \\ 50,000 \\ +\ 100,000 \\ \hline 250,000 \end{array}$ **39.** $\begin{array}{r} 600,000 \\ -\ 100,000 \\ \hline 500,000 \end{array}$ **41.** $\begin{array}{r} 80,000,000 \\ -\ 60,000,000 \\ \hline 20,000,000 \end{array}$ **43.** $\begin{array}{r} 30,000,000 \\ -20,000,000 \\ \hline 10,000,000 \end{array}$ **45.** $\begin{array}{r} 60 \\ \times\ 50 \\ \hline 3000 \end{array}$ **47.** $\begin{array}{r} 2000 \\ \times\ \ \ 7 \\ \hline 14,000 \end{array}$

49. $\begin{array}{r} 600,000 \\ \times\ \ 300 \\ \hline 180,000,000 \end{array}$ **51.** $\begin{array}{r} 1,000 \\ 20\overline{)20,000} \end{array}$ **53.** $\begin{array}{r} 5,000 \\ 40\overline{)200,000} \end{array}$ **55.** $\begin{array}{r} 14,000 \\ 500\overline{)7,000,000} \end{array}$ **57.** Incorrect $\begin{array}{r} 400 \\ 500 \\ 900 \\ +\ 200 \\ \hline 2000 \end{array}$ **59.** Incorrect $\begin{array}{r} 100,000 \\ 50,000 \\ +\ \ 40,000 \\ \hline 190,000 \end{array}$

61. Correct $\begin{array}{r} 300,000 \\ -\ 100,000 \\ \hline 200,000 \end{array}$ **63.** Incorrect $\begin{array}{r} 80,000,000 \\ -\ 50,000,000 \\ \hline 30,000,000 \end{array}$ **65.** Incorrect $\begin{array}{r} 200 \\ \times\ \ 20 \\ \hline 4000 \end{array}$ **67.** Correct $\begin{array}{r} 6000 \\ \times\ \ 70 \\ \hline 420,000 \end{array}$ **69.** $\begin{array}{r} 5000 \\ 80\overline{)400,000} \end{array}$ Correct **71.** $\begin{array}{r} 500 \\ 400\overline{)200,000} \end{array}$ Correct

73. 4000 square yards **75.** $1500 **77.** 5400 sit-ups **79.** Estimate = $170; exact amount = $145 **81.** About $16 per week **83.** **(a)** 400,000 hours **(b)** 20,000 days **84.** **(a)** 300,000 hours **(b)** 15,000 days **85.** 83 **86.** 27 **87.** 28 **88.** 66

1.8 Exercises

1. 25,231 **3.** $4668 **5.** 3750 hors d'oeuvres **7.** 6¢ per ounce **9.** $23 **11.** 9 hours. This is equivalent to 540 minutes. **13.** 800,000 people **15.** $20,382 **17.** $786 **19.** Her balance will be $1482. **21.** $16,405 **23.** 25 miles per gallon **25.** 343 **26.** 21 **27.** 4788 **28.** 258

Chapter 1 Review Problems

1. Three hundred seventy-six **2.** Five thousand, eighty-two **3.** One hundred nine thousand, two hundred seventy-six **4.** Four hundred twenty-three million, five hundred seventy-six thousand, fifty-five **5.** 4000 + 300 + 60 + 4 **6.** 20,000 + 7000 + 900 + 80 + 6 **7.** 1,000,000 + 300,000 + 5000 + 100 + 20 + 8 **8.** 40,000,000 + 2,000,000 + 100,000 + 60,000 + 6000 + 30 + 7 **9.** 924 **10.** 6095 **11.** 1,328,828 **12.** 45,092,651 **13.** 130 **14.** 115 **15.** 690 **16.** 150 **17.** 400 **18.** 1469 **19.** 1598 **20.** 125,423 **21.** 14,703 **22.** 10,582 **23.** 17 **24.** 6 **25.** 27 **26.** 171 **27.** 159 **28.** 4828 **29.** 3026 **30.** 63,146 **31.** 224,757 **32.** 1,230,691 **33.** 36 **34.** 114 **35.** 0 **36.** 24 **37.** 18 **38.** 64 **39.** 105 **40.** 120 **41.** 144 **42.** 0 **43.** 240 **44.** 168 **45.** 2,612,100 **46.** 84,312,000 **47.** 83,200,000 **48.** 563,000,000 **49.** 864 **50.** 1856 **51.** 4050 **52.** 13,680 **53.** 25,524 **54.** 24,096 **55.** 87,822 **56.** 268,513 **57.** 543,510 **58.** 255,068 **59.** 150,000 **60.** 240,000 **61.** 7,200,000 **62.** 7,500,000 **63.** 2,000,000,000 **64.** 12,000,000,000 **65.** 2 **66.** 5 **67.** 14 **68.** 4 **69.** 0 **70.** 12 **71.** 7 **72.** 0 **73.** 7 **74.** 7 **75.** not possible **76.** 4 **77.** 7 **78.** 6 **79.** 8 **80.** not possible **81.** 125 **82.** 125 **83.** 258 **84.** 369 **85.** 25,874 **86.** 3692 **87.** 36,958 **88.** 36,921 **88.** 15,986 R 2 **90.** 35,783 R 4 **91.** 7 R 21 **92.** 4 R 37 **93.** 31 R 15 **94.** 14 R 11 **95.** 38 R 30 **96.** 54 R 38 **97.** 258 **98.** 95 **99.** 54 **100.** 19 **101.** 13^2 **102.** 24^2 **103.** 8^5 **104.** 9^5 **105.** 64 **106.** 81 **107.** 125 **108.** 128 **109.** 49 **110.** 81 **111.** 216 **112.** 256 **113.** 11 **114.** 8 **115.** 22 **116.** 66 **117.** 107 **118.** 76 **119.** 17 **120.** 48 **121.** 44 **122.** 93 **123.** 5670 **124.** 1280 **125.** 15,310 **126.** 42,640 **127.** 12,000 **128.** 23,000 **129.** 676,000 **130.** 202,000 **131.** 5,700,000 **132.** 10,000,000

133. $\begin{array}{r} 600 \\ 600 \\ 900 \\ +\ 900 \\ \hline 3000 \end{array}$ **134.** $\begin{array}{r} 30,000 \\ 40,000 \\ +\ 70,000 \\ \hline 140,000 \end{array}$ **135.** $\begin{array}{r} 30,000 \\ -\ 20,000 \\ \hline 10,000 \end{array}$ **136.** $\begin{array}{r} 4,000,000 \\ -\ 3,000,000 \\ \hline 1,000,000 \end{array}$ **137.** $\begin{array}{r} 2000 \\ \times\ 6000 \\ \hline 12,000,000 \end{array}$ **138.** $\begin{array}{r} 3,000,000 \\ \times\ 900 \\ \hline 2,700,000,000 \end{array}$ **139.** $\begin{array}{r} 4,000 \\ 20\overline{)80,000} \end{array}$

140. $\begin{array}{r} 20,000 \\ 400\overline{)8,000,000} \end{array}$ **141.** Correct $\begin{array}{r} 90 \\ 90 \\ 40 \\ +\ 60 \\ \hline 280 \end{array}$ **142.** Correct $\begin{array}{r} 900,000 \\ -\ 400,000 \\ \hline 500,000 \end{array}$ **143.** Incorrect $\begin{array}{r} 200,000 \\ \times\ 5000 \\ \hline 1,000,000,000 \end{array}$ **144.** Correct $\begin{array}{r} 30,000 \\ 30\overline{)900,000} \end{array}$

145. 175 words **146.** 204 cans **147.** $59,470 **148.** 2397 km **149.** $7028 **150.** 10,301 feet **151.** $64 per share **152.** $1356 **153.** $334 **154.** $278 **155.** 25 miles per gallon **156.** 24 miles per gallon **157.** $5041 **158.** $3456

Chapter 1 Test

1. forty-four million, seven thousand, six hundred thirty-five **2.** 20,000 + 6000 + 800 + 50 + 9 **3.** 3,581,076 **4.** 700 **5.** 1045 **6.** 318,977 **7.** 7419 **8.** 38,341 **9.** 5,225,768 **10.** 378 **11.** 2184 **12.** 91,875 **13.** 129,437 **14.** 3014 R 1 **15.** 2189 **16.** 352 **17.** 11^3 **18.** 64 **19.** 23 **20.** 78 **21.** 79 **22.** 26,450 **23.** 6,460,000 **24.** 3,600,000 **25.** 150,000,000,000 **26.** 18,000 **27.** $2148 **28.** 467 feet **29.** $127 **30.** $292

Chapter 2

Pretest Chapter 2

1. $\frac{3}{8}$ **2.** Answers will vary **3.** $\frac{17}{124}$ **4.** $\frac{1}{6}$ **5.** $\frac{1}{3}$ **6.** $\frac{1}{7}$ **7** $\frac{5}{7}$ **8.** $\frac{4}{11}$ **9.** $\frac{9}{7}$ **10.** $\frac{55}{9}$ **11.** $24\frac{1}{4}$ **12.** $5\frac{4}{5}$ **13.** $2\frac{2}{17}$ **14.** $\frac{5}{44}$ **15.** $\frac{2}{3}$ **16.** $67\frac{5}{6}$ **17.** 1 **18.** $\frac{1}{2}$ **19.** $5\frac{4}{13}$ **20.** $4\frac{2}{3}$ **21.** 8 **22.** 45 **23.** 55 **24.** 72 **25.** $\frac{19}{72}$ **26.** $\frac{49}{72}$ **27.** $\frac{13}{3}$ or $4\frac{1}{3}$ **28.** $3\frac{41}{42}$ **29.** $\frac{77}{18}$ or $4\frac{5}{18}$ **30.** $7\frac{1}{6}$ miles **31.** $7\frac{11}{36}$ tons **32.** 18 students

2.1 Exercises

5. N: 3; D: 5 **7.** N: 2; D: 3 **9.** N: 1; D: 17 **11.** $\frac{1}{2}$ **13.** $\frac{5}{6}$ **15.** $\frac{2}{3}$ **17.** $\frac{5}{6}$ **19.** $\frac{1}{4}$ **21.** $\frac{3}{10}$ **23.** $\frac{5}{8}$ **25.** $\frac{4}{7}$ **27.** $\frac{7}{8}$ **29.** $\frac{2}{5}$ **31.** 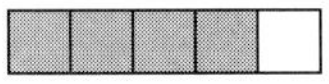 **33.** 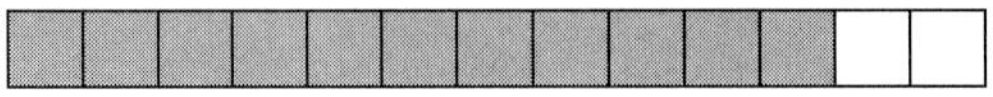 **35.** 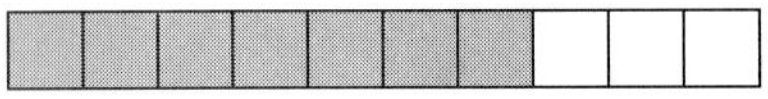 **37.** $\frac{29}{165}$ **39.** $\frac{209}{750}$ **41.** $\frac{89}{211}$ **43.** $\frac{17}{50}$ **45.** $\frac{9}{14}$ **47. (a)** $\frac{90}{195}$ **(b)** $\frac{22}{195}$ **49.** The amount of money each of 6 business owners gets if the business has a profit of $0. **50.** We cannot do it. Division by zero is undefined. **51.** 241 **52.** 11,106 **53.** 119,944 **54.** 177 R 6

2.2 Exercises

7. 3×5 **9.** 2×3 **11.** 7^2 **13.** 2^6 **15.** 5×11 **17.** 5×7 **19.** 3×5^2 **21.** 2×3^3 **23.** $2^2 \times 3 \times 7$ **25.** 2×7^2 **27.** Prime **29.** 3×19 **31.** Prime **33.** 2×31 **35.** Prime **37.** Prime **39.** 11×11 **41.** 7×23 **43.** $\frac{18 \div 9}{27 \div 9} = \frac{2}{3}$ **45.** $\frac{32 \div 16}{48 \div 16} = \frac{2}{3}$ **47.** $\frac{30 \div 6}{48 \div 6} = \frac{5}{8}$ **49.** $\frac{42 \div 6}{48 \div 6} = \frac{7}{8}$ **51.** $\frac{3 \times 1}{3 \times 5} = \frac{1}{5}$ **53.** $\frac{3 \times 13}{2 \times 2 \times 13} = \frac{3}{4}$ **55.** $\frac{2 \times 3 \times 3}{2 \times 2 \times 2 \times 3} = \frac{3}{4}$ **57.** $\frac{3 \times 3 \times 3}{3 \times 3 \times 5} = \frac{3}{5}$ **59.** $\frac{3 \times 11}{3 \times 12} = \frac{11}{12}$ **61.** $\frac{5 \times 13}{13 \times 13} = \frac{5}{13}$ **63.** $\frac{11 \times 8}{11 \times 11} = \frac{8}{11}$ **65.** $\frac{3 \times 50}{4 \times 50} = \frac{3}{4}$ **67.** $\frac{7 \times 17}{7 \times 30} = \frac{17}{30}$ **69.** Yes **71.** Yes **73.** No **75.** No **77.** Yes **79.** $\frac{2}{3}$ **81.** $\frac{6}{19}$ **83.** $\frac{1}{9}$ **85.** 164,050 **86.** 1296 **87.** 1,630,000 **88.** 25,920,000,000

2.3 Exercises

1. $\frac{7}{2}$ **3.** $\frac{14}{3}$ **5.** $\frac{17}{7}$ **7.** $\frac{53}{10}$ **9.** $\frac{29}{6}$ **11.** $\frac{65}{3}$ **13.** $\frac{55}{6}$ **15.** $\frac{169}{6}$ **17.** $\frac{131}{12}$ **19.** $\frac{79}{10}$ **21.** $\frac{201}{25}$ **23.** $\frac{211}{2}$ **25.** $\frac{494}{3}$ **27.** $\frac{119}{15}$ **29.** $\frac{138}{25}$ **31.** $1\frac{2}{5}$ **33.** $2\frac{3}{4}$ **35.** $2\frac{1}{2}$ **37.** $3\frac{3}{8}$ **39.** 5 **41.** $9\frac{5}{9}$ **43.** $2\frac{2}{13}$ **45.** $3\frac{3}{16}$ **47.** $9\frac{1}{3}$ **49.** $17\frac{1}{2}$ **51.** 13 **53.** $22\frac{2}{9}$ **55.** 6 **57.** $36\frac{7}{11}$ **59.** $2\frac{3}{4}$ **61.** $4\frac{1}{6}$ **63.** $12\frac{3}{8}$ **65.** 4 **67.** $\frac{12}{5}$ **69.** $\frac{26}{3}$ **71.** $2\frac{88}{126} = 2\frac{44}{63}$ **73.** $2\frac{138}{424} = 2\frac{69}{212}$ **75.** $2\frac{250}{350} = 2\frac{5}{7}$ **77.** 1,082 **79.** $50\frac{1}{3}$ acres **81.** No. 101 is prime and is not a factor of 5687. **82.** No. 157 is prime and not a factor of 9810. **83.** 28,567 **84.** 260,247 **85.** 16,000,000,000 **86.** 3000

2.4 Exercises

1. $\frac{21}{55}$ **3.** $\frac{15}{52}$ **5.** 1 **7.** $\frac{24}{35}$ **9.** $\frac{5}{12}$ **11.** $\frac{21}{8}$ *or* $2\frac{5}{8}$ **13.** $\frac{24}{7}$ *or* $3\frac{3}{7}$ **15.** $\frac{5}{2}$ *or* $2\frac{1}{2}$ **17.** $\frac{4}{15}$ **19.** $\frac{1}{2}$ **21.** $\frac{14}{5}$ *or* $2\frac{4}{5}$ **23.** $\frac{55}{12}$ *or* $4\frac{7}{12}$ **25.** 15 **27.** $\frac{69}{50}$ *or* $1\frac{19}{50}$ **29.** 0 **31.** $\frac{154}{3}$ *or* $51\frac{1}{3}$ **33.** $\frac{53}{15}$ *or* $3\frac{8}{15}$ **35.** $11\frac{5}{7}$ **37.** $x = \frac{9}{5}$ **39.** $x = \frac{11}{13}$ **41.** $37\frac{11}{12}$ square miles **43.** 134 miles **45.** 146 yards **47.** $9\frac{3}{16}$ ounces **49.** $\$6141\frac{1}{4}$ **51. (a)** $\frac{5}{8}$ of the students **(b)** 5375 students voted **53.** 529 cars **54.** 368 calls **55.** 1752 lines **56.** 173,040 gallons

2.5 Exercises

1. $\frac{4}{35}$ **3.** $\frac{2}{3}$ **5.** $\frac{9}{10}$ **7.** $\frac{3}{2}$ *or* $1\frac{1}{2}$ **9.** $\frac{4}{5}$ **11.** $\frac{16}{7}$ *or* $2\frac{2}{7}$ **13.** $\frac{4}{3}$ *or* $1\frac{1}{3}$ **15.** 1 **17.** $\frac{9}{49}$ **19.** $\frac{36}{7}$ *or* $5\frac{1}{7}$ **21.** 0 **23.** Cannot be done **25.** $\frac{3}{4}$ **27.** $\frac{4}{5}$ **29.** 1 **31.** 16 **33.** $\frac{7}{32}$ **35.** 5000 **37.** $\frac{2}{15}$ **39.** $\frac{7}{40}$ **41.** $\frac{32}{3}$ *or* $10\frac{2}{3}$ **43.** $\frac{7}{18}$ **45.** $\frac{2}{3}$ **47.** 4 **49.** $\frac{15}{7}$ *or* $2\frac{1}{7}$ **51.** 3 **53.** $\frac{63}{32}$ *or* $1\frac{31}{32}$ **55.** $\frac{1}{2}$ **57.** 12 **59.** $\frac{2}{3}$ **61.** $\frac{5}{21}$ **63.** $\frac{32}{15}$ *or* $2\frac{2}{15}$ **65.** $x = \frac{7}{5}$ **67.** $x = \frac{4}{7}$ **69.** $2\frac{1}{4}$ gallons **71.** $37\frac{1}{2}$ miles per hour **73.** 28 flags **75.** $a = 2$, $b = 3$, $c = 4$, $d = 6$. $\frac{2}{3} \div \frac{4}{6} = \frac{4}{6} \div \frac{2}{3}$. In general, if $\frac{a}{b} = \frac{c}{d}$, then it is true. **76.** $3\frac{7}{12} \div \frac{1}{4} = \left(3 + \frac{7}{12}\right) \times 4 = 12 + \frac{28}{12} = 12 + \frac{7}{3} = 14\frac{1}{3}$ **77.** Thirty-nine million, five hundred seventy-six thousand, three hundred four. **78.** 400,000 + 50,000 + 9000 + 200 + 70 + 3 **79.** 1099 **80.** 87,595,631

2.6 Exercises

1. 10 **3.** 28 **5.** 35 **7.** 18 **9.** 48 **11.** 16 **13.** 90 **15.** 60 **17.** 105 **19.** 36 **21.** 6 **23.** 120 **25.** 144 **27.** 84 **29.** 120 **31.** 3 **33.** 45 **35.** 20 **37.** 14 **39.** 96 **41.** 39 **43.** $\frac{21}{36}$ and $\frac{20}{36}$ **45.** $\frac{24}{200}$ and $\frac{35}{200}$ **47.** $\frac{114}{150}$ and $\frac{52}{150}$ **49.** LCD = 42; $\frac{30}{42}$ and $\frac{7}{42}$ **51.** LCD = 48; $\frac{20}{48}$ and $\frac{3}{48}$ **53.** LCD = 80; $\frac{52}{80}$ and $\frac{55}{80}$ **55.** LCD = 60; $\frac{35}{60}$ and $\frac{46}{60}$ **57.** LCD = 72; $\frac{15}{72}, \frac{22}{72}, \frac{3}{72}$ **59.** LCD = 56; $\frac{3}{56}, \frac{49}{56}, \frac{40}{56}$ **61.** LCD = 63; $\frac{5}{63}, \frac{12}{63}, \frac{56}{63}$ **63.** **(a)** 16 **(b)** $\frac{3}{16}, \frac{12}{16}, \frac{6}{16}$ **65.** 178 R 3 **66.** 3624 **72.** 9963

2.7 Exercises

1. $\frac{7}{9}$ **3.** $\frac{11}{15}$ **5.** $\frac{20}{23}$ **7.** $\frac{17}{44}$ **9.** $\frac{7}{12}$ **11.** $\frac{9}{20}$ **13.** $\frac{7}{8}$ **15.** $\frac{3}{2}$ *or* $1\frac{1}{2}$ **17.** $\frac{37}{100}$ **19.** $\frac{26}{175}$ **21.** $\frac{31}{24}$ *or* $1\frac{7}{24}$ **23.** $\frac{27}{40}$ **25.** $\frac{1}{4}$ **27.** $\frac{8}{35}$ **29.** $\frac{9}{16}$ **31.** $\frac{11}{60}$ **33.** $\frac{1}{4}$ **35.** $\frac{1}{2}$ **37.** $\frac{5}{48}$ **39.** 0 **41.** 0 **43.** $\frac{8}{5}$ *or* $1\frac{3}{5}$ **45.** $\frac{11}{30}$ **47.** $1\frac{3}{5}$ **49.** $\frac{3}{14}$ **51.** $\frac{5}{33}$ **53.** $\frac{8}{15}$ **55.** $1\frac{7}{15}$ cup **57.** $1\frac{1}{2}$ pounds **59.** $1\frac{23}{36}$ inches **61.** $7\frac{7}{12}$ cups **63.** $\frac{4}{21}$ of the membership **65.** $\frac{3}{17}$ **66.** $\frac{2}{3}$ **67.** $7\frac{11}{16}$ **68.** $\frac{101}{7}$

2.8 Exercises

1. $9\frac{3}{4}$ **3.** $4\frac{1}{7}$ **5.** $17\frac{1}{2}$ **7.** $16\frac{1}{10}$ **9.** $\frac{4}{7}$ **11.** $2\frac{17}{24}$ **13.** $16\frac{1}{4}$ **15.** $3\frac{1}{2}$ **17.** $7\frac{5}{12}$ **19.** $14\frac{4}{7}$ **21.** $41\frac{4}{5}$ **23.** $6\frac{7}{12}$ **25.** $16\frac{2}{3}$ **27.** $73\frac{37}{40}$ **29.** $5\frac{1}{2}$ **31.** 0 **33.** $4\frac{41}{60}$ **35.** $8\frac{8}{15}$ **37.** $102\frac{5}{8}$ **39.** $13\frac{5}{24}$ **41.** $43\frac{1}{8}$ miles **43.** $2\frac{1}{24}$ pounds **45.** **(a)** $9\frac{19}{24}$ pounds **(b)** $6\frac{5}{24}$ pounds **47.** $\frac{2607}{40}$ *or* $65\frac{7}{40}$ **48.** $\frac{277}{42}$ *or* $6\frac{25}{42}$ **49.** $\frac{941}{21}$ *or* $44\frac{17}{21}$ **50.** $\frac{589}{9}$ *or* $65\frac{4}{9}$ **51.** 111,303 **52.** 38,912 **53.** 512,012 **54.** 8,529,300

Putting Your Skills to Work

1. $16,600 **2.** $5060

2.9 Exercises

1. $16\frac{7}{24}$ tons **3.** $25\frac{11}{24}$ tons **5.** $1\frac{9}{16}$ inches **7.** 17 pieces **9.** $\$93\frac{1}{3}$ **11.** $275\frac{5}{8}$ gallons **13.** $106\frac{7}{8}$ nautical miles **15.** $451 per week is left **17.** **(a)** 42 bags **(b)** $147 **(c)** $145 profit **19.** **(a)** $14\frac{1}{8}$ ounces of bread **(b)** $\frac{5}{8}$ ounce **21.** **(a)** $30\frac{1}{2}$ knots **(b)** 7 hours **23.** **(a)** 5485 bushels **(b)** $11{,}998\frac{7}{16}$ cubic feet **(c)** $9598\frac{3}{4}$ bushels **25.** 44,245 **26.** 27,814 **27.** 45,441 **28.** 278 R 3

Chapter 2 Review Problems

1. $\frac{5}{12}$ **2.** $\frac{3}{8}$ **3.** Answers will vary. **4.** Answers will vary. **5.** $\frac{6}{31}$ **6.** $\frac{87}{100}$ **7.** $2 \times 3 \times 7$ **8.** 2×3^3 **9.** $2^3 \times 3 \times 7$ **10.** Prime **11.** $2 \times 3 \times 13$ **12.** Prime **13.** $\frac{1}{4}$ **14.** $\frac{2}{7}$ **15.** $\frac{7}{12}$ **16.** $\frac{13}{17}$ **17.** $\frac{7}{8}$ **18.** $\frac{17}{35}$ **19.** $\frac{35}{8}$ **20.** $\frac{65}{23}$ **21.** $2\frac{5}{7}$ **22.** $4\frac{11}{13}$ **23.** $3\frac{3}{11}$ **24.** $\frac{117}{8}$ **25.** $1\frac{29}{48}$ **26.** $\frac{20}{77}$ **27.** $\frac{7}{15}$ **28.** 0 **29.** $\frac{4}{63}$ **30.** $\frac{492}{5}$ *or* $98\frac{2}{5}$ **31.** $\frac{817}{40}$ *or* $20\frac{17}{40}$ **32.** $\frac{51}{2}$ *or* $25\frac{1}{2}$ **33.** $24\frac{1}{2}$ **34.** $\$667\frac{1}{4}$ **35.** $486\frac{17}{20}$ square inches **36.** $\frac{15}{14}$ *or* $1\frac{1}{14}$ **37.** $\frac{5}{34}$ **38.** 1920 **39.** $\frac{27}{40}$ **40.** $\frac{25}{6}$ *or* $4\frac{1}{6}$ **41.** $\frac{17}{164}$ **42.** 0 **43.** $\frac{46}{33}$ *or* $1\frac{13}{33}$ **44.** $186\frac{2}{3}$ calories **45.** 7 dresses **46.** 98 **47.** 120 **48.** 90 **49.** $\frac{24}{56}$ **50.** $\frac{33}{72}$ **51.** $\frac{36}{172}$ **52.** $\frac{187}{198}$ **53.** $\frac{1}{14}$ **54.** $\frac{13}{12}$ *or* $1\frac{1}{12}$ **55.** $\frac{85}{63}$ *or* $1\frac{22}{63}$ **56.** $\frac{11}{40}$ **57.** $\frac{23}{70}$ **58.** $\frac{25}{36}$ **59.** $\frac{61}{75}$ **60.** $\frac{19}{48}$ **61.** $\frac{6}{23}$ **62.** $5\frac{4}{9}$ **63.** $8\frac{2}{3}$ **64.** $20\frac{5}{7}$ **65.** $\frac{39}{8}$ *or* $4\frac{7}{8}$ **66.** $\frac{209}{48}$ *or* $4\frac{17}{48}$ **67.** $\frac{310}{3}$ *or* $103\frac{1}{3}$ **68.** $\frac{857}{12}$ *or* $71\frac{5}{12}$ **69.** $8\frac{29}{40}$ miles **70.** $\$1\frac{5}{8}$ **71.** $1\frac{2}{3}$ cups sugar, $2\frac{1}{8}$ cups flour **72.** $206\frac{1}{8}$ miles **73.** 15 lengths **74.** $9\frac{5}{8}$ liters **75.** 30 words per minute **76.** $\$56\frac{1}{4}$ **77.** $\$8\frac{3}{4}$ **78.** $1\frac{1}{16}$ inch **79.** $242

80. **(a)** 25 miles per gallon **(b)** $\$22\frac{2}{25}$

Chapter 2 Test

1. $\frac{3}{5}$ **2.** $\frac{311}{388}$ **3.** $\frac{3}{14}$ **4.** $\frac{3}{7}$ **5.** $\frac{9}{2}$ **6.** $\frac{34}{5}$ **7.** $8\frac{1}{7}$ **8.** 12 **9.** $\frac{14}{45}$ **10.** $31\frac{1}{5}$ **11.** $\frac{77}{40}$ or $1\frac{37}{40}$ **12.** $\frac{12}{7}$ or $1\frac{5}{7}$ **13.** $6\frac{12}{13}$ **14.** $\frac{39}{62}$ **15.** 72 **16.** 48 **17.** 24 **18.** $\frac{30}{72}$ **19.** $\frac{13}{36}$ **20.** $\frac{11}{20}$ **21.** $\frac{25}{28}$ **22.** $14\frac{6}{35}$ **23.** $4\frac{13}{14}$ **24.** $2\frac{4}{9}$ **25.** $21\frac{1}{4}$ **26.** 42 square yards **27.** 8 packages **28.** $\frac{7}{10}$ mile **29.** $14\frac{1}{24}$ miles **30.** **(a)** 40 oranges **(b)** $9.60 or 960¢

Cumulative Test for Chapters 1–2

1. Eighty-four million, three hundred sixty-one thousand, two hundred eight **2.** 869 **3.** 719,220 **4.** 2075 **5.** 17,216 **6.** 4788 **7.** 202,896 **8.** 4307 R 1 **9.** 369 **10.** 49 **11.** 6,037,000 **12.** 38 **13.** $174 **14.** $306 **15.** $\frac{55}{84}$ **16.** $\frac{7}{13}$ **17.** $\frac{75}{4}$ **18.** $14\frac{2}{7}$ **19.** $\frac{527}{48}$ *or* $10\frac{47}{48}$ **20.** $\frac{12}{35}$ **21.** 39 **22.** $\frac{31}{54}$ **23.** $\frac{71}{8}$ *or* $8\frac{7}{8}$ **24.** $\frac{113}{15}$ *or* $7\frac{8}{15}$ **25.** $\frac{13}{28}$ **26.** $23\frac{1}{8}$ tons **27.** $24\frac{3}{5}$ miles per gallon **28.** $8\frac{1}{8}$ cups of sugar; $5\frac{5}{6}$ cups of flour **29.** 60,000,000 miles

Chapter 3

Pretest Chapter 3

1. Forty-seven and eight hundred thirteen thousandths **2.** 0.0567 **3.** $2\frac{11}{100}$ **4.** $\frac{21}{40}$ **5.** 1.59, 1.6, 1.601, 1.61 **6.** 10.5 **7.** 1.053 **8.** 19.45 **9.** 27.191 **10.** 10.59 **11.** 8.892 **12.** 0.3501 **13.** 4780.5 **14.** 5.129 **15.** 0.354 **16.** 0.128 **17.** 0.4375 **18.** $0.2\overline{27}$ or 0.22727 . . . **19.** 0.97 **20.** 17.9 miles per gallon **21.** $489.51 **22.** $8.53

3.1 Exercises

5. Fifty-seven hundredths **7.** Three and eight tenths **9.** One hundred twenty-four thousandths **11.** Twenty-eight and seven ten thousandths **13.** Thirty-six and $\frac{18}{100}$ dollars **15.** One thousand two hundred thirty-six and $\frac{8}{100}$ dollars **17.** Ten thousand and $\frac{76}{100}$ dollars **19.** 0.7 **21.** 0.039 **23.** 0.0065 **25.** 0.000286 **27.** 0.3 **29.** 0.76 **31.** 0.771 **33.** 0.053 **35.** 0.0026 **37.** 8.3 **39.** 84.13 **41.** 1.019 **43.** 126.0571 **45.** $\frac{1}{50}$ **47.** $3\frac{3}{5}$ **49.** $\frac{121}{1000}$ **51.** $12\frac{5}{8}$ **53.** $7\frac{3}{2000}$ **55.** $235\frac{627}{5000}$ **57.** $\frac{187}{10,000}$ **59.** $8\frac{27}{2500}$ **61.** $289\frac{47}{125}$ **63.** $\frac{9889}{10,000}$ **65.** $\frac{1}{50,000,000}$ **67.** 818 **68.** 938 **69.** 56,800 **70.** 8,069,000

3.2 Exercises

1. > **3.** < **5.** < **7.** > **9.** < **11.** > **13.** < **15.** > **17.** 12.6, 12.65, 12.8 **19.** 0.007, 0.0071, 0.05 **21.** 1.1, 1.79, 1.8, 1.81 **23.** 26.003, 26.033, 26.034, 26.04 **25.** 18.006, 18.060, 18.065, 18.066, 18.606 **27.** 5.7 **29.** 29.5 **31.** 578.1 **33.** 2176.8 **35.** 26.03 **37.** 5.77 **39.** 156.12 **41.** 2786.71 **43.** 1.061 **45.** 0.0474 **47.** 5.00761 **49.** 0.007537 **51.** 129 **53.** $2537 **55.** $10,098 **57.** $56.98 **59.** $5783.72 **61.** 0.10 kilogram, 0.07 kilogram **63.** 0.0059, 0.006, 0.0519, $\frac{6}{100}$, 0.0601, 0.0612, 0.062, $\frac{6}{10}$, 0.61 **64.** 0.049, 0.0515, 0.052, 1.05, 1.051, 1.0513, $\frac{15}{10}$, $\frac{151}{100}$, 1.512 **65.** You should consider only one digit to the right of the decimal place that you wish to round to. 86.23498 is closer to 86.23 than to 86.24. **66.** The bank rounds up for any fractional part of a cent. **67.** $12\frac{1}{8}$ **68.** $10\frac{9}{20}$ **69.** 692 miles **70.** $9658

3.3 Exercises

1. 72.8 **3.** 1215.55 **5.** 7.025 **7.** 136.844 **9.** 323.9 **11.** 23.00 **13.** 36.7287 **15.** 67.42 **17.** 1112.21 **19.** 19.86 m **21.** 60.1 gallons **23.** \$78.12 **25.** \$1,411.97 **27.** 3.9 **29.** 27.17 **31.** 64.18 **33.** 508.313 **35.** 0.02151 **37.** 4.6465 **39.** 6.737 **41.** 1189.07 **43.** 1.4635 **45.** \$33.50 **47.** 17.661 kilograms **49.** \$36,947.16 more **51.** \$45.30 **53.** 1.46 cm **55.** \$601,409.07 revenue shortage **57.** 0.008 milligrams, no **59.** \$5.7 billion; 5,700,000,000 dollars **61.** \$13.3 billion; 13,300,000,000 dollars **63.** $x = 8.4$ **65.** $x = 43.7$ **67.** $x = 2.109$ **69.** 20,288 **70.** 18,213 **71.** $\frac{77}{25}$ *or* $3\frac{2}{25}$ **72.** $\frac{35}{4}$ *or* $8\frac{3}{4}$

Putting Your Skills to Work

1. \$83.24 **2.**

	Check Number	Date	Transaction	Amount of Withdrawal (−)		Amount of Deposit (+)		New Balance	
								83	24
(a)		3/20	Service fee	3	50			79	74
(b)		3/22	Deposit			301	64	381	38
(c)	158	3/30	Sears	159	95			221	43
(d)		3/31	Deposit			450	36	671	79
(e)	159	4/2	Gulf Oil	46	52			625	27
(f)	160	4/3	Western Bell	39	64			585	63
(g)	161	4/14	IRS	561	18			24	45
(h)		4/18	New Checks	6	95			17	50

3. \$290.20

3.4 Exercises

1. 0.12 **3.** 0.06 **5.** 0.00288 **7.** 0.00711 **9.** 0.002025 **11.** 0.6582 **13.** 2555.52 **15.** 0.013244 **17.** 768.1517 **19.** 4911.3 **21.** 441 **23.** 43,358.6 **25.** 0.000238 **27.** 6.5237 **29.** \$234.00 **31.** \$382.00 **33.** 235.48 square feet **35.** \$664.20 **37.** \$0.77 **39.** 514.8 miles **41.** 28.6 **43.** 70.1 **45.** 128,650 **47.** 56,098.2 **49.** 28,056,020 **51.** 816,320 **53.** 671.8 **55.** 81.376 **57.** 593.2 centimeters **59.** 2980 meters **61.** \$36,405,000.00 **63.** \$62,279.00 **65.** 98 **66.** 86 **67.** 125 R 4 **68.** 129 R 4

3.5 Exercises

1. 2.1 **3.** 0.0369 **5.** 18.31 **7.** 0.0565 **9.** 0.0029 **11.** 12.2 **13.** 64.3 **15.** 8.01 **17.** 21 **19.** 230 **21.** 5.3 **23.** 2.3 **25.** 33.8 **27.** 65.96 **29.** 12.24 **31.** 24.92 **33.** 31.020 **35.** 12.246 **37.** 50 **39.** 13 **41.** 6.2 ounces in each portion **43.** 20 miles per gallon **45.** \$376.50 **47.** 9 payments **49.** There are 38 blocks of chalk in the box. The error was made by placing 2 fewer blocks of chalk in the box than what is listed on the label. **51.** 0.123 **53.** 69 **55.** 45 **57.** 810 **59.** 74 **60.** 41 **61.** $1\frac{31}{40}$ **62.** $\frac{15}{16}$ **63.** $\frac{91}{12}$ or $7\frac{7}{12}$ **64.** $\frac{5}{3}$ or $1\frac{2}{3}$

3.6 Exercises

5. 0.25 **7.** 0.875 **9.** 0.4375 **11.** 0.35 **13.** 0.62 **15.** 2.25 **17.** 2.125 **19.** 1.4375 **21.** $0.\overline{6}$ **23.** $0.\overline{5}$ **25.** $0.8\overline{3}$ **27.** $0.\overline{90}$ **29.** $0.9\overline{3}$ **31.** $0.\overline{15}$ **33.** $1.\overline{037}$ **35.** $2.2\overline{7}$ **37.** 0.571 **39.** 0.905 **41.** 0.146 **43.** 1.296 **45.** 0.247 **47.** 0.944 **49.** 3.143 **51.** 0.474 **53.** $0.208\overline{3}$ inch thick **55.** It is too small by 0.125 inch. **57.** Too thick, 0.00125 inch **59.** 5.078 **61.** 114 **63.** 374.65 **65.** 156.0664 **67.** 21.414 **69.** 2.5 **71.** 0.325 **73.** 28.6 **75.** 20.836 **77.** 0.586930

79. **(a)** $\begin{array}{r} 0.1616\overline{16} \\ -\,0.0016\overline{16} \\ \hline 0.16 \end{array}$ **(b)** $\begin{array}{r} 0.1616\overline{16} \\ -\,0.01666\overline{6} \\ \hline 0.144949 \end{array}$ **(c)** The repeating patterns line up differently. **80.** **(a)** $\begin{array}{r} 1.8989\overline{89} \\ -\,0.0189\overline{89} \\ \hline 1.88 \end{array}$ **(b)** $\begin{array}{r} 1.898989\overline{89} \\ -\,0.1899999\overline{9} \\ \hline 1.70898989 \end{array}$ **(c)** The repeating patterns line up differently. **81.** 312 square feet **82.** 869 votes **83.** \$377 was deposited **84.** \$83 per employee

3.7 Exercises

1. 621.2 pounds **3.** \$8868.88 **5.** 7889.42 square meters **7.** 46 packages **9.** Yes. After she bought the eggs she had \$0.33. **11.** 11.59 meters of rainfall per year **13.** 12 days **15.** \$1263.09 **17.** \$743.60 **19.** The balance is \$426.55 **21.** \$17,319. He will pay back \$5819 more than the loan. **23.** \$1872 **25.** Yes, by 0.149 milligram **27.** 137 minutes **29.** 17.7 quadrillion Btu **31.** Approximately 47.8 quadrillion Btu. This is the same as 47,800,000,000,000,000 Btu. **33.** $\frac{79}{70}$ or $1\frac{9}{70}$ **34.** $\frac{11}{34}$ **35.** $\frac{5}{9}$ **36.** 22

Chapter 3 Review Problems

1. Thirteen and six hundred seventy-two thousandths **2.** Eighty-four hundred thousandths **3.** 0.7 **4.** 0.81 **5.** 1.523 **6.** 0.0079 **7.** $\frac{17}{100}$ **8.** $\frac{73}{200}$ **9.** $26\frac{22}{25}$ **10.** $1\frac{1}{4000}$ **11.** $>$ **12.** $>$ **13.** 0.901, 0.918, 0.98, 0.981 **14.** 5.2, 5.26, 5.59, 5.6, 5.62 **15.** 0.6 **16.** 19.21 **17.** 1.100 **18.** $156.00 **19.** 6.639 **20.** 77.6 **21.** **(a)** 8.5 gallons **(b)** 9 gallons **22.** 3.894 **23.** 90.739 **24.** 236.9 miles **25.** 0.0228 **26.** 0.90118 **27.** 0.000364 **28.** 887.81 **29.** 86.4 **30.** 2398.02 **31.** 78.6 **32.** 15,637.1 **33.** 0.613 **34.** 123,540 **35.** $0.90 **36.** 24,075 **37.** 0.00258 **38.** 36.8 **39.** 232.9 **40.** 574.4 **41.** 0.059 **42.** $66.75 **43.** $0.2\overline{7}$ **44.** 0.175 **45.** $1.8\overline{3}$ **46.** 0.9375 **47.** 0.786 **48.** 2.294 **49.** 3.6 **50.** 14.1 **51.** 1.152 **52.** 1.301208 **53.** 87.13 **54.** 6 **55.** 0.2128 **56.** 0.55 **57.** 0.6 **58.** 13.122 **59.** 439.19 **60.** 0.402216 **61.** 0.8403 **62.** 69.2 **63.** 20.004 **64.** 1.25 **65.** $9.13 **66.** $84.46 **67.** $368.08 **68.** 24.8 miles **69.** $2170.30 **70.** He will earn more at ABC Company. **71.** no; by 0.0005 mg **72.** 7 test tubes

Chapter 3 Test

1. One hundred fifty-seven thousandths **2.** 0.3977 **3.** $7\frac{3}{20}$ **4.** $\frac{261}{1000}$ **5.** 2.19, 2.9, 2.907, 2.91 **6.** 78.66 **7.** 0.0342 **8.** 43.989 **9.** 37.53 **10.** 0.0979 **11.** 71.155 **12.** 0.5817 **13.** 218.9 **14.** 25.7 **15.** 47 **16.** $1.\overline{2}$ **17.** 0.5625 **18.** 1.487 **19.** 6.1952 **20.** $33.15 **21.** 18.8 miles per gallon **22.** 3.43 centimeters less

Cumulative Test for Chapters 1–3

1. Thirty-eight million, fifty-six thousand, nine hundred fifty-four **2.** 479,587 **3.** 54,480 **4.** 39,463 **5.** 258 **6.** 16 **7.** $\frac{3}{8}$ **8.** $7\frac{1}{2}$ **9.** $\frac{9}{35}$ **10.** $\frac{7}{6}$ or $1\frac{1}{6}$ **11.** 16 **12.** $\frac{33}{10}$ or $3\frac{3}{10}$ **13.** 24,000,000,000 **14.** 0.571 **15.** 2.01, 2.1, 2.11, 2.12, 20.1 **16.** 26.080 **17.** 19.54 **18.** 8.639 **19.** 1.136 **20.** 36,512.3 **21.** 1.058 **22.** 0.8125 **23.** 13.597 **24.** 456 miles **25.** $195.57

Chapter 4

Pretest Chapter 4

1. $\frac{13}{18}$ **2.** $\frac{1}{5}$ **3.** $\frac{9}{2}$ **4.** $\frac{11}{12}$ **5.** **(a)** $\frac{7}{24}$ **(b)** $\frac{11}{120}$ **6.** $\frac{\$41}{12 \text{ cabinets}}$ **7.** $\frac{31 \text{ gallons}}{42 \text{ square feet}}$ **8.** $\frac{30.5 \text{ miles}}{1 \text{ hour}}$ or 30.5 miles/hour **9.** $\frac{\$37}{1 \text{ radio}}$ or $37/radio **10.** $\frac{13}{40} = \frac{39}{120}$ **11.** $\frac{42}{78} = \frac{21}{39}$ **12.** true **13.** false **14.** $n = 17$ **15.** $n = 18$ **16.** $n = 4$ **17.** $n = 7$ **18.** $n = 600$ **19.** 3.5 cups **20.** 364.5 miles **21.** 146 miles **22.** 54 defective bulbs

4.1 Exercises

5. $\frac{3}{4}$ **7.** $\frac{7}{6}$ **9.** $\frac{5}{11}$ **11.** $\frac{2}{3}$ **13.** $\frac{15}{16}$ **15.** $\frac{2}{3}$ **17.** $\frac{8}{5}$ **19.** $\frac{2}{3}$ **21.** $\frac{2}{3}$ **23.** $\frac{10}{17}$ **25.** $\frac{43}{60}$ **27.** $\frac{9}{1}$ **29.** $\frac{13}{19}$ **31.** $\frac{11}{19}$ **33.** $\frac{7}{33}$ **35.** $\frac{41}{245}$ **37.** $\frac{90}{41}$ **39.** $\frac{9}{16}$ **41.** $\frac{\$10}{3 \text{ plants}}$ **43.** $\frac{\$85}{6 \text{ bushels}}$ **45.** $\frac{62 \text{ gallons}}{125 \text{ sq. ft.}}$ **47.** $\frac{410 \text{ revolutions}}{1 \text{ mile}}$ *or* 410 rev/mile **49.** $\frac{9 \text{ miles}}{4 \text{ hours}}$ **51.** $13/hr **53.** 16 mi/gal **55.** 155 gal/hr **57.** 230 words/page **59.** 517 km/hr **61.** $30/share **63.** $14 profit/watch **65.** **(a)** $0.08/oz small box; $0.07/oz large box **(b)** 1¢/oz or $.01/oz **(c)** $0.48 **67.** **(a)** 13 moose **(b)** 12 moose **(c)** North Slope **69.** $24.53 **71.** Increased by Mach 0.2 **72.** Decreased by Mach 0.1 **73.** $2\frac{5}{8}$ **74.** 5 **75.** $\frac{5}{46}$ **76.** $1\frac{1}{48}$ **77.** $12.25/sq yd

4.2 Exercises

3. $\frac{48}{32} = \frac{3}{2}$ **5.** $\frac{8}{3} = \frac{32}{12}$ **7.** $\frac{20}{36} = \frac{5}{9}$ **9.** $\frac{27}{15} = \frac{9}{5}$ **11.** $\frac{22}{30} = \frac{11}{15}$ **13.** $\frac{45}{135} = \frac{9}{27}$ **15.** $\frac{5.5}{10} = \frac{11}{20}$ **17.** $\frac{12 \text{ pounds}}{\$4} = \frac{33 \text{ pounds}}{\$11}$ **19.** $\frac{10 \text{ runs}}{45 \text{ games}} = \frac{36 \text{ runs}}{162 \text{ games}}$ **21.** $\frac{20 \text{ pounds}}{\$75} = \frac{30 \text{ pounds}}{\$112.50}$ **23.** $\frac{2{,}200 \text{ people}}{100 \text{ benches}} = \frac{2{,}750 \text{ people}}{125 \text{ benches}}$ **25.** $\frac{16 \text{ pounds}}{1{,}520 \text{ sq ft}} = \frac{19 \text{ pounds}}{1{,}805 \text{ sq ft}}$ **27.** $525 = 525$ True **29.** $504 = 504$ True **31.** $2160 \neq 2240$ False **33.** $180 = 180$ True **35.** $4950 \neq 4900$ False **37.** $31{,}500 = 31{,}500$ True **39.** $63 = 63$ True **41.** $110 \neq 114$ False **43.** $26 = 26$ True **45.** $38 = 38$ True **47.** $880 = 880$ True **49.** $5148 = 5148$ True **51.** $2160 \neq 2240$ False **53.** Yes **55.** $6570 \neq 6540$ No **57.** **(a)** $\frac{63}{161} = \frac{9}{23}$ $\frac{171}{437} = \frac{9}{23}$ True **(b)** $63 \times 437 \stackrel{?}{=} 161 \times 171$; $27{,}531 = 27{,}531$ True **(c)** For most students it is faster to multiply than to reduce fractions. **58.** **(a)** $\frac{169}{221} = \frac{13}{17}$ $\frac{247}{323} = \frac{13}{17}$ True **(b)** $169 \times 323 \stackrel{?}{=} 221 \times 247$; $54{,}587 = 54{,}587$ True **(c)** For most students it is faster to multiply than to reduce fractions. **59.** 23.1405 **60.** 14.15566 **61.** 402.408 **62.** 25.8

4.3 Exercises

1. $n = 8$ **3.** $n = 38$ **5.** $n = 16$ **7.** $n = 13$ **9.** $n = 26$ **11.** $n = 5.6$ **13.** $n = 5$ **15.** $n = 7$ **17.** $n = 5.5$ **19.** $n = 15$ **21.** $n = 16$ **23.** $n = 7.5$ **25.** $n = 32$ **27.** $n = 24$ **29.** $n = 7$ **31.** $n = 66$ **33.** $n \approx 8.8$ **35.** $n = 20$ **37.** $n = 25$ **39.** $n = 18$ **41.** $n = 14$ **43.** $n = 63$ **45.** $n = 50$ **47.** $n \approx 2.9$ **49.** $n \approx 9.4$ **51.** $n \approx 31.3$ **53.** $n = 2.8$ **55.** $n = 121.60$ **57.** $n = 8.64$ **59.** $n = 4.35$ **61.** $n = 1.36$ **63.** $n = 15.75$ **65.** $n = 3\frac{5}{8}$ **66.** $n = 3\frac{13}{15}$ **67.** $n = 10\frac{8}{9}$ **68.** $n = 4\frac{33}{34}$ **69.** $64 + 4 + 18 - 10 = 76$ **70.** $2.56 - 0.42 + 14.72 = 16.86$ **71.** five hundred sixty-three thousandths **72.** 0.0034

4.4 Exercises

1. 380 desserts **3.** 9 shots of coffee **5.** 916 drops of pigment **7.** \$336 Canadian **9.** 16.9 miles per hour **11.** 197.6 feet **13.** 33 minutes **15.** 356 miles **17.** 42 millimeters **19.** $11\frac{2}{3}$ cups or approximately 11.7 cups **21.** 15 cars **23.** 35 errors **25.** 4898 people **27.** 78 giraffes **29.** \$3570 **31.** 903 pounds **33.** 3410 people **35.** Yes. The units show that the parts of the proportion correspond correctly. **36.** Yes. The units show the parts of the proportion correspond correctly. **37.** No. Look at the denominators: 49 miles per hour does not correspond to 88 feet per second; 60 miles per hour does. This should be $\frac{n \text{ feet per second}}{88 \text{ feet per second}} = \frac{49 \text{ miles per hour}}{60 \text{ miles per hour}}$ **38.** No. 49 miles per hour should be in the numerator on the right. Then the parts of the proportion will correspond. $\frac{60 \text{ miles per hour}}{88 \text{ feet per second}} = \frac{49 \text{ miles per hour}}{n \text{ feet per second}}$ **39.** 56,200 **40.** 196,380,000 **41.** 56.1 **42.** 1.96 **43.** 0.0762 **44.** 598.321

Putting Your Skills to Work

1. There are 4668 people per square mile in Cincinnati and 4014 people per square mile in Austin. Cincinnati is more densely populated.
2. There are 12,209 people per square mile in Chicago and 12,484 people per square mile in Boston. Boston is more densely populated.

Chapter 4 Review Problems

1. $\frac{11}{5}$ **2.** $\frac{5}{3}$ **3.** $\frac{4}{5}$ **4.** $\frac{10}{19}$ **5.** $\frac{25}{62}$ **6.** $\frac{1}{3}$ **7.** $\frac{52}{147}$ **8.** $\frac{40}{93}$ **9.** $\frac{2}{5}$ **10.** $\frac{1}{5}$ **11.** $\frac{3}{4}$ **12.** $\frac{23}{26}$ **13.** $\frac{7}{43}$ **14.** $\frac{4}{43}$ **15.** $\frac{5 \text{ gallons}}{9 \text{ people}}$ **16.** $\frac{4 \text{ revolutions}}{11 \text{ minutes}}$ **17.** $\frac{47 \text{ vibrations}}{4 \text{ seconds}}$ **18.** $\frac{6 \text{ cups}}{19 \text{ people}}$ **19.** \$17.00/share **20.** \$17.00/chair **21.** \$13.50/square yard **22.** \$12.40/ticket **23.** **(a)** \$0.74 **(b)** \$0.58 **(c)** \$0.16 **24.** **(a)** \$0.22 **(b)** \$0.25 **(c)** \$0.03 **25.** $\frac{12}{48} = \frac{7}{28}$ **26.** $\frac{10}{21} = \frac{30}{63}$ **27.** $\frac{7.5}{45} = \frac{22.5}{135}$ **28.** $\frac{8.6}{43} = \frac{17.2}{86}$ **29.** $\frac{136}{17} = \frac{408}{51}$ **30.** $\frac{117}{61} = \frac{351}{183}$ **31.** $\frac{4.50 \text{ dollars}}{15 \text{ pounds}} = \frac{8.10 \text{ dollars}}{27 \text{ pounds}}$ **32.** $\frac{138 \text{ passengers}}{3 \text{ buses}} = \frac{230 \text{ passengers}}{5 \text{ buses}}$ **33.** False **34.** True **35.** True **36.** True **37.** False **38.** False **39.** True **40.** False **41.** True **42.** False **43.** $n = 23$ **44.** $n = 32$ **45.** $n = 31$ **46.** $n = 17$ **47.** $n = 33$ **48.** $n = 42$ **49.** $n = 7$ **50.** $n = 24$ **51.** $n \approx 3.9$ **52.** $n \approx 3.6$ **53.** $n = 3$ **54.** $n = 1$ **55.** $n = 87$ **56.** $n = 324$ **57.** $n = 16.8$ **58.** $n \approx 4.7$ **59.** $n = 12$ **60.** $n = 550$ **61.** $n \approx 12.6$ **62.** $n \approx 4.5$ **63.** 15 gallons **64.** \$231.00 **65.** 9 nurses **66.** 2184 students **67.** 2016 francs **68.** 50 pounds **69.** 600 miles **70.** 16 miles per hour **71.** 67,200 feet **72.** \$3045.69 **73.** 120 feet **74.** **(a)** 7.65 gallons **(b)** \$8.80

Putting Your Skills to Work

1. No, the ratio is slightly less. **2.** $\frac{3}{25}$ The room expense would have been \$2640. **3.** \$21,862

Chapter 4 Test

1. $\frac{9}{26}$ **2.** $\frac{14}{37}$ **3.** $\frac{98 \text{ miles}}{3 \text{ gallons}}$ **4.** $\frac{140 \text{ sq. ft.}}{3 \text{ pounds}}$ **5.** 3.8 tons/day **6.** \$8.28/hour **7.** 245.45 feet/pole **8.** \$85.21/share **9.** $\frac{17}{29} = \frac{51}{87}$ **10.** $\frac{12}{19} = \frac{18}{28.5}$ **11.** $\frac{490 \text{ miles}}{21 \text{ gallons}} = \frac{280 \text{ miles}}{12 \text{ gallons}}$ **12.** $\frac{5 \text{ tablespoons}}{18 \text{ people}} = \frac{15 \text{ tablespoons}}{54 \text{ people}}$ **13.** False **14.** True **15.** True **16.** False **17.** $n = 16$ **18.** $n = 14$ **19.** $n = 12$ **20.** $n = 29.4$ **21.** $n = 120$ **22.** $n = 8.4$ **23.** $n = 120$ **24.** $n = 5.13$ **25.** 6 eggs **26.** 80.95 pounds **27.** 19 miles **28.** \$360 **29.** 4 quarts **30.** 696.67 kilometers
30. 696.67 kilometers

Cumulative Test for Chapters 1–4

1. Twenty-six million, five hundred ninety-seven thousand, eighty-nine **2.** 411 **3.** 13,936 **4.** 68 **5.** $\frac{43}{40}$ *or* $1\frac{3}{40}$ **6.** $\frac{27}{35}$ **7.** $\frac{117}{8}$ *or* $14\frac{5}{8}$ **8.** $\frac{17}{6}$ *or* $2\frac{5}{6}$ **9.** 163.58 **10.** 8.2584 **11.** 0.179586 **12.** 14 **13.** $\frac{3}{1}$ **14.** $\frac{\$0.14}{1 \text{ banana}}$ **15.** $\frac{4 \text{ yen}}{1 \text{ peso}}$ **16.** True **17.** True **18.** $n = 3$ **19.** $n = 2$ **20.** $n = 128$ **21.** $n = 9$ **22.** $n \approx 3.4$ **23.** $n = 21$ **24.** 8.33 inches **25.** \$750.00 **26.** 5 pounds

Chapter 5

Pretest Chapter 5

1. 13% **2.** 21% **3.** 18.5% **4.** 37.2% **5.** 134% **6.** 894% **7.** 0.2% **8.** 0.4% **9.** 17% **10.** 27% **11.** 13.4% **12.** 19.8% **13.** $6\frac{1}{2}$% **14.** $1\frac{3}{8}$% **15.** 60% **16.** 2.5% **17.** 115% **18.** 106.25% **19.** 71.43% **20.** 28.57% **21.** 95.65% **22.** 68.42% **23.** 440% **24.** 275% **25.** 0.33% **26.** 0.25% **27.** $\frac{11}{50}$ **28.** $\frac{19}{50}$ **29.** $\frac{53}{100}$ **30.** $\frac{41}{100}$ **31.** $\frac{3}{2}$ or $1\frac{1}{2}$ **32.** $\frac{8}{5}$ or $1\frac{3}{5}$ **33.** $\frac{19}{300}$ **34.** $\frac{7}{150}$ **35.** $\frac{41}{80}$ **36.** $\frac{7}{16}$ **37.** 55.2 **38.** 22.36 **39.** 94.44% **40.** 81.82% **41.** 3000 **42.** 1850 **43.** They won 63.16% of the games. **44.** \$5400 commission would be paid. **45.** The dinner without the tax was \$14.40. **46.** The percent decrease was 30%.

5.1 Exercises

5. 45% **7.** 7% **9.** 80% **11.** 245% **13.** 5.3% **15.** 0.6% **17.** 29% **19.** 78% **21.** 32% **23.** 0.51 **25.** 0.07 **27.** 0.2 **29.** 0.436 **31.** 0.003 **33.** 0.0072 **35.** 1.26 **37.** 3.66 **39.** 74% **41.** 50% **43.** 8% **45.** 56.3% **47.** 0.2% **49.** 0.57% **51.** 135% **53.** 272% **55.** 27% **57.** 36% **59.** 143% **61.** 30% **63.** 0.5% **65.** 0.62 **67.** 1.28 **69.** 0.005 **71.** 0.8 **73.** 49% **75.** 3.413% **77.** 36% = 36 percent = 36 "per one hundred" = $36 \times \frac{1}{100} = \frac{36}{100} = 0.36$. The rule is using the fact that 36% means 36 per one hundred. **78.** $10.65 = 1065 \times \frac{1}{100}$ = 1065 "per one hundred" = 1065 percent = 1065%. We change 10.65 to $1065 \times \frac{1}{100}$ and use the idea that percent means "per one hundred." **79.** **(a)** 555.62 **(b)** $\frac{55{,}562}{100}$ **(c)** $\frac{27{,}781}{50}$ **81.** $\frac{14}{25}$ **82.** $\frac{18}{25}$ **83.** 0.6875 **84.** 0.8125

5.2 Exercises

1. $\frac{22}{25}$ **3.** $\frac{7}{100}$ **5.** $\frac{33}{100}$ **7.** $\frac{11}{20}$ **9.** $\frac{3}{4}$ **11.** $\frac{1}{5}$ **13.** $\frac{29}{200}$ **15.** $\frac{22}{125}$ **17.** $\frac{81}{125}$ **19.** $\frac{57}{80}$ **21.** $1\frac{19}{25}$ **23.** $3\frac{2}{5}$ **25.** 12 **27.** $\frac{13}{600}$ **29.** $\frac{99}{800}$ **31.** $\frac{11}{125}$ **33.** $\frac{7}{325}$ **35.** $\frac{9}{220}$ **37.** 75% **39.** 33.33% **41.** 31.25% **43.** 28% **45.** 27.5% **47.** 41.67% **49.** 360% **51.** 283.33% **53.** 412.5% **55.** 42.86% **57.** 93.75% **59.** 52% **61.** 34.31% **63.** 34.84% **65.** 2.5% **67.** 1.23% **69.** $83\frac{1}{3}$% **71.** $68\frac{3}{4}$% **73.** $37\frac{1}{2}$% **75.** $7\frac{1}{2}$%

	Fraction	Decimal	Percent
77.	$\frac{5}{12}$	0.4167	41.67%
79.	$\frac{3}{50}$	0.06	6%
81.	$\frac{2}{5}$	0.4	40%
83.	$\frac{69}{200}$	0.345	34.5%
85.	$\frac{3}{200}$	0.015	1.5%
87.	$\frac{5}{9}$	0.5556	55.56%
89.	$\frac{1}{32}$	0.0313	$3\frac{1}{8}$%

	Fraction	Decimal	Percent
78.	$\frac{7}{12}$	0.5833	58.33%
80.	$\frac{2}{25}$	0.08	8%
82.	$\frac{3}{5}$	0.6	60%
84.	$\frac{5}{8}$	0.625	62.5%
86.	$\frac{9}{200}$	0.045	4.5%
88.	$\frac{7}{9}$	0.7778	77.78%
90.	$\frac{21}{800}$	0.0263	$2\frac{5}{8}$%

91. $\frac{463}{1600}$ **93.** 15.375% **95.** $n = 5.625$ **96.** $n = 4$ **97.** $n = 88$ **98.** $n = 96$

5.3A Exercises

1. $n = 38\% \times 500$ **3.** $50\% \times n = 7$ **5.** $17 = n \times 85$ **7.** $n = 128\% \times 4000$ **9.** $n \times 400 = 15$ **11.** $136 = 145\% \times n$ **13.** $n = 150$ **15.** $n = 912$ **17.** \$6.39 **19.** $n = 1200$ **21.** $n = 1300$ **23.** \$150 **25.** 60% **27.** 28% **29.** 5% **31.** 50.4 **33.** 68 **35.** 12% **37.** 3.28 **39.** 64% **41.** 445 **43.** 35% **45.** 39.6 **47.** 1.44 **49.** 902.825 **51.** 59.29% (rounded) **53.** 85% **55.** 554 students **57.** 40 years **59.** 57.6 **61.** 2.448 **62.** 14.892 **63.** 2834 **64.** 2917

5.3B Exercises

	p	b	a
1.	75	660	495
3.	42	400	a
5.	49	b	2450
7.	p	50	30
9.	p	25	10
11.	160	b	400

13. 72 **15.** 75 **17.** 56 **19.** 75 **21.** 60 **23.** 600,000 **25.** 25 **27.** 2.5 **29.** 1.8 **31.** 91 **33.** 300 **35.** 16.4% **37.** 3.64 **39.** 25% **41.** 170 **43.** 43% **45.** 47.5 **47.** 2.8 **49.** 15.82% **51.** 10.91 **53.** 153.62 **55.** $118.80 **57.** $1\frac{31}{45}$ **58.** $\frac{1}{26}$ **59.** $4\frac{1}{5}$ **60.** $1\frac{13}{15}$

5.4 Exercises

1. 180,000 pencils **3.** $94.25 **5.** 20% **7.** Approximately $42.26 **9.** $6.44 **11.** 9% **13.** 333 people **15.** $240 **17.** 25% **19.** $9,600,000 **21.** 1365 M&Ms are blue **23.** $75,000 **25.** 216 babies **27.** 4% **29.** $55 **31.** $12,210,000 for personnel, food, and decorations; $20,790,000 for security facility rental and all other expenses **33.** **(a)** $14.16 **(b)** $250.16 **35.** 36.94% **37.** 60% **39.** **(a)** $1279.20 **(b)** $14,710.80 **41.** **(a)** $255 **(b)** $1870 **43.** **(a)** $200 **(b)** $2700 **45.** **(a)** $118.40 **(b)** $3818.40 **47.** $849.01 **49.** $10.18, $708.62 **51.** $17,108.38 **53.** 1,698,000 **54.** 2,452,400 **55.** 1.63 **56.** 0.793 **57.** 0.0556 **58.** 0.0792

Putting Your Skills to Work

1. An increase of 153.8%; health benefits of reducing amount of fat consumed in milk **2.** Decrease of 21.6%; health benefits of reducing amount of caffeine consumed

Chapter 5 Review Problems

1. 87% **2.** 59% **3.** 27.6% **4.** 32.9% **5.** 7.13% **6.** 6.08% **7.** 252% **8.** 437% **9.** 103.6% **10.** 105.2% **11.** 0.6% **12.** 0.2% **13.** 0.29% **14.** 0.53% **15.** 72% **16.** 61% **17.** 19.5% **18.** 21.6% **19.** 0.24% **20.** 0.98% **21.** $4\frac{1}{12}\%$ **22.** $3\frac{5}{12}\%$ **23.** 317% **24.** 225% **25.** 64% **26.** 88% **27.** 90% **28.** 70% **29.** 45.45% **30.** 44.44% **31.** 225% **32.** 375% **33.** 442.86% **34.** 555.56% **35.** 190% **36.** 183.33% **37.** 0.38% **38.** 0.63% **39.** 0.002 **40.** 0.007 **41.** 0.219 **42.** 0.431 **43.** 1.66 **44.** 1.39 **45.** 0.32125 **46.** 0.26375 **47.** $\frac{41}{50}$ **48.** $\frac{9}{25}$ **49.** $\frac{37}{20}$ **50.** $\frac{9}{4}$ **51.** $\frac{41}{250}$ **52.** $\frac{61}{200}$ **53.** $\frac{5}{16}$ **54.** $\frac{7}{16}$ **55.** $\frac{1}{2000}$ **56.** $\frac{3}{5000}$

	Fraction	Decimal	Percent
57.		0.6	60%
58.		0.875	87.5%
59.	$\frac{3}{8}$	0.375	
60.	$\frac{9}{16}$	0.5625	
61.	$\frac{1}{125}$		0.8%
62.	$\frac{9}{20}$		45%

63. 288 **64.** 340 **65.** 15 **66.** 225 **67.** 40% **68.** 40% **69.** 57.5 **70.** 70.8 **71.** 160 **72.** 140 **73.** 20% **74.** 46.67% **75.** 13 **76.** 28 **77.** 464.29% **78.** 324.32% **79.** 51 students **80.** 96 trucks **81.** 4.17% **82.** 15.69% **83.** $11,200 **84.** $80,200 **85.** 15% **86.** 7.5% **87.** $1200 **88.** $558 **89.** 4.09% **90.** **(a)** $330 **(b)** $1320 **91.** **(a)** $319 **(b)** $1276 **92.** $1800

Chapter 5 Test

1. 42% **2.** 1% **3.** 0.6% **4.** 13.9% **5.** 218% **6.** 71% **7.** 2.7% **8.** $3\frac{1}{7}\%$ **9.** 47.5% **10.** 40% **11.** 140% **12.** 175% **13.** 17.13% **14.** 302.4% **15.** $1\frac{13}{25}$ **16.** $\frac{31}{400}$ **17.** 26.69 **18.** 130 **19.** 55.56% **20.** 200 **21.** 5000 **22.** 46% **23.** 699.6 **24.** 2.29% **25.** $6092 commission **26.** **(a)** $150.81 discount **(b)** They paid $306.19 **27.** 89.29% were not defective **28.** 8.93% is the percent of decrease **29.** 12,000 registered voters **30.** **(a)** $240 in interest in 6 months **(b)** $960 in interest in 2 years

Cumulative Test for Chapters 1–5

1. 2241 **2.** 8444 **3.** 5292 **4.** 89 **5.** $\frac{67}{12}$ *or* $5\frac{7}{12}$ **6.** $\frac{1}{12}$ **7.** $\frac{35}{12}$ *or* $2\frac{11}{12}$ **8.** $\frac{5}{21}$ **9.** 5731.7 **10.** 34.118 **11.** 1.686 **12.** 0.368 **13.** $\frac{3 \text{ pounds}}{5 \text{ square feet}}$ **14.** True **15.** $n = 24$ **16.** 673 faculty **17.** 2.3% **18.** 46.8% **19.** 198% **20.** 3.75% **21.** 2.43 **22.** 0.0675 **23.** 17.76% **24.** 114.58 **25.** 300 **26.** 718.2 **27.** $8370 **28.** 3200 students **29.** 11.31% increase **30.** $352

Chapter 6

Pretest Chapter 6

1. 204 **2.** 56 **3.** 3520 **4.** 6400 **5.** 1320 **6.** 24 **7.** 5320 **8.** 4680 **9.** 98.6 **10.** 0.027 **11.** 529.6 **12.** 0.123 **13.** 2376 m **14.** 94.262 m **15.** 3820 **16.** 3.162 **17.** 0.0563 **18.** 4800 **19.** 0.568 **20.** 8900 **21.** 4.73 **22.** 1.28 **23.** 59.52 **24.** 1826.78 **25.** 39.69 **26.** 103.4 **27.** 55.8 feet **28. (a)** 95°F **(b)** No **29.** 12.2 miles farther **30.** 22.5 gal/hr

6.1 Exercises

1. 12 **3.** 1760 **5.** 2000 **7.** 4 **9.** 2 **11.** 60 **13.** 5 **15.** 7 **17.** 2 **19.** 12,320 **21.** 144 **23.** 123 **25.** 6.25 **27.** 12 **29.** 26,000 **31.** 36 **33.** 28 **35.** 9 **37.** 62 **39.** 64 **41.** 84 **43.** 11 **45.** 264 **47.** 4200 **49.** 64,800 **51.** 3 **53.** 20,265 **55.** $135 **57.** $9.75 **59.** 199,056 feet **61.** $161\frac{3}{4}$ inches **63.** 133,953 tons **65.** 26,280 hours **67.** 14,739 land miles

6.2 Exercises

7. 370 **9.** 3600 **11.** 0.328 **13.** 0.563 **15.** 200,000 **17.** 0.078 **19.** 538,600 **21.** 3.5; 0.035 **23.** 3.582; 0.003582 **25.** 0.0032; 0.0000032 **27.** a **29.** c **31.** 30 **33.** 270 **35.** 1.98 **37.** 0.482 **39.** 5236 **41.** 3255 m **43.** 463 cm **45.** 406.71 m **47.** 2.5464 cm *or* 25.464 mm **49.** False **51.** True **53.** True **55.** False **57.** True **59. (a)** 11.92 km **(b)** 11,920 meters **61. (a)** 481,800 centimeters **(b)** 4.818 kilometers **63.** 0.000000004 kilometer

6.3 Exercises

7. 64,000 **9.** 4700 **11.** 0.0189 **13.** 0.752 **15.** 2,430,000 **17.** 82 **19.** 0.005261 **21.** 74,000 **23.** 0.162 **25.** 0.035 **27.** 6.328 **29.** 2920 **31.** 17,000 **33.** 0.00032 **35.** 0.007896 **37.** 5,900,000 **39.** 0.007; 0.000007 **41.** 0.128; 0.000128 **43.** 0.522; 0.000522 **45.** 3.607; 0.003607 **47.** b **49.** a **51.** 113.922 L **53.** 9.803 t **55.** 260,160 mg **57.** True **59.** False **61.** True **63.** False **65.** True **67.** $358 **69.** $573,000 **71.** $340,000 **73.** $10,102,500 **75.** 0.005632 **76.** 76,182 **77.** 20% **78.** 57.5 **79.** $321.30 **80.** $716.80

6.4 Exercises

1. 2.14 m **3.** 22.86 cm **5.** 15.26 yd **7.** 28.89 yd **9.** 9.3 mi **11.** 21.94 m **13.** 132.02 km **15.** 82 ft **17.** 6.90 in. **19.** 218 yd **21.** 18.95 L **23.** 1061.2 L **25.** 21.76 L **27.** 5.02 gal **29.** 4.77 qt **31.** 14.53 kg **33.** 198.45 g **35.** 35.2 lb **37.** 4.45 oz **39.** 5.45 ft **41.** 502.92 cm **43.** 31 mi/hr **45.** 96.6 km/hr **47.** 0.51 in. **49.** 104°F **51.** 185°F **53.** 53.6°F **55.** 20°C **57.** 75.56°C **59.** 30°C **61.** 143.87 km **63.** 18.85 L **65.** 1397 lb **67.** It is 66.2°F at 4:00 A.M. The temperature may reach 113°F after 7:00 A.M. **69.** 3420°C **71.** 21.773 g **73.** 180.6448 sq cm **75.** 32.7273 mi/hr **77.** 47 **78.** 49 **79.** 91 **80.** 74

Putting Your Skills to Work

1. 1,182,633,000 bytes of space

6.5 Exercises

1. 8 yd **3.** 11 yd **5.** $33.46 **7.** 775 mL per bottle **9.** 310 cm per piece **11.** 1340 cm **13.** 125 samples **15.** The discrepancy is 4.4°F. The sign in the store is 4.4°F less than it should be. **17.** The difference is 6°F. The temperature reading of 180°C is hotter. **19.** 195 miles further **21. (a)** About 105 kilometers per hour **(b)** Probably not. We cannot be sure, but we have no evidence to indicate that they broke the speed limit. **23.** 60 gal/hr **25.** $102 **27.** 85.05 g **29. (a)** She bought 5 qt extra. **(b)** $6.95 **31. (a)** $5.46 **(b)** 132 miles per gallon **33.** $n = 0.64$ **34.** $n = 0.2$ **35.** 15.5 miles **36.** 41.25 yards

Chapter 6 Review Problems

1. 9 **2.** 11 **3.** 5280 **4.** 7040 **5.** 7.5 **6.** 6.5 **7.** 3 **8.** 2 **9.** 8000 **10.** 14,000 **11.** 5.75 **12.** 6.25 **13.** 60 **14.** 84 **15.** 15.5 **16.** 13.5 **17.** 36 **18.** 23 **19.** 840 **20.** 660 **21.** 560 **22.** 290 **23.** 176.3 **24.** 259.8 **25.** 920 **26.** 740 **27.** 5000 **28.** 7000 **29.** 0.285 **30.** 0.473 **31.** 7.93 m **32.** 17.01 m **33.** 35.63 m **34.** 89.59 m **35.** 17,000 **36.** 23,000 **37.** 0.059 **38.** 0.077 **39.** 196,000 **40.** 721,000 **41.** 0.778 **42.** 0.459 **43.** 0.125 **44.** 0.705 **45.** 76,000 **46.** 41,000 **47.** 765 **48.** 423 **49.** 2430 **50.** 1930 **51.** 92.4 **52.** 72.6 **53.** 4.58 **54.** 2.75 **55.** 368.55 **56.** 481.95 **57.** 54.86 **58.** 39.62 **59.** 5.52 **60.** 7.09 **61.** 9.08 **62.** 13.62 **63.** 10.97 **64.** 12.80 **65.** 49.6 **66.** 43.4 **67.** 59 **68.** 77 **69.** 105 **70.** 85 **71.** 0 **72.** 3.12 cm **73.** **(a)** 17 ft **(b)** 204 in. **74.** **(a)** 200 m **(b)** 0.2 km **75.** $2.22 **76.** 32.6 miles farther **77.** Yes 1.67 feet extra **78.** 166.67 milliliters per jar **79.** $581.12 **80.** 380 centimeters **81.** 25°F too hot

Chapter 6 Test

1. 3200 **2.** 228 **3.** 84 **4.** 7 **5.** 0.75 **6.** 30 **7.** 0.273 **8.** 9200 **9.** 4.6 **10.** 0.0988 **11.** 1270 **12.** 9.36 **13.** 0.046 **14.** 127,000 **15.** 0.0289 **16.** 0.983 **17.** 920 **18.** 9420 **19.** 67.62 **20.** 1.63 **21.** 3.55 **22.** 10.03 **23.** 16.06 **24.** 85.05 **25.** **(a)** 20 meters **(b)** 21.8 yards **26.** **(a)** 15°F **(b)** Yes **27.** 82.5 gal/hr **28.** **(a)** 300 km **(b)** 14 miles

Cumulative Test for Chapters 1–6

1. 6028 **2.** 185,440 **3.** 69 **4.** $\frac{19}{42}$ **5.** $1\frac{3}{8}$ **6.** True **7.** $n = 6$ **8.** 209.23 grams **9.** 250% **10.** 64.8 **11.** 20,000 **12.** 9.5 **13.** 5000 **14.** 3.5 **15.** 300 **16.** 3700 **17.** 0.0628 **18.** 790 **19.** 0.05 **20.** 672 **21.** 106.12 **22.** 43.58 **23.** 25.81 **24.** 14.49 **25.** 11.88 meters **26.** 59°F The difference is 44°F. The 15°C temperature is higher. **27.** 7 miles **28.** 1.738 centimeters

Chapter 7

Pretest Chapter 7

1. 18 m **2.** 14 m **3.** 23 sq cm **4.** 2.4 sq cm **5.** 25.6 yd **6.** 52 ft **7.** 135 sq in. **8.** 171 sq in. **9.** 97 sq m **10.** 59° **11.** 15.3 m **12.** 30 sq m **13.** 8 **14.** 12 **15.** 6.782 **16.** 8 ft **17.** 12 ft **18.** 28 in. **19.** 94.2 cm **20.** 153.9 sq m **21.** 27.4 sq m **22.** 240 cu yd **23.** 113 cu ft **24.** 1846.3 cu in. **25.** 4375 cu m **26.** 1130.4 cu m **27.** $n = 120$ cm **28.** $n = 28$ m **29.** **(a)** 3706.5 sq yd **(b)** $555.98

7.1 Exercises

5. 15 mi **7.** 23.6 ft **9.** 49.2 ft **11.** 1.92 mm **13.** 25.46 in. **15.** 17.12 km **17.** 0.0246 cm **19.** 43.7 m **21.** 52 m **23.** 4.8 mi **25.** 0.0208 mm **27.** 31.84 cm **29.** 94 m **31.** 180 cm **33.** 72 in.^{2} **35.** 96.04 ft^2 **37.** 0.288 m^2 **39.** 14,976 yd^2 **41.** 294 m^2 **43.** $132,000 **45.** $100.70 **47.** $953.22 **49.** $598.22 **51.** 223.3 **52.** 7.18 **53.** 21,842.8 **54.** 1.58

7.2 Exercises

5. 40.2 m **7.** 29.8 in. **9.** 1057.5 m^2 **11.** 440.75 cm^2 **13.** 3528 yd^2 **15.** 82 m **17.** 380 cm **19.** 42 m^2 **21.** 86 cm^2 **23.** 344 yd^2 **25.** 550 km^2 **27.** 718 m^2 **29.** 162.5 cm^2 **31.** 345 ft^2 **33.** $80,960.00 **35.** 694.49 in.^{2} (rounded) **37.** Area = $5.2415 \times b^2$ sq units **39.** 30 **40.** 3 **41.** 1800 **42.** 2.6

7.3 Exercises

7. True **9.** True **11.** False **13.** 60° **15.** 30° **17.** 90° **19.** 128° **21.** 104 m **23.** 130 in. **25.** 10.5 mi **27.** 25.5 in. **29.** 22.5 ft^2 **31.** 15.75 in.^{2} **33.** 83.125 cm^2 **35.** 14.07 m^2 **37.** 12.25 yd^2 **39.** 16.4 cm **41.** 188 yd^2 **43.** 1740 ft^2 **45.** $21,060.00 **47.** $n = 12$ **48.** $n = 67.67$ **49.** 716 **50.** 96 magazines

7.4 Exercises

5. 1 **7.** 4 **9.** 5 **11.** 7 **13.** 10 **15.** 11 **17.** 13 **19.** 0 **21.** 10 **23.** 11 **25.** 14 **27.** 4 **29.** 10 **31.** 5 **33.** 22 **35.** **(a)** Yes **(b)** 7 **37.** **(a)** Yes **(b)** 16 **39.** 4.243 **41.** 5.568 **43.** 9.110 **45.** 10.954 **47.** 11.180 **49.** 11 m **51.** 5.099 m **53.** 8.660 m **55.** **(a)** 2 **(b)** 0.2 **(c)** 0.02 **(d)** Each answer is obtained from the previous answer by dividing by 10. **(e)** No, because 0.004 isn't a perfect square. **57.** 205.027 **59.** 39.299 **61.** 4800 in.^{2} **62.** 80,500 meters **63.** 0.92 meter **64.** 0.0989 kilogram

7.5 Exercises

1. 5 in. **3.** 8.544 yd **5.** 15.199 ft **7.** 3.606 **9.** 10.247 km **11.** 11.402 m **13.** 7.071 m **15.** 3 ft **17.** 15 in. **19.** 6.928 yd **21.** 17 ft **23.** 5 mi **25.** 8.7 ft **27.** 4 in., 6.9 in. **29.** 7 m, 12.1 m **31.** 8.5 m **33.** 15.6 cm **35.** 7.1 in. **36.** 10 m **37.** 52 yd **39.** 341 m^2 **40.** 168 m^2 **41.** 441 in.^{2} **42.** 4224 yd^2

7.6 Exercises

5. 58 in. **7.** 5 mm **9.** 22.5 yd **11.** 1.9 cm **13.** 75.4 cm **15.** 69.1 in. **17.** 81.6 in. **19.** 41.87 ft **21.** 78.5 yd^2 **23.** 907.5 m^2 **25.** 803.8 cm^2 **27.** 153.9 ft^2 **29.** 11,304 mi^2 **31.** 0.70057457 m **33.** 628 m^2 **35.** 163.3 m^2 **37.** 189.3 m^2 **39.** 31.0 m^2 **41.** \$1211.20 **43.** 6.28 ft **45.** 256.43 ft **47.** 200 revolutions **49.** 3215.36 $in.^2$ **51.** \$226.08 **53.** **(a)** \$0.75 per slice; 22.1 $in.^2$ **(b)** \$0.67 per slice; 18.8 $in.^2$ **(c)** The 12-in. pizza is \$0.04 per $in.^2$; the 15-in. pizza is \$0.03 per $in.^2$ The 15-in. pizza is a better value. **55.** 13.92 **56.** 0.3 **57.** 6000 **58.** 3000

7.7 Exercises

7. 24 m^3 **9.** 700 mm^3 **11.** 87.9 m^3 **13.** 2260.8 m^3 **15.** 3052.1 yd^3 **17.** 267.9 m^3 **19.** 718.0 m^3 **21.** 1017.4 cm^3 **23.** 261.7 ft^3 **25.** 21 m^3 **27.** 120 m^3 **29.** 170.832816 m^3 **31.** 592.3814 m^3 **33.** 95.6887288 ft^3 **35.** 40 yd^3 **37.** 1004.8 $in.^3$ **39.** 381,510,000,000,000 mi^3 **41.** \$942 **43.** 263,900 yd^3 **45.** $9\frac{7}{12}$ **46.** $6\frac{3}{8}$ **47.** $\frac{135}{16}$ *or* $8\frac{7}{16}$ **48.** $\frac{25}{14}$ *or* $1\frac{11}{14}$

Putting Your Skills to Work

1. 224 pounds **2.** 30 $in.^2$

7.8 Exercises

5. 8 m **7.** 17.5 cm **9.** 1.9 yd **11.** *a* corresponds to *f*; *b* corresponds to *e*; *c* corresponds to *d* **13.** 2.2 in. **15.** 3.36 meters long **17.** 2.1 ft **19.** 36 ft **21.** 81.2 ft **23.** 1.4 km **25.** 16.3 cm **27.** 33.3 m **29.** 11.6 yd^2 **30.** 220.4 m^2 **31.** 12 **32.** 31 **33.** 42 **34.** 62

7.9 Exercises

1. \$22.10 **3.** **(a)** 75 km/hr **(b)** 76 km/hr **(c)** Through Woodville and Palermo **5.** 118 minutes or 1 hour, 58 minutes **7.** \$510.00 **9.** 953.8 gal **11.** \$74.42 **13.** **(a)** 40,820 km **(b)** 20,410 km/hr **15.** \$9880.34 **17.** 128 **18.** 915 **19.** 0.25 **20.** 4.87

Chapter 7 Review Problems

1. 19.8 m **2.** 24.8 cm **3.** 23.2 yd **4.** 9.6 yd **5.** 16.5 cm^2 **6.** 73.5 m^2 **7.** 18.5 $in.^2$ **8.** 51.8 $in.^2$ **9.** 38 ft **10.** 58 ft **11.** 68 m^2 **12.** 63.5 m^2 **13.** 100.4 m **14.** 76.4 cm **15.** 80 in. **16.** 62 mi **17.** 2700 m^2 **18.** 2050 m^2 **19.** 336 yd^2 **20.** 720 yd^2 **21.** 422 cm^2 **22.** 357 cm^2 **23.** 22 ft **24.** 14 ft **25.** 153° **26.** 40° **27.** 24.5 m^2 **28.** 59.4 m^2 **29.** 45.6 cm^2 **30.** 40.0 m^2 **31.** 450 m^2 **32.** 87 m^2 **33.** 9 **34.** 8 **35.** 10 **36.** 11 **37.** 12 **38.** 15 **39.** 6 **40.** 6 **41.** 5 **42.** 12 **43.** ≈ 5.916 **44.** ≈ 6.708 **45.** ≈ 9.381 **46.** ≈ 8.718 **47.** ≈ 13.077 **48.** ≈ 13.416 **49.** 5 m **50.** 5 m **51.** 8.25 cm **52.** 8.06 m **53.** 6.4 cm **54.** 9.2 ft **55.** 6.3 ft **56.** 3.6 ft **57.** 106 cm **58.** 24 cm **59.** 37.7 in. **60.** 44.0 in. **61.** 113.0 m^2 **62.** 254.3 m^2 **63.** 201.0 ft^2 **64.** 153.9 ft^2 **65.** 226.1 $in.^2$ **66.** 201.0 m^2 **67.** 318.5 ft^2 **68.** 126.1 m^2 **69.** 107.4 ft^2 **70.** 80.1 m^2 **71.** 45 ft^3 **72.** 450 m^3 **73.** 14,130 ft^3 **74.** 7.2 ft^3 **75.** 307.7 m^3 **76.** 4578.1 cm^3 **77.** 1728 m^3 **78.** 100 m^3 **79.** 3768 ft^3 **80.** 9074.6 yd^3 **81.** 30 m **82.** 3.3 m **83.** 33.8 cm **84.** 9.6 mi **85.** 324 cm **86.** 175 ft **87.** \$92.40 **88.** **(a)** 210 cm **(b)** 1764 times larger **89.** \$736.00 **90.** $V = 2034.7$ $in.^3$ $W = 32{,}555.2$ g **91.** **(a)** 21,873.2 ft^3 **(b)** 17,498.6 bushels **92.** **(a)** 50 km 100 km/h **(b)** 56 km 70 km/hr **(c)** Through Ipswich

Chapter 7 Test

1. 40 yd **2.** 34 ft **3.** 20 m **4.** 80 m **5.** 24.8 m **6.** 180 yd^2 **7.** 104.0 m^2 **8.** 78 m^2 **9.** 144 m^2 **10.** 12 cm^2 **11.** 52.5 m^2 **12.** 9 **13.** 8 **14.** 12 **15.** 7 **16.** ≈ 7.348 **17.** ≈ 10.954 **18.** ≈ 13.675 **19.** 9.22 **20.** 10 **21.** 5.83 cm **22.** 9 ft **23.** 37.7 in. **24.** 254.3 ft^2 **25.** 107.4 $in.^2$ **26.** 144.3 $in.^2$ **27.** 840 m^3 **28.** 803.8 m^3 **29.** 33.5 m^3 **30.** 508.7 ft^3 **31.** 56 m^3 **32.** 43.2 cm **33.** 46.7 ft **34.** 6456 yd^2 \$2582.40

Cumulative Test for Chapters 1–7

1. 935,760 **2.** 33,415 **3.** $\frac{26}{45}$ **4.** $\frac{4}{21}$ **5.** 56.13 **6.** 7.2272 **7.** 83 **8.** $n = 27$ **9.** 800 students **10.** 75% **11.** 2000 **12.** 40.7 **13.** 5.86 m **14.** 1512 in. **15.** 54.56 mi **16.** 50 m **17.** 208 cm **18.** 56.5 yd **19.** 1.4 cm^2 **20.** 540 m^2 **21.** 192 m^2 **22.** 664 yd^2 **23.** 50.2 m^2 **24.** 2411.5 m^3 **25.** 3052.1 cm^3 **26.** 3136 cm^3 **27.** 2713.0 m^3 **28.** 33.4 m **29.** 4.1 ft **30.** 124 yd^2 \$992.00 **31.** 11 **32.** ≈ 7.550 **33.** 10.440 in. **34.** 4.899 m **35.** 13.9 mi

Chapter 8

Pretest Chapter 8

1. Under age 18 **2.** 43% **3.** 17% **4.** 1650 students **5.** 350 students **6.** 300 people **7.** 600 people **8.** 3rd quarter of both 1993 and 1994 **9.** 4th quarter of 1993 **10.** 100 more people **11.** 100 more people **12.** August and December **13.** December **14.** November **15.** 20,000 sets **16.** 30,000 sets **17.** 55,000 cars **18.** 60,000 cars **19.** 25,000 cars **20.** 20,000 cars **21.** 37 pages per day **22.** 40 pages per day

8.1 Exercises

1. Rent **3.** \$200 **5.** \$500 **7.** $\frac{10}{7}$ **9.** $\frac{3}{10}$ **11.** Hit batters **13.** 294 **15.** 379 pitches **17.** $\frac{72}{325}$ **19.** $\frac{49}{24}$ **21.** 15% **23.** 72% **25.** 44,840 **27.** 1,850,000,000 **29.** 37% **31.** 81% **33.** 1.5% **35.** 42 in.^{2} **36.** 204 in.^{2} **37.** 16 gal **38.** About 3 g

8.2 Exercises

1. 10 million people **3.** 14 million people **5.** 1960–1970 **7.** 50,000 women **9.** 1990 **11.** 25,000 more **13.** 1930 to 1950 **15.** 3.5 million dollars **17.** 1992 **19.** one million dollars **21.** 2.5 in. **23.** October, November, and December **25.** 1.5 in. more

27.

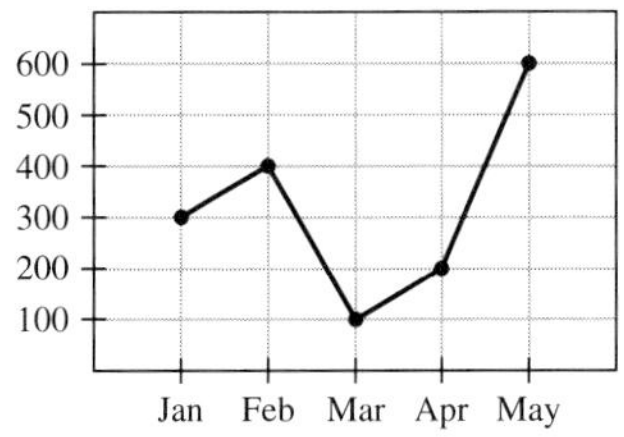

28. increase **29.** 35 **30.** 115

Putting Your Skills to Work

1. 1989 **2.** Trucks that are 12 years old or more

8.3 Exercises

1. 10 cars **3.** 35 cars **5.** 45 cars **7.** 145 cars **9.** 12 days **11.** 3 days **13.** 42 days **15.** 37 days **17.** 63.8%

	TALLY	FREQUENCY
19.	|||	3
21.	卌 |	6
23.	|||	3
25.	||	2

27.

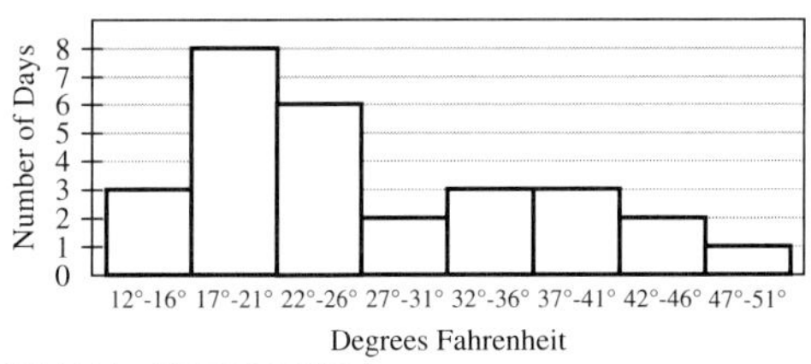

	TALLY	FREQUENCY
29.	|||	3
31.	||	2
33.	|	1
35.	||	2

37. $n = 59.5$ **38.** $n = 7$ **39.** 12 pounds **40.** 13 in.

8.4 Exercises

1. 91.8 **3.** 38 **5.** \$87,000 **7.** 5 hrs **9.** 98,500 **11.** 0.375 **13.** 23.7 mi/gal **15.** 47 **17.** 1011 **19.** 0.58 **21.** \$20,250 **23.** 35.5 min **25.** \$207 **27.** 46 actors **29.** \$69,161.88 **31.** 4850 **33. (a)** 2,340,400,000 barrels of oil per year **(b)** 98,296,800,000 gallons of oil per year **35. (a)** \$2157 **(b)** \$1615 **(c)** The median because the mean is affected by the high amount \$6300. **36. (a)** 12.65 sec **(b)** 11.95 sec **(c)** The median because the large time, 18 seconds, affects the mean. **37.** 60 **39.** 121 and 150 **41.** \$249

Chapter 8 Review Problems

1. 36 computers **2.** 43 computers **3.** 20 computers **4.** 21 computers **5.** $\frac{4}{1}$ **6.** $\frac{2}{1}$ **7.** 7.5% **8.** 30% **9.** 15% **10.** 4% **11.** 57% **12.** 27% **13.** $192 **14.** $360 **15.** $816 **16.** $1008 **17.** 6000 customers **18.** 4000 customers **19.** 4th quarter 1994 **20.** 3rd quarter 1993 **21.** 1000 customers **22.** 1500 customers **23.** Yes **24.** Yes **25.** 400 students **26.** 500 students **27.** 650 students **28.** 450 students **29.** 100 students more **30.** 50 students more **31.** 1992–1993 **32.** 1993–1994 **33.** 45,000 cones **34.** 30,000 cones **35.** 10,000 cones more **36.** 30,000 cones more **37.** 25,000 cones more **38.** 30,000 cones more **39.** The cooler the temperature, the fewer cones sold. **40.** The warmer the temperature, the more cones sold. **41.** 50 bridges **42.** 25 bridges **43.** 20 and 39 **44.** 25 bridges **45.** 150 bridges **46.** 5 bridges more

	TALLY	FREQUENCY
47.	𝍸 𝍸	10
48.	𝍸 III	8
49.	III	3
50.	𝍸	5
51.	II	2

52.

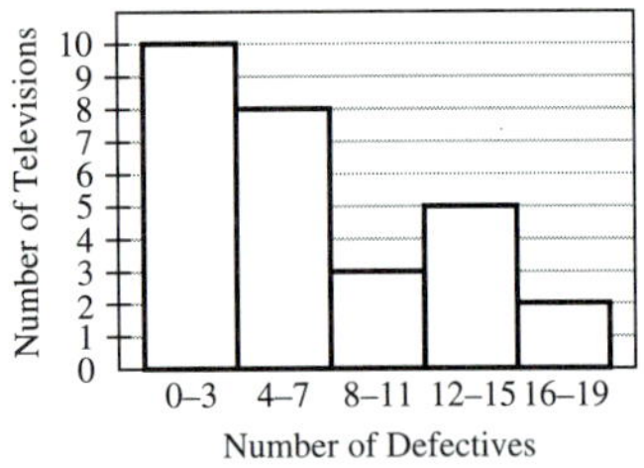

53. 18 times **54.** 90° **55.** $114 **56.** 58 textbooks **57.** 188 women **58.** 1421 cars **59.** 1353 employees **60.** 82 **61.** 83 students **62.** $36,000 **63.** $141,500 **64.** 19.5 cups **65.** 18.5 deliveries **66.** The median, because of one low score, 31. **67.** The median, because of the one high data item, 39.

Chapter 8 Test

1. 37% **2.** 21% **3.** 12% **4.** 90,000 automobiles **5.** 81,000 automobiles **6.** 350 cars **7.** 500 cars **8.** 3rd quarter 1993 **9.** 1st quarter **10.** 50 cars more **11.** 150 cars more **12.** 20 yrs **13.** 26 yrs **14.** 12 yrs **15.** Age 35 **16.** Age 65 **17.** 60,000 televisions **18.** 25,000 televisions **19.** 20,000 televisions **20.** 60,000 televisions **21.** 15.125 **22.** 15.5

Cumulative Test for Chapters 1–8

1. 20,825 **2.** 78,104 **3.** $\frac{153}{40}$ *or* $3\frac{33}{40}$ **4.** $\frac{108}{35}$ *or* $3\frac{3}{35}$ **5.** 2864.37 **6.** 72.65 **7.** 72.23 **8.** $n = 0.6$ **9.** 39 cars **10.** 0.325 **11.** 350 **12.** 1.98 m **13.** 54 ft **14.** 28.3 in.^{2} **15.** 68 in. **16.** 34% **17.** 3840 students **18.** 3 million dollars **19.** 1 million dollars **20.** 16 in. **21.** 1950, 1960 **22.** 8 students **23.** 16 students **24.** $6.50 **25.** $4.95

Chapter 9

Pretest Chapter 9

1. -19 **2.** -9 **3.** 4.5 **4.** 0 **5.** $-\frac{4}{12}$ *or* $-\frac{1}{3}$ **6.** $-\frac{7}{6}$ **7.** -7 **8.** 1.7 **9.** -8 **10.** -31 **11.** $\frac{14}{17}$ **12.** -12 **13.** 1.4 **14.** -2.8 **15.** 42 **16.** $\frac{19}{15}$ *or* $1\frac{4}{15}$ **17.** 24 **18.** 4 **19.** -8 **20.** -10 **21.** -24 **22.** $\frac{15}{16}$ **23.** -64 **24.** -10 **25.** -10 **26.** 31 **27.** -12 **28.** 4.8 **29.** 5 **30.** -29 **31.** $-\frac{1}{3}$ **32.** $\frac{1}{5}$ **33.** 8×10^4 **34.** 5×10^{-4} **35.** 0.0067 **36.** 1,320,000

9.1 Exercises

1. 24 **3.** -17 **5.** -7 **7.** 16.5 **9.** $\frac{17}{35}$ **11.** $-\frac{11}{12}$ **13.** 9 **15.** -7 **17.** -5 **19.** 22 **21.** -2.8 **23.** $-\frac{2}{3}$ **25.** $\frac{5}{9}$ **27.** -22 **29.** -0.7 **31.** $\frac{1}{2}$ **33.** -38 **35.** -7.6 **37.** 4 **39.** $-\frac{1}{6}$ **41.** 18.1 **43.** $-5\frac{1}{2}$ **45.** -4 **47.** -11 **49.** -6 **51.** 10 **53.** $-\frac{184}{225}$ **55.** $-\$94{,}000$ **57.** $9000 **59.** $-\$38{,}000$ **61.** -18°F **63.** -1°F **65.** 904.3 ft^3 **66.** 210 m^3

APPENDIX A
Tables

TABLE OF BASIC ADDITION FACTS

+	0	1	2	3	4	5	6	7	8	9
0	0	1	2	3	4	5	6	7	8	9
1	1	2	3	4	5	6	7	8	9	10
2	2	3	4	5	6	7	8	9	10	11
3	3	4	5	6	7	8	9	10	11	12
4	4	5	6	7	8	9	10	11	12	13
5	5	6	7	8	9	10	11	12	13	14
6	6	7	8	9	10	11	12	13	14	15
7	7	8	9	10	11	12	13	14	15	16
8	8	9	10	11	12	13	14	15	16	17
9	9	10	11	12	13	14	15	16	17	18

TABLE OF BASIC MULTIPLICATION FACTS

×	0	1	2	3	4	5	6	7	8	9
0	0	0	0	0	0	0	0	0	0	0
1	0	1	2	3	4	5	6	7	8	9
2	0	2	4	6	8	10	12	14	16	18
3	0	3	6	9	12	15	18	21	24	27
4	0	4	8	12	16	20	24	28	32	36
5	0	5	10	15	20	25	30	35	40	45
6	0	6	12	18	24	30	36	42	48	54
7	0	7	14	21	28	35	42	49	56	63
8	0	8	16	24	32	40	48	56	64	72
9	0	9	18	27	36	45	54	63	72	81

TABLE OF PRIME FACTORS

Number	Prime Factors	Number	Prime Factors	Number	Prime Factors	Number	Prime Factors
		51	3×17	**101**	prime	**151**	prime
2	prime	**52**	$2^2 \times 13$	**102**	$2 \times 3 \times 17$	**152**	$2^3 \times 19$
3	prime	**53**	prime	**103**	prime	**153**	$3^2 \times 17$
4	2^2	**54**	2×3^3	**104**	$2^3 \times 13$	**154**	$2 \times 7 \times 11$
5	prime	**55**	5×11	**105**	$3 \times 5 \times 7$	**155**	5×31
6	2×3	**56**	$2^3 \times 7$	**106**	2×53	**156**	$2^2 \times 3 \times 13$
7	prime	**57**	3×19	**107**	prime	**157**	prime
8	2^3	**58**	2×29	**108**	$2^2 \times 3^3$	**158**	2×79
9	3^2	**59**	prime	**109**	prime	**159**	3×53
10	2×5	**60**	$2^2 \times 3 \times 5$	**110**	$2 \times 5 \times 11$	**160**	$2^5 \times 5$
11	prime	**61**	prime	**111**	3×37	**161**	7×23
12	$2^2 \times 3$	**62**	2×31	**112**	$2^4 \times 7$	**162**	2×3^4
13	prime	**63**	$3^2 \times 7$	**113**	prime	**163**	prime
14	2×7	**64**	2^6	**114**	$2 \times 3 \times 19$	**164**	$2^2 \times 41$
15	3×5	**65**	5×13	**115**	5×23	**165**	$3 \times 5 \times 11$
16	2^4	**66**	$2 \times 3 \times 11$	**116**	$2^2 \times 29$	**166**	2×83
17	prime	**67**	prime	**117**	$3^2 \times 13$	**167**	prime
18	2×3^2	**68**	$2^2 \times 17$	**118**	2×59	**168**	$2^3 \times 3 \times 7$
19	prime	**69**	3×23	**119**	7×17	**169**	13^2
20	$2^2 \times 5$	**70**	$2 \times 5 \times 7$	**120**	$2^3 \times 3 \times 5$	**170**	$2 \times 5 \times 17$
21	3×7	**71**	prime	**121**	11^2	**171**	$3^2 \times 19$
22	2×11	**72**	$2^3 \times 3^2$	**122**	2×61	**172**	$2^2 \times 43$
23	prime	**73**	prime	**123**	3×41	**173**	prime
24	$2^3 \times 3$	**74**	2×37	**124**	$2^2 \times 31$	**174**	$2 \times 3 \times 29$
25	5^2	**75**	3×5^2	**125**	5^3	**175**	$5^2 \times 7$
26	2×13	**76**	$2^2 \times 19$	**126**	$2 \times 3^2 \times 7$	**176**	$2^4 \times 11$
27	3^3	**77**	7×11	**127**	prime	**177**	3×59
28	$2^2 \times 7$	**78**	$2 \times 3 \times 13$	**128**	2^7	**178**	2×89
29	prime	**79**	prime	**129**	3×43	**179**	prime
30	$2 \times 3 \times 5$	**80**	$2^4 \times 5$	**130**	$2 \times 5 \times 13$	**180**	$2^2 \times 3^2 \times 5$
31	prime	**81**	3^4	**131**	prime	**181**	prime
32	2^5	**82**	2×41	**132**	$2^2 \times 3 \times 11$	**182**	$2 \times 7 \times 13$
33	3×11	**83**	prime	**133**	7×19	**183**	3×61
34	2×17	**84**	$2^2 \times 3 \times 7$	**134**	2×67	**184**	$2^3 \times 23$
35	5×7	**85**	5×17	**135**	$3^3 \times 5$	**185**	5×37
36	$2^2 \times 3^2$	**86**	2×43	**136**	$2^3 \times 17$	**186**	$2 \times 3 \times 31$
37	prime	**87**	3×29	**137**	prime	**187**	11×17
38	2×19	**88**	$2^3 \times 11$	**138**	$2 \times 3 \times 23$	**188**	$2^2 \times 47$
39	3×13	**89**	prime	**139**	prime	**189**	$3^3 \times 7$
40	$2^3 \times 5$	**90**	$2 \times 3^2 \times 5$	**140**	$2^2 \times 5 \times 7$	**190**	$2 \times 5 \times 19$
41	prime	**91**	7×13	**141**	3×47	**191**	prime
42	$2 \times 3 \times 7$	**92**	$2^2 \times 23$	**142**	2×71	**192**	$2^6 \times 3$
43	prime	**93**	3×31	**143**	11×13	**193**	prime
44	$2^2 \times 11$	**94**	2×47	**144**	$2^4 \times 3^2$	**194**	2×97
45	$3^2 \times 5$	**95**	5×19	**145**	5×29	**195**	$3 \times 5 \times 13$
46	2×23	**96**	$2^5 \times 3$	**146**	2×73	**196**	$2^2 \times 7^2$
47	prime	**97**	prime	**147**	3×7^2	**197**	prime
48	$2^4 \times 3$	**98**	2×7^2	**148**	$2^2 \times 37$	**198**	$2 \times 3^2 \times 11$
49	7^2	**99**	$3^2 \times 11$	**149**	prime	**199**	prime
50	2×5^2	**100**	$2^2 \times 5^2$	**150**	$2 \times 3 \times 5^2$	**200**	$2^3 \times 5^2$

TABLE OF SQUARE ROOTS
Square Root Values Are Rounded to the Nearest Thousandth Unless the Answer Ends in .000

n	$\sqrt{n}$	n	$\sqrt{n}$	n	$\sqrt{n}$	n	$\sqrt{n}$	n	$\sqrt{n}$
1	1.000	**41**	6.403	**81**	9.000	**121**	11.000	**161**	12.689
2	1.414	**42**	6.481	**82**	9.055	**122**	11.045	**162**	12.728
3	1.732	**43**	6.557	**83**	9.110	**123**	11.091	**163**	12.767
4	2.000	**44**	6.633	**84**	9.165	**124**	11.136	**164**	12.806
5	2.236	**45**	6.708	**85**	9.220	**125**	11.180	**165**	12.845
6	2.449	**46**	6.782	**86**	9.274	**126**	11.225	**166**	12.884
7	2.646	**47**	6.856	**87**	9.327	**127**	11.269	**167**	12.923
8	2.828	**48**	6.928	**88**	9.381	**128**	11.314	**168**	12.961
9	3.000	**49**	7.000	**89**	9.434	**129**	11.358	**169**	13.000
10	3.162	**50**	7.071	**90**	9.487	**130**	11.402	**170**	13.038
11	3.317	**51**	7.141	**91**	9.539	**131**	11.446	**171**	13.077
12	3.464	**52**	7.211	**92**	9.592	**132**	11.489	**172**	13.115
13	3.606	**53**	7.280	**93**	9.644	**133**	11.533	**173**	13.153
14	3.742	**54**	7.348	**94**	9.695	**134**	11.576	**174**	13.191
15	3.873	**55**	7.416	**95**	9.747	**135**	11.619	**175**	13.229
16	4.000	**56**	7.483	**96**	9.798	**136**	11.662	**176**	13.266
17	4.123	**57**	7.550	**97**	9.849	**137**	11.705	**177**	13.304
18	4.243	**58**	7.616	**98**	9.899	**138**	11.747	**178**	13.342
19	4.359	**59**	7.681	**99**	9.950	**139**	11.790	**179**	13.379
20	4.472	**60**	7.746	**100**	10.000	**140**	11.832	**180**	13.416
21	4.583	**61**	7.810	**101**	10.050	**141**	11.874	**181**	13.454
22	4.690	**62**	7.874	**102**	10.100	**142**	11.916	**182**	13.491
23	4.796	**63**	7.937	**103**	10.149	**143**	11.958	**183**	13.528
24	4.899	**64**	8.000	**104**	10.198	**144**	12.000	**184**	13.565
25	5.000	**65**	8.062	**105**	10.247	**145**	12.042	**185**	13.601
26	5.099	**66**	8.124	**106**	10.296	**146**	12.083	**186**	13.638
27	5.196	**67**	8.185	**107**	10.344	**147**	12.124	**187**	13.675
28	5.292	**68**	8.246	**108**	10.392	**148**	12.166	**188**	13.711
29	5.385	**69**	8.307	**109**	10.440	**149**	12.207	**189**	13.748
30	5.477	**70**	8.367	**110**	10.488	**150**	12.247	**190**	13.784
31	5.568	**71**	8.426	**111**	10.536	**151**	12.288	**191**	13.820
32	5.657	**72**	8.485	**112**	10.583	**152**	12.329	**192**	13.856
33	5.745	**73**	8.544	**113**	10.630	**153**	12.369	**193**	13.892
34	5.831	**74**	8.602	**114**	10.677	**154**	12.410	**194**	13.928
35	5.916	**75**	8.660	**115**	10.724	**155**	12.450	**195**	13.964
36	6.000	**76**	8.718	**116**	10.770	**156**	12.490	**196**	14.000
37	6.083	**77**	8.775	**117**	10.817	**157**	12.530	**197**	14.036
38	6.164	**78**	8.832	**118**	10.863	**158**	12.570	**198**	14.071
39	6.245	**79**	8.888	**119**	10.909	**159**	12.610	**199**	14.107
40	6.325	**80**	8.944	**120**	10.954	**160**	12.649	**200**	14.142

APPENDIX B

Scientific Calculators

This text does *not require* the use of a calculator. However, you may want to consider the purchase of an inexpensive scientific calculator. It is wise to ask your instructor for advice before you purchase any calculator for this course. It should be stressed that students are asked to avoid using a calculator for any of the exercises in which the calculations can be readily done by hand. The only problems in the text that really demand the use of a scientific calculator are marked with the ▦ symbol. Dependence on the use of the scientific calculator for regular exercises in the text will only hurt the student in the long run.

The Two Types of Logic Used in Scientific Calculators

There are two major types of scientific calculators that are popular today. The most common type employs a type of logic known as **algebraic** logic. The calculators manufactured by Casio, Sharp, and Texas Instruments as well as many other companies employ this type of logic. An example of calculation on such a calculator would be the following. To add 14 + 26 on an algebraic logic calculator, the sequence of buttons would be:

14 [+] 26 [=]

The second type of scientific calculator requires the entry of data in **Reverse Polish Notation (RPN).** Calculators manufactured by Hewlett-Packard and a few other specialized calculators are made to use RPN. To add 14 + 26 on a RPN calculator, the sequence of buttons would be:

14 [enter] 26 [+]

Graphing scientific calculators such as the TI-82 and TI-83 have a large liquid display for viewing graphs. To perform the calculation on most graphing calculators, the sequence of buttons would be:

14 [+] 26 [enter]

Mathematicians and scientists do not agree on which type of scientific calculator is superior. However, the clear majority of college students own calculators that employ *algebraic* logic. Therefore this section of the text is explained with reference to the sequence of steps employed by an *algebraic* logic calculator. If you already own or intend to purchase a scientific calculator that uses RPN, you are encouraged to study the instruction booklet that comes with the calculator and practice the problems shown in the booklet. After this practice you will be able to solve the calculator problems discussed in this section.

Performing Simple Calculations

The following example will illustrate the use of the scientific calculator in doing basic arithmetic calculations.

EXAMPLE 1 Add. 156 + 298

We first key in the number 156, then press the [+] key, then enter the number 298, and finally press the [=] key.

156 [+] 298 [=] 454

Practice Problem 1 Add. 3792 + 5896 ■

EXAMPLE 2 Subtract. 1508 − 963

We first enter the number 1508, then press the [−] key, then enter the number 963, and finally press the [=] key.

1508 [−] 963 [=] 545

Practice Problem 2 Subtract. 7930 − 5096 ■

EXAMPLE 3 Multiply. 196 × 358

196 [×] 358 [=] 70168

Practice Problem 3 Multiply. 896 × 273 ■

EXAMPLE 4 Divide. 2054 ÷ 13

2054 [÷] 13 [=] 158

Practice Problem 4 Divide. 2352 ÷ 16 ■

Decimal Problems

Problems involving decimals can be readily done on the calculator. Entering numbers with a decimal point is done by pressing the [•] key at the appropriate time.

EXAMPLE 5 Calculate. 4.56 × 283

To enter 4.56, we press the [4] key, the decimal point key, then the [5] key, and finally the [6] key.

4.56 [×] 283 [=] 1290.48

The answer is 1290.48. Observe how your calculator displays the decimal point.

Practice Problem 5 Calculate. 72.8 × 197 ■

EXAMPLE 6 Add. 128.6 + 343.7 + 103.4 + 207.5

128.6 [+] 343.7 [+] 103.4 [+] 207.5 [=] 783.2

The answer is 783.2. Observe how your calculator displays the answer.

Practice Problem 6 Add. 52.98 + 31.74 + 40.37 + 99.82 ■

Combined Operations

You must use extra caution concerning the order of mathematical operations when you are doing a problem on the calculator that involves two or more different operations.

Any scientific calculator with algebraic logic uses a priority system that has a clearly defined order of operations. It is the same order we use in performing arithmetic operations by hand. In either situation, calculations are performed in the following order:

1. First calculations within a parentheses are completed.
2. Then numbers are raised to a power or a square root is calculated.
3. Then multiplication and division operations are performed from left to right.
4. Then addition and subtraction operations are performed from left to right.

This order is carefully followed on *scientific calculators*. Small inexpensive calculators that do not have scientific functions often do not follow this order of operations.

The number of digits displayed in the answer varies from calculator to calculator. In the following examples, your calculator may display more or fewer digits than the answer we have listed.

EXAMPLE 7 Evaluate. $5.3 \times 1.62 + 1.78 \div 3.51$

This problem requires that we multiply 5.3 by 1.62 and divide 1.78 by 3.51 first and then add the two results. If the numbers are entered directly into the claculator exactly as the problem is written, the calculator will perform the calculations in the correct order.

5.3 [×] 1.62 [+] 1.78 [÷] 3.51 [=] 9.09312251

Practice Problem 7 Evaluate. $0.0618 \times 19.22 - 59.38 \div 166.3$ ■

The Use of Parentheses

In order to perform some calculations on the calculator the use of parentheses is helpful. These parentheses may or may not appear in the original problem.

EXAMPLE 8 Evaluate. $5 \times (2.123 + 5.786 - 12.063)$

The problem requires that the numbers in the parentheses be combined first. By entering the parentheses on the calculator this will be accomplished.

5 [×] [(] 2.123 [+] 5.786 [−] 12.063 [)] [=] −20.77

Note: The result is a negative number.

Practice Problem 8 Evaluate. $3.152 \times (0.1628 + 3.715 - 4.985)$ ■

Negative Numbers

To enter a negative number, enter the number followed by the [+/−] button.

EXAMPLE 9 Evaluate. $(-8.634)(5.821) + (1.634)(-16.082)$

The products will be evaluated first by the calculator. Therefore parentheses are not needed as we enter the data.

8.634 [+/−] [×] 5.821 [+] 1.634 [×] 16.082 [+/−] [=] −76.536502

Note: The result is negative.

Practice Problem 9 Evaluate. $(0.5618)(-98.3) - (76.31)(-2.98)$ ■

Scientific Notation

If you wish to enter a number in scientific notation, you should use the special scientific notation button. On most calculators it is denoted as [EXP] or [EE].

EXAMPLE 10 Multiply. $(9.32 \times 10^6)(3.52 \times 10^8)$

9.32 [EXP] 6 [×] 3.52 [EXP] 8 [=] 3.28064 15

This notation means the answer is 3.28064×10^{15}.

Practice Problem 10 Divide. $(3.76 \times 10^{15}) \div (7.76 \times 10^7)$ ■

Raising a Number to a Power

All scientific calculators have a key for finding powers of numbers. It is usually labeled [y^x]. (On a few calculators the notation is [x^y] or sometimes [^].) To raise a number to a power, first you enter the base, then push the [y^x] key. Then you enter the exponent, then finally the [=] button.

EXAMPLE 11 Evaluate. $(2.16)^9$

2.16 [y^x] 9 [=] 1023.490369

Practice Problem 11 Evaluate. $(6.238)^6$ ■

There is a special key to square something. It is usually labeled [x^2].

EXAMPLE 12 Evaluate. $(76.04)^2$

76.04 [x^2] 5782.0816

Practice Problem 12 Evaluate. $(132.56)^2$ ■

Finding Square Roots of Numbers

There is usually a key to approximate square roots on a scientific calculator labeled [√]. In this example we will need to use a parentheses.

EXAMPLE 13 Evaluate. $\sqrt{5618 + 2734 + 3913}$

[(] 5618 [+] 2734 [+] 3913 [)] [√] 110.747605

Practice Problem 13 Evaluate. $\sqrt{0.0782 - 0.0132 + 0.1364}$ ■

Scientific Calculator Exercises

Use your calculator to complete each of the following. Your answers may vary slightly because of the characteristics of individual calculators.

Complete the following table.

To Do This Operation	Use These Keystrokes	Record Your Answer Here
1. 8963 + 2784	8963 [+] 2784 [=]	
2. 15,308 − 7980	15308 [−] 7980 [=]	
3. 2631 × 134	2631 [×] 134 [=]	
4. 70,221 ÷ 89	70221 [÷] 89 [=]	
5. 5.325 − 4.031	5.325 [−] 4.031 [=]	
6. 184.68 + 73.98	184.68 [+] 73.98 [=]	
7. 2004.06 ÷ 7.89	2004.06 [÷] 7.89 [=]	
8. 1.34 × 0.763	1.34 [×] 0.763 [=]	

Write down the answer for the following and then show what problem you have solved.

9. 123.45 [+] 45.9876 [+] 8765.3 [=]

10. 0.0897 [×] 234.56 [×] 2.5428 [=]

11. 34 [÷] 8 [+] 12.56 [=]

12. 458 [÷] 4 [−] 16.897 [=]

Perform the following calculations using your calculator.

13. 9.467 + 0.563

14. 0.347 + 23.457

15. 34.89 + 39.6 + 214.897

16. 12.567 + 48.31 + 189.38

17. 412,899 − 34,675

18. 87,456 − 2876

19. 3,567,089 − 2,876,805

20. 8,345,802 − 4,985,004

21. 234 × 4.567

22. 1.9876 × 347

23. 0.456 × 3.48

24. 67,876 × 0.0946

25. 3458 ÷ 2.5

26. 9764 ÷ 8

27. 12.107524 ÷ 15.86

28. 16.06513 ÷ 17.98

Perform the following calculations using your calculator.

29. 1.98 + 6.34 + 7.71

30. 8.92 + 9.31 + 7.79

31. \$103.91 + \$2653.82 + \$9804.61

32. \$3986.21 + \$4502.89 + \$989.30

33. 368,781.5 − 283,617.8

34. 571,809.6 − 539,376.8

35. \$1,393,271.86 − \$1,289,663.21

36. \$8,571,300.76 − \$4,098,789.39

37. 345.34 × 45.7

38. 8954.34 × 425.4

39. 0.6314 × 3.96

40. 0.0789 × 12.38

41. $40.36\overline{)36202.92}$

42. $52.98\overline{)172{,}608.84}$

43. $0.7613\overline{)17.12925}$

44. $0.9854\overline{)3.59671}$

Perform the following operations in the proper order using your calculator.

45. $4.567 + 87.89 - 2.45 \times 3.3$

46. $4.891 + 234.5 - 0.98 \times 23.4$

47. $7 \div 8 + 3.56$

48. $9 \div 4.5 + 0.6754$

49. $(9.34)(0.345) + 98.345$

50. $(0.628)(398) + 34.4581$

51. $\dfrac{(95.34)(0.9874)}{381.36}$

52. $\dfrac{(0.8759)(45.87)}{183.48}$

53. $2.56 + 8.98 \times 3.14$

54. $1.62 + 3.81 - 5.23 \times 6.18$

55. $(-4.23)(1.863) - 5.998$

56. $12.34 - (26.314)(-1.856)$

57. $5.62(5 \times 3.16 - 18.12)$

58. $9.356(4.8 - 7.2 - 15.94)$

59. $(3.42 \times 10^8)(0.97 \times 10^{10})$

60. $(6.27 \times 10^{20})(1.35 \times 10^3)$

61. $\dfrac{(2.16 \times 10^3)(1.37 \times 10^{14})}{6.39 \times 10^5}$

62. $\dfrac{(3.84 \times 10^{12})(1.62 \times 10^5)}{7.78 \times 10^8}$

63. $\dfrac{2.3 + 5.8 - 2.6 - 3.9}{5.3 - 8.2}$

64. $\dfrac{(2.6)(-3.2) + (5.8)(-0.9)}{2.614 + 5.832}$

65. $\sqrt{253.12}$

66. $\sqrt{0.0713}$

67. $\sqrt{5.6213 - 3.7214}$

68. $\sqrt{3417.2 - 2216.3}$

69. $(1.78)^3 + 6.342$

70. $(2.26)^8 - 3.1413$

71. $\sqrt{(6.13)^2 + (5.28)^2}$

72. $\sqrt{(0.3614)^2 + (0.9217)^2}$

73. $\sqrt{56 + 83} - \sqrt{12}$

74. $\sqrt{98 + 33} - \sqrt{17}$

Find an approximate value for each of the following. Round your answer to five decimal places.

75. $\dfrac{7}{18} + \dfrac{9}{13}$

76. $\dfrac{5}{22} + \dfrac{1}{31}$

77. $\dfrac{7}{8} + \dfrac{3}{11}$

78. $\dfrac{9}{14} + \dfrac{5}{19}$

Appendix C: Square Roots

Square Roots

MathPro Video 15 SSM

After studying this section, you will be able to:

1 *Evaluate the square root of a number that is a perfect square.*
2 *Approximate the square root of a number that is not a perfect square.*

1 *Evaluating the Square Root of a Number That Is a Perfect Square*

We know that by using the formula $A = s^2$ we can quickly find the area of a square with a side of 3 in. We simply square 3 in. That is, $A = (3 \text{ in.})(3 \text{ in.}) = 9 \text{ in.}^2$. Sometimes we want to ask another kind of question. If a square has an area of 64 in.^{2}, what is the length of its sides?

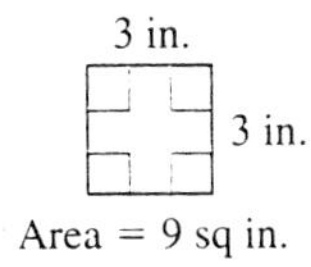

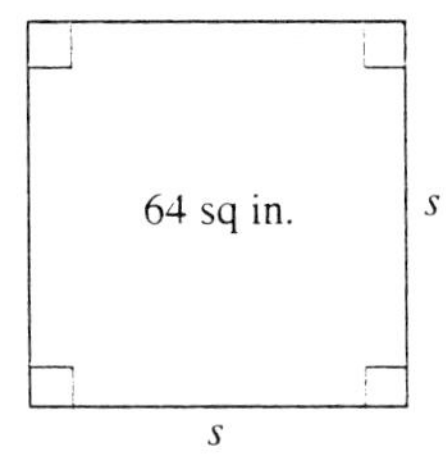

The answer is 8 in. Why? The skill we need to find a number when we are given the square of that number is called *finding the square root.* The square root of 64 is 8.

The **square root** of a number is one of the two identical factors of that number.

The square root of 64 is 8 because $(8)(8) = 64$.

The square root of 9 is 3 because $(3)(3) = 9$.

The symbol for finding the square root of a number is $\sqrt{}$. To write the square root of 64, we write $\sqrt{64} = 8$. Sometimes we speak of finding the square root of a number as *taking* the square root of the number, or we can say that we will *evaluate* the square root of the number. Thus to take the square root of 9, we write $\sqrt{9} = 3$; to evaluate the square root of 9, we write $\sqrt{9} = 3$.

EXAMPLE 1 Find the square roots. **(a)** $\sqrt{25}$ **(b)** $\sqrt{121}$

(a) $\sqrt{25} = 5$ because $(5)(5) = 25$ **(b)** $\sqrt{121} = 11$ because $(11)(11) = 121$

Practice Problem 1 Find the square roots. **(a)** $\sqrt{49}$ **(b)** $\sqrt{169}$ ■

If square roots are added or subtracted, they must be evaluated FIRST, then added or subtracted.

EXAMPLE 2 Find. $\sqrt{25} + \sqrt{36}$

$$\sqrt{25} = 5 \text{ because } (5)(5) = 25$$

$$\sqrt{36} = 6 \text{ because } (6)(6) = 36$$

Thus $\sqrt{25} + \sqrt{36} = 5 + 6 = 11$.

Practice Problem 2 Find. $\sqrt{49} - \sqrt{4}$ ■

When a whole number is multiplied by itself, the number that is obtained is called a **perfect square.**

36 is a perfect square because $(6)(6) = 36$.

49 is a perfect square $(7)(7) = 49$.

The numbers 20 or 48 are *not* perfect squares. There is no *whole number* that when squared—multiplied by itself—yields 20 or 48. Consider 20. $4^2 = 16$, which is less than 20. $5^2 = 25$, which is more than 20. We realize, then, that the square root of 20 is between 4 and 5 because 20 is between 16 and 25. Since there is no whole number between 4 and 5, no whole number squared equals 20. Since the square root of a perfect square is a whole number, we can say that 20 is NOT a perfect square.

EXAMPLE 3

(a) Is 81 a perfect square? **(b)** If so, find $\sqrt{81}$.

(a) Yes. 81 is a perfect square because $(9)(9) = 81$. **(b)** $\sqrt{81} = 9$

Practice Problem 3

(a) Is 144 a perfect square? **(b)** If so, find $\sqrt{144}$. ∎

It is helpful to know the first 15 perfect squares. Take a minute to complete the following table.

Number, n	1	2	3	4	5	6	7	8	9	10	11	12	13	14	15
Number Squared, n^2	1	4	9	16										196	225

2 *Approximating the Square Root of a Number That Is Not a Perfect Square*

If a number is not a perfect square, we can only approximate its square root. This can be done by using a square root table such as the one below. Except for exact values such as $\sqrt{4} = 2.000$, all values are rounded to the nearest thousandth.

Number, n	Square Root of the Number, $\sqrt{n}$	Number, n	Square Root of the Number, $\sqrt{n}$
1	1.000	8	2.828
2	1.414	9	3.000
3	1.732	10	3.162
4	2.000	11	3.317
5	2.236	12	3.464
6	2.449	13	3.606
7	2.646	14	3.742

A square root table is located on page A–4. It gives you the square root of whole numbers up to 200. Square roots can also be found with any calculator that has a square root key. Usually the key looks like this $\boxed{\sqrt{\ }}$ or this $\boxed{\sqrt{x}}$. To find the square root of 8 on a calculator, enter the number 8 and press $\boxed{\sqrt{\ }}$ or $\boxed{\sqrt{x}}$. You will see displayed 2.8284271. (Your calculator may display fewer or more digits.) Remember, no matter how many digits your calculator displays, when we find $\sqrt{8}$, we have only an **approximation.** It is not an exact answer. To emphasize this we use the $\approx$ notation to mean "is approximately equal to." Thus $\sqrt{8} \approx 2.828$.

CALCULATOR

Square Roots

Locate the square root key [$\sqrt{\ }$] on your calculator.

1. To find $\sqrt{289}$ enter

   289 [$\sqrt{\ }$]

   The display should read

   [17]

2. To find $\sqrt{194}$ enter

   194 [$\sqrt{\ }$]

   The display should read

   [13.928388]

   This is just an approximation of the actual square root. We will round the answer to the nearest thousandth.
   $\sqrt{194} \approx 13.928$

EXAMPLE 4 Find approximate values using the square root table or a calculator. Round your answer to the nearest thousandth.

(a) $\sqrt{2}$ (b) $\sqrt{12}$ (c) $\sqrt{7}$

(a) $\sqrt{2} \approx 1.414$ (b) $\sqrt{12} \approx 3.464$ (c) $\sqrt{7} \approx 2.646$

Practice Problem 4 Approximate the square roots to the nearest thousandth.

(a) $\sqrt{3}$ (b) $\sqrt{13}$ (c) $\sqrt{5}$ ■

EXAMPLE 5 Find the length of the side of a square that has an area of 81 in.^{2}.

$$\sqrt{81 \text{ in.}^2} = 9 \text{ in.}$$

The side measures 9 in.

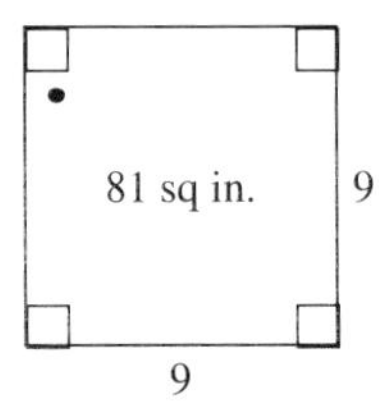

Practice Problem 5 Find the length of the side of a square that has an area of 100 in.^{2}. ■

EXAMPLE 6 Approximate to the nearest thousandth of an inch the length of the side of a square that has an area of 6 in.^{2}.

$$\sqrt{6 \text{ in.}^2} \approx 2.449 \text{ in.}$$

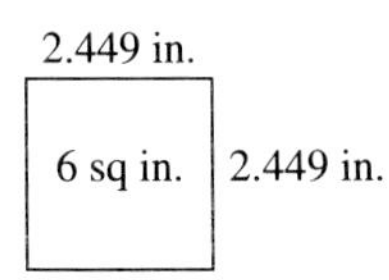

Thus, to the nearest thousandth of an inch, the side measures 2.449 in.

Practice Problem 6 Approximate to the nearest thousandth of a meter the length of the side of a square that has an area of 22 m^2. ■

Exercises

Verbal and Writing Skills

1. Why is the $\sqrt{25} = 5$?
2. 25 is a perfect square because its square root, 5, is a _____ number.
3. Is 32 a perfect square? Why or why not?
4. How can you approximate the square root of a number that is not a perfect square?

Find each square root. Do not *use a calculator. Do* not *refer to a table of square roots.*

5. $\sqrt{1}$
6. $\sqrt{4}$
7. $\sqrt{16}$
8. $\sqrt{9}$
9. $\sqrt{25}$
10. $\sqrt{36}$
11. $\sqrt{49}$
12. $\sqrt{64}$
13. $\sqrt{100}$
14. $\sqrt{81}$
15. $\sqrt{121}$
16. $\sqrt{144}$
17. $\sqrt{169}$
18. $\sqrt{196}$
19. $\sqrt{0}$
20. $\sqrt{225}$

Evaluate the square roots first, then add or subtract the results. Do not *use a calculator or a square root table.*

21. $\sqrt{49} + \sqrt{9}$
22. $\sqrt{25} + \sqrt{64}$
23. $\sqrt{100} + \sqrt{1}$
24. $\sqrt{0} + \sqrt{121}$
25. $\sqrt{36} + \sqrt{64}$
26. $\sqrt{1} + \sqrt{25}$
27. $\sqrt{0} + \sqrt{16}$
28. $\sqrt{36} + \sqrt{81}$
29. $\sqrt{144} - \sqrt{4}$
30. $\sqrt{169} - \sqrt{64}$
31. $\sqrt{121} - \sqrt{36}$
32. $\sqrt{225} - \sqrt{100}$
33. Find an exact value for
$$\sqrt{1} + \sqrt{9} + \sqrt{25} + \sqrt{36} + \sqrt{49}$$
34. Find an exact value for
$$\sqrt{4} + \sqrt{9} + \sqrt{16} + \sqrt{81} + \sqrt{121}$$
35. **(a)** Is 49 a perfect square?
(b) If so, find $\sqrt{49}$.
36. **(a)** Is 25 a perfect square?
(b) If so, find $\sqrt{25}$.
37. **(a)** Is 256 a perfect square?
(b) If so, find $\sqrt{256}$.
38. **(a)** Is 289 a perfect square?
(b) If so, find $\sqrt{289}$.

Use a table of square roots or a calculator with a square root key to approximate the following to the nearest thousandth.

39. $\sqrt{18}$
40. $\sqrt{24}$
41. $\sqrt{31}$
42. $\sqrt{42}$
43. $\sqrt{83}$
44. $\sqrt{76}$
45. $\sqrt{120}$
46. $\sqrt{136}$
47. $\sqrt{125}$
48. $\sqrt{150}$

Find the length of the side of each square. If the area is not a perfect square, approximate by using a square root table or a calculator with a square root key. Round your answer to the nearest thousandth.

49. A square with area 121 m^2
50. A square with area 169 m^2
51. A square with area 26 m^2
52. A square with area 34 m^2
53. A square with area 75 m^2
54. A square with area 90 m^2

To Think About

55. Find each square root.

(a) $\sqrt{4}$

(b) $\sqrt{0.04}$

(c) $\sqrt{0.0004}$

(d) What pattern do you observe?

(e) Can you find $\sqrt{0.004}$ exactly? Why?

56. Find each square root.

(a) $\sqrt{25}$

(b) $\sqrt{0.25}$

(c) $\sqrt{0.0025}$

(d) What pattern do you observe?

(e) Can you find $\sqrt{0.025}$ exactly? Why?

Using a calculator with a square root key, evaluate each of the following and round your answer to the nearest thousandth.

57. $\sqrt{42{,}036}$

58. $\sqrt{25{,}231}$

59. $\sqrt{456} + \sqrt{322}$

60. $\sqrt{578} + \sqrt{984}$

Cumulative Review Problems

61. The Taronga Zoo in Sydney, Australia, has a viewing tank for its platapi (plural for platapus). The tank is 60 in. high, and 80 in. wide. What is the area of the front of the rectangular tank?

62. Ashrita Furman of Jamaica, New York, holds the world record in "joggling"—juggling and running at the same time—over 80.5 km. How many meters did he "joggle"?

63. The gate that leads from the street to the neighborhood basketball courts measures 92 centimeters wide. How many *meters* wide is the door?

64. Goliath beetles in Equatorial Africa can weigh up to 98.9 grams. How many *kilograms* can they weigh?